Advances in Modal Logic

volume 4

Advances in Modal Logic

volume 4

Edited by

Philippe Balbiani, Nobu-Yuki Suzuki, Frank Wolter, and
Michael Zakharyaschev

© Individual authors and King's College 2003. All rights reserved.

ISBN 0-9543006-1-0 (Paperback)
ISBN 0-9543006-2-9 (Hardback)

King's College Publications
Scientific Director: Dov Gabbay
Managing Director: Jane Spurr
Department of Computer Science
Strand, London WC2R 2LS, UK
kcp@dcs.kcl.ac.uk

http://www.dcs.kcl.ac.uk/AiML

Cover design by Richard Fraser, www.avalonarts.co.uk
Printed by Lightning Source, Milton Keynes, UK

All rights reserved. No part of this publication may be reproduced, stored in a retrieval system or transmitted, in any form, or by any means, electronic, mechanical, photocopying, recording or otherwise, without prior permission, in writing, from the publisher.

Advances in Modal Logic

Volume 4

edited by Philippe Balbiani, Nobu-Yuki Suzuki, Frank Wolter,
and Michael Zakharyaschev

CONTENTS

Editorial Preface **P. Balbiani, N.-Y. Suzuki, F. Wolter and M. Zakharyaschev**	vii
Modal Logics with a Linear Hierarchy of Local Propositional Quantifiers **K. Engelhardt, R. van der Meyden, K. Su**	9
Functional Completenes for a Natural Deduction Formulation of Hybridized S5 **T. Brauner**	31
Relativized Action Complement for Dynamic Logics **J. Broersen**	51
How Many Variables Does One Need to Prove PSPACE-hardness of Modal Logics **A. V. Chagrov and M. N. Rybakov**	71
Non-normal Modalisation **R. A. S. Fajardo and M. Finger**	83
Bisimulations and Boolean Vectors **M. Fitting**	97
A Sound and Complete Proof System for QPTL **T. French and M. Reynolds**	127
Equational Logic of Polynomial Coalgebras **R. Goldblatt**	149
Towards Uniform Reasoning via Structured Subset Spaces **B. Heinemann**	185

Controlled Model Exploration 205
Gabriel G. Infante-Lopez, Carlos Areces, and Maarten de Rijke

A Note on Relativised Products of Modal Logics 221
A. Kurucz and M. Zakharyaschev

Notes on the Space Requirements for Checking Satisfiability in Modal Logics 243
M. Kracht

Description Logics with Concrete Domains—A Survey 265
C. Lutz

Restricted Interpolation in Modal Logics 297
L. Maksimova

Binary Logics, Orthologics, and their Relations to Normal Modal Logics 313
Y. Miyazaki

Completions of Algebras and Completeness of Modal and Substructural Logics 335
H. Ono

An Axiomatization of Prior's Ockhamist Logic of Historical Necessity 355
M. Reynolds

Combining Dynamic Logic with Doxastic Modal Logics 371
R. A. Schmidt and D. Tishkovsky

The Complexity of Temporal Logic Model Checking 393
Ph. Schnoebelen

Chronological Future Modality in Minkowski Spacetime 437
I. Shapirovsky and V. Shehtman

An Incompleteness Result for Predicate Extensions of Intermediate Propositional Logics 461
D. Skvortsov

CONTENTS

On IF Modal Logic and its Expressive Power   475
T. Tulenheimo

Modal Definability in Languages with a Finite Number of   499
Propositional Variables and a New Extension of the
Sahlqvist's Class
D. Vakarelov

Index   519

PREFACE

The Fourth Advances in Modal Logic (AiML) conference was held in Toulouse, France, in October 2002. Like the previous conferences in Berlin (1996), Uppsala (1998), and Leipzig (2000), AiML 2002 brought together modal logicians from all over the world to present and discuss the latest developments in modal logic and its applications in mathematics, computer science, philosophy, artificial intelligence, and linguistics. The call for papers attracted 51 submissions out of which 24 were selected for presentation in Toulouse. Invited talks were given by

- Melvin Fitting,
- Robert Goldblatt,
- Ian Hodkinson,
- Carsten Lutz,
- Hiroakira Ono, and
- Philippe Schnoebelen.

This volume contains a selection of best papers presented at the conference; they have been carefully refereed again to ensure journal quality. The *AiML 2002 Best Paper Award* was given to Kai Engelhardt, Ron van der Meyden, and Kaile Su for their contribution "Modal Logics with a linear hierarchy of local propositional quantifiers."

We would like to acknowledge the enormous amount of work put in by the programme committee:

- Philippe Balbiani,
- Giovanna Corsi,
- Luis Farinas del Cerro,
- Larry Moss,
- Maarten de Rijke,
- Mark Reynolds,
- Renate Schmidt,
- Nobu-Yuki Suzuki,
- Heinrich Wansing,

- Frank Wolter,
- Michael Zakharyaschev

and the additional referees Carlos Areces, Sebastian Bauer, Henry Chinaski, Stephane Demri, Hans de Nivelle, Tim French, Olivier Gasquet, Roselle Gennari, Robert Goldblatt, Valentin Goranko, Rajeev Gore, Juan Heguiabehere, Andreas Herzig, Ian Hodkinson, Ullrich Hustadt, Makoto Kanazawa, Satoshi Kobayashi, Roman Kontchakov, Agi Kurucz, Carsten Lutz, Maarten Marx, Fabio Massacci, Stephan Merz, Hiroakira Ono, Ulrike Sattler, Valentin Shehtman, Tatsuya Shimura, Dmitry Tishkovsky.

We are also grateful to Jane Spurr, Head of King's College Publications, for her efficient and dedicated work on this volume.

The organization of the conference was financially supported by the following institutions:

- Conseil régional Midi-Pyrénées,
- Fédération de recherche en informatique et en automatique,
- Institut de recherche en informatique de Toulouse, and
- Université Paul Sabatier.

Information about AiML and related events is available at http://www.aiml.net.

Philippe Balbiani, Nobu-Yuki Suzuki, Frank Wolter,
and Michael Zakharyaschev

Toulouse, Shizuoka, Liverpool, and London

1
Modal Logics with a Linear Hierarchy of Local Propositional Quantifiers

KAI ENGELHARDT, RON VAN DER MEYDEN AND KAILE SU

ABSTRACT. Local propositions arise in the context of the semantics for logics of knowledge in multi-agent systems. A proposition is local to an agent when it depends only on that agent's local state. We consider a logic, LLP, that extends S5, the modal logic of necessity (in which the modality refers to truth at all worlds) by adding a quantifier ranging over the set of all propositions and, for each agent, a propositional quantifier ranging over the agent's local propositions. LLP is able to express a large variety of epistemic modalities, including knowledge, common knowledge and distributed knowledge. However, this expressiveness comes at a cost: the logic is equivalent to second order predicate logic when two independent agents are present [5], hence undecidable and not axiomatizable. This paper identifies a class of multi-agent S5 structures, *hierarchical structures*, in which the agents' information has the structure of a linear hierarchy. All systems with just a single agent are hierarchical. It is shown that LLP becomes decidable with respect to hierarchical systems. The main result of the paper is the completeness of an axiomatization for the hierarchical case.

1 Introduction

Although modal logics are most commonly studied in their propositional forms, which may often be interpreted in first order logic, the topic of modal logics with second order expressive power is almost as old as the field itself. Motivated by earlier applications of quantification over propositions in the context of conditional logic, Kripke's early work on the semantics of modal logic [14] already contains an axiomatization of the logic S5 combined with quantification over propositions.

Kripke's assumptions concerning the range of the propositional quantifiers were somewhat unnatural, however. Further progress in the area was made in the 1960's, with Fine, Bull, and Kaplan [8, 2, 13] establishing axiomatizations and decidability results of several more natural variants. Decidability of S5 plus propositional quantification can be seen from the decidability of monadic second order logic [1].

The above works consider logics with a single modality. Motivated by applications of the logic of knowledge in distributed systems [10], Engelhardt, van der Meyden, and Moses [5] consider propositional quantification in a multi-modal S5 setting. A natural notion in this setting is that of *local propositions*. Local propositions arise in the context of the semantics for *logics of knowledge* in distributed systems due to Halpern and Moses [10]. In this semantics, each agent in a system has in each configuration of the system a local state, representing the information it has about the global state of the system and its history. A proposition is local for the agent when it is a function of this local state.

Engelhardt, van der Meyden, and Moses [5] show that S5 (with the modality interpreted as truth in all possible worlds) when combined with quantification over local propositions provides a framework that is able to express a large variety of epistemic modalities, including knowledge, common knowledge, and distributed knowledge [6].

In the most general case, in which more than one independent agent is involved, the logic obtained by adding local propositional quantifiers to S5 is highly undecidable [5], (indeed, equivalent to second order predicate logic) hence this logic is also not axiomatizable. After introducing syntax and semantics of this logic in Sect. 2, we identify a natural restriction on the semantic structures, expressing that the agents' information has the structure of a linear hierarchy, with respect to which the logic is decidable even in the multi-agent case. The single agent case satisfies this condition. A practical example in which hierarchical structures arise is information-based models for computer security, which frequently assume a linear hierarchy of secrecy levels [4]. We provide an axiomatization for the hierarchical case in Sect. 3. Bounded model property, decidability, and completeness follow from a normal form result presented in Sect. 4. Sect. 5 concludes and indicates directions for future work.

2 Syntax and semantics

We deal with a class of Kripke structures suited to a multi-modal logic, of the kind used in the literature on reasoning about knowledge [6]. Let *Prop* be an infinite set of propositional variables. A *Kripke structure for n agents* $M = (W, R_1, \ldots, R_n, \pi)$ consists of a set W of possible worlds, binary accessibility relations $R_1, \ldots, R_n \subseteq W^2$, and an assignment $\pi : Prop \longrightarrow 2^W$ of propositions to the propositional variables. Each relation R_i represents the information available to agent i. Intuitively, $u\,R_i\,v$ for two worlds u and v when, from agent i's point of view, these worlds are indistinguishable. As is usual in the literature on knowledge, we assume that the accessibility relations R_i are equivalence relations. Structures satisfying this restriction are called $S5_n$ *structures*.

A *proposition* in such a structure is a subset of the set of worlds W, intuitively, just the worlds where the proposition is true. We will be concerned with a special class of propositions. For each agent i, a proposition is called *i-local* if it is a union of R_i-equivalence classes. Intuitively, the i-local propositions are those that depend only upon agent i's information. Agent i can always determine whether a given i-local proposition is true.

While the class of $S5_n$ structures is standard, the language we consider is not the usual one from the knowledge literature. Write p for a typical element of *Prop*. Define the language $\mathcal{L}_{(\forall,\forall_1,\ldots,\forall_n,\Box)}$ with typical element ϕ by:

$$\phi ::= p \mid \neg \phi \mid \phi \wedge \phi \mid \Box \phi \mid \forall p\, (\phi) \mid \forall_i p\, (\phi)$$

We employ parentheses to indicate aggregation and take *true*, $\vee$, $\Diamond$, $\exists p\,()$, and the remaining familiar connectives to be defined in the usual way. Intuitively, $\Box \phi$ says that ϕ is true in all possible worlds. The formula $\forall p\,(\phi)$ says that ϕ is true for all assignments of a proposition to the propositional variable p. The meaning of $\forall_i p\,(\phi)$ is similar, but here p is restricted to range over i-local propositions.

These intuitions are made precise as follows. Another $S5_n$ structure $M' = (W', R'_1, \ldots, R'_n, \pi')$ is a p-variant of M (denoted $M' \simeq_p M$) if it differs from M at most on the interpretation of p. That is, $M' \simeq_p M$ if we have $W' = W$, the relation $R'_i = R_i$ for each $i = 1\ldots n$, and $\pi'(q) = \pi(q)$ for all $q \in Prop \setminus \{p\}$. If, moreover, $\pi'(p)$ is i-local then M' is called an *i-local p-variant* (denoted $M' \simeq^i_p M$). Formulas are interpreted at a world w of a structure M by means of the satisfaction relation $\models$, defined inductively by:

$$\begin{aligned} M, w &\models p & &\text{iff} & w &\in \pi(p) \\ M, w &\models \neg \phi & &\text{iff} & M, w &\not\models \phi \\ M, w &\models \phi \wedge \psi & &\text{iff} & M, w &\models \phi \text{ and } M, w \models \psi \\ M, w &\models \Box \phi & &\text{iff} & \forall v &\in W\, (M, v \models \phi) \\ M, w &\models \forall p\,(\phi) & &\text{iff} & \forall M' &\simeq_p M\, (M', w \models \phi) \\ M, w &\models \forall_i p\,(\phi) & &\text{iff} & \forall M' &\simeq^i_p M\, (M', w \models \phi) \end{aligned}$$

Let $\mathcal{M}$ be a class of Kripke structures for n agents. We say that ϕ is *satisfiable* in $\mathcal{M}$ iff $M, w \models \phi$ for some $M \in \mathcal{M}$ and some possible world w of M.

We write $\mathcal{L}_{(o_1,\ldots,o_m)}$ for the language generated from a set *Prop* of propositional variables by $\vee$, $\neg$, and operators o_i. Fine, Bull, and Kaplan discussed axiomatizations of $\mathcal{L}_{(\forall,\Box)}$ [8, 2, 13] and Fine already claimed decidability.

The local propositional quantifiers were introduced in [5]. These quantifiers are of interest because they are able to express many of the knowledge-like notions, including common knowledge and distributed knowledge, that have been discussed in the literature. This point is discussed at greater length elsewhere [5]. We confine ourselves here to an illustration using the standard knowledge operators K_i and modalities L_i expressing i-*locality* of formulas. The semantics of these operators is defined as follows.

$$M, w \models K_i \phi \quad \text{iff} \quad \forall v \in W \, (w \, R_i \, v \Rightarrow M, v \models \phi)$$
$$M, w \models L_i \phi \quad \text{iff} \quad \{ v \in W \mid M, v \models \phi \} \text{ is } i\text{-local}$$

The expressive power of $\mathcal{L}_{(\forall, \forall_1, \ldots, \forall_n, \Box)}$ is illustrated by the fact that the operators $\forall_i$, K_i, and L_i are interexpressible in the presence of $\Box$ and $\forall$:

$$K_i \phi \equiv \exists_i p \, (p \wedge \Box (p \to \phi)) \quad , \text{ provided } p \text{ not free in } \phi \tag{1}$$
$$L_i \phi \equiv \exists_i p \, (\Box (p \equiv \phi)) \quad , \text{ provided } p \text{ not free in } \phi \tag{2}$$
$$\forall_i p \, (\phi) \equiv \forall p \, (L_i p \to \phi) \tag{3}$$

Note that in $\mathcal{L}_{(\forall, \forall_1, \ldots, \forall_n, \Box)}$ one can express that the set of worlds at which p is true is an R_i-equivalence class, which we abbreviate to $\mathcal{E}_i p$:

$$\mathcal{E}_i p \equiv \Diamond p \wedge L_i p \wedge \forall_i q \, (\Diamond (p \wedge q) \to \Box (p \to q)) \tag{4}$$

Unfortunately, the expressiveness of $\mathcal{L}_{(\forall, \forall_1, \ldots, \forall_n, \Box)}$ comes at a cost. Even if there are just two agents, satisfiability is undecidable for this language [5], when interpreted over all $S5_n$ structures. As the use of two agents appears essential in this result, this motivates the consideration of two variants: (i) the single agent case, and (ii) restricted classes of structures. We focus in this paper on the following class that subsumes both variants. An $S5_n$ structure $(W, R_1, \ldots, R_n, \pi)$ is *hierarchical* if $R_{i-1} \subseteq R_i$, for $1 < i \leq n$. That is, in such structures, i-local propositions are also $(i - 1)$-local, and $\forall_i$-quantification is weaker than $\forall_{i-1}$-quantification. Note that every $S5_1$ structure is hierarchical, so results on hierarchical structures also apply to the single agent case.

3 A proof system

Our axiomatization of hierarchical $\mathcal{L}_{(\forall, \forall_1, \ldots, \forall_n, \Box)}$ below consists of an adaptation of Fine's for $\mathcal{L}_{(\forall, \Box)}$ [8]. Fine's axiomatization consists of the following. For the propositional basis, we have the usual axioms and inference rule:

all substitution instances of propositional tautologies **PC**

$$\phi \to \psi, \phi \vdash \psi \qquad \textbf{MP}$$

The operator $\Box$ is interpreted as necessity, so we have the axioms and inference rules of S5 for this modality:

$$\vdash \Box\phi \to \phi \qquad \textbf{T}_\Box$$
$$\vdash \Diamond\phi \to \Box\Diamond\phi \qquad \textbf{5}_\Box$$
$$\vdash \Box(\phi \to \psi) \to (\Box\phi \to \Box\psi) \qquad \textbf{K}_\Box$$
$$\phi \vdash \Box\phi \qquad \textbf{RN}_\Box$$

For the universal quantification over propositions, we have a set of axioms and a rule of inference that resemble closely the usual rules for predicate logic, the main difference being that we substitute formulas where we would substitute terms in predicate logic:

$$\vdash \forall p\,(\phi) \to \phi[^\psi/_p] \text{ , where } \psi \text{ is free for } p \text{ in } \phi \qquad \textbf{1}_\forall$$
$$\vdash \forall p\,(\phi \to \psi) \to (\forall p\,(\phi) \to \forall p\,(\psi)) \qquad \textbf{K}_\forall$$
$$\vdash \phi \to \forall p\,(\phi) \text{ , where } p \text{ not free in } \phi \qquad \textbf{N}_\forall$$
$$\phi \vdash \forall p\,(\phi) \qquad \textbf{G}$$

Additionally, we have an axiom that captures the fact that the lattice of propositions in a Kripke structure is atomic. The following axiom can be understood as saying that the proposition p, which is true only at the current world, is a subset of all propositions true at the current world.

$$\vdash \exists p\,(p \wedge \forall q\,(q \to \Box(p \to q))) \qquad \textbf{AT}$$

Fine [8] established that the above axioms are sound and complete for $\mathcal{L}_{(\forall,\Box)}$.

We note the following property of the axiomatization. Let ψ be a formula with free variables including e and p. Let $p_1, p_2 \in Prop$ be free for p in ψ. The formula

$$\forall p_1\,(\forall p_2\,(\Box(e \to (p_1 \equiv p_2)) \to (\psi[^{p_1}/_p] \equiv \psi[^{p_2}/_p])))$$

expresses that the truth value of ψ depends only on the values of p in e. Then for formulas $\alpha_1, \ldots, \alpha_k$ free for e in a formula ψ we have

$$\vdash \begin{pmatrix} \forall e\,(\forall p_1, p_2\,(\Box(e \to (p_1 \equiv p_2)) \to (\psi[^{p_1}/_p] \equiv \psi[^{p_2}/_p]))) \\ \wedge \bigwedge_{i \neq j} \Box(\neg(\alpha_i \wedge \alpha_j)) \\ \to \exists p\,\left(\bigwedge_{i=1}^{k} \psi[^{\alpha_i}/_e]\right) \end{pmatrix} \qquad \textbf{Ch}$$

Intuitively, this formula says that if ψ depends only on the values of p in e then we may take a set of propositions that have some desirable properties on a finite, mutually exclusive collection of pieces e of the model (defined by the α_i), and combine these propositions into a single proposition p that satisfies ψ on each piece e.

We now extend the axioms above to an axiomatization for $\mathcal{L}_{(\forall,\forall_1,\ldots,\forall_n,\Box)}$. (In this axiomatization, we take the previous axioms **PC** – **AT** to refer to $\mathcal{L}_{(\forall,\forall_1,\ldots,\forall_n,\Box)}$-formulas.) We add three sets of axioms. The first set consists of axioms relating the local propositional quantifiers to the standard propositional quantifier. The second set concerns the hierarchy between the local quantifiers. The final axiom generalizes the finite choice property.

For each of the universal quantification operators over i-local propositions for $1 \leq i \leq n$, the following axioms characterize the local quantifiers. We use $L_i \phi$ as an abbreviation for $\exists_i p (\Box(p \equiv \phi))$, which expresses that ϕ is an i-local proposition, as noted above. The first axiom characterizes i-local propositional quantification as the restriction of propositional quantification to i-local propositions:

$$\vdash \forall_i p (\phi) \equiv \forall p (L_i p \to \phi) \qquad \textbf{Def}\forall_i$$

The following axiom says that locality of a proposition is independent of the world.

$$\vdash L_i \phi \to \Box L_i \phi \qquad \textbf{NL}_i$$

The following is a variant of the atomicity axiom, but dealing with local propositions.

$$\vdash \exists_i p (p \wedge \Box(p \to \forall_i q (q \to \Box(p \to q)))) \qquad \textbf{AT}_i$$

Intuitively, this axioms says that there exists an i-local proposition p which is, at every world where p holds, the smallest i-local proposition holding at that world. Next, we have an axiom that says that local propositions are closed under union.

$$\vdash \forall p (\theta(p) \to L_i p) \to L_i(\exists q (\theta(q) \wedge q)) \qquad \textbf{U}_i$$

where $\theta(q)$ is some formula with free variable q. To establish the hierarchy on the agents' knowledge, for $1 < i \leq n$, we have the following:

$$\vdash \forall p (\phi) \to \forall_1 p (\phi) \qquad \textbf{H}_\forall$$
$$\vdash \forall_{i-1} p (\phi) \to \forall_i p (\phi) \qquad \textbf{H}_{\forall_i}$$

The final axiom generalizes the finite choice property **Ch** noted above:

$$\vdash \begin{pmatrix} \forall e \left(\forall p_1, p_2 \left(\Box(e \to (p_1 \equiv p_2)) \right) \to (\psi[^{p_1}/_p] \equiv \psi[^{p_2}/_p]) \right) \\ \wedge \, \forall e, e' \left(\theta \wedge \theta[^{e'}/_e] \right) \to (\Diamond(e \wedge e') \to \Box(e \equiv e')) \\ \to (\forall e \, (\theta \to \exists p \, (\psi)) \to \exists p \, (\forall e \, (\theta \to \psi))) \end{pmatrix} \quad \textbf{Choice}$$

where θ is a formula not containing p free. The antecedent of this axiom states that ψ depends only on the values of p in e, and that the formula θ defines a collection of disjoint sets of worlds. The conclusion of the formula says that if for each of these sets of worlds e there exists proposition p making ψ true, then there is a single proposition p that makes ψ true for all the sets e satisfying θ simultaneously. This axiom is valid if the axiom of choice holds in our semantic meta-theory.

This axiom is also valid for $\mathcal{L}_{(\forall,\Box)}$, but does not need to be stated as part of the axiomatization for that language because of its limited expressive power. A difference that emerges in $\mathcal{L}_{(\forall,\forall_1,\ldots,\forall_n,\Box)}$ is that we may take θ to express properties such as $\mathcal{E}_i e$, i.e., "e is an R_i-equivalence class", that define an infinite collection of disjoint sets. This is not possible for $\mathcal{L}_{(\forall,\Box)}$. (This impossibility claim follows from the completeness proof for $\mathcal{L}_{(\forall,\Box)}$, which is given below.)

Write LLPH$_n$ for the above set of axioms and inference rules. We say that ϕ is *derivable*, and write $\vdash \phi$, when the formula ϕ can be derived using the axioms and rules of inference above. The main result of the paper is the following.[1]

THEOREM 1. *Assume the axiom of choice holds meta-theoretically. Then* LLPH$_n$ *is a sound and complete axiomatization of* $\mathcal{L}_{(\forall,\forall_1,\ldots,\forall_n,\Box)}$ *with respect to hierarchical* $S5_n$ *structures. The language has the finite model property with respect to these structures.*

3.1 Some theorems of LLPH$_n$

We list some derived rules and theorems of LLPH$_n$. We first note that a set of axioms and a rule for the local quantifiers, corresponding to the axioms and rule for the global propositional quantifier, can be straightforwardly derived using **Def**$\forall_i$:

$$\vdash \forall_i p \, (\phi) \to (L_i \psi \to \phi[^{\psi}/_p]) \, , \text{ where } \psi \text{ is free for } p \text{ in } \phi \quad \textbf{1}_{\forall_i}$$

$$\vdash \forall_i p \, (\phi \to \psi) \to (\forall_i p \, (\phi) \to \forall_i p \, (\psi)) \quad \textbf{K}_{\forall_i}$$

$$L_i p \to \phi \vdash \forall_i p \, (\phi) \quad \textbf{G}_i$$

$$\vdash \phi \to \forall_i p \, (\phi) \, , \text{ where } p \text{ not free in } \phi \quad \textbf{N}_{\forall_i}$$

---

[1] We have not investigated completeness when we do not assume the axiom of choice semantically.

The Barcan formula is derivable using the same argument used to derive it in predicate S5 [11].

$$\vdash \forall p\,(\Box \phi) \equiv \Box \forall p\,(\phi)$$

Using $\mathbf{NL}_i$, we also obtain the Barcan formula for the i-local quantifiers.

$$\vdash \forall_i p\,(\Box \phi) \equiv \Box \forall_i p\,(\phi)$$

We can also establish some additional properties of the locality predicate. Using $\mathbf{AT}_i$ and $\mathbf{U}_i$, we obtain that the set of local propositions is closed under complementation:

$$\vdash L_i \phi \to L_i \neg \phi \qquad\qquad \mathbf{Comp}_i$$

Defining $K_i \phi$ as an abbreviation for $\exists_i p\,(p \land \Box(p \to \phi))$, we can now derive that K_i satisfies the axioms and rule of S5:

$$\vdash K_i(\phi \to \psi) \to (K_i \phi \to K_i \psi) \qquad \mathbf{K}_{K_i}$$
$$\vdash K_i \phi \to \phi \qquad \mathbf{T}_{K_i}$$
$$\vdash \neg K_i \phi \to K_i \neg \phi \qquad \mathbf{5}_{K_i}$$
$$\phi \vdash K_i \phi \qquad \mathbf{N}_{K_i}$$

The proof of $\mathbf{K}_{K_i}$ uses the fact that the local propositions are closed under intersection, which follows from $\mathbf{Comp}_i$ and $\mathbf{U}_i$. The proof of $\mathbf{5}_{K_i}$ uses $\mathbf{AT}_i$. Finally, we have some derivable formulas that express properties of R_i-equivalence classes and their relation to local propositions:

$$\vdash \mathcal{E}_i e \to \Box \mathcal{E}_i e \qquad \mathbf{EC}_1$$
$$\vdash \mathcal{E}_i e \land \forall e'\,(\Box(e' \to e) \land \mathcal{E}_{i-1} e' \to \Box(e' \to \phi)) \to \Box(e \to \phi) \qquad \mathbf{EC}_2$$
$$\vdash \forall e\,(\mathcal{E}_i e \to \Box(e \to p) \lor \Box(e \to \neg p)) \to L_i p \qquad \mathbf{EC}_3$$
$$\vdash (\mathcal{E}_i e \land \Diamond(e \land p) \land \Diamond(e \land \neg p)) \to \neg L_i p \qquad \mathbf{EC}_4$$

$\mathbf{EC}_1$ says that whether or not a proposition is an equivalence class is independent of the world of evaluation. The fact that an R_i-equivalence class is the union of the R_{i-1}-equivalence classes it contains is expressed by $\mathbf{EC}_2$. The formulas $\mathbf{EC}_3$ and $\mathbf{EC}_4$ allow us to derive conclusions about the i-locality of a proposition. Intuitively, a proposition is i-local if it is determinate within each R_i-equivalence class.

4 A normal form

The proof of the completeness result for LLPH$_n$ is by means of a normal form for $\mathcal{L}_{(\forall,\forall_1,\ldots,\forall_n,\Box)}$ over hierarchical structures. A side effect of the normal form result is a proof that the logic is decidable.

4.1 Completeness for $\mathcal{L}_{(\forall,\Box)}$

Since the construction for $\mathcal{L}_{(\forall,\forall_1,\ldots,\forall_n,\Box)}$ is moderately complex, we first illustrate the idea of the proof on the restricted case of $\mathcal{L}_{(\forall,\Box)}$. (Note that since models for this language have no accessibility relations, all structures are hierarchical.) We discuss the generalizations required to deal with $\mathcal{L}_{(\forall,\forall_1,\ldots,\forall_n,\Box)}$ in the next section. (The axiomatization LLPH$_0$ for $\mathcal{L}_{(\forall,\Box)}$ omits the axioms for the local quantifiers and their hierarchy.)

Decidability and completeness for $\mathcal{L}_{(\forall,\Box)}$ were already established by Fine and Kaplan [8, 13]. They have proofs based on the *graded modalities* [9, 12, 7, 3].[2] These are the unary modalities C_l, for l a natural number. The semantics for the graded modal logic $\mathcal{L}_{(\mathsf{C}_1,\mathsf{C}_2,\ldots)}$ is based on the same set of structures as used for $\mathcal{L}_{(\forall,\Box)}$, i.e., Kripke structures for 0 agents of the form $M = (W, \pi)$. The semantics of the graded modalities is defined by $M, w \models \mathsf{C}_l \phi$ if there are at least l distinct possible worlds $v \in W$ with $M, v \models \phi$.

Note that we may express the graded modal logic formula $\mathsf{C}_l \phi$ in $\mathcal{L}_{(\forall,\Box)}$ as[3]

$$\exists q_1 \ldots \exists q_l \left(\bigwedge_{1 \leq i < j \leq l} \Box[q_i \to \neg q_j] \land \bigwedge_{i=1}^{l} (\Diamond q_i \land \Box[q_i \to \phi]) \right)$$

where q_i is the i-th propositional variable not free in ϕ and $1 \leq i \leq l$. We use $\mathsf{E}_l \phi$ as an abbreviation for the translation of the graded modal logic formula $\mathsf{C}_l \phi \land \neg \mathsf{C}_{l+1} \phi$, which states that there are exactly l worlds satisfying ϕ. We also define the abbreviation $\mathsf{M}_{l,N} \phi$, to be $\mathsf{E}_l \phi$ if $l < N$ and $\mathsf{C}_l \phi$ if $l \geq N$.

Let $\mathbf{p} = p_1, \ldots, p_m$ be a vector of propositional variables. Define a *point atom* for $\mathbf{p}$ to be a formula of the form $l_1 \land \ldots \land l_m$ where each l_i is either p_i or $\neg p_i$. Write $\mathrm{PA}(\mathbf{p})$ for the set of point atoms of $\mathbf{p}$.

Given a point atom a for $\mathbf{p}$ and a number N, define an *N-bounded count* of a to be either a formula of the form $\mathsf{E}_l a$ where $l < N$, or the formula

---

[2] As the proofs in the readily accessible literature are sketchy, we do not know if the completeness arguments used by these authors are the same as those we present. Our expression of the graded modalities is similar to those by Fine and Kaplan. Fine states that decidability can be proved by a quantifier elimination argument using the graded modalities and that this can be used to prove completeness but does not provide details of the normal form.

[3] The approach used here results in an exponential blowup when l is expressed in binary form since the quantifier prefix then has length exponential in the length of l. However, there exists another translation that involves only a linear blowup in this case.

$C_N a$. Define a $(\mathbf{p}, k)$-*atom* to be a formula of the form

$$a \wedge \bigwedge_{b \in PA(\mathbf{p})} c_b$$

where a is a point atom for $\mathbf{p}$ and c_b is an 2^k-bounded count of b for each $b \in PA(\mathbf{p})$, such that c_a is not $E_0 a$. We write $At(\mathbf{p}, k)$ for the set of $(\mathbf{p}, k)$-atoms. These atoms have the following properties.

LEMMA 2.

1. *If* $A, A' \in At(\mathbf{p}, k)$ *are distinct atoms, then* $\vdash \neg(A \wedge A')$.

2. $\vdash \bigvee_{A \in At(\mathbf{p}, k)} A$

3. *If* $A, A' \in At(\mathbf{p}, k)$ *then* $\vdash A \to \Diamond A'$ *or* $\vdash A \to \neg \Diamond A'$

4. *If* $A \in At(\mathbf{p}, k+1)$ *and* $B \in At(\mathbf{p} \cdot q, k)$ *then* $\vdash A \to \exists q (B)$ *or* $\vdash A \to \neg \exists q (B)$.

Sketch of the proof For part 1, note that if A and A' are distinct then they differ either in their point atoms, or they disagree about the count of some atom b. In the first case, we have $\vdash \neg(A \wedge A')$ by propositional reasoning, and in the second case, we get the result by reasoning about the quantificational encoding of the counting modalities.

Part 2 is established by noting that $\vdash \bigvee_{k=0}^{N-1} E_k(b) \vee C_N(b)$ for each point atom b. The result is obtained by conjoining these formulas, distributing, and then eliminating cases containing a conjunct of the form $a \wedge C_0 a$ using the fact that $\vdash a \to \neg C_0 a$.

For part 3, let $A = a \wedge \bigwedge_{b \in PA(\mathbf{p})} c_b$ and $A' = a' \wedge \bigwedge_{b \in PA(\mathbf{p})} c'_b$. We consider two cases, depending on whether there exists a point atom b such that $c_b \neq c'_b$. Suppose first that such an atom b exists. Note that $\vdash c_b \to \Box c_b$. It follows by S5 reasoning that $\vdash A \wedge \Diamond A' \to \Diamond (c_b \wedge c'_b)$, which gives that $\vdash A \to \neg \Diamond A'$ since $\vdash \neg (c_b \wedge c'_b)$. In the second case, we have $c_b = c'_b$ for all point atoms b. In particular, $c_{a'} = c'_{a'}$ is not $E_0 a'$. It follows that $\vdash A \to \Diamond (a')$. Moreover, for each point atom b we have $\vdash c_b \to \Box c_b$, so it follows that $\vdash A \to \Diamond (a' \wedge \bigwedge_{b \in PA(\mathbf{p})} c_b))$. Since $c_b = c'_b$ for each b, this is just $\vdash A \to \Diamond (A')$.

For part 4, let A be a $(\mathbf{p}, k+1)$-atom of the form $a \wedge \bigwedge_{b \in PA(\mathbf{p})} c_b$ where each c_b is a 2^{k+1}-bounded count of b. Let B be a $(\mathbf{p} \cdot q, k+1)$-atom, which we write in the form $a' \wedge \bigwedge_{b \in PA(\mathbf{p})} (c_b^+ \wedge c_b^-)$. Here c_b^+ is a 2^k bounded count of $b \wedge q$ and c_b^- is a 2^k bounded count of $b \wedge \neg q$.

Define a *q-partition* of a 2^{k+1} bounded count c_b of an atom $b \in PA(\mathbf{p})$ to be a formula of the form $c_+ \wedge c_-$ where c_+ is a 2^k bounded count of $b \wedge q$ and c_- is a 2^k bounded count of $b \wedge \neg q$, subject to the following:

1. if c_b is $\mathsf{E}_l b$, then c_+ is $\mathsf{M}_{l_+, 2^k}(b \wedge q)$ and c_- is $\mathsf{M}_{l_-, 2^k}(b \wedge \neg q)$ where $l_+ + l_- = l$;

2. if c_b is $\mathsf{C}_{2^{k+1}} b$ then

   (a) c_+ is $\mathsf{C}_{2^k}(b \wedge q)$ and c_- is $\mathsf{C}_{2^k}(b \wedge \neg q)$, or
   (b) c_+ is $\mathsf{C}_{2^k}(b \wedge q)$ and c_- is $\mathsf{E}_{l_-}(b \wedge \neg q)$ where $l_- < 2^k$, or
   (c) c_+ is $\mathsf{E}_{l_+}(b \wedge q)$ where $l_+ < 2^k$ and c_- is $\mathsf{C}_{2^k}(b \wedge \neg q)$.

   That is, c_+ and c_- can be any combination of 2^k-bounded counts in which at least one has the operator C_{2^k}.

Then we may show that if $c^+ \wedge c^-$ is a q-partition of c_b, then we have $\vdash c_b \to \exists q\, (c^+ \wedge c^-)$. Moreover, we have $\vdash b \wedge c_b \to \exists q\, (b \wedge q \wedge c^+ \wedge c^-)$ and $\vdash b \wedge c_b \to \exists q\, (b \wedge \neg q \wedge c^+ \wedge c^-)$. Conversely, if $c^+ \wedge c^-$ is not a q-partition of c_b, then $\vdash c_b \to \neg \exists q\, (c^+ \wedge c^-)$.

Define B to be q-*compatible* with A if either $b = a \wedge q$ or $b = a \wedge \neg q$ and for all point atoms $b \in \mathrm{PA}(\mathbf{p})$ we have that $c_b^+ \wedge c_b^-$ is a q-partition of c_b. Then, if B is q-compatible with A, it follows from the observations of the previous paragraph that

$$\vdash A \to \exists q\, (a' \wedge c_a^+ \wedge c_a^-) \wedge \bigwedge_{b \in \mathrm{PA}(\mathbf{p}) \setminus \{a\}} \exists q\, (c_b^+ \wedge c_b^-) \ .$$

Using **Ch** and the fact that point atoms correspond to mutually exclusive propositions, we obtain that $\vdash A \to \exists q\, (B)$. Conversely, if B is not q-compatible with A, then we have either that the point atom a' is neither $a \wedge q$ nor $a \wedge \neg q$, or there exists a point atom b such that $c_b^+ \wedge c_b^-$ is not a q-partition of c_b. If a' is neither $a \wedge q$ nor $a \wedge \neg q$, we plainly have $\vdash A \to \neg \exists q\, (B)$. If $c_b^+ \wedge c_b^-$ is not a q-partition of c_b, then $\vdash A \to \neg \exists q\, (B)$ follows using the observation of the previous paragraph. ∎

Using Lemma 2, we may now establish the following result.

LEMMA 3. *If A is a $(\mathbf{p}, k)$-atom and ϕ is a formula of $\mathcal{L}_{(\forall, \Box)}$ of quantification depth at most k with free variables amongst $\mathbf{p}$, then either $\vdash A \to \phi$ or $\vdash A \to \neg \phi$.*

Proof. By induction on k, and, within each k, on the complexity of ϕ. The cases where ϕ is a propositional variable or formed from simpler formulas using negation or conjunction are straightforward, from completeness of the propositional fragment of the logic.

Suppose ϕ is $\Box \psi$. (Here ϕ and ψ have the same quantification depth k.) We consider two cases, depending on whether there exists a point atom $A' \in \mathrm{At}(\mathbf{p}, k)$ such that $\vdash A \to \Diamond A'$ and $\vdash A' \to \psi$. If this is the case, then it follows that $\vdash A \to \Diamond\psi$, i.e. $\vdash A \to \neg\Box\psi$, by S5 reasoning. Otherwise, for all $A' \in \text{At}(\mathbf{p}, k)$, if $\vdash A \to \Diamond A'$ then not $\vdash A' \to \psi$. Let $\tilde{A}$ be the set of $(\mathbf{p}, k)$-atoms A' such that $\vdash A \to \Diamond A'$. By the induction hypothesis, it follows that $\vdash A' \to \neg\psi$ for all $A' \in \tilde{A}$. Now by Lemma 2(3), we have that $\vdash A \to \Box\neg A'$ for all $(\mathbf{p}, k)$-atoms not in $\tilde{A}$. Hence, by Lemma 2(2) and S5 reasoning, we obtain that $\vdash A \to \Box(\bigvee_{A' \in \tilde{A}} A')$. It follows that $\vdash A \to \Box\psi$.

The proof of the case where ψ is $\forall q(\psi)$ is almost identical to the proof of the case for $\Box\psi$, with the following exceptions. In this case, we take ϕ to have quantification depth $k+1$, so A is a $(\mathbf{p}, k+1)$-atom. The formula ψ in this case has quantification depth k, and in place of the $(\mathbf{p}, k+1)$-atoms A' in the preceding paragraph, we use $(\mathbf{p} \cdot q, k)$-atoms B. The rest of the argument follows as above, but we use Lemma 2(4) in place of Lemma 2(3) ∎

For the completeness proof, we now argue as follows. Let ϕ be a consistent formula of quantification depth k with free variables $\mathbf{p}$. By Lemma 2(1), we have $\vdash \bigvee_{A \in \text{At}(\mathbf{p},k)} A$. Hence, by Lemma 3 there exists a $(\mathbf{p}, k)$-atom A such that $\vdash A \to \phi$. It is now straightforward to construct a finite model M with a world w such that $M, w \models A$, since this amounts merely to creating the right number of worlds for each point atom. It now follows that $M, w \models \phi$ by soundness. Clearly, the proof also establishes the finite model property.

4.2 Dealing with $\mathcal{L}_{(\forall, \forall_1, \ldots, \forall_n, \Box)}$

We now show how the proof idea of the previous section can be generalized to give a normal form and completeness proof for $\mathcal{L}_{(\forall, \forall_1, \ldots, \forall_n, \Box)}$. The basic structure of the completeness proof will be the same: we identify a kind of atom generalizing the $(\mathbf{p}, k)$-atoms of the previous section, and prove a result analogous to Lemma 2.

As above, atoms count certain sorts of objects up to a given bound, which depends on the formula we are dealing with. However, where for $\mathcal{L}_{(\forall, \Box)}$ it suffices to count worlds satisfying a point atom, we now also need to count equivalence classes, and to distinguish equivalence classes having different internal structure in this count. We distinguish equivalence classes of the most highly informed agent (agent 1) according to the number of worlds of each propositional type they contain. That is, we relativize the normal form formulas of the previous section to each equivalence class of agent 1, and then count the number of distinct such relativizations that we obtain, just as we did with worlds in the previous section. This gives a normal form for the language $\mathcal{L}_{(\forall, \forall_1, \Box)}$. To deal with additional agents, we recursively apply the construction, using at each level the normal form for the previous level.

As we proceed up the levels, when dealing with a formula ϕ of quantification depth k, the bound up to which we count (which was 2^k in the previous section) increases, depending on i (an agent), $\mathbf{p}$ (the list of free variables of ϕ) and the quantification depth k.

Recall that $\mathcal{E}_i e$ expresses that the set of worlds satisfying e is an R_i-equivalence class. Consequently, just as $\mathcal{L}_{(\forall,\Box)}$ permits counting of worlds, $\mathcal{L}_{(\forall,\forall_1,\ldots,\forall_n,\Box)}$ permits counting of equivalence classes. We write $\mathsf{E}_k^i x(\phi)$ as an abbreviation for the formula expressing that there exists exactly k distinct R_i-equivalence classes x such that ϕ holds. We use $x \subseteq y$ as an abbreviation for $\Box(x \to y)$. We will also write $\mathsf{E}_k^i x \subseteq y(\phi)$ for $\mathsf{E}_k^i x(x \subseteq y \wedge \phi)$. Similarly, we write $\mathsf{C}_k^i x(\phi)$ as an abbreviation for the formula expressing that there exists at least k distinct R_i-equivalence classes x such that ϕ holds, and use $\mathsf{C}_k^i x \subseteq y(\phi)$ for $\mathsf{C}_k^i x(x \subseteq y \wedge \phi)$. For uniformity, it is convenient to treat E_k^0 and C_k^0 as notations for E_k and C_k, respectively.

It is convenient to represent the normal form by means of the following objects, which we call $(i, \mathbf{p}, k)$-*trees*, where $\mathbf{p} = p_1, \ldots, p_m$ is a vector of propositional variables, k a natural number and $0 \le i \le n+1$. We write $\mathbf{p}^+$ for $p_1, \ldots, p_m, p_{m+1}$. The definition of the set $\mathcal{T}_{i,\mathbf{p},k}$ of $(i, \mathbf{p}, k)$-trees is by induction on i. Set $\mathcal{T}_{0,\mathbf{p},k} = 2^{\mathbf{p}}$, i.e., the power set of $\mathbf{p}$. For $i = 1, \ldots, n+1$, we define $\mathcal{T}_{i,\mathbf{p},k}$ to be the set of functions $u : \mathcal{T}_{i-1,\mathbf{p},k} \longrightarrow \{0, \ldots, N_{i-1,\mathbf{p},k}\}$ such that $u(t) \ne 0$ for some $t \in \mathcal{T}_{i-1,\mathbf{p},k}$. Here, the $N_{i,\mathbf{p},k}$ are the numbers defined by the following mutual recursion with the definition of $(i, \mathbf{p}, k)$-trees:

$$N_{0,\mathbf{p},k} = 2^k$$
$$N_{i,\mathbf{p},0} = 1$$
$$N_{i,\mathbf{p},k+1} = |\mathcal{T}_{i,\mathbf{p}^+,k}| \cdot N_{i,\mathbf{p}^+,k}.$$

Note that

$$|\mathcal{T}_{i,\mathbf{p},k}| = (1 + N_{i-1,\mathbf{p},k})^{|\mathcal{T}_{i-1,\mathbf{p},k}|}$$

It is not too difficult to verify that this recursion is well defined. A *branch* in a tree $t \in \mathcal{T}_{i,\mathbf{p},k}$ is a sequence of trees $t_i, t_{i-1}, \ldots, t_0$ where $t_j \in \mathcal{T}_{j,\mathbf{p},k}$ for each $j = 0, \ldots, i$ such that $t = t_i$ and $t_j(t_{j-1}) \ne 0$ for each $j = 1, \ldots, i$.

Each $(i, \mathbf{p}, k)$-tree corresponds to a formula as follows. It is convenient to define the expression $\mathsf{M}_l^{i,\mathbf{p},k} x \subseteq y(\phi)$ to be $\mathsf{E}_l^i x \subseteq y(\phi)$ if $l < N_{i,\mathbf{p},k}$ and to be $\mathsf{C}_{N_{i,\mathbf{p},k}}^i x \subseteq y(\phi)$ if $l \ge N_{i,\mathbf{p},k}$. With every $(i, \mathbf{p}, k)$-tree u, we associate a distinct propositional variable e_u. Given an $(i, \mathbf{p}, k)$-tree u, we define the formula ϕ_u, by induction on i. When $i = 0$, we define ϕ_u to be the point atom $\bigwedge_{j=1}^m l_j$, where $l_j = p_i$ if $p_j \in u$ and $l_j = \neg p_i$ otherwise. When $1 \le i \le n+1$, we define the formula corresponding to an $(i, \mathbf{p}, k)$-tree u to be the formula

$$\phi_u(e_u) = \bigwedge_{t \in \mathcal{T}_{i-1,\mathbf{p},k}} \mathsf{M}_{u(t)}^{i-1,\mathbf{p},k} e_t \subseteq e_u(\phi_t(e_t)).$$

Clearly, $\phi_t(e_t)$ has only the variables e_t and $\mathbf{p}$ free.

The normal form uses the set of distinct propositional variables $\mathbf{c} = \{c_1, \ldots, c_n\}$. Let $A_\mathbf{c}$ be the formula

$$\bigwedge_{j=1}^n (c_j \wedge \mathcal{E}_j c_j).$$

Intuitively, $A_\mathbf{c}$ says that each c_i corresponds to the R_i-equivalence class containing the current world.

From now on, we call a formula a $(\mathbf{p}, k)$-atom if it is of the form

$$\exists \mathbf{c}(A_\mathbf{c} \wedge \phi_{t_0} \wedge \bigwedge_{i=1}^n \phi_{t_i}(c_i) \wedge \phi_{t_{n+1}}(true)),$$

where $t_i \in \mathcal{T}_{i,\mathbf{p},k}$ for each i, and $t_{n+1}, \ldots, t_0$ is a branch in t_{n+1}. Again, we write $\mathrm{At}(\mathbf{p}, k)$ for the set of $(\mathbf{p}, k)$-atoms. These atoms have the following properties. The first four of these are identical to the properties of Lemma 2. However, we have some new properties relating to the local propositional quantifiers.

LEMMA 4.

1. If $A, A' \in \mathrm{At}(\mathbf{p}, k)$ are distinct atoms, then $\vdash \neg(A \wedge A')$.

2. $\vdash \bigvee_{A \in \mathrm{At}(\mathbf{p},k)} A$.

3. If $A, A' \in \mathrm{At}(\mathbf{p}, k)$ then $\vdash A \to \Diamond A'$ or $\vdash A \to \neg \Diamond A'$.

4. If $A \in \mathrm{At}(\mathbf{p}, k+1)$ and $B \in \mathrm{At}(\mathbf{p} \cdot q, k)$ then $\vdash A \to \exists q (B)$ or $\vdash A \to \neg \exists q (B)$.

5. If $A \in \mathrm{At}(\mathbf{p}, k)$ and p is one of the propositions in $\mathbf{p}$ then $\vdash A \to L_i p$ or $\vdash A \to \neg L_i p$.

6. If $A \in \mathrm{At}(\mathbf{p}, k+1)$ and $B \in \mathrm{At}(\mathbf{p} \cdot q, k)$ then $\vdash A \to \exists_i q (B)$ or $\vdash A \to \neg \exists_i q (B)$.

We defer discussion of the proof. Using this result, we may obtain a result directly analogous to Lemma 3.

LEMMA 5. If A is a $(\mathbf{p}, k)$-atom and ϕ is a formula of $\mathcal{L}_{(\forall, \forall_1, \ldots, \forall_n, \Box)}$ of quantification depth at most k, with free variables $\mathbf{p}$ then either $\vdash A \to \phi$ or $\vdash A \to \neg \phi$.

The proof of Lemma 5 is identical to that of Lemma 3, except that we now have a new case for ϕ of the form $\exists_i p\,(\phi')$. The proof for this case follows exactly along the lines of the proof for the case $\exists p\,(\phi')$ in the proof of Lemma 3, but uses Lemma 4(6) in place of Lemma 2(4).

As in the previous section, it then follows straightforwardly that a consistent formula is equivalent to a disjunction of $(\mathbf{p}, k)$-atoms, and we may give a completeness and finite model argument exactly as before.

The bulk of the work of the completeness proof is therefore in the proof of Lemma 4. We now sketch some of the key steps of this proof, focussing on part 4.

As a first observation, we note that the formulas $\phi_u(e)$ have the following basic properties:

LEMMA 6.

1. If $u, v \in \mathcal{T}_{i,\mathbf{p},k}$ and $u \neq v$ then $\vdash \neg(\phi_u(e) \wedge \phi_v(e))$.

2. $\vdash \mathcal{E}_i e \to \bigvee_{v \in \mathcal{T}_{i,\mathbf{p},k}} \phi_v(e)$.

One of the key steps in the proof of Lemma 2(4) is the notion of partition of a count. In the proof of Lemma 2, we dealt with counts of point atoms a, which split into two point atoms $a \wedge q$ and $a \wedge \neg q$ when taking a new proposition q into consideration. We now need to deal with a partition of a formula that counts equivalence classes satisfying a counting property, rather than a type of world. These nests of equivalence classes split into a larger and more complex collection of nests of equivalence classes when we add a new proposition.

To handle this, we introduce a notion of *compatibility* that generalizes the partition of a count of a point atom a into counts of the point atoms $a \wedge q$ and $a \wedge \neg q$. For $u \in \mathcal{T}_{i,\mathbf{p},k+1}$ we define a set $\mathcal{C}(u) \subseteq \mathcal{T}_{i,\mathbf{p}^+,k}$, of $(i, \mathbf{p}^+, k)$-trees. If $v \in \mathcal{C}(u)$ then we say that v is *compatible* with u. The definition is by induction on i, as follows. If $i = 0$ then $v \in \mathcal{C}(u)$ if $u = v \cap \{p_1, \ldots, p_m\}$. For $i > 0$, $v \in \mathcal{C}(u)$ if there exists a function $f : \mathcal{T}_{i-1,\mathbf{p},k+1} \times \mathcal{T}_{i-1,\mathbf{p}^+,k} \to [0, \ldots, N_{i-1,\mathbf{p}^+,k}]$ such that

C1. for all $t \in \mathcal{T}_{i-1,\mathbf{p},k+1}$ and all $t' \in \mathcal{T}_{i-1,\mathbf{p}^+,k}$, if $f(t,t') \neq 0$ then $t' \in \mathcal{C}(t)$,

C2. for all $t' \in \mathcal{T}_{i-1,\mathbf{p}^+,k}$, we have
$$v(t') = \min\left(N_{i-1,\mathbf{p}^+,k}, \sum_{t \in \mathcal{T}_{i-1,\mathbf{p},k+1}} f(t,t')\right), \text{ and}$$

C3. for all $t \in \mathcal{T}_{i-1,\mathbf{p},k+1}$, if $f(t,t') < N_{i-1,\mathbf{p}^+,k}$ for all $t' \in \mathcal{T}_{i-1,\mathbf{p}^+,k}$ then $u(t) = \sum_{t' \in \mathcal{T}_{i-1,\mathbf{p}^+,k}} f(t,t')$ otherwise $u(t) \geq \sum_{t' \in \mathcal{T}_{i-1,\mathbf{p}^+,k}} f(t,t')$.

We may then prove the following results that mirror the observations concerning partitions in the proof of Lemma 2.

LEMMA 7. *Let $u \in \mathcal{T}_{i,\mathbf{p},k+1}$ and $v \in \mathcal{T}_{i,\mathbf{p}+,k}$.*

1. *If $v \in \mathcal{C}(u)$ then $\vdash \phi_u \to \exists p_{m+1}(\phi_v)$.*

2. *If $v \notin \mathcal{C}(u)$ then $\vdash \phi_u \to \neg \exists p_{m+1}(\phi_v)$.*

Rather than give the full proof of this proof-theoretic result, we give the proof of a closely related semantic result, and describe the key steps required to mirror this proof within our proof system. The first part of Lemma 7 corresponds to the following semantic result.

PROPOSITION 8. *Suppose that $u \in \mathcal{T}_{i,\mathbf{p},k+1}$ and $v \in \mathcal{T}_{i,\mathbf{p}+,k}$. If $v \in \mathcal{C}(u)$ then $\models \phi_u \to \exists p_{m+1}(\phi_v)$.*

Proof. The proof is by induction on i. The base case is straightforward. When $i = 0$, if $v \in \mathcal{C}(u)$ then we have either $\phi_v = \phi_u \wedge p_{m+1}$ or $\phi_v = \phi_u \wedge \neg p_{m+1}$ and the claim is immediate from $\mathbf{1}_\forall$.

We now establish the inductive step. Let $i \in \{1, \ldots, n\}$, let $u \in \mathcal{T}_{i,\mathbf{p},k+1}$, let $v \in \mathcal{C}(u)$, and let f be the witness to compatibility. Suppose $M \models \phi_u(e)$. For each $t \in \mathcal{T}_{i-1,\mathbf{p},k+1}$, let $S(t)$ be the set of R_{i-1}-equivalence classes $U \subseteq \pi(e)$ such that $M[U/e] \models \phi_t(e)$. Then $|S(t)| = u(t)$ if $u(t) < N_{i-1,\mathbf{p},k+1}$ and $|S(t)| \geq N_{i-1,\mathbf{p},k+1}$ otherwise. Note that by Lemma 6(1), we have that if $t_1 \neq t_2$ are both $(i, \mathbf{p}, k+1)$-trees, then $S(t_1) \cap S(t_2) = \emptyset$.

For each $t \in \mathcal{T}_{i-1,\mathbf{p},k+1}$, we partition $S(t)$ into a disjoint union

$$S(t) = \bigcup_{t' \in \mathcal{T}_{i-1,\mathbf{p}+,k}} S(t,t')$$

in such a way that $|S(t,t')| = f(t,t')$ if $f(t,t') < N_{i-1,\mathbf{p}+,k}$ and $|S(t,t')| \geq f(t,t')$ if $f(t,t') = N_{i-1,\mathbf{p}+,k}$. The reason we can partition in this way is as follows. By **C3**, there are two possibilities.

1. The first possibility is that $f(t,t') < N_{i-1,\mathbf{p}+,k}$ for all $t' \in \mathcal{T}_{i-1,\mathbf{p}+,k}$. In this case, we have

$$u(t) = \sum_{t' \in \mathcal{T}_{i-1,\mathbf{p}+,k}} f(t,t') < |\mathcal{T}_{i-1,\mathbf{p}+,k}| \cdot N_{i-1,\mathbf{p}+,k} = N_{i-1,\mathbf{p},k+1}.$$

It follows that $|S(t)| = u(t) = \sum_{t' \in \mathcal{T}_{i-1,\mathbf{p}+,k}} f(t,t')$, so we partition $S(t)$ in such a way that $|S(t,t')| = f(t,t')$ for all $t' \in \mathcal{T}_{i-1,\mathbf{p}+,k}$.

2. The second possibility is that $f(t,t'_0) = N_{i-1,\mathbf{p}^+,k}$ for some t'_0. In this case we are still guaranteed that $u(t) \geq \sum_{t' \in \mathcal{T}_{i-1,\mathbf{p}^+,k}} f(t,t')$, and since $M \models \phi_u(e)$ we have that $|S(t)| \geq u(t)$. In this case we first partition $\sum_{t' \in \mathcal{T}_{i-1,\mathbf{p}^+,k}} f(t,t')$ of the elements of $S(t)$ as before, and then place any remaining elements of $S(t)$ in $S(t,t'_0)$. Clearly we then have $|S(t,t')| = f(t,t')$ for $t' \neq t'_0$, and both $|S(t,t')| \geq f(t,t')$ and $f(t,t') = N_{i-1,\mathbf{p}^+,k}$ in the case $t' = t'_0$, so the constraint on the partitioning is satisfied.

For each R_{i-1}-equivalence class U in $S(t,t')$, we have $M[U/e] \models \phi_t(e)$. By the induction hypothesis, we have $\models \phi_t \to \exists p_{m+1}(\phi_{t'})$ for all $t' \in \mathcal{C}(t)$. Since the sets $S(t,t')$ are disjoint, for each such set the equivalence classes $U \in S(t,t')$ are disjoint, the formulas $\phi_{t'}$ depend only on the values of p_m+1 in U, and the propositions witnessing these facts may be aggregated into a single proposition P such that $M[P/p_{m+1}, U/e] \models \phi_{t'}(e)$ for all $U \in S(t,t')$, all $t \in \mathcal{T}_{i-1,\mathbf{p},k+1}$ and all $t' \in \mathcal{T}_{i-1,\mathbf{p}^+,k}$.

Write $S'(t')$ for the set of R_{i-1}-equivalence classes U in $\pi(e)$ such that $M[P/p_{m+1}, U/e] \models \phi_{t'}(e)$. Then

$$S'(t') = \bigcup_{t \in \mathcal{T}_{i-1,\mathbf{p},k+1}} S(t,t')$$

is a partition of $S'(t')$. To see this, note that $S(t,t') \subseteq S'(t')$ for each t by construction of P. By Lemma 6(2), each R_{i-1}-equivalence class U in $\pi(e)$ is in $S(t)$ for some $t \in \mathcal{T}_{i-1,\mathbf{p},k+1}$, hence in $S(t,t'')$ for some $t'' \in \mathcal{T}_{i-1,\mathbf{p}^+,k}$. By Lemma 6(1) and construction of P, we cannot have $t'' \neq t'$ if $M[P/p_{m+1}, U/e] \models \phi_{t'}(e)$. This proves the containment in the other direction. The union is therefore a partition because all the sets $S(t,t')$ are disjoint.

We now claim that $M[P/p_{m+1}] \models \phi_v(e)$. We show this by establishing that $M[P/p_{m+1}] \models \mathsf{M}^{i-1,\mathbf{p}^+,k}_{v(t')} e' \subseteq e(\phi_{t'}(e'))$ for all $t' \in \mathcal{T}_{i-1,\mathbf{p}^+,k}$. There are two cases.

1. First, suppose $v(t') < N_{i-1,\mathbf{p}^+,k}$. Then by **C2**, we have that $v(t') = \sum_{t \in \mathcal{T}_{i-1,\mathbf{p},k+1}} f(t,t') < N_{i-1,\mathbf{p}^+,k}$, so $f(t,t') < N_{i-1,\mathbf{p}^+,k}$ for each $t \in \mathcal{T}_{i-1,\mathbf{p},k+1}$. Thus, for each $t' \in \mathcal{T}_{i-1,\mathbf{p}^+,k}$, we have

$$|S'(t')| = \sum_{t \in \mathcal{T}_{i-1,\mathbf{p},k+1}} |S(t,t')| = \sum_{t \in \mathcal{T}_{i-1,\mathbf{p},k+1}} f(t,t') = v(t'),$$

hence $M[P/p_{m+1}] \models \mathsf{E}^{i-1}_{v(t')} e' \subseteq e(\phi_{t'}(e'))$, as is required to establish $M[P/p_{m+1}] \models \mathsf{M}^{i-1,\mathbf{p}^+,k}_{v(t')} e' \subseteq e(\phi_{t'}(e'))$ in case that $v(t') < N_{i-1,\mathbf{p}^+,k}$.

2. In the other case, where $v(t') = N_{i-1,\mathbf{p}+,k}$, we have by **C2** that $\sum_{t\in\mathcal{T}_{i-1,\mathbf{p},k+1}} f(t,t') \geq N_{i-1,\mathbf{p}+,k}$. Since we always have $|S(t,t')| \geq f(t,t')$, we have

$$|S'(t')| = \sum_{t\in\mathcal{T}_{i-1,\mathbf{p},k+1}} S(t,t') \geq \sum_{t\in\mathcal{T}_{i-1,\mathbf{p},k+1}} f(t,t') \geq N_{i-1,\mathbf{p}+,k}.$$

Thus, $M[P/p_{m+1}] \models \mathsf{C}^{i-1}_{N_{i-1,\mathbf{p}+,k}} e' \subseteq e(\phi_{t'}(e'))$, again as required.

∎

The main difficulty in converting this proof into a proof of Lemma 7(1), is its use of the axiom of choice. To capture this in the proof system, we use axiom **Choice**, with the formula θ as $\mathcal{E}_i p \wedge \Box(p \to e)$, and ψ as a disjunction that describes the mapping from the R_{i-1}-equivalence classes e' in e to the formulas $\phi_{t'}(e')$ that the construction chooses to make them satisfy.

For the second part of Lemma 7, we note the following semantic result.

PROPOSITION 9. *Suppose $u \in \mathcal{T}_{i,m,k+1}$ and $v \in \mathcal{T}_{i,m+1,k}$ is not compatible with u. Then $\models \phi_u \to \neg \exists p_{m+1}(\phi_v)$.*

Proof. Suppose $v \notin \mathcal{C}(u)$. We proceed by induction on i. For $i = 0$, the claim is straightforward, since one of ϕ_v and ϕ_v contains, for some $j = 1\ldots m$, a conjunct p_j whereas the other contains a conjunct $\neg p_j$. For $i > 0$, suppose that v is not compatible with u, and that $M \models \phi_u(e)$. Let M' be a p_{m+1}-variant of M. We suppose that $M' \models \phi_v(e)$ and derive the contradiction that $v \in \mathcal{C}(u)$.

Define $S(t)$ for $t \in \mathcal{T}_{i-1,\mathbf{p},k+1}$ to be the set of R_{i-1}-equivalence classes U in $\pi(e)$ such that $M'[U/e] \models \phi_t(e)$. Similarly, define $S'(t')$ for $t' \in \mathcal{T}_{i-1,\mathbf{p}+,k}$ to be the set of R_{i-1}-equivalence classes U in $\pi(e)$ such that $M'[U/e] \models \phi_{t'}(e)$. Additionally, define $S(t,t') = S(t) \cap S'(t')$. By Lemma 6, each of the collections $\{ S(t) \mid t \in \mathcal{T}_{i-1,\mathbf{p},k+1} \}$, $\{ S'(t') \mid t' \in \mathcal{T}_{i-1,\mathbf{p}+,k} \}$ and $\{ S(t,t') \mid t \in \mathcal{T}_{i-1,\mathbf{p},k+1}, t' \in \mathcal{T}_{i-1,\mathbf{p}+,k} \}$ partition the set of R_{i-1}-equivalence classes U in $\pi(e)$.

Define $f(t,t') = \min(N_{i-1,\mathbf{p}+,k}, |S(t,t')|)$. We show that f satisfies all the conditions of the definition of compatibility in order to witness that $v \in \mathcal{C}(u)$.

C1: If $f(t,t') \neq 0$ then there exists an R_{i-1}-equivalence class U in e such that $M'[U/e] \models \phi_t(e) \wedge \phi_{t'}(e)$, hence $M[U/e] \models \phi_t(e) \wedge \exists p_{m+1}(\phi_{t'}(e))$. Thus, we do not have $\models \phi_t(e) \to \neg \exists p_{m+1}(\phi_{t'}(e))$. By the induction hypothesis, we have $t' \in \mathcal{C}(t)$.

C2: We consider two cases. Suppose first that $v(t') < N_{i-1,\mathbf{p}+,k}$. Since $M' \models \phi_v(e)$, we have $v(t') = |S'(t')| < N_{i-1,\mathbf{p}+,k}$. Consequently, for all

$t \in \mathcal{T}_{i-1,\mathbf{p},k+1}$ we also have $|S(t,t')| < N_{i-1,\mathbf{p}^+,k}$, so $f(t,t') = |S(t,t')|$. Thus,
$$\sum_{t \in \mathcal{T}_{i-1,\mathbf{p},k+1}} f(t,t') = \sum_{t \in \mathcal{T}_{i-1,\mathbf{p},k+1}} |S(t,t')| = |S(t')| = v(t').$$
It follows that $v(t') = \min(\sum_{t \in \mathcal{T}_{i-1,\mathbf{p},k+1}} f(t,t'), N_{i-1,\mathbf{p}^+,k})$.

In the other case, suppose that $v(t') = N_{i-1,\mathbf{p}^+,k}$. We break this case down into two possibilities. Suppose first that there exists $t_0 \in \mathcal{T}_{i-1,\mathbf{p},k+1}$ such that $|S(t_0,t')| \geq N_{i-1,\mathbf{p}^+,k}$. Then $f(t_0,t') = N_{i-1,\mathbf{p}^+,k}$, so
$$\min\left(\sum_{t \in \mathcal{T}_{i-1,\mathbf{p},k+1}} f(t,t'), N_{i-1,\mathbf{p}^+,k}\right) = N_{i-1,\mathbf{p}^+,k} = v(t').$$
Alternately, if $|S(t,t')| < N_{i-1,\mathbf{p}^+,k}$ for all $t \in \mathcal{T}_{i-1,\mathbf{p},k+1}$ then for all $t \in \mathcal{T}_{i-1,\mathbf{p},k+1}$ we have $f(t,t') = |S(t,t')|$. Thus, $\sum_{t \in \mathcal{T}_{i-1,\mathbf{p},k+1}} f(t,t') = \sum_{t \in \mathcal{T}_{i-1,\mathbf{p},k+1}} |S(t,t')| = |S(t')| \geq N_{i-1,\mathbf{p}^+,k}$. Hence
$$\min\left(\sum_{t \in \mathcal{T}_{i-1,\mathbf{p},k+1}} f(t,t'), N_{i-1,\mathbf{p}^+,k}\right) = N_{i-1,\mathbf{p}^+,k} = v(t').$$

C3: Note that we always have $f(t,t') \leq |S(t,t')|$, hence
$$\sum_{t' \in \mathcal{T}_{i-1,\mathbf{p}^+,k}} f(t,t') \leq \sum_{t' \in \mathcal{T}_{i-1,\mathbf{p}^+,k}} |S(t,t')| = |S(t')|.$$
Since $f(t,t') \leq N_{i-1,\mathbf{p}^+,k}$, we have
$$\sum_{t' \in \mathcal{T}_{i-1,\mathbf{p}^+,k}} f(t,t') \leq |\mathcal{T}_{i-1,\mathbf{p}^+,k}| \cdot N_{i-1,\mathbf{p}^+,k} = N_{i-1,\mathbf{p},k+1}$$
Thus $\sum_{t' \in \mathcal{T}_{i-1,\mathbf{p},k+1}} f(t,t') \leq \min(|S(t)|, N_{i-1,\mathbf{p},k+1})$. Since $M \models \phi_u(e)$, we have $\min(|S(t)|, N_{i-1,\mathbf{p},k+1}) = u(t)$. It follows that we always have $\sum_{t' \in \mathcal{T}_{i-1,\mathbf{p},k+1}} f(t,t') \leq u(t)$. In the case that $f(t,t') < N_{i-1,\mathbf{p}^+,k}$ for all $t' \in \mathcal{T}_{i-1,\mathbf{p}^+,k}$, we have also $f(t,t') = |S(t,t')|$. Thus
$$|S(t)| = \sum_{t' \in \mathcal{T}_{i-1,\mathbf{p}^+,k}} |S(t,t')|$$
$$= \sum_{t' \in \mathcal{T}_{i-1,\mathbf{p}^+,k}} f(t,t')$$
$$< |\mathcal{T}_{i-1,\mathbf{p}^+,k}| \cdot N_{i-1,\mathbf{p}^+,k}$$
$$= N_{i-1,\mathbf{p},k+1}.$$

Since $M \models \phi_u(e)$, it follows that $u(t) = |S(t)| = \sum_{t' \in \mathcal{T}_{i-1,\mathbf{p}^+,k}} f(t,t')$. ∎

The main step in converting this semantic proof to a proof-theoretic argument is to replace the use of satisfiability by a consistency argument.

Just as in the proof of Lemma 2(4), we needed to treat the atom a as a special case, establishing that $\vdash a \wedge c_a \to \exists q(a' \to c_a^+ \wedge c_a^-)$, we actually need a slight strengthening of Lemma 7 that deals with branches rather than trees. For this we need a generalized notion of compatibility. Define a branch $v_i, \ldots, v_0$ in a tree $v_i \in \mathcal{T}_{i,\mathbf{p}^+,k}$ to be *compatible* with a branch $u_i, \ldots, u_0$ in a tree $u_i \in \mathcal{T}_{i,\mathbf{p},k+1}$ when v_0 is compatible with u_0, and there exist functions $f_1, \ldots, f_i$ such that for each $j = 1, \ldots, i$, the function f_j witnesses that v_j is compatible with u_j, and $f_j(u_{j-1}, v_{j-1}) \neq 0$. We write Chain$(e_0, \ldots, e_i)$ for $\bigwedge_{j=0}^{i-1} e_j \subseteq e_{j+1} \wedge \bigwedge_{j=0}^{i} \mathcal{E}_j e_j$. That is, Chain$(e_0, \ldots, e_i)$ says that $e_0, \ldots, e_i$ is a nested set of equivalence classes.

LEMMA 10. *Let $u_i \in \mathcal{T}_{i,\mathbf{p},k+1}$ and let $u_i, \ldots, u_0$ be a branch in u_i. Similarly, let $v_i \in \mathcal{T}_{i,\mathbf{p}^+,k}$ and let $v_i, \ldots, v_0$ be a branch in v_i. Then if $v_i, \ldots, v_0$ is compatible with $u_i, \ldots, u_0$, we have*

$$\vdash \text{Chain}(e_0, \ldots, e_i) \wedge \bigwedge_{j=0}^{i} \phi_{u_j}(e_j) \to \exists p_{m+1} \left(\bigwedge_{j=0}^{i} \phi_{v_j}(e_j) \right),$$

else

$$\vdash \text{Chain}(e_0, \ldots, e_i) \wedge \bigwedge_{j=0}^{i} \phi_{u_j}(e_j) \to \neg \exists p_{m+1} \left(\bigwedge_{j=0}^{i} \phi_{v_j}(e_j) \right).$$

The proof of Lemma 10 is very similar to the proof of Lemma 7: we simply need to add some special treatment of the proposition p_{m+1} along the distinguished branch.

Lemma 4(4) now follows easily from Lemma 10. Parts 1 – 3 of Lemma 4 can be established by arguments similar to those in the proof of Lemma 2. For part 5, we use **EC**$_2$ (Section 3.1) and an induction on i to show that for $(i, \mathbf{p}, k)$-trees u and propositions p in $\mathbf{p}$, we have $\vdash \mathcal{E}_i e \wedge \phi_u(e) \to \Box(e \to p)$ or $\vdash \mathcal{E}_i e \wedge \phi_u(e) \to \neg\Box(e \to p)$, and also $\vdash \mathcal{E}_i e \wedge \phi_u(e) \to \Box(e \to \neg p)$ or $\vdash \mathcal{E}_i e \wedge \phi_u(e) \to \neg\Box(e \to \neg p)$. From this we obtain, using **EC**$_3$ and **EC**$_4$, that when u is an $(n+1, \mathbf{p}, k)$-tree and p is one of the propositions in $\mathbf{p}$, either $\vdash \phi_u(true) \to L_i p$ or $\vdash \phi_u(true) \to \neg L_i p$. Part 5 of Lemma 4 follows from this. Part 6 follows from part 4 and part 5 using **Def**$\forall_i$.

5 Conclusion

The results presented in this paper leave open a number of questions concerning $\mathcal{L}_{(\forall, \forall_1, \ldots, \forall_n, \Box)}$. One of these is the complexity of the language with respect to hierarchical structures: we will report on this in the full version.

(Note that decidability follows from the finite model property we have established.) Another question is the effect of restrictions on the range of quantification. We have interpreted the quantifiers to range over all propositions. For $\mathcal{L}_{(\forall,\Box)}$, Fine [8] proved completeness with respect to semantics for quantification that place various restrictions of the range of quantification, such as the assumption that this space forms a boolean algebra. We do not know yet what the effect of such assumptions would be on our logic.

Tenney showed a bounded model property for the monadic second-order theory of an equivalence relation [15]. His language and $\mathcal{L}_{(\forall,\forall_1,\Box)}$ are of equal expressive power and effectively translatable. Consequently our axiomatization LLPH$_1$ translates to an axiomatization of Tenney's language. Moreover, Tenney's bounds on model-sizes are similar to the ones we calculated for $\mathcal{L}_{(\forall,\forall_1,\Box)}$.

The class of models for which we are able to axiomatize the language $\mathcal{L}_{(\forall,\forall_1,\ldots,\forall_n,\Box)}$ can be generalized slightly from the hierarchical models to *locally* hierarchical models. These are models in which for all pairs U, V of propositions, with U an R_i-equivalence class and V an R_j-equivalence class, we have one of $U \cap V = \emptyset$, $U \subseteq V$ or $V \subseteq U$. We leave further discussion of this to the full paper.

Acknowledgments

Research supported by Australian Research Council Large grant A10030007. Work of the third author performed while a visiting fellow at UNSW. Also partially supported by the National Natural Science Foundation of China under Grant 60073056, the Guangdong Provincial Natural Science Foundation under Grant 001174, and the MOE Project of Key Research Institute of Humanities and Social Science in University.

Thanks to Phokion Kolaitis for drawing our attention to Tenney's paper, and several of the AIML participants for suggesting the extension to locally hierarchical models.

BIBLIOGRAPHY

[1] H. Behmann. Beiträge zur Algebra der Logik, insbesondere zum Entscheidungsproblem. *Mathematische Annalen*, 86:163–229, 1922.
[2] R. Bull. On modal logic with propositional quantifiers. *The Journal of Symbolic Logic*, 34:257–263, 1969.
[3] C. Cerratto. Graded modalities V. *Studia Logica*, 53:61–73, 1992.
[4] D. E. Denning. A lattice model of secure information flow. *Communications of the ACM*, 19(5):236–243, May 1976. Papers from the Fifth ACM Symposium on Operating Systems Principles (Univ. Texas, Austin, Tex., 1975).
[5] K. Engelhardt, R. van der Meyden, and Y. Moses. Knowledge and the logic of local propositions. In I. Gilboa, editor, *Theoretical Aspects of Rationality and Knowledge, Proceedings of the Seventh Conference (TARK 1998)*, pages 29–41. Morgan Kaufmann, July 1998.

[6] R. Fagin, J. Y. Halpern, Y. Moses, and M. Y. Vardi. *Reasoning About Knowledge.* MIT-Press, 1995.
[7] M. Fattorosi Barnaba and F. de Caro. Graded modalities I. *Studia Logica*, 44:197–221, 1985.
[8] K. Fine. Propositional quantifiers in modal logic. *Theoria*, 36:336–346, 1970.
[9] K. Fine. In so many possible worlds. *Notre Dame Journal of Formal Logic*, 13:516–520, 1972.
[10] J. Y. Halpern and Y. Moses. Knowledge and common knowledge in a distributed environment. *Journal of the ACM*, 37(3):549–587, July 1990.
[11] G. E. Hughes and M. J. Cresswell. *A New Introduction to Modal Logic.* Routledge, London, New York, 1996.
[12] D. Kaplan. S5 with multiple possibility (Abstract). *The Journal of Symbolic Logic*, 35:355–356, 1970.
[13] D. Kaplan. S5 with quantifiable propositional variables (Abstract). *The Journal of Symbolic Logic*, 35:355, 1970.
[14] S. A. Kripke. A completeness theorem in modal logic. *The Journal of Symbolic Logic*, 24:1–14, 1959.
[15] R. L. Tenney. Second-order Ehrenfeucht games and the decidability of the second-order theory of an equivalence relation. *Journal of the Australian Mathematical Society, Series A*, 20:323–331, 1975.

Kai Engelhardt and Ron van der Meyden
School of Computer Science and Engineering
The University of New South Wales
Sydney 2052, Australia
E-mail: {kaie,meyden}@cse.unsw.edu.au

Kaile Su
Institute of Logic and Cognition
Zhongshan University
Guangzhou 510275, China
E-mail: isskls@zsu.edu.cn

Functional Completeness for a Natural Deduction Formulation of Hybridized S5

TORBEN BRAÜNER

ABSTRACT. In this paper we give a functional completeness result for a natural deduction formulation of hybridized $S5$. Hybridized $S5$ is obtained by adding to ordinary $S5$ further expressive power in the form of so-called satisfaction operators and a second sort of propositional symbols called nominals.

1 Introduction

In this paper we give a functional completeness result for a natural deduction formulation of hybridized $S5$. We shall begin with making a few general remarks on the topics of hybridized $S5$, natural deduction, and functional completeness.

Hybridized $S5$ is a hybrid logical version of $S5$ which is obtained by adding to $S5$ further expressive power in the form of so-called satisfaction operators and a second sort of propositional symbols called nominals. A nominal is assumed to be true at exactly one world, so a nominal can be considered the name of a world. Thus, in hybrid logic a name is a particular sort of propositional symbol whereas in first-order logic it is an argument to a predicate. Beside nominals, a satisfaction operator a : is added for each nominal a. A satisfaction operator a : makes it possible to express that a formula is true at one particular world, namely the world at which the nominal a is true. The history of hybrid logic goes back to Arthur Prior's work, more precisely, it goes back to what he called four grades of tense logical involvement. They were presented in the book [24], Chapter XI (also Chapter XI in the new edition [14]). See also [23] Chapter V.6 and Appendix B.3–4. Prior introduced nominals and satisfaction operators, and moreover, he introduced the so-called binder ∀ which is analogous to the standard first-order universal quantifier. We shall not consider binders in the present paper, however. See [19] for an account of Prior's work. The paper [6] makes a connection to Donald Davidson's notion of a theory of truth.

Hybridized $S5$ has a number of different applications. For example, if hybridized $S5$ is extended with further modal operators, then it can be considered a natural generalisation of the so-called *description logics* used in knowledge representation, see [4]. The description logic called $\mathcal{ALC}$ is a notational variant of ordinary multi-modal logic, that is, propositional logic extended with a finite number of modal operators $\Box_1, \ldots, \Box_m$. The *concept expressions* of $\mathcal{ALC}$ simply corresponds to formulas of multi-modal logic and vice versa. Given a description logic, for example $\mathcal{ALC}$, a knowledge base is a set of metalinguistic statements expressing relationships between concepts and individuals. There are two kinds of metalinguistic statements; they are called *TBox-statements* and *ABox-statements* respectively. A TBox-statement $\phi \sqsubseteq \psi$ expresses that the concept ϕ is subsumed by the concept ψ, that is, that any individual that belongs to the extension of ϕ also belongs to the extension of ψ. An ABox-statement $\phi(a)$ expresses that the individual a belongs to the extension of the concept ϕ and an ABox-statement $R_i(a,c)$ expresses that the individual a is R_i-related to the individual c. This can all be expressed in terms of hybridized $S5$ extended with further modal operators: The TBox-statement above is expressed by the formula $\Box(\phi \to \psi)$, where $\Box$ is the original $S5$ modal operator, and the ABox-statements are expressed by the formulas $a : \phi$ and $a : \Diamond_i c$. Note that no binders are needed. Of course, a nominal is here considered a name of an individual. Thus, hybridized $S5$ generalises description logic such that no distinction between an object language and a metalanguage is needed.

Now, natural deduction style inference rules for ordinary classical first-order logic were originally introduced by Gerhard Gentzen in [12] and later on considered by Dag Prawitz in [20, 21]. With reference to Gentzen's work, Prawitz made the following remarks on the significance of natural deduction.

> ... the essential logical content of intuitive logical operations that can be formulated in the languages considered can be understood as composed of the atomic inferences isolated by Gentzen. It is in this sense that we may understand the terminology *natural* deduction.
>
> Nevertheless, Gentzen's systems are also natural in the more superficial sense of corresponding rather well to informal practices; in other words, the structure of informal proofs are often preserved rather well when formalised within the systems of natural deduction. ([21], p. 245)

Natural deduction systems are characterised by having two different kinds of rules for each non-nullary connective; there is a kind of rules which introduces a connective and there is a kind of rules which eliminates a connective. A maximum formula in a derivation is then a formula occurrence that is both introduced by an introduction rule and eliminated by an elimination rule. A maximum formula can be considered a "detour" in the derivation and it can be removed by rewriting the derivation. Such a step is called a reduction step and it is a notable feature of natural deduction systems that they satisfy a so-called normalisation theorem which says that any derivation can be rewritten to a derivation without maximum formulas by repeated applications of reductions. Another notable feature of natural deduction systems is that they correspond closely to Gentzen systems. It can be proved that our natural deduction formulation of hybridized $S5$ has both of these features. We shall, however, not consider this further in the present paper.

The notion of functional completeness we shall consider is defined as follows. Let a set of natural deduction inference rules involving a set a connectives Σ be given. Moreover, let a set of general rule-schemas involving a new connective $\sharp$ be given. The general rule-schemas can be instantiated to a set of introduction and elimination rules for $\sharp$. A functional completeness result then says that any such new connective $\sharp$ is explicitly definable in terms of the original connectives Σ in the following sense: For any formula ϕ built using the connectives $\Sigma \cup \{\sharp\}$, there exists a formula ψ built using only the original connectives Σ such that ψ can be proved to be equivalent to ϕ using the original inference rules as well as the introduction and elimination rules for the new connective $\sharp$. Thus, such a functional completeness result says that the original natural deduction system involving only the connectives Σ can define explicitly any connective with introduction and elimination rules of a certain form. Note that this notion of functional completeness is purely proof-theoretic, that is, it makes no reference to model-theoretic notions. A different notion of functional completeness which does refer to model-theoretic notions is exemplified by the result in ordinary classical propositional logic which says that any truth-functional connective is definable in terms of the connectives $\{\neg, \wedge, \top\}$. See [15] for a similar result for intuitionistic propositional logic.

With the aim of proving a functional completeness result for our natural deduction formulation of hybridized $S5$, we shall specify a set of general rule-schemas involving the new connective $\sharp$. The connective $\sharp$ takes as arguments an arbitrary, but fixed, number of nominals and an arbitrary, but again fixed, number of formulas. The rule-schemas are obviously not arbitrary; they satisfy certain requirements, one of the requirements being the so-called inversion principle (which is a prerequisite for obtaining a normalisation theorem).

This paper is structured as follows. In the second section of the paper we recapitulate the basics of hybrid logic, in the third section we introduce the natural deduction system for hybridized $S5$, and in the fourth section we prove the functional completeness theorem. In the final section we discuss related and future work.

2 Hybrid logic

In this section we recapitulate the basics of hybrid logic, and moreover, we define models for hybridized $S5$. We shall in many cases adopt the terminology of [3] and [1]. The hybrid logic we consider is obtained by adding a second sort of propositional symbols called *nominals* to ordinary modal logic. It is assumed that a countably infinite set of ordinary propositional symbols and a countably infinite set of nominals are given. The sets are assumed to be disjoint. The metavariables p, q, r, ... range over ordinary propositional symbols and a, b, c, ... range over nominals.

Beside nominals, an operator $a :$ called a *satisfaction operator* is added for each nominal a. The formulas of hybrid modal logic are defined by the grammar

$$S ::= p \mid a \mid S \wedge S \mid S \rightarrow S \mid \bot \mid \Box S \mid a : S$$

where p is an ordinary propositional symbol and a is a nominal. In what follows $\mathcal{H}$ denotes the language, that is, the set of formulas, defined by the grammar. Many hybrid logics involve so-called *binders*, but we shall not consider binders in the present paper. We shall make use of the following conventions. The metavariables ϕ, ψ, θ, ... range over formulas. Negation, nullary conjunction, disjunction, and biimplication are defined by the conventions that $\neg \phi$ is an abbreviation for $\phi \rightarrow \bot$, $\top$ is an abbreviation for $\neg \bot$, $\phi \vee \psi$ is an abbreviation for $\neg(\neg \phi \wedge \neg \psi)$, and $\phi \leftrightarrow \psi$ is an abbreviation for $(\phi \rightarrow \psi) \wedge (\psi \rightarrow \phi)$. Similarly, $\Diamond \phi$ is an abbreviation for $\neg \Box \neg \phi$. If $\overline{a}$ is a list of pairwise distinct nominals and $\overline{c}$ is a list of nominals of the same length as $\overline{a}$, then $\psi[\overline{c}/\overline{a}]$ is the formula ψ where the nominals $\overline{c}$ have been simultaneously substituted for all occurrences of the nominals $\overline{a}$. An analogous definition is obtained if $\overline{a}$ is replaced by a list of ordinary propositional symbols and $\overline{c}$ is replaced by a list of formulas.

We now define models for hybridized $S5$ (note, however, that the functional completeness result we shall prove later do not refer to any model-theoretic notions). A *model* is a tuple (W, V) where W is a non-empty set and V is a *valuation*, that is, V is a function that to each pair consisting of an element of W and an ordinary propositional symbol assigns an element of $\{0, 1\}$. The elements of W are called *worlds*. An *assignment* for a model (W, V) is a function g that to each nominal assigns an element of W. Given a model $\mathcal{M} = (W, V)$, the relation $\mathcal{M}, g, w \models \phi$ is defined by induction, where g is an assignment, w is an element of W, and ϕ is a formula.

$$
\begin{array}{lll}
\mathcal{M}, g, w \models p & \text{iff} & V(w, p) = 1 \\
\mathcal{M}, g, w \models a & \text{iff} & w = g(a) \\
\mathcal{M}, g, w \models \phi \wedge \psi & \text{iff} & \mathcal{M}, g, w \models \phi \text{ and } \mathcal{M}, g, w \models \psi \\
\mathcal{M}, g, w \models \phi \to \psi & \text{iff} & \mathcal{M}, g, w \models \phi \text{ implies } \mathcal{M}, g, w \models \psi \\
\mathcal{M}, g, w \models \bot & \text{iff} & \text{falsum} \\
\mathcal{M}, g, w \models \Box \phi & \text{iff} & \text{for any } v \text{ in } W, \mathcal{M}, g, v \models \phi \\
\mathcal{M}, g, w \models a : \phi & \text{iff} & \mathcal{M}, g, g(a) \models \phi
\end{array}
$$

Of course, it is the quantification over all worlds in the clause for $\Box$ that makes this modality an $S5$ modality. The quantification over all worlds also explains why the modality often is called the *universal* modality. A formula ϕ is said to be *true* at w if $\mathcal{M}, g, w \models \phi$; otherwise it is said to be *false* at w. By convention $\mathcal{M}, g \models \phi$ means $\mathcal{M}, g, w \models \phi$ for every element w of W.

3 Natural deduction for hybrid logic

Before considering natural deduction inference rules for hybridized $S5$, we shall sketch the basics of natural deduction. See [20] and [25] for further details.

A characteristic feature of natural deduction is that derivations have the form of trees. Formula occurrences at the leaves of a tree are called *assumptions* and the formula occurrence at the root of the tree is called the *end-formula*. All assumptions are annotated with numbers. An assumption is either *undischarged* or *discharged*. If an assumption is discharged, then it is discharged at a specified rule-instance which is indicated by annotating the assumption and the rule-instance with identical numbers. We shall often omit this information when no confusion can occur. A rule-instance annotated with some number discharges all undischarged assumptions that are above it and are annotated with the number in question, and moreover, are occurrences of a specified formula. Two assumptions in a derivation belong to the same *parcel* if they are annotated with the same number and are occurrences of the same formula, and moreover, either are both undischarged or have both been discharged at the same rule-instance. Thus, in this terminology rules discharge parcels. We shall make use of the standard notations

which means a derivation π where ψ is the end-formula and $[\phi^r]$ is the parcel consisting of all undischarged assumptions that have the form ϕ^r,

$$\frac{c:\phi \quad c:\psi}{c:(\phi \wedge \psi)}(\wedge I) \qquad \frac{c:(\phi \wedge \psi)}{c:\phi}(\wedge E1) \quad \frac{c:(\phi \wedge \psi)}{c:\psi}(\wedge E2)$$

$$\frac{\begin{array}{c}[c:\phi]\\ \vdots\\ c:\psi\end{array}}{c:(\phi \to \psi)}(\to I) \qquad \frac{c:(\phi \to \psi) \quad c:\phi}{c:\psi}(\to E)$$

$$\frac{a:\phi}{c:a:\phi}(:I) \qquad \frac{c:a:\phi}{a:\phi}(:E)$$

$$\frac{b:\phi}{c:\Box\phi}(\Box I)^* \qquad \frac{c:\Box\phi}{f:\phi}(\Box E)$$

$$\frac{\begin{array}{c}[a:\neg\phi]\\ \vdots\\ a:\bot\end{array}}{a:\phi}(\bot 1)^* \qquad \frac{a:\bot}{c:\bot}(\bot 2)$$

$$\frac{}{a:a}(Ref) \qquad \frac{a:c \quad a:\phi}{c:\phi}(Nom1)^\star$$

* b does not occur in $c:\Box\phi$ or in any undischarged assumptions.
★ ϕ is a propositional symbol.

Figure 1. Inference rules for $\mathbf{N}_\mathcal{H}$

and moreover, a derivation π where ψ is the end-formula and a derivation τ with end-formula ϕ has been substituted for each of the undischarged assumptions indicated by $[\phi^r]$. A derivation in a natural deduction system is generated from a set of inference rules from derivations consisting of a single undischarged assumption.

Another characteristic feature of natural deduction is that there are two different kinds of rules for each non-nullary connective; there is a kind of rules called *introduction rules* which introduce a connective (that is, the connective occurs in the conclusion of the rule, but not in the premises) and there is a kind of rules called *elimination rules* whicheliminate a connective (the connective occurs in the premises of the rule, but not in the conclusion). Introduction rules traditionally have names of the form $(\ldots I \ldots)$, and similarly, elimination rules traditionally have names of the form $(\ldots E \ldots)$.

We shall make use of the following conventions. The metavariables π, τ, ... range over derivations. Formulas of the form $a : \phi$ are called *satisfaction statements*, cf. a similar notion in [2]. The metavariables Γ, Δ, ... range over sets of satisfaction statements. A derivation π is a *derivation of* ϕ if the end-formula of π is an occurrence of ϕ and π is a *derivation from* Γ if each undischarged assumption in π is an occurrence of a formula in Γ (note that numbers annotating undischarged assumptions are ignored). If there exists a derivation of ϕ from $\emptyset$, then we simply say that ϕ is *derivable*. Moreover, $\pi[\bar{c}/\bar{a}]$ is the derivation π where each formula occurrence ψ has been replaced by $\psi[\bar{c}/\bar{a}]$.

Now, natural deduction inference rules for hybridized $S5$ are given in Figure 1. The system will be denoted $\mathbf{N}_\mathcal{H}$. All formulas in the rules are satisfaction statements. Note that the rules ($\perp 1$) and ($\perp 2$) are neither introduction rules nor elimination rules (recall that $\neg \phi$ is an abbreviation for $\phi \to \perp$).

The system $\mathbf{N}_\mathcal{H}$ is sound and complete in the usual model-theoretic sense.

THEOREM 1 *The statements below are equivalent.*

1. *ϕ is derivable from Γ in $\mathbf{N}_\mathcal{H}$.*

2. *For any model $\mathcal{M}$ and any assignment g, if, for any formula $\psi \in \Gamma$, $\mathcal{M}, g \models \psi$, then $\mathcal{M}, g \models \phi$.*

Proof. Soundness and completeness of the system given in the present paper follows from the soundness and completeness result given in [7]. It is straightforward to prove that the system given here is equivalent to a mono-modal version of the system given in [7] extended with the inference rule corresponding to the condition $\forall a \forall b R(a, b)$ on the accessibility relation R. ∎

Below are two examples of derivations in $\mathbf{N}_{\mathcal{H}}$.

$$\dfrac{\dfrac{\dfrac{\dfrac{a : \Box(\phi \to \psi)^2}{b : (\phi \to \psi)}(\Box E) \quad \dfrac{a : \Box\phi^1}{b : \phi}(\Box E)}{b : \psi}(\to E)}{\dfrac{a : \Box\psi}{a : (\Box\phi \to \Box\psi)}(\to I)^1}(\Box I)}{a : (\Box(\phi \to \psi) \to (\Box\phi \to \Box\psi))}(\to I)^2$$

$$\dfrac{\dfrac{\dfrac{a : \Box\neg c^1}{c : \neg c}(\Box E) \quad \dfrac{}{c : c}(Ref)}{\dfrac{c : \bot}{a : \bot}(\bot 2)}(\to E)}{a : \Diamond c}(\to I)^1$$

The end-formula of the derivation on the left hand side is the standard modal axiom K prefixed by a satisfaction operator. The nominal b in this derivation is new. In connection with the right hand side derivation, recall that the formulas $a : \Diamond c$ and $a : (\Box\neg c \to \bot)$ are identical.

4 Functional completeness

In this section we shall prove the functional completeness result: It considers a natural deduction system for the language obtained by extending $\mathcal{H}$ with a new connective $\sharp$ that takes as arguments an arbitrary, but fixed, number of nominals and an arbitrary, but again fixed, number of formulas. The extended language will be denoted $\mathcal{H}(\sharp)$ and the corresponding natural deduction system, which is obtained by extending $\mathbf{N}_{\mathcal{H}}$ with appropriate introduction and elimination rules for the connective $\sharp$, will be denoted $\mathbf{N}_{\mathcal{H}(\sharp)}$.

Now, what does it mean to be an appropriate introduction or elimination rule? We answer this question by requiring that the introduction and elimination rules are instances of general rule-schemas for such rules. These schemas are obviously not arbitrary; they satisfy the following requirements.

1. Besides satisfaction operators, the schemas exhibit no other connectives than $\sharp$.

2. Each schema exhibits exactly one occurrence of the connective $\sharp$ such that the schema for introduction rules exhibits $\sharp$ in the conclusion and the schemas for elimination rules exhibit $\sharp$ in the premises.

3. The schemas satisfy the so-called inversion principle: A derivation of the conclusion of an elimination rule is already "contained" in the derivations of the premises if the major premiss is the conclusion of an introduction rule. (The major premiss of an elimination rule is the premiss that exhibits the connective being eliminated, in this case $\sharp$.)

The history of the inversion principle goes back to [20]. In the formulation of the inversion principle above it is not made explicit what is meant by

$$\frac{[\mathbf{a}_{11}:\boldsymbol{\phi}_{11}]\ldots[\mathbf{a}_{1k_1}:\boldsymbol{\phi}_{1k_1}] \quad [\mathbf{a}_{n1}:\boldsymbol{\phi}_{n1}]\ldots[\mathbf{a}_{nk_n}:\boldsymbol{\phi}_{nk_n}]}{\mathbf{c}:\sharp(\overline{\mathbf{e}_1},\ldots,\overline{\mathbf{e}_n},\overline{\boldsymbol{\phi}_1},\ldots,\overline{\boldsymbol{\phi}_n},\overline{\boldsymbol{\psi}},\overline{\boldsymbol{\theta}})} \; (\sharp I)^*$$

with middle premisses $\mathbf{b}_1:\boldsymbol{\psi}_1 \quad \ldots \quad \mathbf{b}_n:\boldsymbol{\psi}_n$

$$\frac{\mathbf{c}:\sharp(\overline{\mathbf{e}_1},\ldots,\overline{\mathbf{e}_n},\overline{\boldsymbol{\phi}_1},\ldots,\overline{\boldsymbol{\phi}_n},\overline{\boldsymbol{\psi}},\overline{\boldsymbol{\theta}}) \quad \mathbf{a}_{i1}[\overline{\mathbf{f}_i/\mathbf{d}_i}]:\boldsymbol{\phi}_{i1}\ldots\mathbf{a}_{ik_i}[\overline{\mathbf{f}_i/\mathbf{d}_i}]:\boldsymbol{\phi}_{ik_i}}{\mathbf{b}_i[\overline{\mathbf{f}_i/\mathbf{d}_i}]:\boldsymbol{\psi}_i} \; (\sharp Ei)$$

* For each $i \in \{1,\ldots,n\}$, the nominals $\overline{\mathbf{d}_i}$ are pairwise distinct and do not occur in $\mathbf{c}:\sharp(\overline{\mathbf{e}_1},\ldots,\overline{\mathbf{e}_n},\overline{\boldsymbol{\phi}_1},\ldots,\overline{\boldsymbol{\phi}_n},\overline{\boldsymbol{\psi}},\overline{\boldsymbol{\theta}})$ or in any undischarged assumptions other than the specified occurrences of $\mathbf{a}_{i1}:\boldsymbol{\phi}_{i1},\ldots,\mathbf{a}_{ik_i}:\boldsymbol{\phi}_{ik_i}$.

Figure 2. The rule-schemas for $\mathbf{N}_{\mathcal{H}(\sharp)}$

requiring that some derivations "contain" a derivation of a certain formula, but it means that a derivation of the formula in question can be obtained by composition of derivations and by substitution of nominals for nominals in derivations. The rules for the standard connective $\wedge$ is an instructive example of introduction and elimination rules that satisfy the inversion principle: If the premiss of an instance of $(\wedge E1)$ is the conclusion of an introduction rule, necessarily the rule $(\wedge I)$, then the conclusion of $(\wedge E1)$ already occurs as premiss of the instance of $(\wedge I)$. Note that neither composition (of derivations) nor substitution (of nominals for nominals) is needed in this case.

Recall from the introduction of the paper that a formula occurrence that is both introduced by an introduction rule and eliminated by an elimination rule is called a maximum formula, thus, the fact that the inversion principle is satisfied suggests that such maximum formulas can be dispensed with, and this is indeed what a normalisation theorem says. We shall, however, not consider this issue further.

4.1 Rule-schemas for $\mathbf{N}_{\mathcal{H}(\sharp)}$

Below we shall give the technical details of the general rule-schemas for introduction and elimination rules of $\mathbf{N}_{\mathcal{H}(\sharp)}$. In the formulation of the rule-schemas, we shall make use of metametavariables $\mathbf{a}$, $\mathbf{b}$, $\mathbf{c}$, ... and $\boldsymbol{\phi}$, $\boldsymbol{\psi}$, $\boldsymbol{\theta}$, ... that range over respectively the metavariables a, b, c, ... (that in turn range over nominals) and the metavariables ϕ, ψ, θ, ... (that in turn range over formulas). Note that metametavariables are in boldface. The rule-schemas are given in Figure 2 where for each $i \in \{1,\ldots,n\}$,

$\overline{\mathbf{d}_i} = \mathbf{d}_{i1}, \ldots, \mathbf{d}_{iq_i}$, $\overline{\mathbf{e}_i} = \mathbf{e}_{i1}, \ldots, \mathbf{e}_{ir_i}$, and, $\overline{\phi_i} = \phi_{i1}, \ldots, \phi_{ik_i}$, and moreover, where $\overline{\theta} = \theta_1, \ldots, \theta_m$ simply are "dummy" arguments of $\sharp$. In Figure 2 we have made use of substitution of metavariables (that range over nominals) for metavariables (that also range over nominals). This is defined in analogy to the usual substitution operation where nominals are substituted for nominals. The system $\mathbf{N}_{\mathcal{H}(\sharp)}$ is obtained by extending $\mathbf{N}_{\mathcal{H}}$ with introduction and elimination rules for $\sharp$ that are instances of the schemas such that

1. the metavariables $\mathbf{c}, \overline{\mathbf{d}_1}, \ldots, \overline{\mathbf{d}_n}, \overline{\mathbf{f}_1}, \ldots, \overline{\mathbf{f}_n}, \overline{\mathbf{e}_1}, \ldots, \overline{\mathbf{e}_n}$ are pairwise distinct; and

2. $\{\mathbf{a}_{i1}, \ldots, \mathbf{a}_{ik_i}, \mathbf{b}_i\} \subseteq \{\mathbf{d}_{i1}, \ldots, \mathbf{d}_{iq_i}, \mathbf{e}_{i1}, \ldots, \mathbf{e}_{ir_i}, \mathbf{c}\}$ where $i \in \{1, \ldots, n\}$.

Note that the new connective $\sharp$ takes as arguments $r_1 + \ldots + r_n$ nominals and $k_1 + \ldots + k_n + n + m$ formulas, and also, note that there is only one schema for introduction rules but n schemas for elimination rules.

The reason why we make use of metametavariables in the definition of the rule-chemas is that we wish to be able to distinguish between (i) the case where the same metavariable occurs more than once in an inference rule, for example the rule $(\prec I)$ below where the metavariable d occurs twice, and (ii) the case where different metavariables are instantiated to the same nominal or formula, for example if the nominal c is identical to the nominal a in an instance of the rule $(: I)$ given in Figure 1. Thus, metametavariables and rule-schemas are used to speak about metavariables and rules whereas metavariables and rules are used to speak about variables and rule-instances.

The rule-schemas given in Figure 2 obviously satisfy the first two requirements listed above and it is straightforward to check that the inversion principle is satisfied. This follows from the observation that a derivation in which an occurrence of the connective $\sharp$ is both introduced by an instance of $(\sharp I)$ and eliminated by an instance of $(\sharp E i)$ can be rewritten such that the occurrence of $\sharp$ disappears. That is, a derivation of the form

$$\cfrac{\cfrac{[a_{11}{:}\phi_{11}]\ldots[a_{1k_1}{:}\phi_{1k_1}] \quad\quad [a_{n1}{:}\phi_{n1}]\ldots[a_{nk_n}{:}\phi_{nk_n}]}{\cfrac{\begin{array}{c}\vdots \pi_1 \\ b_1 : \psi_1\end{array} \quad \ldots \quad \begin{array}{c}\vdots \pi_n \\ b_n : \psi_n\end{array}}{c : \sharp(\overline{e_1}, \ldots, \overline{e_n}, \overline{\phi_1}, \ldots, \overline{\phi_n}, \overline{\psi}, \overline{\theta})} \quad \begin{array}{c}\vdots \tau_1 \\ a_{i1}[\overline{f_i}/\overline{d_i}]{:}\phi_{i1}\end{array} \ldots \begin{array}{c}\vdots \tau_{k_i} \\ a_{ik_i}[\overline{f_i}/\overline{d_i}]{:}\phi_{ik_i}\end{array}}}{b_i[\overline{f_i}/\overline{d_i}] : \psi_i}$$

can be rewritten to

$$\begin{array}{ccc} \vdots \tau_1 & & \vdots \tau_{k_i} \\ a_{i1}[\overline{f_i/d_i}] : \phi_{i1} & \cdots & a_{ik_i}[\overline{f_i/d_i}] : \phi_{ik_i} \\ & \vdots \pi_i[\overline{f_i/d_i}] & \\ & b_i[\overline{f_i/d_i}] : \psi_i & \end{array}$$

where the metavariables for nominals $\overline{\mathbf{d}_i}$ and $\overline{\mathbf{f}_i}$ are instantiated to the nominals $\overline{d_i}$ and $\overline{f_i}$ (see Figure 2).

We shall need the convention and the proposition below. First the convention. The *degree* of a formula is the number of occurrences of connectives in it which is different from $\bot$. Then the proposition.

PROPOSITION 2 *The rules*

$$\dfrac{\begin{array}{c}[a : \neg\phi]\\ \vdots \\ a : \bot\end{array}}{a : \phi}\,(\bot) \qquad \dfrac{a : d \quad a : \phi}{d : \phi}\,(Nom)$$

are admissible in $\mathbf{N}_{\mathcal{H}(\sharp)}$.

Proof. The proof, which makes use of the notion of the degree of a formula, is along the lines of a similar proof given in [7]. ∎

Note in the proposition above that ϕ can be any formula; not just a propositional symbol. Thus, the rule $(\bot)$ generalises the rule $(\bot 1)$ whereas (Nom) generalises $(Nom1)$. The side-conditions on the rules $(\bot 1)$ and $(Nom1)$ enables us to prove a normalisation theorem such that normal derivations satisfy a version of the subformula property. We shall, however, not consider this issue further in the present paper. In any case, the versions of the rules $(\bot 1)$ and $(Nom1)$ without side-conditions, that is, the rules $(\bot)$ and (Nom), are admissible according to the proposition above. We shall also need the replacement lemma below.

LEMMA 3 *(Replacement) If for any nominal* c, $c : (\psi \leftrightarrow \theta)$ *is derivable in* $\mathbf{N}_{\mathcal{H}(\sharp)}$, *then for any formula* ϕ *and any nominal* d, $d : (\phi[\psi/q] \leftrightarrow \phi[\theta/q])$ *is derivable in* $\mathbf{N}_{\mathcal{H}(\sharp)}$.

Proof. Induction on the degree of ϕ. ∎

4.2 Examples of rules

It is straightforward to check that the ordinary introduction and elimination rules for the connectives $\wedge$, $\rightarrow$, $:$, and $\square$ given in Figure 1 are instances of the rule-schemas.

Another example of instances of the rule-schemas are the rules

$$\frac{\begin{array}{c}[d:\phi]\\ \vdots\\ d:\psi\end{array}}{c:(\phi \prec \psi)}(\prec I) \qquad \frac{c:(\phi \prec \psi) \quad f:\phi}{f:\psi}(\prec E)$$

for the binary connective $\prec$ where the rule $(\prec I)$ is equipped with the side-condition that the nominal d does not occur in $c : (\phi \prec \psi)$ or in any undischarged assumptions other than the specified occurrences of $d : \phi$ (note that we have used infix notation for the new connective instead of the usual prefix notation). The connective $\prec$ is called *strict implication* and $\phi \prec \psi$ is provably equivalent to $\square(\phi \rightarrow \psi)$ in the sense of Lemma 5, that is, $c : ((\phi \prec \psi) \leftrightarrow \square(\phi \rightarrow \psi))$ is derivable in $\mathbf{N}_{\mathcal{H}(\prec)}$ for any nominal c.

An example of rules that do not fit into the rule-schemas are the rules

$$\frac{c:\phi}{c:(\phi \vee \psi)}(\vee I1) \qquad \frac{c:\psi}{c:(\phi \vee \psi)}(\vee I2) \qquad \frac{c:(\phi \vee \psi) \quad \begin{array}{c}[c:\phi]\\ \vdots\\ \theta\end{array} \quad \begin{array}{c}[c:\psi]\\ \vdots\\ \theta\end{array}}{\theta}(\vee E)$$

for the connective $\vee$ which have been obtained by adapting the standard introduction and elimination rules for disjunction to hybrid logic. This is not a loss since the rules are not proof-theoretically well-behaved in the context of classical logic (in contrast to intuitionistic logic). To be precise: In the present system disjunction is taken as defined by the convention that $\phi \vee \psi$ is an abbreviation for $\neg(\neg\phi \wedge \neg\psi)$, but if disjunction instead is taken as a primitive connective, and $\mathbf{N}_{\mathcal{H}}$ is extended with the rules above, then the rule $(\bot 1)$ cannot be restricted to propositional symbols as it is in the present version of $\mathbf{N}_{\mathcal{H}}$, and this is a prerequisite for obtaining a subformula property; an important property of a natural deduction system. We shall not consider this issue further, see [20] and [7]. Note, however, that the rules above are admissible in $\mathbf{N}_{\mathcal{H}}$ as it stands.

4.3 The functional completeness theorem

We now come to the proof of functional completeness. To this end we shall prove a lemma. Before we can state and prove the lemma, we need a definition.

Functional Completeness for a Natural Deduction Formulation of Hybridized S5

DEFINITION 4 Consider a connective $\sharp$ together with introduction and elimination rules which are instances of the rule-schemas given in Figure 2. For any formula
$$\alpha = \sharp(\overline{e_1}, \ldots, \overline{e_n}, \overline{\phi_1}, \ldots, \overline{\phi_n}, \overline{\psi}, \overline{\theta})$$
of $\mathcal{H}(\sharp)$, a formula χ_α of $\mathcal{H}(\sharp)$ is defined as follows. Firstly, for each $i \in \{1, \ldots, n\}$ and $l \in \{1, \ldots, q_i\}$, let ξ_{il} be the conjunction of the formulas
$$\{\phi_{ih} \mid \mathbf{a}_{ih} = \mathbf{d}_{il}\} \cup \{\neg \psi_i \mid \mathbf{b}_i = \mathbf{d}_{il}\}$$
and let ρ_{il} be the formula $\neg\Box\neg\xi_{il}$. (Note that the set $\{\neg\psi_i \mid \mathbf{b}_i = \mathbf{d}_{il}\}$ has at most one element: It contains the formula $\neg\psi_i$ if $\mathbf{b}_i = \mathbf{d}_{il}$; otherwise it is empty.) Secondly, for each $i \in \{1, \ldots, n\}$, if $\mathbf{b}_i \notin \{\mathbf{d}_{i1}, \ldots, \mathbf{d}_{iq_i}, \mathbf{c}\}$, then let $b_i = e_{ij}$ where j is such that $\mathbf{b}_i = \mathbf{e}_{ij}$ (exactly one such j is guaranteed to exist), and moreover, for each $h \in \{1, \ldots, k_i\}$, if $\mathbf{a}_{ih} \notin \{\mathbf{d}_{i1}, \ldots, \mathbf{d}_{iq_i}, \mathbf{c}\}$, then let $a_{ih} = e_{ij}$ where j is such that $\mathbf{a}_{ih} = \mathbf{e}_{ij}$ (again, exactly one such j is guaranteed to exist). Thirdly, for each $i \in \{1, \ldots, n\}$, let ρ_i be the conjunction of the formulas
$$\{a_{ih} : \phi_{ih} \mid \mathbf{a}_{ih} \notin \{\mathbf{d}_{i1}, \ldots, \mathbf{d}_{iq_i}, \mathbf{c}\}\} \cup \{\phi_{ih} \mid \mathbf{a}_{ih} = \mathbf{c}\} \cup$$
$$\{b_i : \neg\psi_i \mid \mathbf{b}_i \notin \{\mathbf{d}_{i1}, \ldots, \mathbf{d}_{iq_i}, \mathbf{c}\}\} \cup \{\neg\psi_i \mid \mathbf{b}_i = \mathbf{c}\}.$$
(Note that the set $\{\neg\psi_i \mid \mathbf{b}_i = \mathbf{c}\}$ has at most one element.) Given the definitions of ρ_i and ρ_{il}, let $\sigma_i = \neg(\rho_i \wedge \rho_{i1} \wedge \ldots \wedge \rho_{iq_i})$, and finally, let $\chi_\alpha = \sigma_1 \wedge \ldots \wedge \sigma_n$.

Note that the definition of χ_α does not depend on the identity of the nominals $\overline{e_1}, \ldots, \overline{e_n}$ and the formulas $\overline{\phi_1}, \ldots, \overline{\phi_n}, \overline{\psi}, \overline{\theta}$ in the sense that
$$\chi_\alpha = \chi_\gamma[\overline{e_1}, \ldots, \overline{e_n}/\overline{g_1}, \ldots, \overline{g_n}][\overline{\phi_1}, \ldots, \overline{\phi_n}, \overline{\psi}, \overline{\theta}/\overline{p_1}, \ldots, \overline{p_n}, \overline{q}, \overline{r}]$$
if $\gamma = \sharp(\overline{g_1}, \ldots, \overline{g_n}, \overline{p_1}, \ldots, \overline{p_n}, \overline{q}, \overline{r})$. Also, note that the number of occurrences of the connective $\sharp$ in χ_α is identical to the total number of occurrences of $\sharp$ in the formulas $\overline{\phi_1}, \ldots, \overline{\phi_n}, \overline{\psi}$.

It is instructive to take a look at some examples: If we consider the connective strict implication together with the rules $(\prec I)$ and $(\prec E)$ given in the previous subsection, and apply the definition above to the formula $\phi \prec \psi$, then $\chi_{\phi \prec \psi} = \neg(\top \wedge \neg\Box\neg(\phi \wedge \neg\psi))$. This can be compared to the rules $(\to I)$ and $(\to E)$ for the usual implication connective where $\chi_{\phi \to \psi} = \neg((\phi \wedge \neg\psi) \wedge \top)$ and the rules $(: I)$ and $(: E)$ for the satisfaction operator where $\chi_{e:\psi} = \neg((e : \neg\psi) \wedge \top)$. Now the lemma.

LEMMA 5 Let a formula $\alpha = \sharp(\overline{e_1}, \ldots, \overline{e_n}, \overline{\phi_1}, \ldots, \overline{\phi_n}, \overline{\psi}, \overline{\theta})$ of $\mathcal{H}(\sharp)$ be given. For any nominal c, $c : (\alpha \leftrightarrow \chi_\alpha)$ is derivable in $\mathbf{N}_{\mathcal{H}(\sharp)}$.

Proof. For each $i \in \{1,\ldots,n\}$, let $\overline{f_i} = f_{i1},\ldots,f_{iq_i}$ be a list of pairwise distinct new nominals and let

$$b'_i = \begin{cases} b_i & \text{if } \mathbf{b}_i \notin \{\mathbf{d}_{i1},\ldots,\mathbf{d}_{iq_i},\mathbf{c}\}, \\ f_{il} & \text{if } \mathbf{b}_i = \mathbf{d}_{il}, \\ c & \text{if } \mathbf{b}_i = \mathbf{c}, \end{cases}$$

and moreover, for each $h \in \{1,\ldots,k_i\}$, let

$$a'_{ih} = \begin{cases} a_{ih} & \text{if } \mathbf{a}_{ih} \notin \{\mathbf{d}_{i1},\ldots,\mathbf{d}_{iq_i},\mathbf{c}\}, \\ f_{il} & \text{if } \mathbf{a}_{ih} = \mathbf{d}_{il}, \\ c & \text{if } \mathbf{a}_{ih} = \mathbf{c}. \end{cases}$$

(Recall that b_i and a_{ih} were defined together with the formula χ_α.) We first prove the implication from right to left: It follows from the rule $(\sharp I)$ that $c : \alpha$ is derivable from $\{c : \chi_\alpha\}$ if for each $i \in \{1,\ldots,n\}$, the formula $b'_i : \psi_i$ is derivable from $\{a'_{i1} : \phi_{i1},\ldots,a'_{ik_i} : \phi_{ik_i}, c : \sigma_i\}$. But this is the case if $c : \bot$ is derivable from $\Gamma_i \cup \{c : \sigma_i\}$ where

$$\Gamma_i = \{a'_{i1} : \phi_{i1},\ldots,a'_{ik_i} : \phi_{ik_i}, b'_i : \neg\psi_i\},$$

that is, if the formulas $c : \rho_{i1},\ldots,c : \rho_{iq_i}$, and $c : \rho_i$ all are derivable from Γ_i. Clearly, $c : \rho_i$ is derivable from Γ_i. Consider now any $l \in \{1,\ldots,q_i\}$. The formula $c : \rho_{il}$ is derivable from Γ_i if $c : \bot$ is derivable from $\Gamma_i \cup \{c : \Box\neg\xi_{il}\}$. Note that $f_{il} : \neg\xi_{il}$ is derivable from $\{c : \Box\neg\xi_{il}\}$ by the rule $(\Box E)$. On the other hand, it is straightforward to check that $f_{il} : \xi_{il}$ is derivable from Γ_i. We therefore conclude that $c : \bot$ is derivable from $\Gamma_i \cup \{c : \Box\neg\xi_{il}\}$.

We now prove the implication from left to right: The formula $c : \chi_\alpha$ is derivable from the formula $\{c : \alpha\}$ if for each $i \in \{1,\ldots,n\}$, $c : \bot$ is derivable from $\{c : \alpha, c : \rho_i, c : \rho_{i1},\ldots,c : \rho_{iq_i}\}$. Note that for any $h \in \{1,\ldots,k_i\}$, if $\mathbf{a}_{ih} \notin \{\mathbf{d}_{i1},\ldots,\mathbf{d}_{iq_i},\mathbf{c}\}$, then $a'_{ih} = a_{ih}$ and $a_{ih} : \phi_{ih}$ is derivable from $\{c : \rho_i\}$. Moreover, if $\mathbf{a}_{ih} = \mathbf{c}$, then $a'_{ih} = c$ and $c : \phi_{ih}$ is derivable from $\{c : \rho_i\}$. Furthermore, for any $l \in \{1,\ldots,q_i\}$, if $\mathbf{a}_{ih} = \mathbf{d}_{il}$, then $a'_{ih} = f_{il}$ and $f_{il} : \phi_{ih}$ is derivable from $\{f_{il} : \xi_{il}\}$. Thus, by the rule $(\sharp Ei)$, $b'_i : \psi_i$ is derivable from

$$\Delta_i = \{c : \alpha, c : \rho_i, f_{i1} : \xi_{i1},\ldots,f_{iq_i} : \xi_{iq_i}\}.$$

We now split up in three cases. The first case is where $\mathbf{b}_i = \mathbf{d}_{il_1}$ for some $l_1 \in \{1,\ldots,q_i\}$. Then $b'_i = f_{il_1}$ and $f_{il_1} : \neg\psi_i$ is derivable from $\{f_{il_1} : \xi_{il_1}\}$. Thus, $f_{il_1} : \bot$ is derivable from Δ_i wherefore the formula $f_{il_1} : \neg\xi_{il_1}$ is derivable from $\Delta_i - \{f_{il_1} : \xi_{il_1}\}$ and so is $c : \Box\neg\xi_{il_1}$ by the rule $(\Box I)$. Therefore $c : \bot$ is derivable from $(\Delta_i - \{f_{il_1} : \xi_{il_1}\}) \cup \{c : \rho_{il_1}\}$. We are done if $q_i = 1$. Otherwise, let $l_2 \in \{1, \ldots, q_i\}$ be such that $l_2 \neq l_1$. Then $f_{il_2} : \neg \xi_{il_2}$ is derivable from $(\Delta_i - \{f_{il_1} : \xi_{il_1}, f_{il_2} : \xi_{il_2}\}) \cup \{c : \rho_{il_1}\}$ and so is $c : \Box \neg \xi_{il_2}$ by the rule $(\Box I)$. Therefore $c : \bot$ is derivable from $(\Delta_i - \{f_{il_1} : \xi_{il_1}, f_{il_2} : \xi_{il_2}\}) \cup \{c : \rho_{il_1}, c : \rho_{il_2}\}$. We are done if $q_i = 2$. Otherwise, etc. This step is carried out q_i times in total. The second case is where $\mathbf{b}_i \notin \{\mathbf{d}_{i1}, \ldots, \mathbf{d}_{iq_i}, \mathbf{c}\}$. Then $b_i' = b_i$ and $b_i : \neg \psi_i$ is derivable from $\{c : \rho_i\}$. Thus, $c : \bot$ is derivable from Δ_i. We are done if $q_i = 0$. Otherwise, etc. The rest of this case is analogous to the first case. The third case is where $\mathbf{b}_i = \mathbf{c}$. Then $b_i' = c$ and $c : \neg \psi_i$ is derivable from $\{c : \rho_i\}$. Again, $c : \bot$ is derivable from Δ_i. The rest of the third case is analogous to the second case. ∎

We are now ready for functional completeness.

THEOREM 6 (Functional completeness) *For any formula ϕ in $\mathcal{H}(\sharp)$, there exists a formula ψ in $\mathcal{H}$ such that for any nominal c, $c : (\phi \leftrightarrow \psi)$ is derivable in $\mathbf{N}_{\mathcal{H}(\sharp)}$.*

Proof. Induction on the number of occurrences of the connective $\sharp$ in ϕ where we use Lemma 5 and Lemma 3. Note that replacing an occurrence of $\alpha = \sharp(\overline{e_1}, \ldots, \overline{e_n}, \overline{\phi_1}, \ldots, \overline{\phi_n}, \overline{\psi}, \overline{\theta})$ in a formula by an occurrence of the formula χ_α decreases the number of occurrences of the connective $\sharp$ by at least one, cf. the definition of χ_α. ∎

5 Related and further work

There is a vast literature on the proof-theory of the modal logic $S5$. It has turned out to be difficult to formulate natural deduction and Gentzen systems for $S5$ without introducing metalinguistic machinery. The history of this problem goes back to the papers [17, 18] where a counter-example to cut-elimination is given for an otherwise very natural and straightforward Gentzen formulation of $S5$. See [5] for an example of a cut-free Gentzen system for $S5$ where metalinguistic machinery is avoided (at the expense of having to make use of a "non-local" side-condition on an inference rule, that is, a side-condition that does not just refer to the premiss of the rule, but to the whole derivation of the premiss). See the book [16] for an example of a cut-free Gentzen formulation of $S5$ that makes use of metalinguistic machinery, namely indexed formulas. This system can be considered a reformulation in Gentzen style of Fitting's prefixed tableau system for $S5$ given in [9]. Fitting's metalinguistic approach is at a general level in line with Gabbay's labelled deduction systems, see [10].

It should be mentioned that various other Gentzen and natural deduction style formulations for $S5$ have been suggested, notable here are formulations in terms of Belnap's display logic and Dösen's higher-order sequents. However, these formulations differ considerably from Gentzen's original sequent and natural deduction systems for classical logic, and moreover, they are relatively complicated from a technical point of view. (Although it has to be acknowledged that display sequents as well as higher-order sequents were introduced as natural generalisations of Gentzen's notion of a sequent. They are intended to allow a uniform sequent-style formulation of many different logics.) An overview can be found in [28].

So what is it that enables us to formulate natural deduction rules for hybridized $S5$ without having to turn to metalinguistic machinery when it is difficult in the context of ordinary $S5$? One feature of hybridization appears to be crucial in the formulation of the introduction and elimination rules for the $S5$ modality: We can express that a formula ϕ is true at a world a, indeed, this is what is expressed by the formula $a : \phi$. But this is exactly the interpretation of the indexes used in the metalingustic approach, so by hybridizing $S5$, we have in the the terminology of [2] internalized the metalanguage in the object language.

Our functional completeness results are in line with a result given in [22] for natural deduction formulations of intuitionistic and classical propositional logic. See also [30] where a functional completeness result is given which is similar to the result of [22] but within a somewhat different framework. Furthermore, see [27]. In [29] functional completeness is proved for a number of classical modal and tense logics which are presentable in terms of Belnap's display logic.

It is straightforward to check that hybridized $S5$ has the same expressive power over models as monadic first-order logic with equality, one variable, and a countably infinite set of constants but no function symbols. That is, a model and an assignment in the usual sense for this first-order logic correspond to a model, an assignment, and a world as appropriate for hybridized $S5$ and there exists truth preserving translations in both directions between the logics. So our functional completeness result for hybridized $S5$ can be seen as a functional completeness result for the mentioned first-order logic, just couched in modal-logical terms. It is, by the way, notable that the expressive power at the level of frames obtained by adding nominals as well as the universal modality to ordinary mono-modal logic is the same as the expressive power obtained by adding the difference operator, this is shown in [11] (the difference operator is the modal operator which expresses that a formula is true in all worlds which are different from the actual world). See [8] for a further investigation of the difference operator as an additional operator and see [13] for an investigation of the universal modality as an additional operator. Also, see [26] for relevant work in connection with counting modalities (a counting modality expresses that a formula is true in at least n distinct worlds, where n is some fixed number).

5.1 Further work

We are currently investigating the extent to which the approach for obtaining the functional completeness result of the present paper can be extended to other hybrid logics.

One line of work is the following. We have very recently proved a functional completeness result for hybridized $S5$ where the rule-schemas for the connective ♯ strongly generalises those considered in the present paper. We have moreover proved a similar functional completeness result for a natural deduction formulation of the hybrid logic of inequality (the hybrid logic of inequality is identical to hybridized $S5$ except that the universal modality has been replaced by the difference operator). This work will be published elsewhere. The hybrid logic of inequality is more expressive than hybridized $S5$ since the universal modality is definable in terms of the difference operator. Given the strongly generalised rule-schemas, different conditions are imposed on the rules for ♯ depending on which logic is under consideration. Of course, the more expressive power the logic has, the weaker the conditions on the rules are. One possible continuation of this line of work is to consider modal operators with stronger expressive power than that of the difference operator (in combination with weaker conditions on the rules for ♯). One such option is a version of the counting modalities mentioned above.

Another possible line of work concerns the natural deduction formulation of multi-modal hybrid logic given in [7]. The system considered in that paper is a hybridized version of multi-modal K which can be extended with additional inference rules corresponding to conditions on the accessibility relations expressed by so-called geometric theories. One technical obstacle is that the introduction rule for a modal operator given in [7] does not only exhibit the modal operator in the conclusion, but also in the discharged assumptions. Thus, it is implicit, rather than explicit, in the terminology of [22], adapted to our natural deduction formulations of hybrid logic.

Acknowledgements

Thanks to Maarten Marx for discussing this paper. The research presented is partially supported by The IT University of Copenhagen.

BIBLIOGRAPHY

[1] C. Areces, P. Blackburn, and M. Marx. Hybrid logics: Characterization, interpolation and complexity. *Journal of Symbolic Logic*, 66:977–1010, 2001.
[2] P. Blackburn. Internalizing labelled deduction. *Journal of Logic and Computation*, 10:137–168, 2000.

[3] P. Blackburn, M. de Rijke, and Y. Venema. *Modal Logic*, volume 53 of *Cambridge Tracts in Theoretical Computer Science*. Cambridge University Press, 2001.
[4] P. Blackburn and M. Tzakova. Hybridizing concept languages. *Annals of Mathematics and Artificial Intelligence*, 24:23–49, 1998.
[5] T. Braüner. A cut-free Gentzen formulation of the modal logic S5. *Logic Journal of the IGPL*, 8:629–643, 2000. Full version of paper in *Proceedings of 6th Workshop on Logic, Language, Information and Computation*.
[6] T. Braüner. Modal logic, truth, and the master modality. *Journal of Philosophical Logic*, 31:359–386, 2002. Revised and extended version of paper in *Advances in Modal Logic, Volume 3*.
[7] T. Braüner. Natural deduction for hybrid logic. 2003. Revised and extended version of paper in *Workshop Proceedings of Methods for Modalities 2*. Submitted for journal publication, hard-copies available from the author.
[8] M. de Rijke. The modal logic of inequality. *Journal of Symbolic Logic*, 57:566–584, 1992.
[9] M. Fitting. Basic modal logic. In D. Gabbay et al., editor, *Handbook of Logic in Artificial Intelligence and Logic Programming, Vol. 1, Logical Foundations*, pages 365–448. Oxford University Press, Oxford, 1993.
[10] D. Gabbay. *Labelled Deductive Systems*. Oxford University Press, 1996.
[11] G. Gargov and V. Goranko. Modal logic with names. *Journal of Philosophical Logic*, 22:607–636, 1993.
[12] G. Gentzen. Investigations into logical deduction. In M. E. Szabo, editor, *The Collected Papers of Gerhard Gentzen*, pages 68–131. North-Holland Publishing Company, 1969.
[13] V. Goranko and S. Passy. Using the universal modality: Gains and questions. *Journal of Logic and Computation*, 2:5–30, 1992.
[14] P. Hasle, P. Øhrstrøm, T. Braüner, and J. Copeland, editors. *Revised and Expanded Edition of Arthur N. Prior: Papers on Time and Tense*. Oxford University Press, 2002.
[15] D.P. McCullough. Logical connectives for intuitionistic propositional logic. *Journal of Symbolic Logic*, 36:15–20, 1969.
[16] G. Mints. *A Short Introduction to Modal Logic*. CSLI, 1992.
[17] M. Ohnishi and K. Matsumoto. Gentzen method in modal calculi. *Osaka Mathematical Journal*, 9:113–130, 1957.
[18] M. Ohnishi and K. Matsumoto. Gentzen method in modal calculi, II. *Osaka Mathematical Journal*, 11:115–120, 1959.
[19] P. Øhrstrøm and P. Hasle. A. N. Prior's rediscovery of tense logic. *Erkenntnis*, 39:23–50, 1993.
[20] D. Prawitz. *Natural Deduction. A Proof-Theoretical Study*. Almqvist and Wiksell, Stockholm, 1965.
[21] D. Prawitz. Ideas and results in proof theory. In J. E. Fenstad, editor, *Proceedings of the Second Scandinavian Logic Symposium*, volume 63 of *Studies in Logic and The Foundations of Mathematics*, pages 235–307. North-Holland, 1971.
[22] D. Prawitz. Proofs and the meaning and completeness of the logical constants. In J. Hintikka et al., editor, *Essays on Mathematical and Philosophical Logic*, Studies in Logic and The Foundations of Mathematics, pages 25–40. Reidel, 1978.
[23] A. Prior. *Past, Present and Future*. Oxford, 1967.
[24] A. Prior. *Papers on Time and Tense*. Clarendon/Oxford University Press, 1968.
[25] A.S. Troelstra and H. Schwichtenberg. *Basic Proof Theory*, volume 43 of *Cambridge Tracts in Theoretical Computer Science*. Cambridge University Press, 1996.
[26] W. van der Hoek and M. de Rijke. Counting objects. *Journal of Logic and Computation*, 5:325–345, 1995.
[27] F. von Kutschera. Die Vollständigkeit des Operatorsystems $\{\neg, \wedge, \vee, \supset\}$ für die intuitionistische Aussagenlogik im Rahmen der Gentzensemantik. *Archiv für Mathematische Logik und Grundlagenforschung*, 11:3–16, 1968.

[28] H. Wansing. Sequent calculi for normal modal propositional logics. *Journal of Logic and Computation*, 4:125–142, 1994.
[29] H. Wansing. A proof-theoretic proof of functional completeness for many modal and tense logics. In H. Wansing, editor, *Proof Theory of Modal Logic*, volume 2 of *Applied Logic Series*, pages 123–136. Kluwer, 1996.
[30] J. Zucker and R.S. Tragesser. The adequacy problem for inferential logic. *Journal of Philosophical Logic*, 7:501–516, 1978.

Torben Braüner
Department of Computer Science
Roskilde University
P.O. Box 260
DK-4000 Roskilde
Denmark
E-mail: torben@ruc.dk
URL: http://www.dat.ruc.dk/~torben/

3
Relativized Action Complement for Dynamic Logics
JAN BROERSEN

ABSTRACT. Dynamic logic is becoming more and more popular as a logic of action, and is used for knowledge representation and reasoning for such distinct domains as action and change, action and time, and action and norms. We argue that to each of these three application areas the notion of 'action complement' is crucial, but that the notions of action complement found in the literature do not suffice, because of counter intuitive aspects of their semantics. As a solution to this problem we propose a range of dynamic logics with a 'relativized' action complement. These relativized complement dynamic logics have a semantics that is better suited for reasoning about action. Other results concern their relative strength and their expressiveness in terms of definability of classes of frames and models. This paper gives motivations, definitions and some initial results.

1 Introduction

Variants of dynamic logic are used more and more for representation and reasoning in the domains of action and change [13, 26], action and time [5, 30], and action and norms [24, 23, 9, 8]. In all of these domains, the notion of 'action complement' in the interpretation of reference to *'alternative action'* arises as a crucial concept. For instance, in the domain of reasoning about action and change, it is natural to consider the property that an effect is brought about exclusively by a certain action [27, 4]. Another way of saying the same thing is that alternative action cannot have that effect. This observation is central to the proposed solutions to the problem of how to express frame properties in dynamic (modal action) logics [13]. If, for the moment, we denote action complement with the symbol −, we can capture the property of 'exclusive effect' by a dynamic logic formula of the form: $\neg \varphi \rightarrow [-\alpha]\neg \varphi$. Intuitively it reads 'performing an action other than α from a state for which $\neg \varphi$ holds, cannot result in a state where φ holds'. So if we want to perform an action that changes φ from false to true, we can only achieve this by performing an action α (if in the given situation it is actually possible to perform such an action). The second area in which action complement, in the interpretation of alternative action, arises as a natural concept is that of temporal reasoning over action. For instance, if we consider the (liveness) property that over all possible futures eventually an action a is inevitable, we should be able to conclude that this is logically equivalent with the property that it is not possible to perform actions alternative to a forever. An important third application area is that of deontic action reasoning. Traditionally, in deontic logics that were not so much concerned with normative statements about actions α as with normative conditions about states of affairs φ, the relation between *obligation* ($O(\varphi)$) and *permission* ($P(\varphi)$) is described as $O(\varphi) \leftrightarrow \neg P(\neg\varphi)$. In the work on (dynamic) deontic action logic, initiated by Meyer [24], this property reappears as $O(\alpha) \leftrightarrow \neg P(-\alpha)$, for actions α. This identification only makes sense under a suitable interpretation of the operation $(-)$ as an action complement that refers to actions alternative to that referred to by α. Under such an interpretation the identification is read as 'obligation to perform an action α bi-implies absence of permission to perform any action alternative to α'.

In the literature, three notions of complement for dynamic (modal action) logics can be found: the above mentioned action complement used by Meyer to define his dynamic deontic logic [24], the dynamic negation defined by Van Benthem [2], and the complement with respect to the universal relation, as imported from relation algebra [14, 11, 15]. The action complement defined by Meyer does not have a straightforward interpretation in terms of standard modal action models. The complement is defined as part of an action algebra, whose relation to modal action models is not made explicit in detail. Furthermore, the complement is designed to meet some specific deontic requirements, and does not count as a solution that is applicable to the action and change or temporal domain. The dynamic negation of Van Benthem [2] does not meet the requirement concerning the intended interpretation of the complement as referring to 'alternative action'. The dynamic negation $\sim_d$ is defined as the dynamic logic test ([10]) for non-possibility to perform an action: $\sim_d \alpha \equiv_{def} ([\alpha]\bot)?$. So the dynamic negation refers to non-activity, and not to alternative action. This leaves the complement with respect to the universal relation as the most serious candidate for the modeling of the notion of alternative action. But, like we do in this paper, some authors have put some question to this view. The following is a quote from Krister Segerberg [29] that addresses the central issue:

> 'By contrast with intersection, the question concerning complement (negative action) is intricate and involves much extra-theoretical consideration: do we humans really think in terms

of complements? Does the analysis of human languages suggest that we do? Is it not the case that the choice between two actions a and $U \setminus a$ is often a choice between a and some action b that is a proper subset of $U \setminus a$? Before these questions have been answered, this author feels a certain unease about the unrestricted acceptance of closure under complement.'

Segerberg uses the symbol U here to refer to the universal relation $S \times S$. He makes several points. One is that he is not convinced of the naturalness of a notion of action complement. We do not agree, as follows from the above discussion concerning the naturalness of the action complement for several independent application domains. The second important point made is that it is in a way counter-intuitive to define action complement with respect to the universal relation, since intuitively, choices concerning the performance of an action or its complement decide on a division of a *subset* of the universal relation space. In section 3, we try to make the uneasiness felt by Segerberg explicit by showing that the universal complement does not fit well with the three mentioned application domains. In section 4 we show how to define weaker notions of complement, that should take Segerberg's feelings of uneasiness away, and that result in logics that are more fitted to the action reasoning domain. But first we define the global setting for our logics.

2 Dynamic logics

The languages of the dynamic logics in this paper are of the form given in the following definition.

DEFINITION 1 (Modal syntax). Taking 'a' to represent arbitrary elements of a given countable set of atomic action symbols $\mathcal{A}$, and taking 'P' to represent arbitrary elements of a given countable set of proposition symbols $\mathcal{P}$, and given a finite set AC of action combinators, well formed formulas $\varphi, \psi, \ldots$ of a dynamic logic language $\mathcal{L}_{DL}(AC)$ are defined by:

$$\varphi, \psi, \ldots \quad := \quad P \mid \neg \varphi \mid \varphi \wedge \psi \mid \langle \alpha \rangle \varphi$$

where α represents arbitrary actions constructed using atomic actions a and the action combinators from the set AC.

We assume the usual syntactic extensions, including $[\alpha]\varphi \equiv_{def} \neg\langle\alpha\rangle\neg\varphi$. The above definition leaves open how complex actions α are built from atomic actions a. We consider several different sets AC of action combinators. We now define the most general action language we consider in this paper, where the symbol '$-$' functions as a 'place holder' for the range of symbols for different notions of action complement we will introduce in subsequent sections.

DEFINITION 2 (Action language).

$$\alpha, \beta, \ldots := a \mid \alpha \cup \beta \mid \alpha; \beta \mid -\alpha \mid \alpha^* \mid \alpha^{\leftarrow}$$

The dynamic logic languages $\mathcal{L}_{DL}(AC)$ we consider are all conceived of as sub-languages of the most general language $\mathcal{L}_{DL}(\cup,;,-,^*,^{\leftarrow})$ by restriction to a subset $AC \subseteq \{\cup,;,-,^*,^{\leftarrow}\}$ of the action combinators. We now turn to the semantics.

DEFINITION 3 (Modal action models).

A modal action model $\mathcal{M} = (S, R^{\mathcal{A}}, V^{\mathcal{P}})$ is defined as follows:

- S is a nonempty set of possible states

- $R^{\mathcal{A}}$ is an action interpretation function $R^{\mathcal{A}} : \mathcal{A} \to 2^{(S \times S)}$, assigning a binary relation over $S \times S$ to each atomic action a in $\mathcal{A}$.

- $V^{\mathcal{P}}$ is a valuation function $V^{\mathcal{P}} : \mathcal{P} \to 2^S$ assigning to each proposition P of $\mathcal{P}$ the subset of states in S for which P is valid.

In the following definition of the semantics of formulas of a dynamic logic language $\mathcal{L}_{DL}(AC)$, we implicitly assume an extension of the interpretation function $R^{\mathcal{A}}$ for atomic actions a, to an interpretation function R for complex actions α.

DEFINITION 4 (Modal semantics, validity, logic). The modal semantics is determined by the notion of validity of a formula φ in a state s of a model $\mathcal{M} = (S, R^{\mathcal{A}}, V^{\mathcal{P}})$, denoted $\mathcal{M}, s \models \varphi$. We assume that the interpretation $R(\alpha)$ of general actions α is given.

$\mathcal{M}, s \models P$ iff $s \in V^{\mathcal{P}}(P)$
$\mathcal{M}, s \models \neg\varphi$ iff   not $\mathcal{M}, s \models \varphi$
$\mathcal{M}, s \models \varphi \land \psi$ iff $\mathcal{M}, s \models \varphi$ and $\mathcal{M}, s \models \psi$
$\mathcal{M}, s \models \langle \alpha \rangle \varphi$ iff   there is a t such that $(s, t) \in R(\alpha)$ and $\mathcal{M}, t \models \varphi$

Validity on a model $\mathcal{M}$ is defined as validity in all states of the model. If φ is valid on a model $\mathcal{M}$, we say that $\mathcal{M}$ is a model for φ. General validity of a formula φ is defined as validity on all modal action models. The subset of all generally valid formulas in $\mathcal{L}_{DL}(AC)$ is defined to be the dynamic logic over AC. This logic is symbolically referred to as $DL(AC)$.

The definition of R in terms of $R^\mathcal{A}$ depends on the set AC of action combinators in the modal action language. Below we give this definition of R in terms of $R^\mathcal{A}$ for all combinators except the action complement, since this will be the subject of subsequent sections.

DEFINITION 5. The relational interpretation R of actions α in terms of the relation $R^\mathcal{A}$ is given by:

$$
\begin{array}{ll}
R(a) = R^\mathcal{A}(a) \quad \text{for } a \in \mathcal{A} & \\
R(\alpha \cup \beta) = R(\alpha) \cup R(\beta) & R(\alpha^*) = (R(\alpha))^* \\
R(\alpha;\beta) = R(\alpha) \circ R(\beta) & R(\alpha^\leftarrow) = \{(s,t) \mid (t,s) \in R(\alpha)\}
\end{array}
$$

The binary operation '$\circ$' denotes the sequential composition of relations, and the unary operation '$*$' denotes the reflexive transitive closure of a relation.

This completes the definition of the semantics of dynamic logics. The semantics enforces that $[\alpha]\varphi$ holds in a state whenever all states reachable by α obey φ, which reflects 'execution of the action α results in validity of the condition φ'. For the notion of 'logic entailment' we can choose between the local and the global variant. See [3] for details.

3 Complement with respect to the universal relation

In this section we show that in particular the extension of PDL with the complement as inherited from relation algebra [31], has properties that although useful and intuitive for relations in general, do not apply to the types of action reasoning we mentioned in the introduction. In the next section we propose weakenings that are better suited to model the action reasoning in these domains. It is straightforward to import the standard relation algebraic complement in the dynamic logics we study. We denote this complement with the symbol '$\sim$'.

DEFINITION 6 (Semantics of $\sim$-logics). $\sim$-logics are defined as modal action logics whose set AC of action combinators includes the unary action operator '$\sim$', which we call the 'universal complement'. The semantics of $\sim$-logics follows from the modal semantics of definition 4, the relational definitions for $;, \cup,^*$ and $\leftarrow$ in definition 5, and the clauses:

$$
\begin{array}{ll}
R(U) = S \times S \\
R(\sim\alpha) = R(U) \setminus R(\alpha)
\end{array}
$$

In almost all the literature known to us concerning the complement in dynamic logics [14, 11, 25, 28, 15, 19] and related formalisms (dynamic algebras, Peirce algebras, boolean modules), this is the version of the complement that is studied. This is not surprising given that the universal complement directly corresponds to standard negation in the first-order view on modal languages. It is straightforward to define a translation T from iteration-free $\sim$-logics involving a set of propositions $\{P, Q, \ldots\}$, and a set of atomic actions $\{a, b, \ldots\}$ to a first-order logic with a set of unary predicates $\{P, Q, \ldots\}$, a set of binary predicates $\{R^a, R^b, \ldots\}$, and a set of variables $\{x, y, \ldots\}$:

$$
\begin{array}{lcl}
T_x(P) & := & Px \\
T_x(\bot) & := & x \neq x \\
T_x(\neg \varphi) & := & \neg T_x(\varphi) \\
T_{xy}(\alpha^{\leftarrow}) & := & T_{yx}(\alpha) \\
T_{xy}(\alpha \cup \beta) & := & T_{xy}(\alpha) \vee T_{xy}(\beta)
\end{array}
\qquad
\begin{array}{lcl}
T_x(\varphi \wedge \psi) & := & T_x(\varphi) \wedge T_x(\psi) \\
T_x(\langle \alpha \rangle \varphi) & := & \exists y (T_{xy}(\alpha) \wedge T_y(\varphi)) \\
T_{xy}(a) & := & R^a xy \\
T_{xy}(\alpha; \beta) & := & \exists z (T_{xz}(\alpha) \wedge T_{zy}(\beta)) \\
T_{xy}(\sim \alpha) & := & \neg T_{xy}(\alpha)
\end{array}
$$

With induction we can prove that for any $\varphi \in \mathcal{L}_{DL}(^{\leftarrow}, ;, \cup, \sim)$ it holds that $\mathcal{M}, s \models \varphi$ iff $\mathcal{M} \models_{FOL} T_x(\varphi)[s]$ (s is assigned to the free variable x). Iteration free $\sim$-logics are thus subsumed by first-order logic. This means that they inherit the compactness property from FOL: any non-satisfiable infinite set has a non-satisfiable finite subset.

We now give a set of action connectives that can be introduced as syntactic abbreviations in terms of the complement ($\sim$) and the choice ($\cup$).

DEFINITION 7. The 'universal relation U', the 'impossible action $fail$', the 'concurrent action $\alpha \cap \beta$', the 'subsumption relation $\alpha \subseteq \beta$', and the 'equivalence action $\alpha \doteq \beta$' are defined through the following syntactic extensions on $\cup$ and $\sim$ (We assume that unary operators bind stronger than binary ones, and freely add parentheses whenever ambiguity in the reading of relational formulas may arise.):

$$
\begin{array}{lcl}
\alpha \cap \beta & \equiv_{def} & \sim(\sim\alpha \cup \sim\beta) \\
\alpha \subseteq \beta & \equiv_{def} & \sim\alpha \cup \beta \\
\alpha \doteq \beta & \equiv_{def} & (\alpha \subseteq \beta) \cap (\beta \subseteq \alpha)
\end{array}
\qquad
\begin{array}{lcl}
fail & \equiv_{def} & \alpha \cap \sim\alpha \\
U & \equiv_{def} & \alpha \cup \sim\alpha
\end{array}
$$

Thus, an action operation like $\cap$ does not have to be introduced explicitly in a modal action logic whenever complement and choice are already included[1].

Introduction of the universal complement in modal logic enhances expressiveness considerably ([14]). However, we argue from semantic intuitions

---

[1] If we have the dynamic logic test in the language, the action $fail$ can also be defined as: $fail \equiv_{def} \bot?$.

that the universal complement is inappropriate for a modal logic of action. Of course this does not mean that it cannot be a useful construct in other applications of modal logic (e.g. epistemic reasoning).

Our first point is that the universal relation U does not correspond to the notion of 'any action', an interpretation it should have if the complement were to reflect the notion of 'alternative action'. The intuition for the notion of 'any action' is that it subsumes all possible actions, but not *more*. But the relation U reaches any state, including the ones that are not reachable by any action. In our view, this makes the U too strong to represent the notion of 'any action'. Nevertheless, some authors on action reasoning in PDL refer to the relation U as the 'any action' [26].

We now go on to show what counter intuitive consequences the use of the complement with respect to the universal relation would have in the three application domains we mentioned.

For temporal reasoning over action, it is intuitive to assume that time evolves by the performance of actions. It follows that only if states of affairs can be achieved through action, they belong to the possible futures that have to be considered. Then it is clear that the universal relation U that reaches all possible states cannot function as an action, since then many states that are not reachable should have to be considered as possible future states of affairs.

The universal complement is also not suited to reason about action (non-) effects. A suitable notion of action complement ('−') should enable the expression of frame properties. In particular we claimed that it should be possible to use formulas of the form $\neg \varphi \rightarrow [-\alpha]\neg\varphi$ to express that the condition φ can only be changed by α. But his cannot be achieved by using the universal complement. Again, it is too strong. The universal complement in the expression $\neg\varphi \rightarrow [\sim\alpha]\neg\varphi$ does not only say that actions other than (not involving) α cannot have φ as a result, it says in addition that all states in the state-space that embody the result φ can actually be reached by α ($[|\alpha|]\varphi \equiv_{def} [\sim\alpha]\neg\varphi$ is also called 'the window operator' [1, 17]). This follows the semantics of the universal complement: any state can be reached from any other state by either α or $\sim\alpha$, and reachability by $\sim\alpha$ contradicts $[\sim\alpha]\neg\varphi$. This means that the universal complement is not suited to encode Reiter's approach to specifying frame properties [13, 27] into modal action logic.

In the deontic context, the universal complement is not useful for similar reasons. We do not give the arguments here, but only mention that they were reason for Meyer to define his version of the action complement [24].

Note that we do not claim that a modal *formula* of the form $[\sim\alpha]\varphi$ does not have an intuitive reading in terms of the action α. We only claim that the operator $\sim$ does not have an intuitive reading as an action operation. An intuitive reading of $[\sim \alpha]\varphi$ in terms of action is 'φ holds in all states that are not the result of performing the action α'. But this does not give a reading of $\sim$ as an action combinator, because $\sim \alpha$ is *not* an action: it may include 'transitions' that do not correspond to any action at all.

4 Relativized complement dynamic logics

The universal complement introduces an aspect that is not in the spirit of modal logic: globalness. This aspect of the complement is inherited from relation algebra, where the semantic view is also global. The complement we consider in this section is faithful to the idea of locality in modal semantics. This results in a better fit with the interpretation of the modal language as a logic of action. As a side-effect we also expect better complexities for the logics with a relativized action complement. The action language comprising the new, relativized complement is only concerned with the relations over the part of the state space that is reachable. It assumes that in general, from any given state at least part of the state space cannot be reached through any action, and that the part of the state space that *is* reachable may vary from state to state. Then, the general intuition for the alternative complement, denoted $\wr^I$, is that it is taken with respect to all possible relations over this reachable state space. But what may be considered reachable, depends on what action combinators are in the action language. This explains the term 'relativized'. The complement operation is relativized with respect to the part of the state space in a modal action model that (1) is the minimal relation space containing all atomic actions and that (2) is closed under application of the action combinators of the dynamic logic language (the complement operation itself excluded). Thus, if we allow iteration in the action language, the complement space is reflexive, and if we allow converse, the complement space is symmetric. Therefore we cannot give, as for the universal complement, a general definition of the new complement for all dynamic logics; each dynamic logic comes with its own interpretation for the complement. All in all we define 6 versions of the relativized complement: $\wr^K, \wr^B, \wr^{S4}, \wr^{K4}, \wr^{B4}, \wr^{S5}$, one for each dynamic logic involving a specific set of action combinators. The annotations give information about the nature of the relation space with respect to which the complement is taken, where we adopt standard terminology from modal logic to refer to transitivity, reflexivity etc. Intuitively, this relation space with respect to which the relativized complement is taken reflects the space of possible alternative complex actions. That this space is relative to the action combinators in the action language makes sense, since what can be done 'alternatively' depends on the way in which the action syntax allows us to combine actions into more complex ones.

As a consequence of this scattering of interpretations, some action operators that can be syntactically defined in terms of the complement also get alternative interpretations (any^I, $\subseteq^I$). Other combinators undergo no effective change in interpretation ($\cup, \cap, ;, *, \leftarrow, fail$).

To our knowledge the notion of relativized complement has not been studied before in modal logic. Only in the work of De Giacomo [12] a complement appears that is equivalent with $\wr^K$.

4.1 Semantics of $\wr^I$-logics

DEFINITION 8 (Semantics of $\wr^I$-logics). Let $\wr^K, \wr^B, \wr^{S4}, \wr^{K4}, \wr^{B4}, \wr^{S5}$, be action complement connectives for the logics $DL(\cup, \wr^K)$, $DL(\leftarrow, \cup, \wr^B)$, $DL(;, \cup, \wr^{K4})$, $DL(*, ;, \cup, \wr^{S4})$, $DL(\leftarrow, ;, \cup, \wr^{B4})$, $DL(*, \leftarrow, ;, \cup, \wr^{S5})$ respectively. We refer to this type of logics as $\wr^I$-logics. The semantics of each $\wr^I$-logic is determined by the modal semantics of definition 4, together with a relational interpretation function R for actions, that follows by selection of the relevant clauses of definition 5 in combination with the relevant clause for the relativized action from the following list (the unary operation '+' denotes the transitive closure of a relation):

$$
\begin{aligned}
\text{for } DL(\cup, \wr^K) : & \quad R(any^K) = \bigcup_{a \in \mathcal{A}} R(a) \\
\text{for } DL(\leftarrow, \cup, \wr^B) : & \quad R(any^B) = \bigcup_{a \in \mathcal{A}} (R(a) \cup R(a^\leftarrow)) \\
\text{for } DL(;, \cup, \wr^{K4}) : & \quad R(any^{K4}) = (\bigcup_{a \in \mathcal{A}} R(a))^+ \\
\text{for } DL(*, ;, \cup, \wr^{S4}) : & \quad R(any^{S4}) = (\bigcup_{a \in \mathcal{A}} R(a))^* \\
\text{for } DL(\leftarrow, ;, \cup, \wr^{B4}) : & \quad R(any^{B4}) = (\bigcup_{a \in \mathcal{A}} (R(a) \cup R(a^\leftarrow)))^+ \\
\text{for } DL(*, \leftarrow, ;, \cup, \wr^{S5}) : & \quad R(any^{S5}) = (\bigcup_{a \in \mathcal{A}} (R(a) \cup R(a^\leftarrow)))^*
\end{aligned}
$$

$$R(\wr^I \alpha) = R(any^I) \setminus R(\alpha)$$

The logics of definition 8 are clearly not the only possible $\wr^I$-logics. But many $\wr^I$-logics are rather eccentric. Examples are $DL(\wr^K)$ (the complement is only applicable to atomic actions), which has the same complement as $DL(\cup, \wr^K)$, and $DL(*, \cup, \wr^{S4})$, which has the same complement as $DL(*, ;, \cup, \wr^{S4})$. We do not consider these logics, since they seem to have less relevancy for the application to reasoning about action.

We introduce the relativized *any* and the relativized *subsumption action* as syntactic extensions by defining them in terms of the relativized complement ($\wr^I$) and the choice ($\cup$). Note that we already defined the relativized any as part of the definition of the relativized action complement above. But we identify dynamic logics by reference to their base-combinators, that is, sets of action combinators that are not definable in terms of others. We can define the 'any' in terms of the complement, but not the other way round.

DEFINITION 9. *The relativized any, and the relativized subsumption action are defined through the following syntactic extensions on $\cup$ and $\wr^I$:*

$$\alpha \subseteq^I \beta \equiv_{def} \wr^I \alpha \cup \beta$$
$$any^I \equiv_{def} \alpha \cup \wr^I \alpha \text{ for an arbitrarily chosen } \alpha$$

The relativized versions of the *any* and the *subsumption action* differ from their non-relativized counterparts. The relativized *any* does not reach the complete state space, but only the part that is reachable through (complex) action, as determined by the action language. The next propositions states that the *intersection, equivalence action* and *fail* have the same interpretation as their relativized counterparts.

PROPOSITION 10.
For any two actions α and β in the action language of a $\wr^I$-logic the following holds for the relational interpretations $R(\alpha)$ and $R(\beta)$ on a modal action model $\mathcal{M} = (S, R^{\mathcal{A}}, V^{\mathcal{P}})$:

$$
\begin{array}{llll}
R(\alpha \cap \beta) = R(\alpha \cap^I \beta) & \text{with} & \alpha \cap^I \beta \equiv_{def} \wr^I (\wr^I \alpha \cup \wr^I \beta) \\
R(\alpha \doteq \beta) = R(\alpha \doteq^I \beta) & \text{with} & \alpha \doteq^I \beta \equiv_{def} (\alpha \subseteq^I \beta) \cap (\beta \subseteq^I \alpha) \\
R(fail) = R(fail^I) & \text{with} & fail^I \equiv_{def} \alpha \cap \wr^I \alpha \text{ for an} \\
& & & \text{arbitrarily chosen } \alpha
\end{array}
$$

Proof. We prove the first equivalence. From the action semantics for the universal complement and the relativized complements it follows that $R(\wr^I \alpha) = R(\sim(\alpha \cup \sim any^I))$. Therefore, in the definition for $\alpha \cap^I \beta$ we can substitute the action $\sim(\alpha \cup \sim any^I)$ for all occurrences of the action $\wr^I \alpha$. This returns $\sim(\sim(\alpha \cup \sim any^I) \cup \sim(\beta \cup \sim any^I) \cup \sim any^I)$. By applying standard boolean properties we arrive at the equivalent $\sim(\sim((\alpha \cup \sim any^I) \cap any^I) \cup \sim((\beta \cup \sim any^I) \cap any^I))$. By again applying standard boolean properties we arrive at (1) $\sim(\sim((\alpha \cap any^I) \cup (\sim any^I \cap any^I)) \cup \sim((\beta \cap any^I) \cup (\sim any^I \cap any^I)))$. Now we focus on the actions (2) $\alpha \cap any^I$ and (3) $\sim any^I \cap any^I$. Action 3 is equivalent with the impossible action $fail$, as follows from the meaning of $\sim$. Action 2 is equivalent with α, since any^I is the action that subsumes any other action: for any α it holds that $R(\alpha) \subseteq R(any^I)$. Substitution of these equivalent actions into action 1 results in $\sim(\sim \alpha \cup \sim \beta)$, which is equivalent with $\alpha \cap \beta$. ∎

For any specific $\wr^I$-logic, we talk of its $\wr^I$-reduced sub-logic when we mean the logic that results by removal of the complement, and we talk of its $\sim$-logic counterpart, when we mean the corresponding logic where the $\wr^I$-operator is replaced by the $\sim$-operator. In the same way we talk about $\sim$-logic counterpart formulas, validities, validity schemes, etc.

We will prove that $\wr^I$-logics are subsumed by their $\sim$-logic counterparts (under syntactic replacement of the complements). In this sense they are 'weaker'. But (with one exception) it is not the case that $\wr^I$-logics can be syntactically defined by their $\sim$-logic counterparts. We will prove this for the two weakest $\wr^I$-logics.

PROPOSITION 11.

For the logics $DL(\cup, \wr^K)$ and $DL(\cup, \leftarrow, \wr^B)$, the relativized complement operator $\langle \wr^I \alpha \rangle \varphi$ is not syntactically definable in their $\sim$-logic counterparts.

Proof. From the translation to first-order logic we gave in section 3 follows compactness for iteration-free $\sim$-logics. The relativized complements $\wr^K$ and $\wr^B$ are not compact. For the logic $DL(\cup, \wr^K)$, it is straightforward to verify that the set $\{\langle \wr^K a \rangle \neg P, [b]P, [c]P, \ldots\}$ (where $\{a, b, c, \ldots\} = \mathcal{A}$), is not satisfiable while all of its finite sub-sets are. For the logic $DL(\cup, \leftarrow, \wr^B)$, the same holds for $\{\langle \wr^B(a \cup a^\leftarrow) \rangle \neg P, [b]P, [b^\leftarrow]P, [c]P, [c^\leftarrow]P, \ldots\}$. ∎

4.2 Validities

In this section we study the relations between validities of $\wr^I$-logics and validities of their $\sim$-logic counterparts.

THEOREM 12. *Under syntactic replacement of $\wr^I$ with $\sim$, each validity for a given $\wr^I$-logic turns into a validity for its $\sim$-logic counterpart.*

Proof. Through negative demonstration. Assume that the formula $\varphi^{\wr^I}$ is a $\wr^I$-validity (so it is not a validity scheme). Denote its $\sim$-logic counterpart formula by $\varphi^\sim$. Now assume that $\varphi^\sim$ is not a validity. Under this assumption, we show how to construct a model on which $\varphi^{\wr^I}$ is not valid, thereby proving the theorem. If $\varphi^\sim$ is not a validity, it follows that there is a model $\mathcal{M}$ and a state s such that $\mathcal{M}, s \not\models \varphi^\sim$. Define $aa(\varphi^\sim)$ to be the set of atomic actions occurring in the formula $\varphi^\sim$. Now construct the model $\mathcal{M}'$, by adding to $\mathcal{M}$ the interpretation of an atomic action r such that $r \notin aa(\varphi^\sim)$ and $R^\mathcal{A}(r) = S \times S \setminus any^I$. It follows that $\mathcal{M}', s \not\models \varphi^\sim$, because the addition of the new relation does not alter the truth-condition for $\varphi^\sim$: (1) all actions in $aa(\varphi^\sim)$ are equally interpreted in $\mathcal{M}$ and $\mathcal{M}'$, and (2) all action connectives are equally interpreted, in particular, addition of $R^\mathcal{A}(r)$ does not in any way change the interpretation of actions of the form $\sim \alpha$, because the universal relation is not affected, (3) all valuations of propositions remain as they are. But now, due to the addition of the extra relation, for any a such that $a \in aa(\varphi^\sim)$, the interpretation of the relation $\wr^I a$ on $\mathcal{M}'$, is exactly the same as the interpretation of $\sim a$ on $\mathcal{M}$. It is easy to see that this does not only hold for the interpretation of negated atomic actions, but that the interpretation of any complex action $\wr^I \alpha$ on $\mathcal{M}'$, is exactly the same as the interpretation of $\sim \alpha$ on $\mathcal{M}$. But then, from the information that $\mathcal{M}'$ and $\mathcal{M}$ agree on all valuations of atomic proposition, it follows that the truth conditions for a formula $\varphi^{\wr^I}$ on $\mathcal{M}'$ is necessarily equal to the truth condition for $\varphi^\sim$ on $\mathcal{M}$. So it holds that $\mathcal{M}', s \not\models \varphi^{\wr^I}$. So there is a model, namely $\mathcal{M}'$, for which the formula $\varphi^{\wr^I}$ is not valid. This contradicts the assumption we started off with. ∎

A second theorem gives inclusion relations between separate $\wr^I$-logics.

THEOREM 13.

Under replacement of complements in formulas, the following inclusion relation between $\wr^I$-logics holds. In the picture, logics are represented by the type of complement they endorse, and arrows denote inclusion of validities.

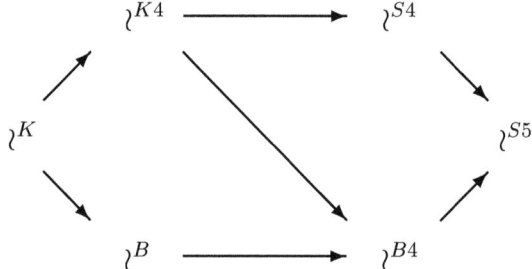

Proof. We can use the same proof strategy as for theorem 12. If we want to prove that for a logic encompassing the complement $\wr^I$, the validities are included in a logic encompassing the complement $\wr^j$, we make use of the addition of a relation $R^A(r)$ for an atomic action r, defined as $R^A(r) = any^j \setminus any^I$. On all other aspects the proofs are completely analogous to the proof of theorem 12. This implies that inclusion of logics is determined by inclusion of the respective complement spaces. So the above partial order simply follows definition 8 by inspecting the inclusion relations between the spaces with respect to which the respective complements are defined. ∎

It is clear right away that the above inclusions are strict, since stronger logics encompass supersets of non-syntactically definable action combinators. But it is illustrative to see how the logics can also be distinguished by validity schemes concerning action combinators that are available in (almost) all relativized complement logics, such as the ∪, the $\wr^I$ and the ;. First we note that many validity schemes are shared by all $\wr^I$-logics and $\sim$-logics. One example is the property K-R:

PROPOSITION 14.
For all $\wr^I$-logics we have the following validity scheme.

$$K\text{-}R \qquad \models \langle \alpha \rangle \varphi \wedge [\wr^I \beta] \neg \varphi \to \langle \alpha \cap \beta \rangle \varphi$$

Proof. Consider an arbitrary model $\mathcal{M}$ and an arbitrary state s. In case that $\mathcal{M}, s \not\models \langle \alpha \rangle \varphi$ or $\mathcal{M}, s \not\models [\wr^I \beta] \neg \varphi$ it holds trivially that $\mathcal{M}, s \models \langle \alpha \rangle \varphi \wedge [\wr^I \beta] \neg \varphi \to \langle \alpha \cap \beta \rangle \varphi$. So, we assume that (1) $\mathcal{M}, s \models \langle \alpha \rangle \varphi$ and (2) $\mathcal{M}, s \models [\wr^I \beta] \neg \varphi$. From (1) it follows that $\alpha \neq fail$ and that there is a state t such that $(s,t) \in R(\alpha)$ and (1') $\mathcal{M}, t \models \varphi$. Now either $(s,t) \in R(\beta)$ or $(s,t) \in R(\wr^I \beta)$ ($\wr^I \beta \cup \beta$ covers all states reachable through complex action). But from $(s,t) \in R(\wr^I \beta)$ and property 2 we would have to conclude that $\mathcal{M}, t \models \neg \varphi$, which contradicts 1'. Then, from $(s,t) \in R(\beta)$ and $(s,t) \in R(\alpha)$ it follows that $(s,t) \in R(\alpha \cap \beta)$, which together with 1' gives $\mathcal{M}, s \models \langle \alpha \cap \beta \rangle \varphi$. ∎

Other properties are valid for some $\wr^I$-logics and invalid for others. We mention the following relevant validity schemes for $\wr^I$-logics:

Trans-any	$\models \langle any^i \rangle \langle any^i \rangle \varphi \to \langle any^i \rangle \varphi$
Symm-any	$\models \varphi \to [any^i] \langle any^i \rangle \varphi$
Refl-any	$\models \varphi \to \langle any^i \rangle \varphi$
NegSeq-R	$\models \langle \alpha \rangle [\wr^i \beta] \varphi \to [\wr^i (\alpha; \beta)] \varphi$

Now the following proposition lists a table saying which properties hold for which logic.

PROPOSITION 15 (validity schemes of $\wr^I$-logics).

MAL variant(s)	Trans-any	Symm-any	Refl-any	NegSeq-R
$(\cup, \wr^K)$	no	no	no	no
$(\leftarrow, \cup, \wr^B)$	no	yes	no	no
$(;, \cup, \wr^{K4})$	yes	no	no	no
$(*, ;, \cup, \wr^{S4})$	yes	no	yes	no
$(\leftarrow, ;, \cup, \wr^{B4})$	yes	yes	no	yes
$(*, \leftarrow, ;, \cup, \wr^{S5})$	yes	yes	yes	yes

Proof. We only verify the soundness of the scheme NegSeq-R for the strongest logic $DL(*, \leftarrow, ;, \cup, \wr^{S5})$. We prove NegSeq-R from negative demonstration: assume there is a model $\mathcal{M}$ with a state s such that (1) $\mathcal{M}, s \models$

$\langle\alpha\rangle[\imath^{S5}\beta]\varphi$ and (2) $\mathcal{M}, s \models \neg[\imath^{S5}(\alpha;\beta)]\varphi$. From (1) it follows that $\alpha \neq fail$ and that there is a state t such that $(s,t) \in R(\alpha)$ and (1') $\mathcal{M}, t \models [\imath^{S5}\beta]\varphi$. Property 2 is equivalent with (2') $\mathcal{M}, s \models \langle \imath^{S5}(\alpha;\beta)\rangle\neg\varphi$. Now there are two cases: $\beta = fail$ or $\beta \neq fail$.

If $\beta = fail$ property 2' reduces to $\mathcal{M}, s \models \langle any^{S5}\rangle\neg\varphi$. This is in direct contradiction with semantic property 1', that for $\beta = fail$ reduces to $\mathcal{M}, t \models [any^{S5}]\varphi$. Since the any^{S5} reaches over all connected states, including the ones reachable by converse action, the contradiction strikes notwithstanding the circumstance that this second validity holds for another state. This is also why the property NegSeq-R does not hold for $DL(*,;,\cup,\imath^{S4})$: the action any^{S4} does not have this 'global' power.

If $\beta \neq fail$, there is a state u such that $(s,u) \in R(\imath^{S5}(\alpha;\beta))$ and (3) $\mathcal{M}, u \models \neg\varphi$. Now either $(t,u) \in R(\beta)$ or $(t,u) \in R(\imath^{S5}\beta)$ (again, $\imath^{S5}\beta \cup \beta$ reaches all connected states). But from $(t,u) \in R(\imath^{S5}\beta)$ and property 1' we would have to conclude that $\mathcal{M}, u \models \varphi$, which contradicts 3. And from $(t,u) \in R(\beta)$ together with $(s,t) \in R(\alpha)$ we get $(s,u) \in R(\alpha;\beta)$, which directly contradicts $(s,u) \in R(\imath^{S5}(\alpha;\beta))$. ∎

It might come as a surprise that the scheme $\varphi \rightarrow \langle any^i\rangle\varphi$, representing reflexivity, only holds in the logic with $\imath^{S4}$ and $\imath^{S5}$, and not for the logic with $\imath^{B4}$. The complement space for the logic with $\imath^{B4}$ is both transitive and symmetric. Transitivity and symmetry together imply reflexivity. But in the present setting, transitivity and symmetry only holds for the reachable relation space, which results in a relativized reflexivity with respect to the complete state space: reflexivity only holds for worlds from which some action is possible.

The table not only lists properties that distinguish $\imath^I$-logics mutually, it also forms a comparison of $\imath^I$-logics with their $\sim$-logic counterparts. For instance, it shows that $DL(\cup,\imath^K)$ is considerably weaker than its sibling $DL(\cup,\sim)$, because it does not support transitivity, or reflexivity or symmetry for the any^K. But the logics lower in the table are closer to their $\sim$-logic counterparts. And for the strongest logic we can prove the following theorem.

THEOREM 16. *Under syntactic interchange of occurrences of $\imath^{S5}$ and $\sim$, the logics $DL(\imath^{S5},\cup,;,*,\leftarrow)$ and $DL(\sim,\cup,;,*,\leftarrow)$ are equivalent, that is, they encompass the same set of validities.*

Proof. Theorem 12 states that under syntactic interchange of occurrences of $\imath^{S5}$ and $\sim$, it holds that $DL(\imath^{S5},\cup,;,*,\leftarrow) \subseteq DL(\sim,\cup,;,*,\leftarrow)$. Here we have to prove that also $DL(\imath^{S5},\cup,;,*,\leftarrow) \supseteq DL(\sim,\cup,;,*,\leftarrow)$. Again we rely on a proof through negative demonstration. Assume that $\varphi^\sim$ is a $\sim$-validity. Denote its $\imath^{S5}$-logic counterpart formula by $\varphi^{\imath^{S5}}$. Now assume that $\varphi^{\imath^{S5}}$ is not a validity. Under this assumption, we show how to construct a model on which $\varphi^\sim$ is not valid, thereby proving the theorem. If $\varphi^{\imath^{S5}}$ is not a validity, it follows that there is a model $\mathcal{M}$ and a state s such that $\mathcal{M}, s \not\models \varphi^{\imath^{S5}}$. Define $aa(\varphi^{\imath^{S5}})$ to be the set of atomic actions occurring in the formula $\varphi^{\imath^{S5}}$. Now construct the model $\mathcal{M}'$, by contraction to the generated submodel of state s (simply remove all non-reachable states, where reachability also accounts for the converse direction of relations). It follows that $\mathcal{M}', s \not\models \Phi^{\imath^{S5}}$, since the truth function does not get another value by this contraction. But now, for any a such that $a \in aa(\varphi^{\imath^{S5}})$, the interpretation of the relation $\sim a$ on $\mathcal{M}'$, corresponds one to one to the interpretation of $\imath^{S5}a$ on $\mathcal{M}$. It is easy to see that actually the interpretation of any complex relation and well-formed formula using $\sim$ on $\mathcal{M}'$ corresponds one to one to the interpretation of a complex relation and well-formed formula using $\imath^{S5}$ on $\mathcal{M}$. But then it holds that $\mathcal{M}', s \not\models \varphi^\sim$, where $\varphi^\sim$ corresponds to $\varphi^{\imath^{S5}}$ with $\sim$ in place for $\imath^{S5}$. It follows that there is a model $\mathcal{M}'$ for which the scheme $\varphi^\sim$ is not valid. This contradicts the assumption we started off with. ■

4.3 Expressiveness

We developed $\imath^I$-logics specifically to reason about concurrency, equivalence and complement of action. And we want the logics to be strong enough to capture all the relevant aspects of the reasoning with these concepts. It is well-known that intersection frames are not definable in standard modal logics (see e.g. [16]). This points to an inadequacy of standard modal logics for reasoning about concurrency, because it shows that not all consequences of the decision to model concurrency through intersection are 'felt' at the level of validities. $\imath^I$-logics do not suffer this inadequacy since they are strong enough to define intersection, action equivalence and action complement both at the level of frames *and* at the level of models. In particular, action equivalence is defined at the level of models by the formula $[\imath^I(\alpha \doteq \beta)]\bot$ saying that any two actions α and β whose relational interpretations do not coincide, are not possible.

4.4 Reasoning about action using the relativized complement

To stress the intuitive adequacy of the complement to reason about action, we show how it can be used in the three application domains we mentioned in the introduction. First we show that it can be used to encode Reiter's approach to the representation of frame formulas. We argued that the universal complement is not suitable for the encoding of Reiter's solution to the frame problem in modal action logics, because the expression $\neg\varphi \to [\sim\alpha]\neg\varphi$ does not only say that actions other than (not involving) α cannot have φ as a result, it says in addition that all states in the state-space that embody the result φ can actually be reached by α. The relativized complement version of this expression, i.e. $\neg\varphi \to [\wr^I \alpha]\neg\varphi$ does not have this unintended additional consequence: only for the states that are reachable through complex action it holds that if φ holds in them, they can be reached by α. Note that we can introduce a relativized version of the window operator ([1, 17]) to abbreviate $[\wr^I \alpha]\neg\varphi$.

To show that the defined complements can be suitably applied to the temporal domain, we define mixed action / time properties using the relativized action complement. To accomplish this, we add modal fixed-point operators to the logic $DL(\wr^K, \cup)$. The resulting μ-calculus (see [6] for more on μ-calculi) enables the expression of mixed dynamic / temporal modalities. We identify 'on all paths the condition φ is preserved until action α is performed' (notation $A(\varphi U \alpha)$), with 'after any finite number of performances of $\wr^K \alpha$ the condition φ holds'. Through similar identifications we define 'on all paths we keep on performing αs until the condition φ is met' (notation $A(\alpha U \varphi)$), and 'on all paths we keep on performing αs until the action β is performed' (notation $A(\alpha U \beta)$). Note that we define non-reflexive versions of these operators.

DEFINITION 17 (Mixed action / time until operators).

$$
\begin{aligned}
A(\varphi U \alpha) &\equiv_{def} [\wr^K \alpha]\mu Z.\ \varphi \wedge [\wr^K \alpha]Z \\
A(\alpha U \varphi) &\equiv_{def} [\wr^K \alpha]\bot \wedge [\alpha]\mu Z.\ \neg\varphi \to ([\wr^K \alpha]\bot \wedge [\alpha]Z) \\
A(\alpha U \beta) &\equiv_{def} [\wr^K \alpha]\bot \wedge [\alpha]\mu Z.\ [\wr^K (\alpha \cup \beta)]\bot \wedge [\alpha]Z
\end{aligned}
$$

Finally, we show how to define deontic modal action logics using the relativized complement (see [7] for details). We will only consider, what we call, 'goal norms'. The class of goal norms takes an action norm to be a norm about the result / goal / effect of an action, while normative repercussions for sequential sub-actions are absent. Thus, goal norms may only be violated in the result state of an action; violations cannot occur in the process of reaching this state. We denote norms of this types with the symbols $O_\odot(\alpha), P_\odot(\alpha)$ and $F_\odot(\alpha)$, where O, P and F stand for Obligation, Permission and Prohibition ('Forbiddenness') respectively, and the subscript '$\odot$' depicts that norms concern action goals. For goal norms it holds, for instance, that each of the formulas $P_\odot(a; b) \wedge \neg P_\odot(a)$, $F_\odot(a) \wedge \neg F_\odot(a; b)$, and $O_\odot(a; b) \wedge \neg O_\odot(a)$ is consistent. Thus, it can be permitted to perform $a; b$, while at the same time a is forbidden. The point is that it is the result of the action that matters: action $a; b$ is permitted, because it brings about a result that is allowed. This view on norms as compliance to action goals was first adopted by Meyer [24]. Meyer identifies the violation of an action goal norm with the validity of a violation proposition V in the resulting state of the action. With the introduction of this special proposition, we may define the following reductions of the modal operators for permission, prohibition and obligation, where (a) free choice permission for α is identified as the absence of a possibility to do α in a way that results in a violation, (b) free choice prohibition for α is identified as the possibility to do α in a way that results in a violation, and (c) free choice obligation for α is identified as the conjunction of free choice permission for α and the occurrence of a violation after any action performance not involving α:

DEFINITION 18 (A reduction to $\wr^I$-logics).

(a) $\quad P_\odot(\alpha) \equiv_{def} \neg \langle \alpha \rangle V$
(b) $\quad F_\odot(\alpha) \equiv_{def} \langle \alpha \rangle V$
(c) $\quad O_\odot(\alpha) \equiv_{def} P_\odot(\alpha) \wedge [\wr^I \alpha] V$

The reduction of definition 18 is easily adapted in order to satisfy other requirements. For instance, we can adapt the reduction in such a way that we exchange free choice norms for imposed choice norms (see e.g. [18, 7] for the distinction). But the interesting aspect of the above reduction is that it gives a consistent logic of free choice permission. In the deontic logic literature it is generally assumed that the introduction of free choice permission leads to inconsistent or degenerate logics [20].

5 Conclusions and future work

In this paper we argued that the more traditional action complement with respect to the universal relation is not suited for many action representation tasks that use dynamic logics. As example domains of such representation and reasoning tasks, we mentioned the domains of (1) action and time, (2) action and norms (deontic logic), and (3) action and change. For all of these domains, the notion of action complement was argued to be important. As an alternative to the traditional complement we defined a series of dynamic logics exhibiting a relativized action complement. Logics with this complement feature a number of attractive properties: (1) semantic intuitiveness, (2) syntactic definability of the standard intersection connective, and (3) definability of intersection, complement and action equivalence at the level of models.

We also expect better complexity properties for the relativized complement logics (as compared to the complexities of their universal complement logic counterparts). But so far we have not studied complexity issues in any depth. We base our expectation on the general insight that relativization may result in better complexity properties. This is demonstrated by the work of Marx [21] on relativized relation algebras. Relativization can be seen as an adaptation of the semantics of formulas that allows for more models, which should make the satisfaction problem easier. Having more models corresponds with having less validities. The validities of proposition 15 are examples of properties that hold for $\sim$-logics but not necessary for their $\wr^I$-logic counterparts.

To prove complexity results we expect that we can use the local character of the relativized complement. This local character enables us to define generalized tree properties such as defined by Marx and Venema [22]. A particular interesting question is whether the logic $DL(;,\cup,\wr^{K4})$ is decidable. The presence of sequence is likely to cause undecidability. Transitivity introduces undecidability in first order generalizations of modal logic such as the two variable and guarded fragments. But for some traditional incarnations of modal logic that were developed for the temporal and process domains, transitivity is a common feature that does not cause undecidability.

BIBLIOGRAPHY

[1] J.F.A.K. van Benthem. Minimal deontic logics. *Bulletin of the Section of Logic*, 8(1):36–42, 1979.

[2] J.F.A.K. van Benthem. Programming operations that are safe for bisimulation. *Studia Logica*, 60(2):311–330, 1998.

[3] P. Blackburn, M. de Rijke, and Y. Venema. *Modal Logic*, volume 53 of *Cambridge Tracts in Theoretical Computer Science*. Cambridge University Press, 2001.

[4] A. Borgida, J. Mylopoulos, and R. Reiter. On the frame problem in procedure specifications. *IEEE Transactions on Software Engineering*, 21:785–798, 1995.

[5] J.C. Bradfield. *Verifying Temporal Properties of Systems*. Birkhuser Boston, Massachussetts, 1992.

[6] J.C. Bradfield and C. Stirling. Modal logics and mu-calculi: An introduction. In J.A. Bergstra, A. Ponse, and S.A. Smolka, editors, *Handbook of Process Algebra*, pages 291–330. Elsevier Science, 2001.

[7] J.M. Broersen. A new action-base for dynamic deontic logics. In J. Horty and A.J.I. Jones, editors, *Pre-proceedings 6th International Workshop on Deontic Logic in Computer Science (DEON'02)*, pages 21–37, 2002.

[8] J.M. Broersen, M. Dastani, Z. Huang, and L.W.N. van der Torre. Trust and commitment in dynamic logic. In H. Shafazand and A. Min Tjoa, editors, *Eurasia-ICT 2002: Information and Communication Technology*, volume 2510 of *Lecture Notes in Computer Science*, pages 677–684. Springer, 2002.

[9] J.M. Broersen, R.J. Wieringa, and J.-J.Ch. Meyer. A fixed-point characterization of a deontic logic of regular action. *Fundamenta Informaticae*, 48(2, 3):107–128, 2001. Special Issue on Deontic Logic in Computer Science.

[10] M.J. Fischer and R.E. Ladner. Propositional dynamic logic of regular programs. *Journal of Computer and System Sciences*, 18(2):194–211, 1979.

[11] G. Gargov and S. Passy. A note on boolean modal logic. In P. Petkov, editor, *Mathematical Logic. Proceedings of The Summer School and Conference "Heyting'88"*, pages 311–321. Plenum Press, 1990.

[12] G. De Giacomo. *Decidability of class-based knowledge representation formalisms*. PhD thesis, Universita' di Roma "La Sapienza", 1995.

[13] G. De Giacomo and M. Lenzerini. PDL-based framework for reasoning about actions. In *Proceedings of the 4th Congress of the Italian Association for Artificial Intelligence (AI*IA'95)*, volume 992 of *Lecture Notes in Artificial Intelligence*, pages 103–114. Springer, 1995.

[14] V. Goranko. Modal definability in enriched languages. *Notre Dame Journal of Formal Logic*, 31(1):81–105, 1990.
[15] D. Harel, D. Kozen, and J. Tiuryn. *Dynamic Logic*. The MIT Press, 2000.
[16] W. van der Hoek. *Modalities for Reasoning about Knowledge and Quantities*. PhD thesis, Faculteit der Wiskunde en Informatica, Vrije Universiteit Amsterdam, 1992.
[17] L. Humberstone. Inaccessible worlds. *Notre Dame Journal of Formal Logic*, 24:346–352, 1983.
[18] H. Kamp. Free choice permission. *Aristotelian Society Proceedings N.S.*, 74:57–74, 1973-1974.
[19] C. Lutz and U. Sattler. The complexity of reasoning with boolean modal logic. In *Advances in Modal Logic*, volume 3, pages 365–387. World Scientific, 2002.
[20] D. Makinson. Stenius' approach to disjunctive permission. *Theoria*, 50:138–147, 1984.
[21] M. Marx. Relativized relation algebras. *Algebra Universalis*, 41:23–45, 1999.
[22] M. Marx and Y. Venema. Local variations on a loose theme: modal logic and decidability. In M.Y. Vardi and S. Weinstein, editors, *Finite Model Theory and its Applications*. Springer. forthcoming.
[23] R. van der Meyden. The dynamic logic of permission. *Journal of Logic and Computation*, 6(3):465–479, 1996.
[24] J.-J.Ch. Meyer. A different approach to deontic logic: Deontic logic viewed as a variant of dynamic logic. *Notre Dame Journal of Formal Logic*, 29:109–136, 1988.
[25] E. Orlowska. Dynamic logic with program specification and its relational proof system. *Journal of Applied Non-Classical Logics*, 3(2):147–171, 1993.
[26] H. Prendinger and G. Schurz. Reasoning about action and change, a dynamic logic approach. *Journal of Logic, Language and Information*, 5:209–245, 1996.
[27] R. Reiter. The frame problem in the situation calculus: A simple solution (sometimes) and a completeness result for goal regression. In V. Lifschitz, editor, *Artificial Intelligence and Mathematical Theory of Computation: Papers in Honor of John McCarthy*. Academic Press, 1991.
[28] M. de Rijke. A system of dynamic modal logic. *Journal of Philosophical Logic*, 27:109–142, 1998.
[29] K. Segerberg. Outline of a logic of action. In *Advances in Modal Logic*, volume 3, pages 365–387. World Scientific, 2002.
[30] C. Stirling. Modal and temporal logics for processes. In *Banff Higher Order Workshop*, volume 1043 of *Lecture Notes in Computer Science*, pages 149–237. Springer, 1996.
[31] A. Tarski. On the calculus of relations. *Journal of Symbolic Logic*, 6:73–89, 1941.

Jan Broersen
Intelligent Systems Group
Institute of Information and Computing Science
Universiteit Utrecht
PO Box 80.089
3508 TB Utrecht
The Netherlands
E-mail: broersen@cs.uu.nl

4
How Many Variables Does One Need to Prove PSPACE-hardness of Modal Logics?

A. V. CHAGROV AND M. N. RYBAKOV

ABSTRACT. The main result of the paper is that the decision problem for the variable-free fragments of both **K** and **K4** is PSPACE-complete. It is also proved that the decision problem for one-variable fragments of **S4**, **Grz**, and **GL** is PSPACE-complete too. The questions about complexity of variable-free, one-variable, and two-variable fragments of some closed logics are discussed.

1 Introduction

It is well known that the decision problem for propositional modal logics **K**, **T**, **S4**, and many others is PSPACE-complete [6] (a good reference on complexity theory is [9]). Ladner's [6] proof of PSPACE-completeness of the decision problem for these logics was essentially based upon the fact that the modal language contains infinitely many propositional variables; he did not try to use as few of them as possible. On the other hand, the fragments with only finitely many variables are rather natural, and in many cases such a restriction may result in a lower algorithmic complexity: for example, classical propositional (non-modal) logic with finitely many variables is decidable in polynomial time.

About 15 years ago, an analysis of Ladner's [6] and other similar proofs gave the first author some (shaky) grounds to conjecture[2] that an infinite set of variables is really needed to prove PSPACE-completeness of modal logics, that finite variable fragments can really be of lower complexity. However, in the 1990s this conjecture was refuted: Halpern [5] showed that a sole variable is enough to prove PSPACE-hardness of the satisfiability problem for modal logics considered in [6]. And even before that Spaan [8] proved the 'Single Variable Reduction Theorem' (Theorem 5.4.6 of [8]) claiming that the satisfiability problem for modal formulas is polynomially reducible to the satisfiability problem for one-variable modal formulas.

---

[2]This conjecture was never published, but it was discussed at several conferences.

In this connection, it is natural to raise the question on the complexity of the *variable-free* fragments of standard modal logics: is it possible to obtain optimal lower complexity bounds (PSPACE-hardness or NP-hardness) for the decision problem for modal logics using only *variable-free formulas*? Some results of this sort were proved by Spaan [8] (see Theorem 5.4.7 of [8]) — but for logics with two modal operators. On the other hand, note that Halpern's [5] language contains no constants ($\bot$ and $\top$), and so the obtained results for the one-variable fragment may be regarded as optimal (although the formulas from [5] that establish PSPACE-hardness of the satisfiability problem for **K** can be easily transformed into variable-free formulas with the same property by some substitution).

Our main aim in this paper is to find out the *minimal number* of variables one needs to prove PSPACE-hardness of the decision problem for standard modal logics. Since this problem (even in the full infinite language) is in PSPACE, we thereby shall get PSPACE-completeness of the corresponding finite-variable (or even variable-free) fragments. For some of the logics considered — **T**, **S4**, **Grz**, **GL** — these fragments contain a sole variable, while variable-free fragments of these logics are decidable by polynomial algorithms. On the other hand, **K** and **K4** do have very expressive PSPACE-hard variable-free fragments.

In conclusion, we consider the satisfiability problem for some other logics. We also obtain a corollary on the frame complexity function for the considered fragments.

Our notation and notions are mostly standard: for modal logics and their complexity problems the reader can consult [3] or [10], for complexity theory see [4] and [9].

2 Main results

In this section we consider normal modal logics located between **K** and **K4**. For our aims it is enough to show PSPACE-hardness of the satisfiability (or, equivalently, the decision) problem for fragments we are interested in. First, we will use only variable-free formulas for proving PSPACE-hardness of the satisfiability problem for **K4**.

To begin with, we prove PSPACE-hardness of the satisfiability problem for **K4** in the *full language* that contains the propositional variables $p_1, p_2, p_3, \ldots$, the Boolean constant $\bot$ ('falsehood'), the Boolean connectives $\land, \lor, \rightarrow$, and the modal operator $\square$. The constant $\top$ and the connectives $\neg$ and $\Diamond$ are used as standard abbreviations.

To prove PSPACE-hardness of the satisfiability problem for **K4**, we construct a reduction of the satisfiability problem for quantified Boolean formulas (which is known to be PSPACE-complete, see [9]) to the satisfiability problem for **K4**. Without loss of generality, we will confine ourselves to considering quantified Boolean formulas of the form $\varphi = Q_1 p_1 \ldots Q_n p_n \varphi'$, where $Q_1, \ldots, Q_n \in \{\forall, \exists\}$ and φ' is a quantifier-free formula with $p_1, \ldots, p_n$ as its variables. Our reduction is a modification of the transformation considered in [6] (it can also be found in [2]). Given a formula $\varphi = Q_1 p_1 \ldots Q_n p_n \varphi'$, we take additional variables $q_0, \ldots, q_{n+1}$ and set

$$A = \bigwedge_{i=1}^{n+1} (q_i \to q_{i-1}),$$

$$B = \bigwedge_{i=1}^{n} [q_i \to (p_i \to \Box(q_i \to p_i)) \land (\neg p_i \to \Box(q_i \land \neg q_{n+1} \to \neg p_i))],$$

$$C = \bigwedge_{\{i\,:\,Q_{i+1}=\forall\}} [q_i \land \neg q_{i+1} \to \Diamond(q_{i+1} \land \neg q_{i+2} \land p_{i+1}) \land \Diamond(q_{i+1} \land \neg q_{i+2} \land \neg p_{i+1})],$$

$$D = \bigwedge_{\{i\,:\,Q_{i+1}=\exists\}} [q_i \land \neg q_{i+1} \to \Diamond(q_{i+1} \land \neg q_{i+2})].$$

Let

$$\varphi^* = q_0 \land \neg q_1 \land \Box^+ A \land \Box^+ B \land \Box^+ C \land \Box^+ D \land \Box^+ (q_n \land \neg q_{n+1} \to \varphi'),$$

where $\Box^+ \psi = \psi \land \Box \psi'$. Note that φ^* is constructed from φ by a polynomial time algorithm in the length of φ. Moreover,

(1) φ is true $\iff$ φ^* is **K4**-satisfiable,

and hence we obtain PSPACE-hardness of the satisfiability problem for **K4**; accordingly to Savitch's theorem [7] coPSPACE=PSPACE therefore we also obtain PSPACE-hardness of the decision problem for **K4**.

To prove ($\Rightarrow$) of (1), observe that the structure of the true formula φ clearly shows how to construct the required **K4**-model. As the underlying **K4**-frame we take the transitive reflexive tree of height $n+1$ such that if $Q_i = \forall$ then from every point of level $i-1$ (the root is of level 0) two points of level i are accessible and if $Q_i = \exists$ then one point of level i is accessible. The valuation is defined in the following way: q_i is true at all points of level greater than or equal to i, p_i is false at all points of level less than i; if a and b are two distinct points of level i accessible from some point of level $i-1$, then p_i is true at one of them, say a, and all points accessible from it, and p_i is false at the other — i.e., b — and all points accessible from it; if a is a point of level $i-1$ and b is the unique point of level i accessible from a, then if for the truth-values of $p_1, \ldots, p_{i-1}$ at a for $\exists p_i Q_{i+1} p_{i+1} \ldots Q_n p_n \varphi'$ to be true the 'true' was chosen for p_i, then we define p_i to be true at b

and at all points accessible from b, and if 'false' was chosen for p_i, then we make p_i false at b and at all points accessible from b as well. It should be clear that the formula φ' is true at every point of level n. Therefore, $\Box(q_n \wedge \neg q_{n+1} \to \varphi')$ is true at the root. It is easy to check that all other conjuncts of φ^* are true at the root. Hence φ^* is satisfied in the constructed **K4**-model.

($\Leftarrow$) If φ^* is true at a point a of some **K4**-model, then the conjuncts of φ^* allow us to extract in a 'step-by-step' manner a submodel of the given model in almost the same way as in the proof of ($\Rightarrow$), except that some points may be irreflexive. The formula φ' will be true at all points of level n in this model and the collections of the truth-values of $p_1, \ldots, p_n$ at points of level n provide us with a representative sample showing the truth of φ.

For any m, let

$$\alpha_m = \Box(\Diamond^m \Box \bot \wedge \neg \Diamond^{m+1} \Box \bot \to \Box(\Diamond \top \to \Diamond \Box \bot))$$

and denote by φ_α^* the formula obtained from φ^* by substituting $\alpha_1, \ldots, \alpha_{2n+2}$ for $p_1, \ldots, p_n, q_0, \ldots, q_{n+1}$, respectively.

LEMMA 1 *The formula φ_α^* is computed from φ^* by a polynomial time algorithm, and φ_α^* is **K4**-satisfiable iff φ^* is **K4**-satisfiable.*

Proof. Suppose φ^* is not **K4**-satisfiable. Then $\neg \varphi^* \in \mathbf{K4}$. But $\neg \varphi_\alpha^*$ is a substitution of $\neg \varphi^*$, from which it follows that $\neg \varphi_\alpha^* \in \mathbf{K4}$, and hence φ_α^* is not **K4**-satisfiable.

Let φ^* be **K4**-satisfiable. It means (see the proof of (1)) that the quantified Boolean formula φ is true. Hence φ^* is satisfied at the root w_0 of a reflexive and transitive tree $M = \langle W, R, v \rangle$ of height $n+1$. It is easy to see that the valuation v is hereditary: for every variable p and for any $w', w'' \in W$ such that $w'Rw''$ and $(M, w') \models p$, we have $(M, w'') \models p$.

Now we extend the model $M = \langle W, R, v \rangle$ to a model $M' = \langle W', R', v' \rangle$ in such a way that $(M', w_0) \models \varphi_\alpha^*$ will hold.

First observe that for α_m to be refuted at a point of a (reflexive and transitive) model, it is sufficient that the frame, depicted on the left-hand side of Fig. 1, be accessible from that point (black points of the frame are irreflexive, white ones are reflexive, and the accessibility relation is transitive). Denote this frame by F_m.

It is clear that if α_m is true at a point of a transitive model then it is true at all points accessible from it (the main connective of α_m is $\Box$). But if $k \neq m$ then $F_m \models \alpha_k$. Using these observations, we construct the required model M'. The construction takes $(2n+3)$ steps.

Step 0. Put $W_0 = W$, $R_0 = R$.

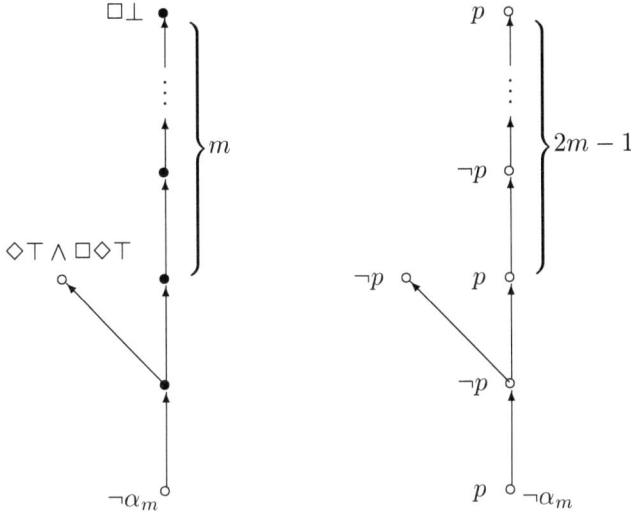

Figure 1. Frame F_m and model M_m.

Step m, where $1 \leq m \leq 2n+2$. Consider the set X_m of points in M where p_m is false. For every $w \in X_m$, we extend W_{m-1} and R_{m-1} so that a copy of F_m be accessible from w. Denote the resultant set of points by W_m and the transitive closure of the obtained accessibility relation by R_m.

Let $W' = W_{2n+2}$, $R' = R_{2n+2}$ and let v be an arbitrary valuation. For every ψ with $p_1, \ldots, p_n, q_0, \ldots, q_{n+1}$ we denote by ψ_α the formula obtained from ψ by substituting $\alpha_1, \ldots, \alpha_{2n+2}$ for $p_1, \ldots, p_n, q_0, \ldots, q_{n+1}$, respectively. Then for every subformula ψ of the (quantifier-free) Boolean formula φ' and every $w \in W$ of level n, the following equivalence holds:

$$(M', w) \models \psi_\alpha \iff (M, w) \models \psi.$$

This claim is proved by induction on the construction of ψ.

The most difficult case is the basis of induction. If $\psi = p_m$ then $\psi_\alpha = \alpha_m$ and we must prove that for every $w \in W$ of level n

$$(M', w) \models \alpha_m \iff (M, w) \models p_m.$$

Let w be a point of level n such that $(M, w) \not\models p_m$. Then at step m of the construction of M', model M was extended with a copy of F_m accessible from w. But then $(M', w) \not\models \alpha_m$.

Let $(M', w) \not\models \alpha_m$, where w is a point of M of level n. It is easy to see that a copy of F_m must be accessible from w. This is possible only if $(M, w) \not\models p_m$.

The induction step is trivial.

Thus, if w is a point in M of level n then $(M', w) \models \varphi'_\alpha$. Observe that the formula $\alpha_{2n+1} \wedge \neg \alpha_{2n+2}$ is true at some point w of M' only if w is a point of level n in M, because points in M validating $q_n \wedge \neg q_{n+1}$ are points of level n. Therefore, they are points of M' from which copies of F_{2n+2} are accessible, but copies of F_{2n+1} are not. So, at every other point of M' the formula $\alpha_{2n+1} \wedge \neg \alpha_{2n+2}$ is not true. Hence $(M', w_0) \models \Box(\alpha_{2n+1} \wedge \neg \alpha_{2n+2} \to \varphi'_\alpha)$.

It is easy to check that

$$(M', w_0) \models \alpha_{n+1} \wedge \neg \alpha_{n+2} \wedge \Box^+ A_\alpha \wedge \Box^+ B_\alpha \wedge \Box^+ C_\alpha \wedge \Box^+ D_\alpha.$$

It follows that $(M', w_0) \models \varphi^*_\alpha$. ∎

Now we show how to modify φ^* and $\alpha_1, \ldots, \alpha_{2n+2}$ in order to prove similar results for some other logics. Consider a logic L such that $\mathbf{K} \subseteq L \subseteq \mathbf{K4}$. For any ψ, let

$$\Box^{\leq 1} \psi = \Box \psi; \quad \Box^{\leq k+1} \psi = \Box^{\leq k} \psi \wedge \Box^{\leq k+1} \psi.$$

Denote by $tr_k(\psi)$ the formula obtained from ψ by replacing each subformula $\Box \delta$ of ψ with $\Box^{\leq k} \delta$. Then one can readily prove that

$$\varphi \text{ is true} \iff \varphi^* \text{ is } \mathbf{K4}\text{-satisfiable}$$
$$\iff tr_n(\varphi^*) \text{ is } L\text{-satisfiable}.$$

Given any natural m, we take again the variable-free formula

$$\alpha_m = \Box(\Diamond^m \Box \bot \wedge \neg \Diamond^{m+1} \Box \bot \to \Box(\Diamond \top \to \Diamond \Box \bot)).$$

Denote by $tr_n(\varphi^*)_\alpha$ the formula obtained from $tr_n(\varphi^*)$ by substituting $\alpha_1, \ldots, \alpha_{2n+2}$ for $p_1, \ldots, p_n, q_0, \ldots, q_{n+1}$, respectively.

LEMMA 2 *The formula $tr_n(\varphi^*)_\alpha$ is constructed from φ^* by a polynomial time algorithm, and $tr_n(\varphi^*)_\alpha$ is L-satisfiable iff φ^* is $\mathbf{K4}$-satisfiable.*

Using the complexity properties of quantified Boolean formulas and their $*$-images, we obtain from Lemmas 1 and 2 the following:

THEOREM 3 *For every logic L from the interval $\mathbf{K} \subseteq L \subseteq \mathbf{K4}$, the satisfiability (and so the decision) problem for its variable-free fragment is PSPACE-hard.*

Ladner [6] gave polynomial space decision algorithms for **K**, **T**, and **S4**. A slight modification of Ladner's algorithm for **S4** provides a similar decision algorithm for **K4**. Hence we obtain

COROLLARY 4 *The satisfiability (and so the decision) problem for the variable-free fragments of* **K** *and* **K4** *are PSPACE-complete.*

Since the variable-free fragments of **S4**, **Grz**, **GL** are decidable in polynomial time (see Theorem 9 below), let us consider their one-variable fragments.

REMARK Halpern [5] proved that the decision problem for the one-variable fragment of **S4** is PSPACE-hard. But [5] contains some inaccuracies. Namely, the proof uses the following one-variable formulas:

$$\hat{q}_1 = \Diamond\Box p,$$
$$\hat{q}_{m+1} = \Diamond(p \land \Diamond(\neg p \land \hat{q}_m)),$$
$$\hat{q}'_j = \hat{q}_j \land \neg\hat{q}_{j+1},$$
$$\hat{r}_n = \Diamond\hat{q}'_n \land \Box(\hat{q}'_n \to \Diamond\Box\neg p),$$

where $\hat{r}_1, \hat{r}_2, \hat{r}_3, \ldots$ must have the property that 'any atom of them (finite conjunction of $\hat{r}_i$'s and negations of $\hat{r}_i$'s) is satisfiable.' However, $\hat{r}_1$ is not **S4**-satisfiable; each of the formulas $\hat{r}_2, \hat{r}_3, \hat{r}_4, \ldots$ is **S4**-satisfiable, but the atom $\hat{r}_2 \land \neg\hat{r}_3 \land \hat{r}_4$ is not.

Further, it is assumed that the formulas $\hat{r}_1, \hat{r}_2, \hat{r}_3, \ldots$ are substituted for variables in some formula, denoted in [5] by ψ_A^{S4}, with its satisfiability (or non-satisfiability) being preserved. But the structure of ψ_A^{S4} is such that, even if we suppose that $\hat{r}_1, \hat{r}_2, \hat{r}_3, \ldots$ do have all of the required properties, one can show that in some cases ψ_A^{S4} is satisfiable, while its substitution instance is not. Note that our formula φ^* is a modification of ψ_A^{S4} (we add the conjunct $\neg q_{n+1}$ to a number of subformulas of ψ_A^{S4}).

Nevertheless, the general approach to the one-variable fragment of **S4** in [5] is sound. Here we show how to cure the defects of Halpern's construction.

So, let us consider **S4** and **Grz**. As in the case of **K4**

$$\varphi \text{ is true} \iff \varphi^* \text{ is \textbf{S4}-satisfiable}$$
$$\iff \varphi^* \text{ is \textbf{Grz}-satisfiable}.$$

For every m, let

$$\delta_1 = \Diamond\Box p, \quad \delta_{m+1} = \Diamond(p \land \Diamond(\neg p \land \delta_m)),$$
$$\alpha_m = \Box(p \to \Box(\neg p \land \delta_m \land \neg\delta_{m+1} \to \Box\Diamond p)).$$

As before, denote by φ_α^* the formula that is obtained from φ^* by substituting $\alpha_1, \ldots, \alpha_{2n+2}$ for $p_1, \ldots, p_n, q_0, \ldots, q_{n+1}$, respectively.

LEMMA 5 *The formula φ_α^* is constructed from φ^* by a polynomial time algorithm, and φ_α^* is **S4**-satisfiable iff φ_α^* is **Grz**-satisfiable iff φ^* is **S4**-satisfiable iff φ^* is **Grz**-satisfiable.*

The proof is similar to the proof of Lemma 1. The only difference is that when constructing M' we use the models M_m instead of the F_m; see Fig. 1 (the accessibility relation of M_m is reflexive and transitive).

Thus, we obtain

THEOREM 6 *The satisfiability (and so the decision) problem for the one-variable fragments of **S4** and **Grz** are PSPACE-complete.*

Now consider the Gödel–Löb provability logic **GL**. Besides the equivalences of Lemma 5 we also have

$$\varphi \text{ is true} \iff \varphi^* \text{ is } \mathbf{GL}\text{-satisfiable.}$$

As is well-known, **GL** is similar to **Grz**. So it is easy to modify the formulas $\alpha_1, \ldots, \alpha_{2n+2}$ for the case of **GL**. In fact, it is enough to put, for any m,

$$\delta_1 = \Diamond\Box^+ p, \quad \delta_{m+1} = \Diamond(p \wedge \Diamond(\neg p \wedge \delta_m)),$$

$$\alpha_m = \Box(p \to \Box(\neg p \wedge \delta_m \wedge \neg\delta_{m+1} \to \Box\Diamond^+ p)),$$

where $\Box^+\psi = \psi \wedge \Box\psi$, $\Diamond^+\psi = \psi \vee \Box\psi$.

An algorithm deciding the satisfiability problem for **GL** in polynomial space can be found in [3], see Theorem 18.29. Therefore, we obtain

THEOREM 7 *The satisfiability (and so the decision) problem for the one-variable fragment of **GL** is PSPACE-complete.*

Moreover, using the argument above one can prove the following more general

THEOREM 8 *Let $\mathbf{K} \subseteq L \subseteq \mathbf{Grz}$ or $\mathbf{K} \subseteq L \subseteq \mathbf{GL}$. Then the satisfiability (and so the decision) problem for the one-variable fragment of L is PSPACE-hard.*

3 Some related problems

3.1 NP-hardness

Theorem 8 does not solve all our problems for **GL**: there are infinitely many non-equivalent pairs of formulas in **GL**, for example, $\Box\bot, \Box\Box\bot, \Box\Box\Box\bot$, etc. Nevertheless, the variable-free fragment of **GL** is quite simple: we describe an algorithm deciding this fragment in polynomial time. We will use two well-known facts: (1) the variable-free fragment of **GL** is complete with respect to all finite linear **GL**-frames (see the description of $\mathfrak{F}_{\mathbf{GL}}^{<\infty}(0)$, i.e., universal frame of rank 0, in [3], p. 275) and (2) if φ is refuted in some **GL**-frame then φ is refuted in its subframe of height not exceeding the length of φ, see [1] or the proof of Theorem 18.13 in [3].

So, let φ be a variable-free formula of length k and let $\varphi_1, \ldots, \varphi_n = \varphi$ be all subformulas of φ such that if φ_i is a subformula of φ_j then $i \leq j$. We construct the table

	φ_1	φ_2	$\ldots$	φ_{n-1}	φ_n
0	f	t			
1	f	f			
$\ldots$					
$k-1$	f	f			

where $\{0, \ldots, k-1\}$ is supposed to be the set of points of a **GL**-frame and aRb iff $b < a$.

If $i \models \varphi_j$ then we write 't' in the (i, φ_j) square of the table; otherwise we write 'f' (for example, in the table above we have $\varphi_1 = \bot$, $\varphi_2 = \Box\bot$). It is not hard to see that this table can be constructed by a polynomial time algorithm in the length of φ.

When the construction is complete we check the last column: if it contains at least one 'f' then $\varphi \notin \mathbf{GL}$; otherwise $\varphi \in \mathbf{GL}$.

Therefore, we obtain

THEOREM 9 *The satisfiability and the decision problems for the variable-free fragment of* **GL** *are decidable in polynomial time.*

The algorithm presented in the proof of Theorem 9 is so simple, and it may seems it is possible to use the ideas of this algorithm to prove similar results for other logics with other restrictions on the number of variables. But in fact we cannot do too much. A first reasonable step in this direction is to consider the logic **GLLin** (= **GL.3**, i.e., the logic of all linear frames for **GL**): **GLLin** is the greatest logic among those having the same variable-free fragment as **GL** — and so the simplest of them. But such an attempt does not seem to be successful.

THEOREM 10 *The satisfiability problem for the one-variable fragment of* **GLLin** *is* NP-*complete.*

Indeed, the satisfiability problem for full **GLLin** (with countably infinite number of variables) is known to be in NP, see [3], Theorem 18.25, where **GLLin** is **GL** $\oplus$ bw_1. So it is sufficient to reduce some NP-complete problem to the one-variable fragment of **GLLin**. We will reduce the satisfiability problem for Boolean formulas (in other words, the validity problem for formulas of the form $\exists p_1 \ldots \exists p_n \varphi$, where φ is quantifier-free formula with the variables $p_1, \ldots, p_n$).

Let φ be a Boolean formula with n variables. Without a loss of generality we will assume that the variables of φ are $p_1, \ldots, p_n$. Replace every p_i in φ with $\Diamond(\Box^{i+1}\bot \wedge \Diamond^i \top \wedge p)$ and take the negation of the obtained formula. Denote the resulting formula by $\tilde{\varphi}$. It is easy to see that $\tilde{\varphi}$ is constructed in polynomial time and that

$$\varphi \text{ is satisfiable} \iff \tilde{\varphi} \notin \textbf{GLLin}.$$

However, for some logics related to **GLLin** the algorithm of Theorem 9 can be modified. For example consider extensions of **S4**: **S4.3** and **Grz.3** (the first is determined by chains of clusters, while the second is characterised by finite linear frames with partial (i.e., reflexive) order).

THEOREM 11 *The one-variable fragments of* **S4.3** *and* **Grz.3** *are decidable in polynomial time.*

THEOREM 12 *The satisfiability problems for the two-variable fragments of* **S4.3** *and* **Grz.3** *are* NP-*complete.*

To prove Theorem 11 one can use the analigies of above-mentioned facts on **GL**: (1) the description of $\mathfrak{F}_{\textbf{S4.3}}^{<\infty}(1)$ and $\mathfrak{F}_{\textbf{Grz.3}}^{<\infty}(1)$ see in [3], p. 275, (2) the statement on linear approximability of these logics is contained in [3], Theorem 18.13. To prove Theorem 12 it is sufficient for any Boolean formula φ with n variables $p_1, \ldots, p_n$ to consider the formula $\tilde{\varphi}$ obtained from φ by replacing every p_i with $\Diamond(\delta_i \wedge \neg \delta_{i+1} \wedge q)$, where δ_n is defined on p. 77.

3.2 Complexity function

The *complexity function* $f_L(n)$ for a logic (a set of formulas) L is defined as follows (see [3]):

$$f_L(n) = \max_{\substack{|\varphi| \leq n \\ \varphi \notin L}} \min_{\substack{F \models L \\ F \not\models \varphi}} |F|,$$

where $|F|$ is the number of points in F and $|\varphi|$ is the length of φ. For all PSPACE-hard logics L considered above, $f_L(n)$ cannot be bounded by any polynomial function. Indeed, take the formulas $\varphi_n = \forall p_1 \ldots \forall p_n \bot$. It is easy to see that any model refuting φ_n^* must have at least $2^n + 1$ points. Since the length of the φ_n is a linear function of n, the length of the φ_n^* is a polynomial function of n, and so we can conclude that the complexity functions for **K**, **K4**, **S4**, **GL**, **Grz** cannot be bounded by any polynomial function (see [3] for a similar argument).

Given a logic L, denote by $L(m)$ the m-variable fragment of L. Then it follows from the discussion above that we have

THEOREM 13 *If $L \in \{\mathbf{K}(0), \mathbf{K4}(0), \mathbf{S4}(1), \mathbf{GL}(1), \mathbf{Grz}(1)\}$ then $f_L(n)$ cannot be bounded by any polynomial function.*

4 Acknowledgements

We are grateful to anonymous referees and our collegues, especially to Michael Zakharyaschev whose remarks and suggestions substantially helped to impove presentation.

BIBLIOGRAPHY

[1] S. N. Artemov. Modal logics axiomatizing provability. *Mathematics of the USSR, Izvestia*, 49:1123–1154, 1985.
[2] P. Blackburn, M. de Rijke, and Y. Venema. *Modal Logic*. Cambridge University Press, 2001.
[3] A. Chagrov and M. Zakharyaschev. *Modal Logic*. Oxford University Press, 1997.
[4] M. R. Garey and D. S. Johnson. *Computers and Intractability, A Guide to the Theory of NP-completeness*. San Francisco, 1979.
[5] J. Y. Halpern. The effect of bounding the number of primitive propositions and the depth of nesting on the complexity of modal logic. *Artificial Intelligence*, 75(2):361–372, 1995.
[6] R. E. Ladner. The computational complexity of provability in systems of modal logic. *SIAM Journal on Computing*, 6:467–480, 1977.
[7] W. J. Savitch. Relationships between nondeterministic and deterministic tape complexity. *Journal of Computer and System Science*, 4:177–192, 1970.
[8] E. Spaan. *Complexity of Modal Logics*. PhD thesis, University of Amsterdam, Department of Mathematics and Computer Science, 1993.
[9] L. Stockmeyer. Classifying the computational complexity of problems. *The Journal of Symbolic Logic*, 52(1):1–43, 1987.
[10] M. Zakharyaschev, F. Wolter, and A. Chagrov. Advanced modal logic. In D. M. Gabbay and F. Guenthner, editors, *Handbook of Philosophical Logic*, volume 3, pages 83–266. Kluwer Academic Publishers, 2nd edition, 2001.

A. V. Chagrov
Department of Mathematics
Tver State University
Zhelyabova Street, 33
Tver, 170000, Russia
E-mail: chagrov@mail.ru

M. N. Rybakov
Department of Mathematics
Tver State University
Zhelyabova Street, 33
Tver, 170000, Russia
E-mail: m_rybakov@mail.ru, s000125@tversu.ru

5
Non-Normal Modalisation
ROGERIO A. S. FAJARDO AND MARCELO FINGER

ABSTRACT. We study the external application of a non-normal modal logic M to a generic logic L, thus generating a modalised logic M(L). We prove the transference of completeness and decidability from M and L to M(L). Previous existing techniques that show the transference of properties in combinations of modal logics all relied on some form of normality of the modal system. Our proof is based on a technique that provides a consistency preserving mapping of L-formulas into propositional classical formulas; this technique allows us to do without normality assumptions in the modal system.

1 Introduction

In this paper we study a method of combining non-normal modal logics called *modalisation* or the external application of a modal logic M to a generic logic system L. This combination generates the modalised logic system M(L), and we analyse the transference of basic logical properties from the component logics to the combined system.

The novelty here is the technique for proving the transference of completeness and decidability, which uses a *consistency preserving mapping* into *propositional classical logic*. This mapping takes formulas of a generic logic L and generates a set of formulas in propositional classical logic, with the following property: any finite set of L-formulas is mapped into a consistent set of propositional classical formulas iff the set of L-formulas is L-consistent. This mapping allows us to show the transference of properties for non-normal modalisation.

Several combinations of modal and temporal logics have been studied in the literature, as discussed below. However, almost all of them relied on the existence of some form of normality, which may be defined either in algebraic terms or in proof-theoretical terms. In a mono-modal logic with modality $\Box$, normality implies that the following formulas are valid:

$$\Box(p \to q) \to (\Box p \to \Box q) \qquad \Box(p \land q) \leftrightarrow (\Box p \land \Box q)$$

as well as the admission of the inference rule: From $\vdash A$ infer $\vdash \Box A$.

None of these is assumed to hold in a non-normal modal logic. It turns out that the existing analysis of transference of logical properties from the component logic systems to combined ones in the literature relied fundamentally on the assumption of normality.

Non-normal modal systems are starting to attract the attention of research community. For example, in deontic logic with an obligation modality O [1], it has been noted that the axiom $OA \wedge OB \to O(A \wedge B)$ needs not always be respected. Furthermore, recent works in the domain of belief revision have extended the basic ontology to deal not only with belief modalities, but also with intension and feasibility modalities [9]; the latter modalities are intrinsically non-normal, and can naturally interact with other modalities. So it is highly desirable to have a theory of transference of logical properties for the combination of non-normal modal logics.

This paper starts to investigate how modal logics can be combined without assuming normality. We concentrate on a not so strong way of combining logics, known as *modalisation*, in which an *external modal logic* M is applied to a generic logic L, where the external logic is assumed to be a non-normal mono-modal logic with a 1-place connective $\Box$. We believe this is a promising step in the study of stronger combinations of non-normal modalities, as has been the case in the study of combination of normal temporal logics [5, 8].

Modalisation is a direct generalisation of the *temporalisation* process, which was previously developed for temporal logics with binary connectives U ("until") and S ("since"), initially restricted over linear time only [4], then extended to any class of flows of time [7] and then to any number of normal modal/temporal operators with any arity [8]. Surprisingly, however, the strategy for proving the transference of completeness and decidability remained the same throughout all these generalisations, as well as in the present work. The proof details, however, differed significantly and have become increasingly more complex.

Apart from the temporalisation/modalisation method, several other combinations of logics have been previously analysed in the literature. The conservativity of independently combined modal logics was presented by Thomason in [11]. Fine and Schurz [6] and Kracht and Wolter [10] have studied the transfer properties of systematically combining independently axiomatisable mono-modal systems, also called *fusion* of modal logics. The work of Fine and Schurz [6] deals with more than two independent normal modalities. A generalisation of such results for many-place multi-modal systems is presented by Wolter in [12]. The independent combination (fusion) of temporal logics was studied in [5, 7] initially for linear time and then for any class of flows of time. Wolter's approach in [10, 12] is algebraic, while all others are based on Kripke frames; even with an algebraic approach, some notion of normality was used; such a notion forbids connectives such as U ('until') and S ('since') to be used in combinations. This restriction was later weakened in [2], so as to allow for connectives temporal U and S as well as other modalities found is description logics, which are not strictly normal; however, some notion of normality, even if presented in a weaker format, still had to be imposed to obtain the transference of logical properties.

The organisation of this paper. The rest of this paper is organised as follows. We first present the modalised system M(L) in Section 2; its language, semantics and inference system will be derived from those of M and L. We then prove the transference of completeness in Section 3; the crucial step is the definition of the consistency preserving mapping in Section 3.1, which leads directly to the transference of completeness in Section 3.2. The transference of decidability is also a consequence of the consistency preserving mapping, as shown in Section 4. We conclude with some remarks on possible application of the results here for stronger combination of logics.

2 The non-normal modalisation M(L)

In this section we describe the system M(L), which is based on the temporalisation process introduced in [4]. By a *logic system* we mean a *tuple* $S = (\mathcal{L}_S, \vdash_S, \mathcal{K}_S, \models_S)$, where $\mathcal{L}_S$ is the system's language, $\vdash_S$ is an inference system, $\mathcal{K}_S$ is the system's associated class of models and $\models_S$ is the system's semantical relation between models and formulas.

The language of M(L)

The language $\mathcal{L}_\mathsf{M}$ of the mono-modal system M is built from a denumerable set of atoms $\mathcal{P} = \{p_0, p_1, \ldots\}$, applying the one-place modality $\Box$ and the Boolean connectives $\neg$ (*negation*) and $\wedge$ (*conjunction*). We use A, B, C for formulas of M, φ for formulas of L and lower Greek letters for formulas of M(L), possibly with subscripts.

Very little is required of the internal logic L, except that its language is a denumerable set of finite formulas and that it has the classical Boolean connectives $\neg$ and $\wedge$ with its usual semantics. Apart from that, any other type of construct is acceptable in the language; for example, it may contain other (normal or non-normal) modalities, or it may possess quantifiers and predicates.

To avoid double parsing of modalised formulas, we partition the language of L into the sets:

- Bool$_\mathsf{L}$, the set of *Boolean combinations* consists of the formulas built up from *any* other formulas with the use of the Boolean connectives $\neg$ or $\wedge$;

- Mono$_L$, the set of *monolithic formulas* is the complementary set of Bool$_L$ in the language of L.

If the internal logic L does not contain the classical connectives $\neg$ and $\wedge$, we assume that Mono$_L = \mathcal{L}_L$ and Bool$_L = \varnothing$, so every formula in L is considered monolithic. As an example, consider the atoms $p, q \in$ L and ■ a modal symbol in L, then ■p and ■$(p \wedge q)$ are monolithic formulas whereas $\neg$■p and ■$p \wedge$ ■q are Boolean combinations.

The set of modalised formulas, $\mathcal{L}_{\mathsf{M(L)}}$, is defined as the smallest set closed under the rules:

1. If $\varphi \in$ Mono$_L$, then $\varphi \in \mathcal{L}_{\mathsf{M(L)}}$;
2. If $\psi_1, \psi_2 \in \mathcal{L}_{\mathsf{M(L)}}$, then $\neg\psi_1 \in \mathcal{L}_{\mathsf{M(L)}}$ and $\psi_1 \wedge \psi_2 \in \mathcal{L}_{\mathsf{M(L)}}$;
3. If $\psi \in \mathcal{L}_L$, then $\square\psi \in \mathcal{L}_{\mathsf{M(L)}}$.

Note that the atoms of M are not elements of $\mathcal{L}_{\mathsf{M(L)}}$. We will use the connectives $\vee$, $\rightarrow$ and $\leftrightarrow$, and the constants $\top$ and $\bot$, in their usual meaning. Also, the formula $\Diamond A$ abbreviate $\neg\square(\neg A)$. The *size* of a formula A is number of symbols it contains.

The semantics of M(L)

The semantics of non-normal modal logics is here based on *minimal models* of possible worlds [3]. A minimal model for modal logic is a structure $\mathcal{M} = (W, N, V)$ such that W is a set of worlds; N is a mapping $N : W \to 2^{2^W}$, that is N associate a set of sets of worlds to each world; and $V : \mathcal{L} \to 2^W$ is a valuation that associates a set of worlds to each formula according to the following restrictions:

- $V(\neg A) = W \setminus V(A)$.
- $V(A \wedge B) = V(A) \cap V(B)$.
- $V(\square A) = \{w \in W | V(A) \in N(w)\}$.

We write $\mathcal{M}, w \models A$ iff $w \in V(A)$. Under this view, a formula is V-associated with the set of worlds in which it holds, and the function N associates a world $w \in W$ with a set of propositions that are *necessary* at w.

Let $\mathcal{K}_{\mathsf{M}}$ be a class of models of logic M, usually defined by placing some restriction on the mapping N. Let $\mathcal{K}_L$ be the class of models for valid formulas of L; we specify some restrictions on the semantic relation $\models_L$ for the logic L, whose class of models will be called $\mathcal{K}_L$. The basic restriction imposes that, for each $\mathcal{M} \in \mathcal{K}_L$ and $\varphi \in \mathcal{L}_L$ we have

$$\text{either } \mathcal{M} \models \varphi \text{ or } \mathcal{M} \models \neg\varphi.$$

To satisfy this condition in some logics, it may be necessary to adapt the notion of a class of models. For instance, if L is some modal logic with a modality ■, an element of $\mathcal{K}_L$ is a *pair* $\langle \mathcal{M}_L, x \rangle$, where $\mathcal{M}_L = (W', N', V')$ and $x \in W'$, such that either $\mathcal{M}_L, x \models \varphi$ or $\mathcal{M}_L, x \models \neg\varphi$; if the class of models were simply defined in terms of $\mathcal{M}_L$, we could have a formula, say, a propositional symbol p, such that neither $\mathcal{M}_L \models p$ nor $\mathcal{M}_L \models \neg p$.

Finally, we can define a minimal model for the modalised logic M(L) as a structure $\mathcal{M}_{\mathsf{M(L)}} = (W, N, g)$, where W and N are as above, and $g : W \to \mathcal{K}_L$ associates to each $w \in W$ a model of L. The satisfaction relation $\models$ is then defined recursively over the structure of modalised formulas:

(i) $\mathcal{M}_{\mathsf{M(L)}}, w \models \alpha$, $\alpha \in \mathrm{Mono}_L$ iff $g(w) = \mathcal{M}_L$ and $\mathcal{M}_L \models \alpha$ (denoted $g(w) \models \alpha$).

(ii) $\mathcal{M}_{\mathsf{M(L)}}, w \models \neg\alpha$ iff $\mathcal{M}_{\mathsf{M(L)}}, w \not\models \alpha$.

(iii) $\mathcal{M}_{\mathsf{M(L)}}, w \models (\alpha \wedge \beta)$ iff $\mathcal{M}_{\mathsf{M(L)}}, w \models \alpha$ and $\mathcal{M}_{\mathsf{M(L)}}, w \models \beta$.

(iv) $\mathcal{M}_{\mathsf{M(L)}}, w \models \Box\alpha$ iff $\{w' \in W \mid \mathcal{M}_{\mathsf{M(L)}}, w' \models \alpha\} \in N(w)$.

A class of modalised models $\mathcal{K}_{\mathsf{M(L)}}$ is obtained from $\mathcal{K}_L$ and $\mathcal{K}_M$ by placing over modalised models $\mathcal{M}_{\mathsf{M(L)}} = (W, N, g)$ the same restrictions over N that are placed on the class $\mathcal{K}_M$.

A formula is *valid* in a class $\mathcal{K}$ if it is verified at all worlds at all models over that class.

The inference system of M(L)

We assume that an *inference system*, $\vdash$, for a generic logic system is a mechanism capable of recursively enumerating the set of all provable formulas of the system, here called *theorems* of the logic system.

An inference system is *sound* with respect to a class of models $\mathcal{K}$ if all its theorems are valid over $\mathcal{K}$. Conversely, it is *complete* if all valid formulas are theorems. We assume that logic L's inference system, $\vdash_L$, is sound and complete with respect to a class $\mathcal{K}_L$.

We will assume that the modal logic M inference system, $\vdash_M$, is given in an axiomatic form, consisting of a set of *axioms* and a set of inference rules. In fact, all we have to assume of $\vdash_M$ is the validity of propositional classical tautologies, the admissibility of Modus Ponens and the following inference rule: if $\vdash_M A \leftrightarrow B$ then $\vdash_M \Box A \leftrightarrow \Box B$.

We include those rules for they are valid in any minimal model, and they define a system that is sound and complete with respect to the class of all minimal models. On the other hand, by forcing other inference rule and axioms on logic M, some restriction is imposed on the structure of the class of models; see [3].

The combined inference system of M(L) is denoted by $\vdash_{M(L)}$ and consists of the following elements:

- The axioms and inference rules of $\vdash_M$;

- The inference rule *Preserve*: For every formula φ in L, if $\vdash_L \varphi$ then $\vdash_{M(L)} \varphi$.

As usual, a formula φ is *consistent* if $\not\vdash \neg\varphi$.

In [4] it was shown that if L is sound over a class $\mathcal{K}_L$ and M is sound over a class $\mathcal{K}_M$, then the inference system of $\vdash_{M(L)}$ is sound over the combined class $\mathcal{K}_{M(L)}$; no extra restrictions were made on the nature of M. That is, soundness transfers over modalisation. In the following, we investigate the transference of completeness and decidability.

3 Completeness of M(L)

In this section we show that the non-normal modalisation of sound and complete logics preserves completeness. The proof strategy is the same that has been used in our previous works of temporalisation [4, 7, 8] and is illustrated in Figure 1. We start with a consistent M(L)-formula φ, translate it into a consistent modal formula A in M; then completeness of M over $\mathcal{K}_M$ gives us a model for A; after some model manipulation using the completeness of L, we obtain a model for φ in $\mathcal{K}_{M(L)}$, thus deriving the transference of completeness from L and M to M(L).

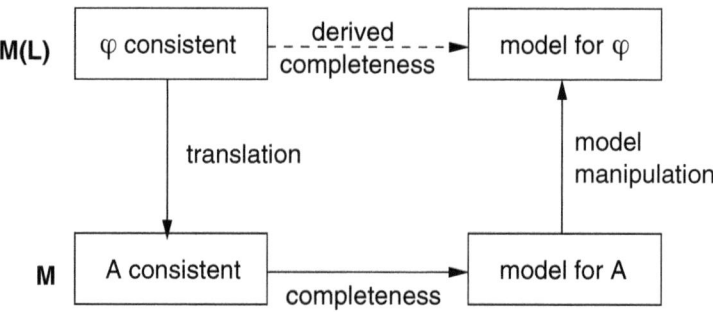

Figure 1. Completeness proof strategy

The difficulty in our proof lies in finding an adequate translation σ such that φ is M(L)-consistent iff $A = \sigma(\varphi)$ is M-consistent and such that any model for $\sigma(\varphi)$ can be easily transformed into a model for φ. The simple idea of mapping each monolithic formula of M(L) to a propositional variable does not work. To see that, suppose the logic L is a modal logic with modality ■ and suppose that the axiom **T** holds for L:

$$\vdash_\mathsf{L} \blacksquare p \to p$$

Now suppose that $\varphi = \blacksquare p \to p$ is a M(L) formula; by the inference rule Preserve, φ is a theorem. If we simply map all monolithic subformulas to atoms, we end up with a formula in M of the form $\sigma(\varphi) = q_1 \to q_2$, where q_1, q_2 are "new" atoms, and M-models of $\sigma(\varphi)$ may make the atom q_1 true but q_2 false at some world, so not all M-models of $\sigma(\varphi)$ can be directly transformed into a M(L)-model of φ. A translation mapping has to guarantee that the mapped formulas behaves exactly the same way as the original formulas in terms of satisfiability and validity.

3.1 The consistency preserving map

Let $\Phi = \{\varphi_1, \ldots, \varphi_n\}$ be a set of monolithic L-formulas. We will map them into propositional classical formulas $A_1, \ldots, A_n$ as defined below built from propositions $q_1, \ldots, q_n$; since M extends classical logic, this will be the basic step for mapping a M(L) formula ψ built from the elements of Φ into a M formula $\sigma(\psi)$. The translation σ from M(L)-formulas to M-formulas is then recursively defined as:

- $\sigma(\varphi_i) = A_i$;
- $\sigma(\neg \psi) = \neg \sigma(\psi)$;
- $\sigma(\psi_1 \wedge \psi_2) = \sigma(\psi_1) \wedge \sigma(\psi_2)$;
- $\sigma(\Box \psi) = \Box \sigma(\psi)$.

It remains to define the propositional classical formulas $A_1, \ldots, A_n$. This mapping of L formulas into classical ones is in fact independent of the modalisation process; we will show it guarantees that consistency is preserved through this mapping. Before we do that, we need some definitions.

If $\Phi = \{\varphi_1, \ldots, \varphi_n\}$, we write $\sigma(\Phi)$ for the set $\{\sigma(\varphi) | \varphi \in \Phi\}$. Let $\mathrm{Lit}(\Phi) = \Phi \cup \{\neg \varphi | \varphi \in \Phi\}$ be the set of *literals* of Φ. Let $\mathrm{Cons}(\Phi) = \{\bigwedge \varphi_i | \varphi_i \in \mathrm{Lit}(\Phi) \text{ and } \not\vdash_\mathsf{L} \neg \bigwedge \varphi_i\}$.

Fix an enumeration of Φ, $\epsilon = \varphi_1, \ldots, \varphi_n$, and suppose we have variables $q_1, \ldots, q_n$ in M. We then recursively define $A_1, \ldots, A_n$ in the following way. Each $A_i = \sigma(\varphi_i)$ will be a boolean combination of $q_1, \ldots, q_i$.

For the base case, we define A_1 as follows:

- $A_1 = q_1 \wedge \neg q_1$, if $\varphi_1 \notin \mathrm{Cons}(\Phi_1)$, i.e. $\vdash_\mathsf{L} \neg \varphi_1$;
- $A_1 = q_1 \vee \neg q_1$, if $\neg \varphi_1 \notin \mathrm{Cons}(\Phi_1)$, i.e. $\vdash_\mathsf{L} \varphi_1$;

- $A_1 = q_1$, otherwise.

Suppose $A_1, \ldots, A_m$ is defined for $m < n$ and define $\Phi_m = \{\varphi_1, \ldots, \varphi_m\}$. As we did for the base case, the definition of A_{m+1} will have to analyse three cases of combinations based on Φ_m, namely those that imply φ_{m+1}, those that imply $\neg\varphi_{m+1}$, and those of which φ_{m+1} is independent.

We start by defining the (transformation of) combinations that imply φ_{m+1}:

$$B^+(\Phi_m) = \bigvee \{\sigma(\psi) | \psi \in \text{Cons}(\Phi_m) \text{ and } \vdash_L \psi \to \varphi_{m+1}\}.$$

$B^+(\Phi_m)$ is a disjunction of the σ-mapped elements of Φ_m that imply φ_{m+1}. $B^+(\Phi_m)$ may have several disjuncts, a single one or even none (in which case $B^+(\Phi_m) = \bot$). Similarly, define:

$$B^-(\Phi_m) = \bigvee \left\{ \begin{array}{c} \sigma(\psi) | \psi \in \text{Cons}(\Phi_m) \text{ and} \\ \vdash_L \psi \to \neg\varphi_{m+1} \end{array} \right\}.$$

$B^-(\Phi_m)$ is a disjunction of the σ-mapped elements of Φ_m that imply $\neg\varphi_{m+1}$. Finally, for the elements of $\text{Cons}(\Phi_m)$ that imply neither φ_{m+1} nor $\neg\varphi_{m+1}$, define:

$$B^0(\Phi_m) = \bigvee \{\sigma(\psi) \wedge q_{m+1} | \psi \in \text{Cons}(\Phi_m)$$
$$\text{and}$$
$$\psi \wedge \varphi_{m+1}, \psi \wedge \neg\varphi_{m+1} \in \text{Cons}(\Phi_{m+1})\}.$$

The presence of q_{m+1} in $B^0(\Phi_m)$ provides the freedom to choose the truth value of A_{m+1}. So, finally, we define A_{m+1} as:

$$A_{m+1} = \neg B^-(\Phi_m) \wedge (B^+(\Phi_m) \vee B^0(\Phi_m)).$$

Note that some of the B-elements may be absent from the formula in case it is an empty disjunction. In particular, if $B^0(\Phi_m)$ is empty we will not have q_{m+1}, meaning that the value of φ_{m+1} is totally determined by $\varphi_1, \ldots, \varphi_m$.

With this construction, if a non-normal modal model satisfies a disjunct of $B^+(\Phi_m)$, it will not satisfy $B^-(\Phi_m)$, and hence it satisfies A_{m+1}. Conversely, by satisfying $B^-(\Phi_m)$, A_{m+1} is falsified. If neither $B^+(\Phi_m)$ nor $B^-(\Phi_m)$ is satisfied, the new atom q_{m+1} determines the value of A_{m+1}.

We now revisit the previous example of mapping the axiom T, $\blacksquare p \to p$, from L to classical propositional logic. We have $\Phi = \{\blacksquare p, p\}$, and $\text{Cons}(\Phi) = \{\blacksquare p \wedge p, \neg\blacksquare p \wedge p, \neg\blacksquare p \wedge \neg p\}$. Fix an enumeration $\epsilon = \blacksquare p, p$; in this case $\text{Cons}(\Phi_1) = \{\blacksquare p, \neg\blacksquare p\}$ and by the definition above $A_1 = \sigma(\blacksquare p) = q_1$. To find A_2, note that $B^+(\Phi_1) = q_1$ (for $\vdash_L \blacksquare p \to p$), $B^-(\Phi_1) = \bot$ and

$B^0(\Phi_1) = \neg q_1 \wedge q_2$. Hence $A_2 = q_1 \vee (\neg q_1 \wedge q_2)$, which is classically equivalent to $\neg q_1 \to q_2$; note that the consistent combinations of A_1 and A_2 are exactly the consistent combinations of $\blacksquare p$ and p. In this case, $\sigma(\blacksquare p \to p) = q_1 \to (\neg q_1 \to q_2)$, which is a classical tautology, as desired for the mapping of an L-theorem.

Had we chosen the other enumeration of monolithic formulas ($\epsilon' = p, \blacksquare p$) we would have obtained different values: $A_1 = \sigma(p) = q_1$, $A_2 = \sigma(\blacksquare p) = q_1 \wedge q_2$; however, this leads to $\sigma(\blacksquare p \to p) = (q_1 \wedge q_2) \to q_1$, which is also a tautology, as desired.

The following lemma shows that, indeed, this construction always works and will be useful for the completeness proof.

LEMMA 1 *Let ψ be a conjunction of literals of a set of L-formulas $\Phi = \{\varphi_1, \ldots, \varphi_n\}$. Then ψ is L-consistent iff $\sigma(\psi)$ is classically consistent.*

Proof. By induction on $m \leq n$. For $m = 1$, the result follows immediately from the definition of A_1. Now suppose the result is valid for Φ_m; we will prove it for Φ_{m+1}. All we have to care for are formulas ψ in which φ_{m+1} occurs, that is $\psi = \mu \wedge \varphi_{m+1}$ or $\psi = \mu \wedge \neg \varphi_{m+1}$.

Suppose $\sigma(\psi)$ is classically consistent. By the definition of σ, $\sigma(\mu)$ is classically consistent so the induction hypothesis gives us that μ is L-consistent. Hence $\sigma(\mu)$ is an element of one of the disjunctions $B^+(\Phi_m)$, $B^-(\Phi_m)$ or $B^0(\Phi_m)$.

If $\sigma(\mu)$ is in $B^+(\Phi_m)$, then $\vdash_\mathsf{L} \mu \to \varphi_{m+1}$. We want to show that $\psi = \mu \wedge \varphi_{m+1}$, in which case ψ is clearly consistent. For contradiction, suppose that $\psi = \mu \wedge \neg \varphi_{m+1}$. Suppose v is a classical valuation that satisfies $\sigma(\mu)$. By the definition of $A_{m+1} = \sigma(\varphi_{m+1})$, $\sigma(\mu)$ is a disjunct in $B^+(\Phi_m)$, thus satisfied by v. We show that v does not satisfy $B^-(\Phi_m)$. Indeed, if there is a conjunct in $B^-(\Phi_m)$ satisfied by v, then there is a conjunction of literals μ' in Φ_m such that $\vdash_\mathsf{L} \mu' \to \neg \varphi_{m+1}$. Since μ is consistent, we also have that $\vdash_\mathsf{L} \mu \to \neg \mu'$, so the induction hypothesis gives us that v cannot satisfy μ', and hence v cannot satisfy $B^-(\Phi_m)$. It follows that v satisfies $\sigma(\varphi_{m+1})$ and thus falsifies $\sigma(\psi)$. On the other hand, any valuation that falsifies $\sigma(\mu)$ also falsifies $\sigma(\psi)$. So any valuation falsifies $\sigma(\psi)$, contradicting its consistency. We have thus proved that $\psi = \mu \wedge \varphi_{m+1}$, and ψ is consistent.

If $\sigma(\mu)$ is in $B^-(\Phi_m)$, by an analogous argument, if $\sigma(\mu)$ is satisfied, $\sigma(\psi_{m+1})$ is falsified, so the only possibility is that $\psi = \mu \wedge \neg \varphi_{m+1}$. But then $\vdash_\mathsf{L} \mu \to \neg \varphi_{m+1}$ and ψ is consistent.

Finally, if $\sigma(\mu)$ is in $B^0(\Phi_m)$, then by definition both $\mu \wedge \varphi_{m+1}$ and $\mu \wedge \neg \varphi_{m+1}$ are consistent.

Suppose now that ψ is L-consistent. Then μ is also L-consistent and by induction hypothesis, $\sigma(\mu)$ is classically consistent. Also, $\sigma(\mu)$ is an element of one of the disjunctions $B^+(\Phi_m)$, $B^-(\Phi_m)$ or $B^0(\Phi_m)$ and we have three possibilities to analyse in a manner totally analogous as in the previous direction. Details omitted. ∎

We now generalise Lemma 1 to full M(L).

LEMMA 2 *Let ψ be a M(L)-formula and let Φ be the set of its monolithic subformulas, over which the mapping σ is defined. Then ψ is M(L)-consistent iff $\sigma(\psi)$ is M-consistent.*

Proof. ($\Rightarrow$) Suppose, for contradiction, that $\sigma(\psi)$ is inconsistent, that is, $\vdash_M \neg\sigma(\psi)$. If φ is a subformula of ψ and a formula of L, then by Lemma 1 $\sigma(\psi)$ is a classical tautology iff φ is a theorem of L and of M(L) by the inference rule *Preserve*. Furthermore, all other M inference step can be copied in M(L). Therefore, the deduction of $\vdash_M \neg\sigma(\psi)$ can be simulated in M(L), so as to prove $\vdash_{M(L)} \neg\psi$, contradicting the consistency of ψ.

($\Leftarrow$) Suppose now that ψ is M(L)-inconsistent. Consider a deduction of $\vdash_{M(L)} \neg\psi$; suppose it contains an instance of the rule preserve, in which for a formula φ built from the elements of Φ, $\vdash_L \varphi$ is inferred. Then, by Lemma 1, $\sigma(\varphi)$ is a classical tautology, that is also a theorem of M. All other inference steps of M(L) can be copied as inference steps in M, so that the deduction of $\vdash_{M(L)} \neg\psi$ can be transformed into a deduction of $\vdash_M \neg\sigma(\psi)$. ∎

This result leads us to a proof of transference of completeness.

3.2 Transference of completeness

THEOREM 3 *If L and M are complete logics, so is M(L).*

Proof. Let ψ be a consistent M(L)-formula and let Φ be the set of all monolithic subformulas of ψ. By Lemma 2, $\sigma(\psi)$ is M-consistent, so by the completeness of M there is a model $\mathcal{M}_M = (W, N, V)$ with $\mathcal{M}_M \models \sigma(\psi)$. We build a M(L) model $\mathcal{M}_{M(L)} = (W, N, g)$ in the following way: for all $w \in W$ and $\varphi \in \Phi$, $g(w) \models_L \varphi$ iff $w \in V(\sigma(\varphi))$. To see that this is indeed a model, consider for each $w \in W$ the elements $\varphi_i \in \Phi$ such that $\sigma(\varphi)$ are satisfied by V in w; by Lemma 1 all such φ_i are simultaneously L-consistent and by completeness of L there is a model $\mathcal{M}_L = g(w)$ that simultaneously satisfies all of them.

A simple structural induction now shows that $\mathcal{M}_{M(L)} \models \psi$ iff $\mathcal{M}_M \models \sigma(\psi)$. Therefore $\mathcal{M}_{M(L)} = (W, N, g)$ is a model for ψ and M(L) is complete over the same class of models that M is. ∎

4 Decidability

We now show the transference of decidability. A logic is *decidable* if there is an algorithm that, given any formula in that logic, decides in a finite number of steps if that formula is a theorem of the logic.

We first note that Lemma 2 gives us immediately that a formula ψ is a M(L)-theorem iff $\sigma(\psi)$ is a M-theorem.

THEOREM 4 *If the logics* L *and* M *are complete and decidable, so is* M(L).

Proof. Give a M(L)-formula ψ, we can algorithmically construct the set Φ of all its monolithic formulas. Since Φ is finite and L is decidable, we can construct the set Cons(Φ) and the M-formula $\sigma(\psi)$. We then apply the decision procedure for logic M over $\sigma(\psi)$ and by Lemma 2 we know that ψ is a M(L)-theorem iff $\sigma(\psi)$ is a M-theorem. ∎

4.1 Complexity

As for complexity, the decision procedure above gives us the following. Let $c_L(n)$ and $c_M(n)$ be the complexities of the decision procedure in L and M, respectively.

If n is the size of the M(L)-formula ψ, the size of the set Φ is $O(n)$ and the number of elements potentially needed in the construction of Cons(Φ) is $O(2^n)$. Then the construction of $\sigma(\psi)$ can be done in time $O(2^n \times c_L(n))$. Note, however, that by the construction of Cons(Φ), each element of Cons(Φ) will generate a disjuct in $\sigma(\psi)$, so the size of $\sigma(\psi)$ is $O(2^n)$. As a consequence, the decision procedure of M is applied to $\sigma(\psi)$ in time $O(c_M(2^n))$.

As a result, we have the following.

LEMMA 5 *The time complexity of the decision procedure in Theorem 4 is:*

$$O(2^n \times c_L(n) + c_M(2^n)).$$

With regards to the space complexity, because the size of Cons(Φ) is $O(2^n)$, so even if the complexity of the decision procedure of logics M and L are in PSPACE, the space of the decision procedure for M(L) above will be in EXPSPACE.

LEMMA 6 *If the logics* L *and* M *are in PSPACE, then* M(L) *is in EX-PSPACE.*

Of course, this does not rule out the possibility of existing a different decision procedure for M(L) that places it in PSPACE. In other words, the discussion above does not provide a lower bound for the complexity of M(L).

5 Conclusion

We have presented an original method that maps formulas of a logic L into boolean combinations of propositional classical formulas, so as to preserve validity and consistency. This mapping was then used to prove the transference of completeness and decidability of the modalisation M(L) that applies non-normal modal logic M to a generic logic L.

For the future, we hope to be able to explore such mapping for other logics, and for translating a logic into another. Also, as done previously, we want to explore the use of successive modalisations as a means to study the *independent combination* or *fusion* of two non-normal modal logics. Another possible path of development is the generalisation of the non-normal modalisation process to modal logics with many connectives and with connectives with any finite arity, following the steps of [12, 8].

Acknowledgements

Marcelo Finger was partly supported by the Brazilian Research Council (CNPq), grant PQ 300597/95-5. Rogerio Fajardo is supported by a Brazilian Research Council (CNPq) undergraduate studentship IC 580800/2001-6.

BIBLIOGRAPHY

[1] Lennart Aqvist. Deontic logic. In D. Gabbay and F. Guenthner, editors, *Handbook of Philosophical Logic*, volume Volume II, pages 605–714. Reidel, 1984.

[2] F. Baader, C. Lutz, H. Sturm, and F. Wolter. Fusions of description logics and abstract description systems. *Journal of Artificial Intelligence Research (JAIR)*, 16:1–58, 2002.

[3] B. F. Chellas. *Modal Logic — an Introduction*. Cambridge University Press, 1980.

[4] M. Finger and D. M. Gabbay. Adding a Temporal Dimension to a Logic System. *Journal of Logic Language and Information*, 1:203–233, 1992.

[5] M. Finger and D. Gabbay. Combining Temporal Logic Systems. *Notre Dame Journal of Formal Logic*, 37(2):204–232, Spring 1996.

[6] K. Fine and G. Schurz. Transfer theorems for stratified multimodal logics. To appear in the Proceedings of Arthur Prior Memorial Conference, Christchurch, New Zeland, 1991.

[7] M. Finger and M. Angela Weiss. The unrestricted addition of a temporal dimension to a logic system. In *3rd International Conference on Temporal Logic (ICTL2000)*, Leipzig, Germany, 4–7 October 2000.

[8] Marcelo Finger and M. Angela Weiss. The unrestricted combination of temporal logic systems. *Logic Journal of the IGPL*, 10(2):165–190, March 2002.

[9] Andreas Herzig and Dominique Longin. Belief dynamics in cooperative dialogues. *J. of Semantics*, 17(2), 2000. vol. published in 2001.

[10] M. Kracht and F. Wolter. Properties of independently axiomatizable bimodal logics. *Journal of Symbolic Logic*, 56(4):1469–1485, 1991.

[11] S. K. Thomason. Independent Propositional Modal Logics. *Studia Logica*, 39:143–144, 1980.

[12] Frank Wolter. Fusions of modal logics revisited. In Marcus Kracht, Maarten de Rijke, Heinrich Wansing, and Michael Zakharyaschev, editors, *Advances in Modal Logic*, volume Volume 1 of *Lecture Notes 87*, pages 361–379. CSLI Publications, Stanford, CA, 1996.

Marcelo Finger
Department of Computer Science
IME/USP – Institute of Mathematics and Statistics
University of Sao Paulo
Rua do Matao, 1010
05508-090, Sao Paulo, SP, Brazil
E-mail: mfinger@ime.usp.br

Rogerio A. S. Fajardo
Department of Mathematics
IME/USP – Institute of Mathematics and Statistics
University of Sao Paulo
Rua do Matao, 1010
05508-090, Sao Paulo, SP, Brazil
E-mail: fajardo@ime.usp.br

6
Bisimulations and Boolean Vectors
MELVIN FITTING

ABSTRACT. A modal accessibility relation is just a transition relation, and so can be represented by a $\{0, 1\}$ valued transition matrix. Starting from this observation, I first show that the machinery of matrices, over Boolean algebras more general than the two-valued one, is appropriate for investigating multi-modal semantics. Then I show that bisimulations have a rather elegant theory, when expressed in terms of transformations on Boolean vector spaces. The resulting theory is a curious hybrid, fitting between conventional modal semantics and conventional linear algebra. I don't know where the investigations begun here will ultimately wind up, but in the meantime the approach has a kind of curious charm that others may find appealing.

1 Introduction

Bisimulations are to Kripke, or transition, structures as homomorphisms are to groups. Yet while this sounds right, it cannot quite be so, since bisimulations are not maps between frames, but are relations between them. In this paper I show that a shift in the point of view, from Kripke frames to closely related Boolean vector spaces, turns bisimulations from relations to linear mappings having rather nice properties. Mono-modal frames give rise to Boolean vector spaces over the familiar two-valued Boolean algebra, while multi-modal frames bring more complex Boolean algebras into the picture. Still, the basic ideas remain the same for both the mono- and the multi-modal cases. The point of this approach is not to prove new results, but to look at well-known results in a new way, hoping that a fresh perspective will lead to fresh insights.

The paper begins with several sections discussing background. The formal work on Boolean valued vector spaces begins with Section 5.

2 Familiar background

As will be seen shortly, there is little formal difference in the algebraic treatment presented here between mono- and multi-modal logics, so I'll begin with the general case from the start. Let $\mathcal{K}$ be a non-empty set, finite or countable. We might consider the members of $\mathcal{K}$ to be *knowers*, but nothing depends on this. For convenience, if $\mathcal{K}$ is finite, I'll assume it is $\{1, 2, \ldots, n\}$, and if it is infinite, $\{1, 2, \ldots\}$.

A $\mathcal{K}$-*language* is a propositional modal language, built up from propositional letters (typically P, Q, ...) using the propositional connectives $\wedge$, $\vee$, $\neg$ (with $\supset$ taken as a defined connective) and the modal operators $\Diamond_k$ for each $k \in \mathcal{K}$ (with $\Box_k$ taken as a defined connective). I'll skip the obvious language details. If $\mathcal{K} = \{1\}$ I'll say the language is *mono-modal*, and otherwise, *multi-modal*.

A $\mathcal{K}$-*frame* is a tuple, $\langle \mathcal{G}, \mathcal{R}_k : k \in \mathcal{K} \rangle$, where $\mathcal{G}$ is a non-empty set and each $\mathcal{R}_k$ is a binary relation on $\mathcal{G}$. As usual, members of $\mathcal{G}$ will be referred to as possible worlds, and each $\mathcal{R}_k$ as an accessibility relation, or a transition.

If $\mathcal{F} = \langle \mathcal{G}, \mathcal{R}_k : k \in \mathcal{K} \rangle$ is a $\mathcal{K}$-frame then the structure $\langle \mathcal{F}, v \rangle$ is a $\mathcal{K}$-*model* based on this frame provided v is a mapping from members of $\mathcal{G}$ and formulas to truth values $\{\mathbf{0}, \mathbf{1}\}$ (with the usual Boolean structure) such that, for each $\Gamma \in \mathcal{G}$:

1. $v(\Gamma, \neg X) = \neg v(\Gamma, X)$

2. $v(\Gamma, X \wedge Y) = v(\Gamma, X) \wedge v(\Gamma, Y)$

3. $v(\Gamma, X \vee Y) = v(\Gamma, X) \vee v(\Gamma, Y)$

4. $v(\Gamma, \Diamond_k X) = \mathbf{1}$ if and only if $v(\Delta, X) = \mathbf{1}$ for some $\Delta \in \mathcal{G}$ such that $\Gamma \mathcal{R}_k \Delta$

Of course the behavior of v is completely determined by its behavior on propositional letters.

Next is the notion of bisimulation. I divide this into two parts, one concerning frames, the other concerning models. Customarily these are combined, but it is more convenient in the present treatment to separate the notions. In addition, I give a version that is more general than usual, allowing the bisimulation relation to be parametrized by modal operator. As it happens, it is no more work to treat this version than the usual one in the context of the paper. The usual version becomes a special case, which I designate with the terminology *standard*.

DEFINITION 1 Let $\mathcal{F}_{\mathcal{G}} = \langle \mathcal{G}, \mathcal{R}_k : k \in \mathcal{K} \rangle$ and $\mathcal{F}_{\mathcal{H}} = \langle \mathcal{H}, \mathcal{S}_k : k \in \mathcal{K} \rangle$ be two $\mathcal{K}$-frames. Also let $\mathcal{A} = \langle \mathcal{A}_k : k \in \mathcal{K} \rangle$ be a family of relations between $\mathcal{G}$ and $\mathcal{H}$. $\mathcal{A}$ is a *frame bisimulation* between $\mathcal{F}_{\mathcal{G}}$ and $\mathcal{F}_{\mathcal{H}}$ provided:

1. For all $\Gamma_1, \Gamma_2 \in \mathcal{G}$ and $\Delta_1 \in \mathcal{H}$, if $\Gamma_1 \mathcal{A}_k \Delta_1$, and $\Gamma_1 \mathcal{R}_k \Gamma_2$, then there is some $\Delta_2 \in \mathcal{H}$ such that $\Gamma_2 \mathcal{A}_k \Delta_2$ and $\Delta_1 \mathcal{S}_k \Delta_2$,

2. For all $\Delta_1, \Delta_2 \in \mathcal{H}$ and $\Gamma_1 \in \mathcal{G}$, if $\Gamma_1 \mathcal{A}_k \Delta_1$, and $\Delta_1 \mathcal{S}_k \Delta_2$, then there is some $\Gamma_2 \in \mathcal{G}$ such that $\Gamma_2 \mathcal{A}_k \Delta_2$ and $\Gamma_1 \mathcal{R}_k \Gamma_2$.

If $\mathcal{A}_j = \mathcal{A}_k$ for all $j, k \in \mathcal{K}$, I'll say $\mathcal{A}$ is a *standard* frame bisimulation. For standard frame bisimulations, I'll identify $\mathcal{A}$ with any relation in its family (all of which are the same).

If $\mathcal{K}$ consists of one element, there is no distinction between what we are calling a frame bisimulation and a standard frame bisimulation. I'll call such a case a *mono-modal* frame bisimulation. Historically it is the earliest notion of bisimulation to appear in modal logic. In the more general setting there is actually no interaction between modalities, so we have the following principle.

THEOREM 2 *Let $\mathcal{F}_\mathcal{G} = \langle \mathcal{G}, \mathcal{R}_k : k \in \mathcal{K} \rangle$ and $\mathcal{F}_\mathcal{H} = \langle \mathcal{H}, \mathcal{S}_k : k \in \mathcal{K} \rangle$ be two $\mathcal{K}$-frames, and let $\mathcal{A} = \langle \mathcal{A}_k : k \in \mathcal{K} \rangle$ be a family of relations between $\mathcal{G}$ and $\mathcal{H}$. $\mathcal{A}$ is a frame bisimulation between $\mathcal{F}_\mathcal{G}$ and $\mathcal{F}_\mathcal{H}$ if and only if, for each $k \in \mathcal{K}$, $\mathcal{A}_k$ is a mono-modal frame bisimulation between $\langle \mathcal{G}, \mathcal{R}_k \rangle$ and $\langle \mathcal{H}, \mathcal{S}_k \rangle$.*

DEFINITION 3 Let $\langle \mathcal{F}_\mathcal{G}, v_\mathcal{G} \rangle$ and $\langle \mathcal{F}_\mathcal{H}, v_\mathcal{H} \rangle$ be two $\mathcal{K}$-models, where $\mathcal{F}_\mathcal{G} = \langle \mathcal{G}, \mathcal{R}_k : k \in \mathcal{K} \rangle$ and $\mathcal{F}_\mathcal{H} = \langle \mathcal{H}, \mathcal{S}_k : k \in \mathcal{K} \rangle$. Also let $\mathcal{A}$ be a standard frame bisimulation between $\mathcal{F}_\mathcal{G}$ and $\mathcal{F}_\mathcal{H}$. $\mathcal{A}$ is a *bisimulation* if, in addition, for all $\Gamma \in \mathcal{G}$ and $\Delta \in \mathcal{H}$ with $\Gamma \mathcal{A} \Delta$ we have $v_\mathcal{G}(\Gamma, P) = v_\mathcal{H}(\Delta, P)$ for all propositional letters P.

The well-known key fact about bisimulations is the following, proved by an easy induction on formula complexity. If $\mathcal{A}$ is a bisimulation between the $\mathcal{K}$-models $\langle \mathcal{F}_\mathcal{G}, v_\mathcal{G} \rangle$ and $\langle \mathcal{F}_\mathcal{H}, v_\mathcal{H} \rangle$, and $\Gamma_1 \in \mathcal{G}$ and $\Gamma_2 \in \mathcal{H}$, then if $\Gamma_1 \mathcal{A} \Gamma_2$, $v_\mathcal{G}(\Gamma_1, X) = v_\mathcal{H}(\Gamma_2, X)$ for every $\mathcal{K}$-formula X. Much is known about bisimulations—I refer you to the standard literature for details, [1] among others.

3 Introducing Boolean algebra use

If $\mathcal{K}$ has cardinality greater than 1, we are dealing with multi-modal frames and models. These can be collapsed to mono-modal versions, provided we are willing to complicate the underlying truth-value space. I'll first introduce the notion of a Boolean valued modal model, then discuss connections with multi-modal frames.

3.1 Boolean valued models

I assume the definition and basic properties of Boolean algebras are known—[7] is a very thorough reference. When working in a Boolean algebra I will use $\wedge$, $\vee$, and $\neg$ to denote the operations of meet, join, and complement. I will write $a \Rightarrow b$ for $\neg a \vee b$. I will also use $\leq$ for the standard ordering relation, $a \leq b$ iff $a \wedge b = a$ iff $a \vee b = b$, and I will use $<$ for strict ordering; $a < b$ if $a \leq b$ and $a \neq b$. The bottom element of the algebra will be denoted by $\mathbf{0}$, and the top by $\mathbf{1}$. Finally, I will use $\bigwedge$ and $\bigvee$ for the infinitary meet and join operations, when they exist.

General Assumption 1 For the rest of this paper, $\mathcal{B}$ is a complete Boolean algebra, where *complete* means that infinite, as well as finite, meets and joins exist.

In the definition below, assume we are using a mono-modal language— there is a single possibility operator.

DEFINITION 4 ($\mathcal{B}$-*frames and models*) A $\mathcal{B}$-frame is a pair $\mathcal{F} = \langle \mathcal{G}, \mathcal{R} \rangle$, where $\mathcal{G}$ is a non-empty set of possible worlds, as usual, and $\mathcal{R}$ is a $\mathcal{B}$-valued accessibility relation: a mapping from pairs of worlds to $\mathcal{B}$. That is, $\mathcal{R} : \mathcal{G} \times \mathcal{G} \to \mathcal{B}$.

If $\mathcal{F} = \langle \mathcal{G}, \mathcal{R} \rangle$ is a $\mathcal{B}$-frame, the structure $\mathcal{M} = \langle \mathcal{F}, v \rangle$ is a $\mathcal{B}$-model based on this frame, provided v maps members of $\mathcal{G}$ and formulas to $\mathcal{B}$ such that, for each $\Gamma \in \mathcal{G}$:

1. $v(\Gamma, \neg X) = \neg v(\Gamma, X)$
2. $v(\Gamma, X \wedge Y) = v(\Gamma, X) \wedge v(\Gamma, Y)$
3. $v(\Gamma, X \vee Y) = v(\Gamma, X) \vee v(\Gamma, Y)$
4. $v(\Gamma, \Diamond X) = \bigvee \{ \mathcal{R}(\Gamma, \Delta) \wedge v(\Delta, X) \mid \Delta \in \mathcal{G} \}$

As usual, the action of v at the atomic level completely determines it for all formulas. If we assume $\Box X$ is defined to be $\neg \Diamond \neg X$, as usual, then the condition for $\Box$ becomes the following.

$$v(\Gamma, \Box X) = \bigwedge \{ \mathcal{R}(\Gamma, \Delta) \Rightarrow v(\Delta, X) \mid \Delta \in \mathcal{G} \}$$

A modal model in the usual sense is simply a $\mathcal{B}$-model where $\mathcal{B}$ is the usual two-element Boolean algebra.

3.2 Connections

I will establish a connection between multi-modal models and $\mathcal{B}$-models, after which we can confine our discussion to the Boolean valued case. Suppose we have a finite or countable set $\mathcal{K}$, $\mathcal{F}_\mathcal{K} = \langle \mathcal{G}, \mathcal{R}_k : k \in \mathcal{K} \rangle$ is a $\mathcal{K}$-frame, and $\mathcal{M}_\mathcal{K} = \langle \mathcal{F}_\mathcal{K}, v_\mathcal{K} \rangle$ is a $\mathcal{K}$-model. I'll use this to create a related Boolean-valued modal model.

First, let $\mathcal{B}$ be the powerset algebra of $\mathcal{K}$. This is an atomic Boolean algebra, with the atoms being members of the form $\{n\}$, for $n \in \mathcal{K}$.

Next, instead of the $\mathcal{K}$-language appropriate for $\mathcal{M}_\mathcal{K}$ we want a monomodal language. But, I enlarge this language by introducing propositional constants: for each $n \in \mathcal{K}$ let P_n be a distinct propositional constant.

I now create a $\mathcal{B}$-model as follows. Let $\mathcal{G}$ be the same set of possible worlds as in the multi-modal frame $\mathcal{F}_\mathcal{K}$, and let $\mathcal{R}$ be the $\mathcal{B}$-valued accessibility relation given by: $\mathcal{R}(\Gamma, \Delta) = \{k \in \mathcal{K} \mid \Gamma \mathcal{R}_k \Delta\}$. This gives us a $\mathcal{B}$-frame, $\mathcal{F} = \langle \mathcal{G}, \mathcal{R} \rangle$. I'll define a valuation v by specifying it for atomic formulas. If A is atomic and not one of the propositional constants P_n, set $v(\Gamma, A)$ to be the **1** of $\mathcal{B}$ if $v_\mathcal{K}(\Gamma, A)$ is true, and set $v(\Gamma, A)$ to be **0** otherwise. Finally, set $v(\Gamma, P_n) = \{n\}$. Extend v to non-atomic formulas as usual. We thus have a $\mathcal{B}$-model $\mathcal{M} = \langle \mathcal{F}, v \rangle$.

Now, define a map θ from formulas of the $\mathcal{K}$-language to formulas of the mono-modal language enlarged with the propositional constants P_n.

1. for an atomic formula A (which can not be any P_n), set $\theta(A) = A$.

2. θ is a homomorphism with respect to propositional connectives, that is $\theta(X \wedge Y) = \theta(X) \wedge \theta(Y)$, and so on.

3. $\theta(\Diamond_k X) = P_k \supset \Diamond \theta(X)$

PROPOSITION 5 *For a formula X in the $\mathcal{K}$-language, $v_\mathcal{K}(\Gamma, X)$ is true in the multi-modal model $\mathcal{M}_\mathcal{K}$ if and only if $v(\Gamma, \theta(X)) = \mathbf{1}$ in the Boolean valued model $\mathcal{M}$.*

I'll leave the proof of this to you. It is a straightforward induction on formula degree, and makes use of the observation that in Boolean valued modal modals, $v(\Gamma, X \supset Y) = \mathbf{1}$ if and only if $v(\Gamma, X) \leq v(\Gamma, Y)$.

The result above will not be needed in what follows. It serves as motivation for the consideration of Boolean valued models in place of multi-modal ones. The switch to the Boolean valued case makes an algebraic approach much simpler and more natural. In [2, 4, 3, 5] Heyting algebras were used, which are more general than Boolean algebras. Using them allowed consideration, not just of multiple modalities, as above, but also of dependencies between them. This is more than is needed here, however.

4 Introducing vector spaces

A (two-valued) Kripke frame is just a directed graph. Transition matrices are a common way of representing edges in a graph, and these relate well to modal machinery. As an example, consider the frame depicted in Figure 1. The accessibility relation is represented by the matrix

$$\begin{bmatrix} 0 & 1 & 1 \\ 0 & 0 & 0 \\ 0 & 1 & 0 \end{bmatrix}$$

where the entry in position (i, j) is **1** if there is an edge from Γ_i to Γ_j, and otherwise is **0**. This is material that is familiar from many books—[6] is a recommended source.

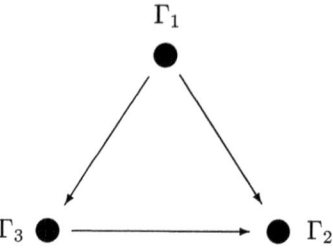

Figure 1. A classical mono-modal frame

Standard terminology is to call a set of possible worlds in a frame a *proposition*, so that in any model based on a frame the set of worlds in which a formula is true is a proposition. For our purposes, instead of working with propositions as sets, we work with propositions as *Boolean vectors*: for instance, using the frame of Figure 1 the vector $\langle \mathbf{1}, \mathbf{1}, \mathbf{0} \rangle$ corresponds to the set $\{\Gamma_1, \Gamma_2\}$; having Γ_i in the set corresponds to having **1** as the ith component of the vector. Of course the introduction of vectors for this purpose depends on an arbitrary ordering of possible worlds, but any two ways of ordering will result in vector spaces that are isomorphic in obvious ways, so we can ignore this point.

The introduction of a Boolean vector space meshes well with the usual Kripke semantics. Propositional connectives correspond to straightforward Boolean operations on vectors. More interestingly, suppose matrix multiplication is defined in the usual way, but with meet (∧) replacing multiplication, and join (∨) replacing addition. In the frame of Figure 1, if the vector $\langle \mathbf{1}, \mathbf{1}, \mathbf{0} \rangle$ represents the worlds at which a formula X is true (that is, X is true at Γ_1 and Γ_2), then the product of the frame transition matrix and the column vector corresponding to $\langle \mathbf{1}, \mathbf{1}, \mathbf{0} \rangle$ is a (column) vector that represents the worlds at which $\Diamond X$ is true.

$$\begin{bmatrix} 0 & 1 & 1 \\ 0 & 0 & 0 \\ 0 & 1 & 0 \end{bmatrix} \begin{bmatrix} 1 \\ 1 \\ 0 \end{bmatrix} = \begin{bmatrix} 1 \\ 0 \\ 1 \end{bmatrix}$$

Rather nicely, this extends directly to Boolean valued frames as well. Figure 2 shows such a frame in which the underlying space of truth values is the power set of $\{1, 2, 3\}$. Using this, we can still identify propositions with vectors, but now they are vectors over the powerset space. So, for instance, the vector $\langle \{1, 2\}, \{3\} \rangle$ could represent the status of a formula X in a particular model over this frame: at Γ_1 the truth value of X is $\{1, 2\}$, and at Γ_2 it is $\{3\}$.

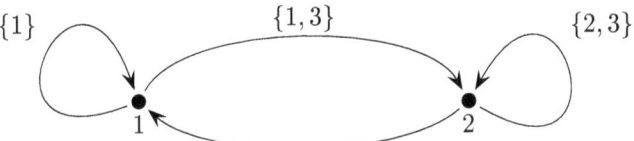

Figure 2. A Boolean-valued mono-modal frame

As we might expect, the accessibility relation can be represented by a matrix, but now with values in the powerset space. For the frame of Figure 2 the matrix is the following.

$$\begin{bmatrix} \{1\} & \{1,3\} \\ \emptyset & \{2,3\} \end{bmatrix}$$

And once again, such a matrix corresponds to the possibility operator, with matrix multiplication as application. Then, if X is represented by the vector $\langle \{1,2\}, \{3\} \rangle$ as above, the following represents the status of $\Diamond X$.

$$\begin{bmatrix} \{1\} & \{1,3\} \\ \emptyset & \{2,3\} \end{bmatrix} \begin{bmatrix} \{1,2\} \\ \{3\} \end{bmatrix} = \begin{bmatrix} \{1,3\} \\ \{3\} \end{bmatrix}$$

Now it is possible to say a little more clearly what is to come in the rest of the paper. The Boolean vector methodology sketched above will be rigorously introduced. Propositions will be identified with vectors. Just as accessibility relations within frames can be represented by matrices, so too for relations *between* frames. These matrices can be thought of as mapping propositions in one frame to propositions in another. Among such matrices are those corresponding to bisimulations. What are the properties of these Boolean algebra valued matrices in general, and what is special about those that are bisimulations?

5 Boolean terminology and notation

Now the primary formal development starts. Think of the preceding sections as establishing a context—why we might be interested—but at this point I'll start fresh, with Boolean valued vector spaces themselves as the topic. Underlying everything will be a complete Boolean algebra—in the example of Figure 2 it was the powerset of $\{1,2,3\}$—Boolean vector spaces will be built on such a Boolean algebra.

5.1 Boolean vector spaces

Recall, $\mathcal{B}$ is a complete Boolean algebra. I do not take an abstract approach to the subject of Boolean vector spaces; a concrete representation is fine for present purposes: $\mathcal{B}^n$ is the space of n-tuples over $\mathcal{B}$. It too constitutes a Boolean algebra using pointwise operations and relations. Overloading notation, I will also denote these by $\wedge$, $\vee$, and $\neg$, and $\leq$, with $\mathbf{0}_n$ and $\mathbf{1}_n$ for bottom and top, respectively $\langle 0, 0, \ldots, 0 \rangle$ and $\langle 1, 1, \ldots, 1 \rangle$ (with n components). For vectors in $\mathcal{B}^n$, a dot product is defined by: $\langle b_1, \ldots, b_n \rangle \cdot \langle c_1, \ldots, c_n \rangle = (b_1 \wedge c_1) \vee \ldots \vee (b_n \wedge c_n)$. Obviously $v \cdot \mathbf{0}_n = \mathbf{0}$ and $v \cdot \mathbf{1}_n \neq \mathbf{0}$ iff $v \neq \mathbf{0}_n$. We also have $v \cdot \mathbf{1}_n = v \cdot v$, incidentally.

I will also consider $\mathcal{B}^\infty$, whose members are infinite tuples over $\mathcal{B}$, $\langle b_1, b_2, b_3, \ldots \rangle$. This corresponds to Kripke frames with a countable set of worlds. Boolean operations and orderings are still defined pointwise and are no problem. Also the dot product operation extends to $\mathcal{B}^\infty$ (recall infinite joins exist): the dot product of $\langle b_1, b_2, b_3, \ldots \rangle$ and $\langle c_1, c_2, c_3, \ldots \rangle$ is the join of the set $\{(b_1 \wedge c_1), (b_2 \wedge c_2), (b_3 \wedge c_3), \ldots\}$.

5.2 Boolean matrices

Boolean matrices are matrices in the usual sense, but with entries from $\mathcal{B}$. I'll use $\mathcal{B}_{n,m}$ to denote the collection of all $n \times m$ Boolean matrices, where either or both of n and m could be ∞. Once again overloading notation, $\wedge$, $\vee$, $\neg$ and $\leq$ are defined componentwise. It is easy to see that $\mathcal{B}_{n,m}$ itself is a Boolean algebra with respect to these operations. Generalizing notation from above, $\mathbf{1}_{n,m}$ is the Boolean $n \times m$ matrix with every entry $\mathbf{1}$, and $\mathbf{0}_{n,m}$ is the Boolean $n \times m$ matrix with every entry $\mathbf{0}$. I will use I_n for the $n \times n$ identity matrix.

For a matrix $A \in \mathcal{B}_{n,m}$, $[A]_i$ denotes row vector i of A, thought of as a member of $\mathcal{B}^m$, so that $[A]_i \in \mathcal{B}^m$. Likewise $[A]^j$ denotes column vector j of A, taken to be a member of $\mathcal{B}^n$. And $[A]_i^j$ denotes the entry in row i and column j. Then Boolean matrix multiplication is characterized in the usual way: if $B \in \mathcal{B}_{m,k}$, AB is the $n \times k$ matrix such that $[AB]_i^j = [A]_i \cdot [B]^j$. I also introduce one non-standard piece of notation.

DEFINITION 6 *If β is a member of $\mathcal{B}$, by $\beta_{n,m}(a,b)$ I mean the $n \times m$ matrix all of whose entries are $\mathbf{0}$ except for the entry in row a, column b, which is β.*

The following will be useful in proving some inequalities involving Boolean matrices. Item 1 is overly generous, in a sense, but using it is no more work and is often easier to apply than a stricter version would be.

LEMMA 7 *Let $A, B \in \mathcal{B}_{n,m}$.*

1. *$A \leq B$ provided $\beta_{n,m}(a,b) \leq A$ implies $\beta_{n,m}(a,b) \leq B$ for every $\beta \in \mathcal{B}$ and every $a \leq n$ and $b \leq m$.*

2. *$\beta_{m,n}(a,b)\beta_{n,k}(b,c) = \beta_{m,k}(a,c)$.*

Proof. Suppose $\beta_{n,m}(a,b) \leq A$ implies $\beta_{n,m}(a,b) \leq B$ for every $\beta \in \mathcal{B}$ and every $a \leq n$ and $b \leq m$. Let $\beta = [A]_i^j$. Then obviously $\beta_{n,m}(i,j) \leq A$, so $\beta_{n,m}(i,j) \leq B$, and thus $[A]_i^j \leq [B]_i^j$. Since i and j were arbitrary, $A \leq B$. Part 2 follows directly from the definition of Boolean matrix multiplication. ∎

6 Elementary matrix multiplication properties

For use later on, we need some basic properties of Boolean matrix multiplication. Some of this is analogous to matrix multiplication over a field, some is rather different. As might be expected, Boolean matrix multiplication is associative but not generally commutative. Also, multiplication distributes over $\vee$, that is, $A(B \vee C) = AB \vee AC$ and $(B \vee C)A = BA \vee CA$. If $B \leq C$ then $AB \leq AC$ and $BA \leq CA$. (All this is under the assumption that dimensions are such that the products displayed are defined, of course.) A^T denotes the transpose of A and, as usual, $(AB)^T = B^T A^T$. In addition, Boolean matrices have a number of special features not shared by matrices over a field.

THEOREM 8 *Assume $A \in \mathcal{B}_{m,n}$. Then:*

1. *$A \leq AA^T A$*
   $A^T \leq A^T AA^T$

2. *$A\mathbf{1}_{n,k} = AA^T A\mathbf{1}_{n,k}$*
   $A^T \mathbf{1}_{m,k} = A^T AA^T \mathbf{1}_{m,k}$

3. $(A^T A) \leq (A^T A)^2 \leq (A^T A)^3 \leq \ldots$
   $(AA^T) \leq (AA^T)^2 \leq (AA^T)^3 \leq \ldots$

Proof. The items to be proved come in pairs. I'll show one half, the other half is obviously similar.

1. By Lemma 7, part 1, it is enough to show that if $\beta_{m,n}(a,b) \leq A$ then $\beta_{m,n}(a,b) \leq AA^T A$ for an arbitrary $\beta \in \mathcal{B}$, a and b. So, suppose $\beta_{m,n}(a,b) \leq A$. Then $\beta_{n,m}(b,a) \leq A^T$ so by Lemma 7, part 2, $\beta_{m,m}(a,a) = \beta_{m,n}(a,b)\beta_{n,m}(b,a) \leq AA^T$, and then by a similar calculation, $\beta_{m,n}(a,b) = \beta_{m,m}(a,a)\beta_{m,n}(a,b) \leq AA^T A$.

2. $A\mathbf{1}_{n,k} \leq AA^T A\mathbf{1}_{n,k}$ by part 1. Conversely, $A^T A\mathbf{1}_{n,k} \leq \mathbf{1}_{n,k}$ so $AA^T A\mathbf{1}_{n,k} \leq A\mathbf{1}_{n,k}$.

3. By part 1 we have $A \leq AA^T A$ so $A^T A \leq A^T AA^T A$. And so on.

∎

Part 3 above is a strictly increasing sequence of inequalities, in the sense that for any n one can find a finite Boolean matrix A such that the sequence strictly grows for the first n steps. If one uses infinite dimension matrices, the growth can be made to continue indefinitely.

EXAMPLE 9 Let

$$A = \begin{bmatrix} 1 & 0 & 0 & 0 \\ 1 & 1 & 0 & 0 \\ 0 & 1 & 1 & 0 \\ 0 & 0 & 1 & 1 \end{bmatrix}$$

Then

$$(A^T A) = \begin{bmatrix} 1 & 1 & 0 & 0 \\ 1 & 1 & 1 & 0 \\ 0 & 1 & 1 & 1 \\ 0 & 0 & 1 & 1 \end{bmatrix} \text{ and } (A^T A)^2 = \begin{bmatrix} 1 & 1 & 1 & 0 \\ 1 & 1 & 1 & 1 \\ 1 & 1 & 1 & 1 \\ 0 & 1 & 1 & 1 \end{bmatrix} \text{ and }$$

$$(A^T A)^3 = \begin{bmatrix} 1 & 1 & 1 & 1 \\ 1 & 1 & 1 & 1 \\ 1 & 1 & 1 & 1 \\ 1 & 1 & 1 & 1 \end{bmatrix}$$

Thus, $(A^T A) < (A^T A)^2 < (A^T A)^3 = (A^T A)^4 = \ldots$

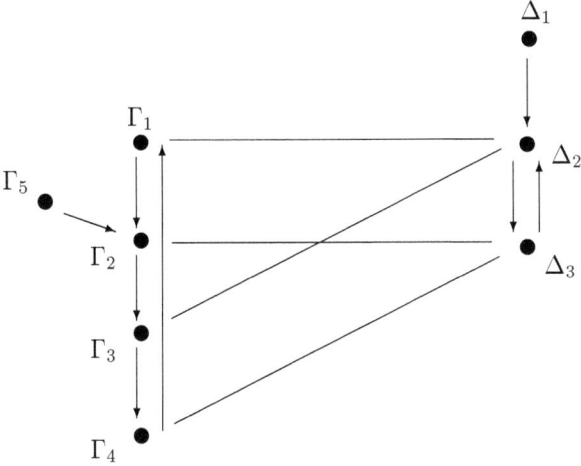

Figure 3. A bisimulation, two-valued setting

7 Bisimulation motivation

In order to motivate the work on matrices in the next several sections, a bisimulation example will be useful. In fact, one in which the underlying logic is two-valued will suffice for this purpose.

EXAMPLE 10 In Figure 3 two frames are displayed, a left-hand one with possible worlds Γ_1, Γ_2, Γ_3, Γ_4, and Γ_5, and a right-hand one with possible worlds Δ_1, Δ_2, and Δ_3. The two (mono-modal) accessibility relations are indicated by arrows.

A relation, call it $\mathcal{A}$, between the two frames is indicated by lines (not arrows) connecting worlds. It meets the conditions for being a standard frame bisimulation as given in Definition 1. It is not, of course, a bisimulation since we do not have models, but only frames, here. The additional condition for bisimulations in Definition 3 can be thought of as a restriction on the models we can base on these frames. For instance, since both Γ_1 and Γ_3 are related by $\mathcal{A}$ to Δ_2, any model based on the left-hand frame must assign the same truth values to propositional letters at both Γ_1 and Γ_3 if the condition for being a bisimulation is to apply.

Although $\mathcal{A}$ is a relation, it can be thought of as a function as well, mapping propositions in one frame to propositions in the other frame. For instance, the proposition $\{\Gamma_1, \Gamma_2, \Gamma_5\}$ maps to the proposition $\{\Delta_2, \Delta_3\}$, the set of worlds in the right frame related to the worlds of the proposition in the left frame. But since I am representing propositions, not as sets, but as vectors, I prefer to say that the vector $\langle 1, 1, 0, 0, 1\rangle$ maps to the vector $\langle 0, 1, 1\rangle$. The relation $\mathcal{A}$ can itself be represented as a matrix, A; I use a representation in which column j is the vector corresponding to the set of worlds related to Γ_j. Stated differently, $[A]_i^j = 1$ just in case $\Gamma_j \mathcal{A} \Delta_i$. For the present example, this gives the following matrix.

$$A = \begin{bmatrix} 0 & 0 & 0 & 0 & 0 \\ 1 & 0 & 1 & 0 & 0 \\ 0 & 1 & 0 & 1 & 0 \end{bmatrix}$$

And once again, matrix multiplication provides appropriate machinery. It was noted above that the vector $\langle 1, 1, 0, 0, 1\rangle$ maps to the vector $\langle 0, 1, 1\rangle$, and in fact we have the following:

$$A \begin{bmatrix} 1 \\ 1 \\ 0 \\ 0 \\ 1 \end{bmatrix} = \begin{bmatrix} 0 \\ 1 \\ 1 \end{bmatrix}$$

Now we return to the point raised above, that if we want to base a model on one of these frames, and respect the frame bisimulation $\mathcal{A}$, we are restricted in our assignments of truth values to atomic formulas—not every proposition can be used. For instance, requiring an atomic formula P to be true only at Γ_1 will not do; since Γ_1 is related by $\mathcal{A}$ to Δ_2 which in turn is related to Γ_3, if P is taken to be true at Γ_1 we must also take it to be true at Γ_3. Rephrasing this in terms of vectors, $\langle 1, 0, 0, 0, 0\rangle$ is not an appropriate vector for us to be working with in the setup of Figure 3. In fact, if we map it from the left-hand frame to the right-hand one using A, then back using A^T, we get the following:

$$A^T A \begin{bmatrix} 1 \\ 0 \\ 0 \\ 0 \\ 0 \end{bmatrix} = \begin{bmatrix} 1 \\ 0 \\ 1 \\ 0 \\ 0 \end{bmatrix}$$

This is what we should expect. But here things stop. If we apply the mapping $A^T A$ to $\langle 1, 0, 1, 0, 0\rangle$, we simply get $\langle 1, 0, 1, 0, 0\rangle$ back again. Vectors that are left unchanged in this way are appropriate candidates for propositional letter assignments. I call such vectors *stable*, and the next section starts a formal investigation of them.

Example 10 was mono-modal, but multi-modal examples can be treated in the same way. Recall that Definition 1 allowed multi-modal models and a parameterized bisimulation relation. Converted to an algebraic setting, this just amounts to allowing bisimulations to be represented by matrices over Boolean algebras that are more complex than the two-valued one, just as we have represented multi-modal accessibility relations by such matrices. This is the setting you should have in mind for the following sections.

8 Linear mappings and stability

The previous section sketched background ideas informally—now it is time for the mathematical details. For this section, let $A \in \mathcal{B}_{m,n}$ (with ∞ allowed as values for m or n). A can be thought of as defining a mapping from $\mathcal{B}^n$ to $\mathcal{B}^m$ (informally, from propositions to propositions). We identify vectors with column vectors in the obvious way so that, properly speaking, A maps $\mathcal{B}_{n,1}$ to $\mathcal{B}_{m,1}$. Then, for $v \in \mathcal{B}_{n,1}$, its image under A is Av. More generally, for each k we can think of A as defining a mapping from $\mathcal{B}_{n,k}$ to $\mathcal{B}_{m,k}$, that is, $A : \mathcal{B}_{n,k} \to \mathcal{B}_{m,k}$. For a matrix $V \in \mathcal{B}_{n,k}$ its image is defined, using matrix multiplication, as AV. Of course the transpose of A maps in the reverse direction, $A^T : \mathcal{B}_{m,k} \to \mathcal{B}_{n,k}$.

In general, the transpose of a Boolean matrix won't be its inverse. In fact, having an inverse is a rare property for a Boolean matrix to possess. But as was noted in the previous section, the collection of things on which the transpose behaves like an inverse will be a collection of special interest.

DEFINITION 11 *I will say a matrix $V \in \mathcal{B}_{n,k}$ is A-stable if $A^T A V = V$. Likewise a matrix $W \in \mathcal{B}_{m,k}$ is A^T-stable if $A A^T W = W$. If no confusion is likely to result, I may just use the term stable.*

THEOREM 12 *The following are properties of stability:*

1. *If V is A-stable then AV is A^T-stable, and if W is A^T-stable then $A^T W$ is A-stable.*

2. *$\mathbf{0}_{n,k}$ is the smallest A-stable matrix in $\mathcal{B}_{n,k}$, and $\mathbf{0}_{m,k}$ is the smallest A^T-stable matrix in $\mathcal{B}_{m,k}$.*

3. *The smallest stable matrices map to each other. That is, $A\mathbf{0}_{n,k} = \mathbf{0}_{m,k}$, and $A^T \mathbf{0}_{m,k} = \mathbf{0}_{n,k}$.*

4. *$A^T A \mathbf{1}_{n,k}$ is the largest A-stable matrix in $\mathcal{B}_{n,k}$, and $AA^T \mathbf{1}_{m,k}$ is the largest A^T-stable matrix in $\mathcal{B}_{m,k}$.*

5. *The largest stable matrices map to each other. That is, $A(A^T A \mathbf{1}_{n,k}) = AA^T \mathbf{1}_{m,k}$ and $A^T (AA^T \mathbf{1}_{m,k}) = A^T A \mathbf{1}_{n,k}$.*

Proof. For each item I'll show one half; the other is similar.

1. Suppose V is A-stable, $A^T AV = V$. Then trivially $AA^T AV = AV$ so AV is A^T-stable.

2. Clearly $A^T A \mathbf{0}_{n,k} = A^T \mathbf{0}_{m,k} = \mathbf{0}_{n,k}$, so $\mathbf{0}_{n,k}$ is A-stable. It is obviously smallest.

3. Already used in previous item.

4. $A^T A(A^T A \mathbf{1}_{n,k}) = A^T (AA^T A \mathbf{1}_{n,k}) = A^T A \mathbf{1}_{n,k}$ by Theorem 8 part 2, so we have A-stability. And if V is A-stable, $V = A^T AV \leq A^T A \mathbf{1}_{n,k}$, so $A^T A \mathbf{1}_{n,k}$ is largest.

5. In one direction, $A \mathbf{1}_{n,k} \leq \mathbf{1}_{m,k}$ so $AA^T A \mathbf{1}_{n,k} \leq AA^T \mathbf{1}_{m,k}$. In the other direction, $A^T \mathbf{1}_{m,k} \leq \mathbf{1}_{n,k}$ so using Theorem 8 part 2, $AA^T \mathbf{1}_{m,k} = AA^T AA^T \mathbf{1}_{m,k} \leq AA^T A \mathbf{1}_{n,k}$.

∎

I will eventually show the stable matrices themselves form a Boolean algebra, but an additional assumption will be needed. Without that, we still have the following.

THEOREM 13 *Let* $V, W \in \mathcal{B}_{n,k}$, *and assume both are A-stable.*

1. $A(V \vee W) = AV \vee AW$

2. $V \vee W$ *is A-stable*

Similar results obtain for A^T as well.

Proof. The arguments are as follows

1. This is immediate since Boolean matrix multiplication always distributes over $\vee$.

2. $A^T A(V \vee W) = A^T(AV \vee AW) = A^T AV \vee A^T AW = V \vee W$.

∎

9 The final general assumption

So far we have been working with arbitrary (complete) Boolean algebras. But algebras arising from multi-modal logics are rather special. One way of saying it is that they are powerset algebras, as the discussion in Section 3 illustrated. Another way of saying it is that they are *atomic*.

DEFINITION 14 For the Boolean algebra $\mathcal{B}$:

1. $a \in \mathcal{B}$ is an *atom* if $a \neq \mathbf{0}$ and there is no $b \in \mathcal{B}$ such that $\mathbf{0} < b < a$.

2. $\mathcal{B}$ is *atomic* if for each $x \in \mathcal{B}$ other than $\mathbf{0}$ there is an atom a such that $a \leq x$.

Atoms have many useful features, several of which we will need. If β is an atom and $\beta \leq x \vee y$ then $\beta \leq x$ or $\beta \leq y$. If α and β are different atoms, $\alpha \wedge \beta = \mathbf{0}$, while of course $\alpha \wedge \alpha = \alpha$. If $\mathcal{B}$ is finite, it is automatically complete and atomic. Whether finite or not, if $\mathcal{B}$ is atomic and complete, each member is the join of the collection of atoms below it. This implies that a complete, atomic Boolean algebra is isomorphic to the collection of all subsets of a set, namely the set of atoms.

General Assumption 2 From now on, $\mathcal{B}$ is both complete and atomic.

LEMMA 15 *Assume $A \in \mathcal{B}_{m,n}$. Then for any $V \in \mathcal{B}_{n,k}$ and $W \in \mathcal{B}_{k,n}$*

1. $V \wedge A^T W \leq A^T A V$

2. $W \wedge A V \leq A A^T W$

Proof. I'll only show item 1. Fix a and b—I'll show $[V \wedge A^T W]_a^b \leq [A^T A V]_a^b$. And since each member of $\mathcal{B}$ is the join of the family of atoms below it, it is enough to show that for any atom β if $\beta \leq [V \wedge A^T W]_a^b$ then $\beta \leq [A^T A V]_a^b$. Assume $\beta \leq [V]_a^b$ and $\beta \leq [A^T W]_a^b$. From the second of these, $\beta \leq [A^T]_a \cdot [W]^b$, that is, $\beta \leq \bigvee_c \{[A^T]_a^c \wedge [W]_c^b\}$. Since β is an atom, for some c, $\beta \leq [A^T]_a^c \wedge [W]_c^b$, so in particular, $\beta \leq [A^T]_a^c$. It follows that $\beta_{n,m}(a,c) \leq A^T$. Then $\beta_{m,n}(c,a) \leq A$, so by Lemma 7, part 2, $\beta_{n,n}(a,a) \leq A^T A$. Then again, since $\beta_{n,k}(a,b) \leq V$, $\beta_{n,k}(a,b) \leq A^T A V$, so $\beta \leq [A^T A V]_a^b$. ∎

Now Theorem 13 for joins can have a companion for meets.

THEOREM 16 *Let $A \in \mathcal{B}_{m,n}$, $V, W \in \mathcal{B}_{n,k}$, and assume both V and W are A-stable.*

1. $A(V \wedge W) = AV \wedge AW$

2. $V \wedge W$ is A-stable

Similar results obtain for A^T as well.

Proof. The arguments are as follows

1. First, $V \wedge W \leq V$, so $A(V \wedge W) \leq AV$. Similarly $A(V \wedge W) \leq AW$. So $A(V \wedge W) \leq AV \wedge AW$.

   Next, V is A-stable so AV is A^T-stable, and consequently $AV \leq AA^T \mathbf{1}_{m,k}$ by Theorem 12 part 4. Similarly $AW \leq AA^T \mathbf{1}_{m,k}$. Now

   $$\begin{align} AV \wedge AW &= AV \wedge AW \wedge AA^T \mathbf{1}_{m,k} \tag{1}\\ &\leq AA^T(AV \wedge AW) \tag{2}\\ &\leq A(A^T AV \wedge A^T AW) \tag{3}\\ &= A(V \wedge W) \tag{4} \end{align}$$

   Here (2) is by Lemma 15 part 2, taking W of that Lemma to be $AV \wedge AW$ and V of that Lemma to be $A^T \mathbf{1}_{m,k}$. For (3) we apply the first half of the argument, using A^T.

2. Similar to part 2 of Theorem 13.

∎

10 Negation and stability

Throughout this section $A \in \mathcal{B}_{m,n}$, so that for each k, $A : \mathcal{B}_{n,k} \to \mathcal{B}_{m,k}$ and $A^T : \mathcal{B}_{m,k} \to \mathcal{B}_{n,k}$. Also I'll systematically use V as a member of $\mathcal{B}_{n,k}$. The goal of this section is to show that A-stable matrices have a 'natural' notion of negation. First, a few minor preliminary items. Recall the notation of Definition 6.

LEMMA 17 *If $\beta_{m,k}(a,b) \leq A\beta_{n,k}(c,b)$ then $\beta_{n,k}(c,b) \leq A^T \beta_{m,k}(a,b)$.*

Proof.
All entries of $\beta_{m,k}(a,b)$ are $\mathbf{0}$ except for β in row a, column b. Consequently if $\beta_{m,k}(a,b) \leq A\beta_{n,k}(c,b)$ it must be that $\beta \leq [A]_a^c$. This in turn implies $\beta \leq [A^T]_c^a$, or $\beta_{n,m}(c,a) \leq A^T$. Consequently $\beta_{n,k}(c,b) = \beta_{n,m}(c,a)\beta_{m,k}(a,b) \leq A^T \beta_{m,k}(a,b)$. ∎

LEMMA 18 *Let V be A-stable. Then*

1. $AV \wedge A(\neg V) = \mathbf{0}_{m,k}$
2. $A^T A(\neg V) \leq \neg V$
3. $A^T A(\neg V) = \neg V \wedge A^T A \mathbf{1}_{n,k}$

Proof. The arguments are as follows.

1. This makes use of an infinite distributive law that holds in complete Boolean algebras, see [7, Vol. 1, Lemma 1.33]. I'll suppose $AV \wedge A(\neg V) \neq \mathbf{0}_{m,k}$ and derive a contradiction. By the supposition, for some a and b the entry in row a and column b is non-zero. Then, using the distributive law:

$$\begin{aligned}
\mathbf{0} &< [AV \wedge A(\neg V)]_a^b \\
&= [AV]_a^b \wedge [A(\neg V)]_a^b \\
&= [A]_a \cdot [V]^b \wedge [A]_a \cdot [\neg V]^b \\
&= \bigvee_c \{[A]_a^c \wedge [V]_c^b\} \wedge \bigvee_d \{[A]_a^d \wedge [\neg V]_d^b\} \\
&= \bigvee_{c,d} \{[A]_a^c \wedge [V]_c^b \wedge [A]_a^d \wedge [\neg V]_d^b\}
\end{aligned}$$

Of course not every member of this join can be $\mathbf{0}$. Let us say that for a, b, c, d,
$$[A]_a^c \wedge [V]_c^b \wedge [A]_a^d \wedge [\neg V]_d^b = \beta > \mathbf{0}$$

From this we have

$$\beta_{m,n}(a,c) \leq A \tag{5}$$
$$\beta_{n,k}(c,b) \leq V \tag{6}$$
$$\beta_{m,n}(a,d) \leq A \tag{7}$$
$$\beta_{n,k}(d,b) \leq \neg V \tag{8}$$

Now, from (5) and (7), and Lemma 7 part 2 we have

$$\beta_{m,k}(a,b) = \beta_{m,n}(a,c)\beta_{n,k}(c,b) \leq A\beta_{n,k}(c,b) \tag{9}$$
$$\beta_{m,k}(a,b) = \beta_{m,n}(a,d)\beta_{n,k}(d,b) \leq A\beta_{n,k}(d,b) \tag{10}$$

From (10) and Lemma 17 we have

$$\beta_{n,k}(d,b) \leq A^T \beta_{m,k}(a,b) \tag{11}$$

And then

$$\beta_{n,k}(d,b) \leq A^T \beta_{m,k}(a,b) \quad \text{by (11)} \quad (12)$$
$$\leq A^T A \beta_{n,k}(c,b) \quad \text{by (9)} \quad (13)$$
$$\leq A^T AV \quad \text{by (6)} \quad (14)$$
$$= V \quad \text{by stability} \quad (15)$$

But this contradicts (8).

2. By part 1, $AV \wedge A(\neg V) = \mathbf{0}_{m,k}$ so $A(\neg V) \leq \neg(AV)$ and hence $A^T A(\neg V) \leq A^T(\neg AV)$. Also AV is A^T-stable so by part 1 again (for A^T), $A^T(AV) \wedge A^T(\neg AV) = \mathbf{0}_{n,k}$ so $A^T(\neg AV) \leq \neg A^T(AV)$. Combining things, $A^T A(\neg V) \leq \neg A^T(AV) = \neg V$.

3. $V \vee \neg V = \mathbf{1}_{n,k}$ so $A^T AV \vee A^T A(\neg V) = A^T A \mathbf{1}_{n,k}$. But V is stable, so $V \vee A^T A(\neg V) = A^T A \mathbf{1}_{n,k}$, and hence $\neg V \wedge A^T A \mathbf{1}_{n,k} \leq A^T A(\neg V)$. In the other direction, by part 2 $A^T A(\neg V) \leq \neg V$, and of course $A^T A(\neg V) \leq A^T A \mathbf{1}_{n,k}$ and so $A^T A(\neg V) \leq \neg V \wedge A^T A \mathbf{1}_{n,k}$.

∎

Now we head to the main material. If V is A-stable, it does not follow that $\neg V$ will also be. But there is a suitable candidate for a negation that is stable. The form is suggested by the following—recall, $A^T A \mathbf{1}_{n,k}$ is the largest A-stable $n \times k$ matrix.

THEOREM 19 *If V is A-stable, $V = V \wedge A^T A \mathbf{1}_{n,k}$*

Proof. Trivially $V \wedge A^T A \mathbf{1}_{n,k} \leq V$. The other direction follows from Theorem 12 part 4. ∎

Now, the following definition should seem reasonable. The Theorem that follows it justifies this sense of reasonableness.

DEFINITION 20 $\overline{V} = \neg V \wedge A^T A \mathbf{1}_{n,k}$. I'll call $\overline{V}$ the *stable negation* of V.

THEOREM 21 *Let V be A-stable. Then*

1. $\overline{V} = A^T A(\neg V)$

2. $\overline{V}$ is A-stable

3. $\overline{\overline{V}} = V$

4. $V \wedge \overline{V} = \mathbf{0}_{n,k}$

5. $V \vee \overline{V} = A^T A \mathbf{1}_{n,k}$

6. $A\overline{V} = \overline{AV}$

7. $A(\neg V) = A(\overline{V})$

8. $\overline{V}$ is the largest matrix W such that W is A-stable and $W \leq \neg V$

Proof. The various parts are as follows.

1. This is a restatement of Lemma 18, part 3.

2. In one direction:

$$A^T A(\neg V \wedge A^T A \mathbf{1}_{n,k}) \leq A^T A \neg V \wedge A^T A A^T A \mathbf{1}_{n,k} \quad (16)$$
$$\leq \neg V \wedge A^T A A^T A \mathbf{1}_{n,k} \quad (17)$$
$$= \neg V \wedge A^T A \mathbf{1}_{n,k} \quad (18)$$

In this, (16) is a standard item (see the beginning of the proof of Theorem 16 part 1); then (17) is by Lemma 18 part 2; and (18) is by Theorem 8 part 2.

In the other direction, $A^T A(\neg V) \leq A^T A A^T A(\neg V)$ by Theorem 8 part 3, then use part 1.

3. $\overline{\overline{V}} = \neg \overline{V} \wedge A^T A \mathbf{1}_{n,k} = \neg(\neg V \wedge A^T A \mathbf{1}_{n,k}) \wedge A^T A \mathbf{1}_{n,k} = (V \vee \neg A^T A \mathbf{1}_{n,k}) \wedge A^T A \mathbf{1}_{n,k} = V \wedge A^T A \mathbf{1}_{n,k} = V$. The last step is by Theorem 19.

4. Straightforward.

5. $V \vee \overline{V} = V \vee (\neg V \wedge A^T A \mathbf{1}_{n,k}) = (V \vee \neg V) \wedge (V \vee A^T A \mathbf{1}_{n,k}) = \mathbf{1}_{n,k} \wedge A^T A \mathbf{1}_{n,k} = A^T A \mathbf{1}_{n,k}$. Note: along the way we used Theorem 12 part 4.

6. $V \wedge \overline{V} = \mathbf{0}_{n,k}$ and both V and $\overline{V}$ are A-stable, so $AV \wedge A\overline{V} = A(V \wedge \overline{V}) = A\mathbf{0}_{n,k} = \mathbf{0}_{m,k}$ and hence $A\overline{V} \leq \neg(AV)$. Also $A\overline{V} \leq AA^T \mathbf{1}_{m,k}$ (Theorem 12 part 4), so $A\overline{V} \leq \neg(AV) \wedge AA^T \mathbf{1}_{m,k} = \overline{AV}$.

In the other direction, $V \vee \overline{V} = A^T A \mathbf{1}_{n,k}$ so using Theorem 12 part 5, $AV \vee A\overline{V} = A(V \vee \overline{V}) = AA^T A \mathbf{1}_{n,k} = AA^T \mathbf{1}_{m,k}$, so $\neg(AV) \wedge AA^T \mathbf{1}_{m,k} \leq A\overline{V}$, or $\overline{AV} \leq A\overline{V}$.

7. From the definition of stable negation, $\overline{V} \leq \neg V$, so $A(\overline{V}) \leq A(\neg V)$. By Lemma 18 part 2, $A^T A(\neg V) \leq \neg V$, so $AA^T A(\neg V) \leq A(\neg V)$. And by Theorem 8 part 1, $A(\neg V) \leq AA^T A(\neg V)$, so $AA^T A(\neg V) = A(\neg V)$. Then $A(\neg V)$ is A^T-stable, so $A(\neg V) \leq AA^T \mathbf{1}_{n,k}$ by Theorem 12 part 4. By Lemma 18 part 1, $AV \wedge A(\neg V) = \mathbf{0}_{n,k}$ so $A(\neg V) \leq \neg(AV)$. Thus $A(\neg V) \leq \neg(AV) \wedge AA^T \mathbf{1}_{n,k}$, so using the definition of stable negation with respect to A^T instead of A, $A(\neg V) \leq \overline{AV} = A\overline{V}$ by part 6 of this Theorem.

∎

11 Summary so far

Once again, let $A \in \mathcal{B}_{m,n}$. Various results about A-stability have been shown, and it is time to collect them together.

1. The A-stable members of $\mathcal{B}_{n,k}$ constitute a Boolean algebra with meet and join being the $\wedge$ and $\vee$ of $\mathcal{B}_{n,k}$, with complementation being stable negation, and with bottom and top being $\mathbf{0}_{n,k}$ and $A^T A \mathbf{1}_{n,k}$.

2. Likewise the A^T-stable members of $\mathcal{B}_{m,k}$ constitute a Boolean algebra.

3. A is an isomorphism from the Boolean algebra of A-stable members of $\mathcal{B}_{n,k}$ onto the A^T-stable members of $\mathcal{B}_{m,k}$, with A^T as its inverse.

12 Boolean bisimulations

Consider again Example 10. There are two mono-modal frames displayed, each with its accessibility relation. Representing these by transition matrices we get

$$R = \begin{bmatrix} 0 & 1 & 0 & 0 & 0 \\ 0 & 0 & 1 & 0 & 0 \\ 0 & 0 & 0 & 1 & 0 \\ 1 & 0 & 0 & 0 & 0 \\ 0 & 1 & 0 & 0 & 0 \end{bmatrix} \quad S = \begin{bmatrix} 0 & 1 & 0 \\ 0 & 0 & 1 \\ 0 & 1 & 0 \end{bmatrix}$$

for the left and right frames respectively. Earlier I also specified a matrix, A, corresponding to the frame bisimulation relation $\mathcal{A}$. Question: what conditions on A, R, and S tell us that $\mathcal{A}$ is, in fact, a (standard) frame bisimulation? Answer: the conditions are $AR \leq SA$ and $SA^T \leq RA^T$. It is easily checked that these inequalities hold in the special case of Example 10. In Section 13 I'll prove these are the inequalities that characterize bisimulations not only in the mono-modal case but also in the multi-modal one. Until then, I'll investigate matrices satisfying the inequalities for their own sakes.

DEFINITION 22 Let $A \in \mathcal{B}_{m,n}$, $R \in \mathcal{B}_{n,n}$, and $S \in \mathcal{B}_{m,m}$. I'll call A a *Boolean bisimulation* from R to S if:

1. $AR \leq SA$

2. $A^T S \leq RA^T$

There are some elementary properties of Boolean bisimulations, whose proofs are immediate and so are omitted.

THEOREM 23 *Boolean bisimulation is an equivalence relation, in the following sense.*

1. *The identity matrix I_n in $\mathcal{B}_{n,n}$ is a Boolean bisimulation from R to R.*

2. *If A is a Boolean bisimulation from R to S, then A^T is a Boolean bisimulation from S to R.*

3. *If A is a Boolean bisimulation from R to S and A' is a Boolean bisimulation from S to U, then $A'A$ is a Boolean bisimulation from R to U.*

Boolean bisimulations always exist, in an uninteresting way, because $\mathbf{0}_{m,n}$ is a Boolean bisimulation. Also there is always a largest Boolean bisimulation, the disjunction of all Boolean bisimulations. This is an immediate consequence of the fact that multiplication of Boolean matrices distributes over disjunction.

13 Connections

In the previous section the notion of Boolean bisimulation was defined, and there is also the usual notion of bisimulation, extended somewhat in Definition 1. It is time to connect these. I'll begin with frame bisimulations in a mono-modal setting (which are trivially standard), then move on to the more general situation.

THEOREM 24 *Let $\mathcal{F}_\mathcal{G} = \langle \mathcal{G}, \mathcal{R} \rangle$ and $\mathcal{F}_\mathcal{H} = \langle \mathcal{H}, \mathcal{S} \rangle$ be two mono-modal frames, and let $\mathcal{A}$ be a relation between $\mathcal{G}$ and $\mathcal{H}$. Let $\mathcal{B}$ be the two-member Boolean algebra, with elements $\{\mathbf{0}, \mathbf{1}\}$. Assume fixed enumerations $\{\Gamma_1, \Gamma_2, \ldots\}$ of $\mathcal{G}$ and $\{\Delta_1, \Delta_2, \ldots\}$ of $\mathcal{H}$. Let R be the transition matrix for $\mathcal{F}_\mathcal{G}$, that is $[R]_i^j = \mathbf{1}$ iff $\Gamma_i \mathcal{R} \Gamma_j$, and similarly let S be the transition matrix for $\mathcal{F}_\mathcal{H}$. Finally let A be the matrix corresponding to the relation $\mathcal{A}$, so that $[A]_i^j = \mathbf{1}$ iff $\Gamma_j \mathcal{A} \Delta_i$.*

$\mathcal{A}$ is a frame bisimulation between $\mathcal{F}_\mathcal{G}$ and $\mathcal{F}_\mathcal{H}$ iff A is a Boolean bisimulation from R to S.

Proof. First assume $\mathcal{A}$ is a frame bisimulation between $\mathcal{F}_\mathcal{G}$ and $\mathcal{F}_\mathcal{H}$. To show $\mathcal{A}$ is a Boolean bisimulation, I'll show $\mathcal{A}\mathcal{R} \leq \mathcal{S}\mathcal{A}$; the other inequality is similar. Suppose $[\mathcal{A}\mathcal{R}]_i^j = \mathbf{1}$, I'll show $[\mathcal{S}\mathcal{A}]_i^j = \mathbf{1}$. Since $[\mathcal{A}\mathcal{R}]_i^j = \mathbf{1}$ we have $[\mathcal{A}]_i \cdot [\mathcal{R}]^j = \mathbf{1}$, and so for some k, $[\mathcal{A}]_i^k = [\mathcal{R}]_k^j = \mathbf{1}$. Since $[\mathcal{A}]_i^k = \mathbf{1}$, $\Gamma_k \mathcal{A} \Delta_i$, and since $[\mathcal{R}]_k^j = \mathbf{1}$, $\Gamma_k \mathcal{R} \Gamma_j$. Then by the definition of frame bisimulation, there must be a world Δ_n with $\Delta_i \mathcal{S} \Delta_n$ and $\Gamma_j \mathcal{A} \Delta_n$. But then $[\mathcal{S}]_i^n = \mathbf{1}$ and $[\mathcal{A}]_n^j = \mathbf{1}$. It follows that $[\mathcal{S}\mathcal{A}]_i^j = \mathbf{1}$.

Next, assume $\mathcal{A}$ is a Boolean bisimulation from $\mathcal{R}$ to $\mathcal{S}$; I'll show $\mathcal{A}$ is a frame bisimulation—actually I'll show one of the two bisimulation conditions, the other is similar The argument is essentially that of the previous paragraph, reversed. So, suppose $\Gamma_j, \Gamma_k \in \mathcal{G}$, $\Delta_i \in \mathcal{H}$, $\Gamma_k \mathcal{A} \Delta_i$, and $\Gamma_k \mathcal{R} \Gamma_j$. From the first of these relation instances, $[\mathcal{A}]_i^k = \mathbf{1}$ and from the second, $[\mathcal{R}]_k^j = \mathbf{1}$. But then $[\mathcal{A}\mathcal{R}]_i^j = \mathbf{1}$ and, since $\mathcal{A}$ is a Boolean bisimulation, $\mathcal{A}\mathcal{R} \leq \mathcal{S}\mathcal{A}$, so $[\mathcal{S}\mathcal{A}]_i^j = \mathbf{1}$. It follows that for some n, $[\mathcal{S}]_i^n = [\mathcal{A}]_n^j = \mathbf{1}$, and hence for the world $\Delta_n \in \mathcal{H}$, $\Gamma_j \mathcal{A} \Delta_n$ and $\Delta_i \mathcal{S} \Delta_n$, which is what was to be shown. ∎

Before extending this Theorem to the multi-modal setting, a small but useful detour is needed. We need a notion of scalar multiplication for Boolean matrices.

DEFINITION 25 If $b \in \mathcal{B}$ and M is a matrix, by bM I mean the matrix such that $[bM]_i^j = b \wedge [M]_i^j$. That is, in bM each component of M has been replaced by its meet with b.

Using scalar multiplication, there is a kind of normal form for matrices over $\mathcal{B}$. Recall, $\mathcal{B}$ is atomic and complete; take $\{a_1, a_2, \ldots\}$ to be the set of atoms. If $A \in \mathcal{B}_{m,n}$, there are matrices $A_{a_1}, A_{a_2}, \ldots$, whose entries are all in $\{\mathbf{0}, \mathbf{1}\}$, such that $A = a_1 A_{a_1} \vee a_2 A_{a_2} \vee \ldots$. Rather than a formal proof of this, an example should suffice. Consider the following matrix from Section 4, where the Boolean algebra is all subsets of $\{1, 2, 3\}$ and the atoms are $\{1\}$, $\{2\}$, and $\{3\}$. The matrix is:

$$\begin{bmatrix} \{1\} & \{1,3\} \\ \emptyset & \{2,3\} \end{bmatrix}$$

This can be written as follows.

$$\{1\} \begin{bmatrix} 1 & 1 \\ 0 & 0 \end{bmatrix} \vee \{2\} \begin{bmatrix} 0 & 0 \\ 0 & 1 \end{bmatrix} \vee \{3\} \begin{bmatrix} 0 & 1 \\ 0 & 1 \end{bmatrix}$$

With normal forms available, we can move to the main item—the multi-modal case.

THEOREM 26 *Let $\mathcal{K}$ be finite or countable. Let $\mathcal{F}_\mathcal{G} = \langle \mathcal{G}, \mathcal{R}_i : i \in \mathcal{K} \rangle$ and $\mathcal{F}_\mathcal{H} = \langle \mathcal{H}, \mathcal{S}_i : i \in \mathcal{K} \rangle$ be two $\mathcal{K}$-frames, and let $\mathcal{A} = \langle \mathcal{A}_i : i \in \mathcal{K} \rangle$ be a family of relations between $\mathcal{G}$ and $\mathcal{H}$. Assume $\mathcal{G} = \{\Gamma_1, \Gamma_2, \ldots\}$ and $\mathcal{H} = \{\Delta_1, \Delta_2, \ldots\}$. Let $\mathcal{B}$ be the powerset Boolean algebra whose elements are the subsets of $\mathcal{K}$. Let R be the $\mathcal{B}$-valued matrix with $[R]_i^j = \{k \in \mathcal{K} \mid \Gamma_i \mathcal{R}_k \Gamma_j\}$, and similarly let S be the matrix corresponding to $\mathcal{S}$. Finally let A be the $\mathcal{B}$-valued matrix such that $[A]_i^j = \{k \in \mathcal{K} \mid \Gamma_j \mathcal{A}_k \Delta_i\}$.*

$\mathcal{A}$ is a frame bisimulation between $\mathcal{F}_\mathcal{G}$ and $\mathcal{F}_\mathcal{H}$ iff A is a Boolean bisimulation from R to S.

Proof. By Theorem 2, $\mathcal{A}$ is a frame bisimulation between $\mathcal{F}_\mathcal{G}$ and $\mathcal{F}_\mathcal{H}$ if and only if $\mathcal{A}_k$ is a mono-modal frame bisimulation between $\langle \mathcal{G}, \mathcal{R}_k \rangle$ and $\langle \mathcal{H}, \mathcal{S}_k \rangle$, for each $k \in \mathcal{K}$. Let R_k be the transition matrix corresponding to $\mathcal{R}_k$, with entries from $\{\mathbf{0}, \mathbf{1}\}$, let S_k similarly correspond to $\mathcal{S}_k$, and A_k correspond to $\mathcal{A}_k$. Then, by Theorem 24, $\mathcal{A}$ is a frame bisimulation if and only if $A_k R_k \leq S_k A_k$ and $A_k^T S_k \leq R_k A_k^T$, for each $k \in \mathcal{K}$. It remains to show these families of inequalities are equivalent to $AR \leq SA$ and $A^T S \leq R A^T$.

It is easy to see that the normal forms for R, S, and A are:

$$R = \bigvee_{k \in \mathcal{K}} \{k\} R_k \text{ and } S = \bigvee_{k \in \mathcal{K}} \{k\} S_k \text{ and } A = \bigvee_{k \in \mathcal{K}} \{k\} A_k$$

Now, $AR \leq SA$ if and only if

$$\bigvee_{j \in \mathcal{K}} \{j\} A_j \bigvee_{k \in \mathcal{K}} \{k\} R_k \leq \bigvee_{j \in \mathcal{K}} \{j\} S_j \bigvee_{k \in \mathcal{K}} \{k\} A_k$$

Making use of distributivity, this can be shown equivalent to

$$\bigvee_{j,k \in \mathcal{K}} (\{j\} \wedge \{k\}) A_j R_k \leq \bigvee_{j,k \in \mathcal{K}} (\{j\} \wedge \{k\}) S_j A_k$$

This in turn is equivalent to

$$\bigvee_{k \in \mathcal{K}} \{k\} A_k R_k \leq \bigvee_{k \in \mathcal{K}} \{k\} S_k A_k$$

Finally, this is equivalent to

$$A_k R_k \leq S_k A_k \text{ for each } k \in \mathcal{K}$$

In a similar way, $A^T S \leq R A^T$ is equivalent to $A_k^T S_k \leq R_k A_k^T$ for every $k \in \mathcal{K}$. These equivalences complete the proof. ∎

14 The modal operator

Suppose we have two Boolean valued frames and a bisimulation between them. Then various results of earlier sections ensure that stable vectors are well-behaved with respect to $\wedge$, $\vee$, and stable negation. But what about $\Diamond$? Its application can turn stable vectors into non-stable ones, as the following shows.

EXAMPLE 27 The diagram below shows two mono-modal Kripke frames, one on the left and one on the right, with a relation between them which is, in fact, a standard frame bisimulation.

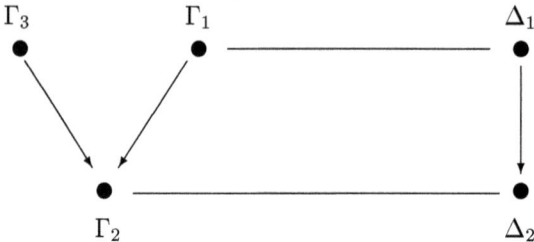

Suppose, in the left frame, we assign the atomic formula P to be true just at Γ_2 and in the right frame just at Δ_2. This meets the conditions of Definition 3 for bisimulation—the worlds at which P is true are related by the frame bisimulation. But, in the left frame $\Diamond P$ will be true at both Γ_1 and Γ_3, and in the right at Δ_1, but only Γ_1 and Δ_1 are related by the bisimulation. This is awkward, but not a serious problem—Γ_3 does not take part in the bisimulation so we can, in effect, ignore it. The question is, how to do that in a way consistent with the theory developed so far.

To make better connection with the present paper, I'll convert this example to an algebraic one. Let R be the transition matrix corresponding to the accessibility relation of the left frame above, and let S correspond to the right frame accessibility relation. Also, let A represent the bisimulation. Then we have the following.

$$R = \begin{bmatrix} 0 & 1 & 0 \\ 0 & 0 & 0 \\ 0 & 1 & 0 \end{bmatrix} \text{ and } S = \begin{bmatrix} 0 & 1 \\ 0 & 0 \end{bmatrix} \text{ and } A = \begin{bmatrix} 1 & 0 & 0 \\ 0 & 1 & 0 \end{bmatrix}$$

A is a Boolean bisimulation from R to S; in fact $AR = SA$, while $A^T S < RA^T$. Above we took P to be true in the left frame just at Γ_2. Algebraically,

P is assigned the vector (proposition) $V = \langle 0, 1, 0 \rangle$. It is easy to check that this is A-stable. But,

$$RV = \begin{bmatrix} 0 & 1 & 0 \\ 0 & 0 & 0 \\ 0 & 1 & 0 \end{bmatrix} \begin{bmatrix} 0 \\ 1 \\ 0 \end{bmatrix} = \begin{bmatrix} 1 \\ 0 \\ 1 \end{bmatrix}$$

and $\langle 1, 0, 1 \rangle$ is *not* A-stable.

In Example 27, Γ_3 is not really relevant to the bisimulation, so what we want to do is eliminate it from consideration. Our solution parallels the treatment of negation in Section 10—instead of RV, we work with the largest stable matrix below it. And this has the same familiar form it did when negation was involved.

Section assumption For the rest of this section, $A \in \mathcal{B}_{m,n}$ so that for each k, $A : \mathcal{B}_{n,k} \to \mathcal{B}_{m,k}$ and $A^T : \mathcal{B}_{m,k} \to \mathcal{B}_{n,k}$. Further, I'll assume A is a Boolean bisimulation from $R \in \mathcal{B}_{n,n}$ to $S \in \mathcal{B}_{m,m}$.

DEFINITION 28 $R \circ V = RV \wedge A^T A \mathbf{1}_{n,k}$.

THEOREM 29 *Let V be A-stable. Then*

1. $R \circ V = A^T A(RV)$

2. $A(R \circ V) = A(RV)$

3. $A(R \circ V) = S \circ (AV)$

4. $R \circ V$ *is A-stable*

5. $R \circ V$ *is the largest matrix W such that W is A-stable and $W \leq RV$*

Proof. Since A is a Boolean bisimulation and V is A-stable, $A^T A(RV) \leq A^T SAV \leq RA^T AV = V$. Of course $A^T ARV \leq A^T A \mathbf{1}_{n,k}$, and so $A^T ARV \leq RV \wedge A^T A \mathbf{1}_{n,k}$. We also have $RV \wedge A^T A \mathbf{1}_{n,k} \leq A^T ARV$ by Lemma 15. Combining things, we have item 1.

Using item 1, $A(R \circ V) = AA^T ARV \geq A(RV)$ by Theorem 8. Also, $A(R \circ V) = A(RV \wedge A^T A \mathbf{1}_{n,k}) \leq A(RV)$, and we have item 2.

Using items 1 and 2 (including their counterparts for S) and the fact that A is a Boolean bisimulation, $A(R \circ V) = AA^T ARV \leq AA^T SAV = S \circ (AV)$. Also using the fact that V is stable, $S \circ (AV) = AA^T SAV \leq ARA^T AV = ARV = A(R \circ V)$. These give us item 3.

$A^T A(R \circ V) = A^T A(RV \wedge A^T A\mathbf{1}_{n,k}) \leq A^T A(RV) = R \circ V$. Also, $R \circ V = A^T A(RV) \leq A^T A A^T A R V = A^T A(R \circ V)$ by Theorem 8 again. So we have item 4.

It is trivial that $R \circ V \leq RV$. And, if W is A-stable and $W \leq RV$, then $W = A^T AW \leq A^T ARV = R \circ V$, so we have item 5. ∎

Note that part 2 of this theorem says that, with respect to bisimulation, $R \circ V$ and RV behave alike. This is analogous to a similar result concerning stable negation: $A\overline{V} = A(\neg V)$, which was established earlier.

15 Bisimulations, models, and formulas

The definition of Boolean bisimulation corresponds to *frame* bisimulation. To turn a frame bisimulation into a bisimulation, as in Definition 3, models must be based on the frames involved—that is, assignments of truth values to atomic formulas at worlds must be given. But not just any assignment will do, since the bisimulation relation must be respected. In the usual two-valued, mono-modal setting, worlds related by a bisimulation must make the same atomic formulas true, and an analogous condition is needed in the Boolean valued case too.

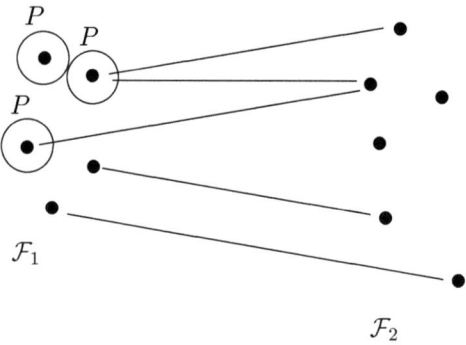

Figure 4. Bisimulations and models

Consider the example displayed in Figure 4, where two mono-modal frames, $\mathcal{F}_1$ and $\mathcal{F}_2$, and a frame bisimulation are indicated (accessibility relations have not been shown). Suppose we want to make the atomic formula P true at the worlds indicated in frame $\mathcal{F}_1$. Then we are required to have P true at the worlds related to them in $\mathcal{F}_2$ (there are two such worlds in this example), and in turn worlds related to these in $\mathcal{F}_1$ must have P

true, that is, they must be part of the original set of worlds where P was to be true. Similar considerations apply to more general Boolean valued cases as well. All this corresponds to an easily expressible condition on vectors. We must require that values assigned to atomic formulas be what I will call *weakly stable*.

DEFINITION 30 V is *weakly stable* (with respect to A) if $A^T AV \leq V$.

Weak stability is too weak to have attractive algebraic features on its own. But note again the example in Figure 4. The set of worlds at which P is true in $\mathcal{F}_1$ is weakly stable, but it is not stable. One of the three worlds is, in an obvious sense, irrelevant to the bisimulation. But if we shift our attention to the set of worlds in $\mathcal{F}_2$ that are related to these three, we have a two-element set that is, in fact, not just weakly stable, but stable. Indeed, this always happens. Suppose V is weakly stable. Then $A^T AV \leq V$, and so $AA^T AV \leq AV$. But also $AV \leq AA^T AV$ by Theorem 8, and so $AA^T AV = AV$, that is, AV is stable. Since we are interested in behavior under bisimulation, and since weak stability turns into stability after one shift from a model to another, we simplify things by requiring stability from the start.

From now on, given two Boolean valued frames and a bisimulation between them, in basing models on these frames, *we require that the values assigned to atomic formulas must be stable vectors*.

Finally, it is time to return to the original reason for introducing bisimulations into modal logic—they preserve formula truth. In an algebraic setting this becomes especially simple, as the formulation in this section will show. For the rest of this section I assume the following.

1. R and S are (transition) matrices, with $R \in \mathcal{B}_{n,n}$ and $S \in \mathcal{B}_{m,m}$.

2. $A \in \mathcal{B}_{m,n}$ is a Boolean bisimulation from R to S.

L is a mono-modal language, but as we have seen, since $\mathcal{B}$ can be a more complex Boolean algebra than $\{\mathbf{0}, \mathbf{1}\}$, we have the effect of a multi-modal language and semantics.

Let v be a mapping from propositional letters of L to A-*stable* vectors in $\mathcal{B}^n$, and let w likewise be a mapping from propositional letters of L to A^T-*stable* vectors in $\mathcal{B}^m$. ($\mathcal{B}^n$ is identified with $\mathcal{B}_{n,1}$ and $\mathcal{B}^m$ with $\mathcal{B}_{m,1}$, as usual.) Both v and w are extended to mappings from arbitrary formulas in L as follows.

1. $v(X \wedge Y) = v(X) \wedge v(Y)$
   $w(X \wedge Y) = w(X) \wedge w(Y)$

2. $v(X \vee Y) = v(X) \vee v(Y)$
   $w(X \vee Y) = w(X) \vee w(Y)$

3. $v(\neg X) = \overline{v(X)}$ (using stable negation in $\mathcal{B}^n$)
   $w(\neg X) = \overline{w(X)}$ (using stable negation in $\mathcal{B}^m$)

4. $v(\Diamond X) = R \circ v(X)$
   $w(\Diamond X) = S \circ w(X)$

Using results from Theorems 13, 16, 21, and 29, for any formula X, $v(X)$ is A-stable and $w(X)$ is A^T-stable. And further, we have a central result on bisimulations, in the following form.

THEOREM 31 *If $A(v(P)) = w(P)$ and $A^T(w(P)) = v(P)$ for propositional letters P of L, then for any formula X, $A(v(X)) = w(X)$ and $A^T(w(X)) = v(X)$.*

16 Conclusion

An algebraic approach to modal semantics, and bisimulation in particular, has been presented. It can be carried further, but details must remain for another paper. I'll sketch a few items, to illustrate the possibilities.

There is a special class of bisimulations known as *P-morphisms* that, historically, were investigated before the more general notion of bisimulation was introduced. These can be characterized easily in the algebraic setting: a matrix A is a *P-morphism onto* if $I_n = A^T A$ and $I_m \leq AA^T$. The expected results follow easily from this characterization.

It is possible to 'multiply' frames, by forming Cartesian products. The algebraic counterpart of this is the tensor product, a standard operation in other contexts. It relates quite well to Boolean bisimulations. In a similar way, the disjoint union operation on frames corresponds to the notion of direct sum, familiar from linear algebra. That is to say, we are seeing operations that are long-known and well-understood, but in a less familiar context.

Bisimulations also arise in other contexts, of course. One such place is automata theory. For instance, to show that the usual algorithm for converting a non-deterministic automaton to a deterministic version is correct, one essentially shows the resulting automaton bisimulates the original one. For some time Dexter Kozen has been developing an approach to automata theory that makes central use of matrices, in ways that are strikingly similar to what was presented here. An extensive set of notes presenting this work can be found on his web site, www.cs.cornell.edu/kozen/, under the heading *CS786 S02 Introduction to Kleene Algebra*.

It is not unreasonable to hope that by looking at modal machinery from an algebraic point of view we will achieve additional insight and understanding, based on the work of generations of mathematicians who have gone before.

Acknowledgements

Research partly supported by PSC-CUNY Grant 64393-00 33. I want to thank an anonymous referee for making suggestions that improved the contents of this paper.

BIBLIOGRAPHY

[1] Patrick Blackburn, Maarten de Rijke, and Yde Venema. *Modal Logic*. Tracts in Theoretical Computer Science. Cambridge University Press, Cambridge, UK, 2001.
[2] Melvin C. Fitting. Many-valued modal logics. *Fundamenta Informaticae*, 15:235–254, 1991.
[3] Melvin C. Fitting. Many-valued modal logics, II. *Fundamenta Informaticae*, 17:55–73, 1992.
[4] Melvin C. Fitting. Many-valued non-monotonic modal logics. In Anil Nerode and Mikhail Taitslin, editors, *Logical Foundations of Computer Science — Tver '92*, pages 139–150. Springer Lecture Notes in Computer Science, 620, 1992.
[5] Melvin C. Fitting. Tableaus for many-valued modal logic. *Studia Logica*, 55:63–87, 1995.
[6] Ki Hang Kim. *Boolean Matrix Theory and Applications*. Marcel Dekker, 1982.
[7] J. Donald Monk and Robert Bonnet, editors. *Handbook of Boolean Algebras*. North-Holland, 1989. (Three volumes).

Melvin Fitting
Department of Mathematics and Computer Science
Lehman College
250 Bedford Park Boulevard West
Bronx, NY 10468-1589, USA
E-mail: fitting@lehman.cuny.edu

7
A Sound and Complete Proof System for QPTL

TIM FRENCH AND MARK REYNOLDS

ABSTRACT. In this paper we give sound and complete proof systems for the useful and expressive *quantified propositional temporal logic*, both with and without past temporal operators. Until now an axiomatization has only existed for the version with the inclusion of the past operators and this axiomatization relied on the past operators in subtle and complicated ways. In certain situations, such as in branching time extensions of the linear time logic, it is important to avoid using the past time operators. Our completeness proof proceeds mostly by using deterministic Rabin automata but the main step is via a new and interesting proof of correctness for an optimal complementation procedure for nondeterministic Büchi automata.

1 Introduction

Quantified propositional temporal logic, QPTL [16], extends traditional propositional temporal logic, PLTL [12], and has the ability to quantify over interpretations of the propositional atoms. It is a logic used for reasoning about structures with atoms whose truth values vary along a discrete, linearly ordered flow of time. The set of formulas of QPTL is built up recursively from the atoms via classical negation and disjunction, temporal next-time and future-always connectives (from the standard PLTL) and quantification over atoms. We also have $\forall x \alpha$ being true at a moment of time in a structure if and only if all other structures which differ only in the interpretation of the atom x make α true at that moment.

Standard PLTL is often (see for example [19]) extended to a logic which we here call PLTLP (in view of a lack of standard terminology) by the addition of temporal operators which refer to the past sequence of states leading up to the point of evaluation (in addition to the usual future time operators). A complete axiomatization is given in [5] (see also [6]) for an extension of QPTL that involves these past-time operators. We shall call this extension QPTLP although it is called QPTL in [5]. The sketch of the completeness proof given in that paper shows that for every non-deterministic Büchi automaton, A, there is a formula of QPTLP, χ_A that is satisfied by exactly the ω-sequences that A accepts and that every formula of QPTLP is equivalent to a non-deterministic Büchi automaton. It should be noted that the proof makes essential use of the past-time operators and restricting the proof system to only the future axioms does not give a complete axiomatization for QPTL.

In our paper we produce a complete axiomatization for pure (future only) QPTL that does not rely on past operators. This is important in various circumstances. For example, when the linear time logic is extended to a branching time version then there are many complications with the semantics for the past time operators [2]. In any case, the past-time operators can, in a certain useful way, be easily defined in terms of the future only operators and quantification.

Our axiom system is a very simple extension of the traditional axiom system [4] for PLTL via a natural induction rule. The completeness proof is also considerably simpler (albeit a little too long to appear in full here). We use deterministic Rabin automata to show that every QPTL formula is provably equivalent to an existentially quantified formula (a Σ_1 formula). Completeness follows from the equivalence between satisfiability and consistency in PLTL and QPTL.

We also show that we obtain a sound and complete axiomatization for QPTLP by the simple addition of the standard PLTLP axioms for the past time operators. We consider the expressiveness of QPTLP versus QPTL: are the past operators important when we do not need them for our completeness proof?

There are important applications of QPTL in verification of complex systems because of the two main recognized advantages of quantification: extra expressivity and the ability to reason about refinement relations between programs [9]. There are well known inadequacies in the expressiveness of PLTL [17] and it is also well known [15] that QPTL is one of several ways of achieving an improvement to reach the important standard of expressiveness of automata and the second order monadic logic of one successor (S1S). If one program produces the same observable behaviour as another but perhaps uses extra variables with perhaps some extra computation steps in between the observed steps then we say that the second program *refines* or *implements* the first. As shown in [10], refinement can be defined via the existence of temporal propositions which satisfy certain temporal properties, i.e. via a formula of QPTLP. There are also important applications of propositional quantification in work on logics of knowledge [1] where we want to quantify over propositions which are local to a particular agent in the sense that their truth value only depends on the state of the agent.

We define the logics in the next section before presenting our proof system. In section 5 we introduce automata and in section 6 we show that every QPTL formula is provably equivalent to a Σ_1 formula. The next section sketches the remainder of the completeness proof: it will appear in full in a longer version of this paper later. We conclude with some discussion of extensions of the results.

2 Syntax and semantics

The language QPTL consists of an infinite set of atomic variables $\mathcal{V} = \{x_0, x_1, \ldots\}$, the boolean operations $\neg, \vee$, (**not** and **or** respectively) and the future temporal operators $\bigcirc, \square$ (**next** and **generally** respectively), along with the quantifier $\forall$ (**for all**). Later in the paper we will also consider an extended version of the language, QPTLP that also includes the past temporal operators $\ominus, \boxminus$ (**weak previous** and **past generally** respectively). We will use the convention that all unary operators (including $\forall$) bind weaker than binary operators, unless otherwise indicated by brackets. The set of formulas of QPTL contains x, for all $x \in \mathcal{V}$ and is closed under the usual temporal operators $\neg, \vee, \bigcirc,$ and $\square$ as well as containing $\forall x \alpha$ for all $x \in \mathcal{V}$.

We will next present the definition of a structure σ that evaluates a formula in the language QPTL. A structure $\sigma = (\sigma_0, \sigma_1, \ldots)$ is a countably infinite sequence of elements of $\wp(\mathcal{V})$, (for each $i \in \mathbb{N}$, $\sigma_i \subseteq \mathcal{V}$). We say $\sigma' = (\sigma'_0, \sigma'_1, \ldots)$ is an X-variant of σ for some finite $X \subset \mathcal{V}$, if $\forall i\ \sigma_i \backslash X = \sigma'_i \backslash X$. For the singleton $\{x\}$, we refer to an x-variant, rather than an $\{x\}$-variant. Given some structure σ we define the jth moment of σ be the tuple (σ, j) for $j \in \mathbb{N}$ and inductively define a formula α to "hold" at the jth moment of σ (denoted $(\sigma, j) \models \alpha$) as follows.

$(\sigma, j) \models x \Leftrightarrow x \in \sigma_j$, for all $x \in \mathcal{V}$. $\quad (\sigma, j) \models \neg \alpha \Leftrightarrow (\sigma, j) \not\models \alpha$.
$(\sigma, j) \models \alpha_1 \vee \alpha_2 \Leftrightarrow$ either $(\sigma, j) \models \alpha_1$ or $(\sigma, j) \models \alpha_2$.
$(\sigma, j) \models \bigcirc \alpha \Leftrightarrow (\sigma, j+1) \models \alpha$. $\quad (\sigma, j) \models \square \alpha \Leftrightarrow \forall k \geq j\ (\sigma, k) \models \alpha$.
$(\sigma, j) \models \forall x \alpha \Leftrightarrow (\sigma', j) \models \alpha$, for all x–variants σ' of σ.

We say σ is an initial model of α (written $\sigma \models \alpha$) if $(\sigma, 0) \models \alpha$. If for every structure σ and for every $i \in \mathbb{N}$ we have $(\sigma, i) \models \alpha$, we say α is valid (written $\models \alpha$), and if $\neg \alpha$ is not valid, we say α is satisfiable.

We complete this section by defining additional operators in terms of those already defined. The propositional operators $\wedge, \rightarrow, \leftrightarrow$ are well known to be expressible in terms of $\neg$ and $\vee$. We will also consider the formulas $\top, \bot$ (respectively "true" and "false") to be abbreviations for, respectively $(x_0 \vee \neg x_0)$ and $\neg(x_0 \vee \neg x_0)$. Given the sets of propositions $X = \{x_0, \ldots, x_m\}$ and $Y = \{y_0, \ldots, y_m\}$ we use the abbreviations $\exists X$ and $X \leftrightarrow Y$ for $\exists x_0 \ldots \exists x_m$ and $\bigwedge_{i=0}^{m} x_i \leftrightarrow y_i$ respectively. For $i \in \mathbb{N}$, and any formula α we define $\bigcirc^i \alpha$ inductively by $\bigcirc^0 \alpha = \alpha$ and $\bigcirc^{i+1} \alpha = \bigcirc \bigcirc^i \alpha$. We will also use the abbreviation $\bigcirc^{\leq n}\alpha$ for $\bigwedge_{i=0}^{n}\bigcirc^{i}\alpha$. The other abbreviations in QPTL are:

$$\exists x\alpha = \neg\forall x\neg\alpha \qquad \qquad \Diamond\alpha = \neg\Box\neg\alpha$$
$$\alpha\mathcal{W}\beta = \exists u(u \wedge \Box(u \rightarrow ((\alpha \wedge \bigcirc u) \vee \beta))) \qquad \alpha\mathcal{U}\beta = \Diamond\beta \wedge (\alpha\mathcal{W}\beta)$$

where $u = x_i$ for the least i such that x_i does not appear in α or β. These operators will be referred to as, respectively: existential quantification, future, waiting, and until. Intuitively, $\Diamond\alpha$ is true if α is true at some future moment, and $\alpha\mathcal{U}\beta$ is true if α is true for all moments until β is true.

The above syntax and semantics without quantification define a language we will refer to as PLTL. Full PLTL also includes the $\mathcal{U}$ operator, but this is not necessary in QPTL.

3 The proof system

In this section we will give the axioms and rules of inference for QPTL. We will also derive some basic theorems using the system, which will be used in the proof of completeness. We let $var(\alpha)$ be set of variables occurring in α. If α and β are formulas of QPTL, then we define the substitution $\alpha[\beta\backslash x]$ to be the formula α with all occurrences of x replaced by the sub-formula β. We say x is *free for* β in α if

1. α does not contain a sub-formula of the form $\forall x\gamma$ and

2. for all $y \in var(\beta)$, if α contains a sub-formula $\forall y\gamma$ then $x \notin var(\gamma)$.

We now give the axiom system for QPTL which we will denote **QF**. It is a direct extension of the axiom system **DX** which was given in [4] and was shown to be sound and complete for PLTL (without the $\mathcal{U}$ operator). The axioms are substitution instances of the following:

C0	any propositional tautology	C1	$\Box(\alpha \rightarrow \beta) \rightarrow (\Box\alpha \rightarrow \Box\beta)$
C2	$\Box\alpha \rightarrow (\alpha \wedge \bigcirc(\Box\alpha))$	C3	$\bigcirc\neg\alpha \leftrightarrow \neg\bigcirc\alpha$
C4	$\Box(\alpha \rightarrow \beta) \rightarrow (\bigcirc\alpha \rightarrow \bigcirc\beta)$	C5	$\Box(\alpha \rightarrow \bigcirc\alpha) \rightarrow (\alpha \rightarrow \Box\alpha)$
QX0	$\exists x(x \wedge \bigcirc\Box\neg x)$	QX1	$\forall x\bigcirc\alpha \leftrightarrow \bigcirc\forall x\alpha$
QX2	$\forall x\alpha \rightarrow \alpha[\beta\backslash x]$	where	x is free for β in α.

There are also four rules of inference, Modus Ponens and Temporal Generalization (from DX), ∀-Generalization (from [5]) and *Quantified Induction*, which is new. The first three rules are given as:

$$\text{MP } \frac{\alpha \rightarrow \beta,\ \alpha}{\beta} \qquad \text{T-Gen } \frac{\alpha}{\Box\alpha} \qquad \forall\text{-Gen } \frac{\alpha \rightarrow \beta}{\alpha \rightarrow \forall x\beta} \text{ when } x \notin var(\alpha).$$

To state the Quantified Induction rule we require the following definitions. Let $X = \{x_0, \ldots, x_m\}$ and $Y = \{y_0, \ldots, y_m\}$ be disjoint sets of variables. We say a formula β has *temporal depth k in X* if every occurrence of $x \in X$ in β does not occur in the scope of a $\square$ operator, or more than $k \bigcirc$ operators. The rule is:

$$\text{QI} \quad \frac{\beta \to \exists Y(\bigcirc^{\leq 1}(Y \leftrightarrow X) \wedge \bigcirc \beta[Y\backslash X])}{\beta \to \exists X \square \beta},$$

where β has temporal depth 1 in X and the variables of Y do not appear in β.

3.1 Some theorems

Using the rule ∀-Gen and the axiom QX2 we can generate the substitution rule. Since the axiom system **QF** directly extends the sound and complete axiom system **DX** [4], any validity of PLTL can be assumed to be provable in **QF**, and formulas of QPTL can be substituted for free variables. This will be referred to as *semantic validification (SV)*. In most cases we will leave the proof of semantic validity for such a formula to the reader, and denote such derivations as PLTL, in the proofs.

We also note that the axioms C0, QX2 and the rule ∀-Gen give a sound and complete axiom system for quantified propositional logic (QPL), so semantic validification also applies to any theorem not containing temporal operators (denoted as QPL in the proofs). We will now prove two theorems that involve both temporal operators and quantifiers. Note that we use $\vdash$ to mean provable in **QF**.

(1) $\square(\alpha \vee \beta) \to \exists z \square((\alpha \wedge z) \vee (\beta \wedge \neg z))$.

PLTL	$\vdash$	$(\alpha \vee \beta) \to ((\alpha \wedge \alpha) \vee (\beta \wedge \neg \alpha))$
T-Gen	$\vdash$	$\square((\alpha \vee \beta) \to ((\alpha \wedge \alpha) \vee (\beta \wedge \neg \alpha)))$
C1	$\vdash$	$\square(\alpha \vee \beta) \to \square((\alpha \wedge \alpha) \vee (\beta \wedge \neg \alpha))$
QX2	$\vdash$	$\square(\alpha \vee \beta) \to \exists z \square((\alpha \wedge z) \vee (\beta \wedge \neg z))$

(2) $\vdash \square \forall x \alpha \leftrightarrow \forall x \square \alpha$.

	($\longrightarrow$)				($\longleftarrow$)	
QX2	$\vdash$	$\forall x \alpha \to \alpha$		C2, SV	$\vdash$	$\forall x \square \alpha \to \forall x \bigcirc \square \alpha$
T-Gen	$\vdash$	$\square(\forall x \alpha \to \alpha)$		QX1	$\vdash$	$\forall x \square \alpha \to \bigcirc \forall x \square \alpha$
C1	$\vdash$	$\square \forall x \alpha \to \square \alpha$		C5	$\vdash$	$\forall x \square \alpha \to \square \forall x \square \alpha$
∀-Gen	$\vdash$	$\square \forall x \alpha \to \forall x \square \alpha$		C1, SV	$\vdash$	$\forall x \square \alpha \to \square \forall x \alpha$

We will now show how the axiom QX0 allows additional variables to be added to a proof. As an example we prove

(3) $\vdash \exists x(\neg x \wedge \bigcirc x \wedge \bigcirc\bigcirc\square\neg x)$

QX0, T-Gen, C2 $\vdash$ $\exists x(x \wedge \bigcirc\square\neg x) \wedge \bigcirc\exists y(y \wedge \bigcirc\square\neg y)$
QX1, PLTL, QPL $\vdash$ $\exists x \exists y(\neg(y \wedge \neg x) \wedge \bigcirc(y \wedge \neg x) \wedge \bigcirc\bigcirc\square\neg(y \wedge \neg x))$
QX2, PLTL, QPL $\vdash$ $\exists x(\neg x \wedge \bigcirc x \wedge \bigcirc\bigcirc\square\neg x)$

A similar proof can be found for all tautologies of the form

(4) $\exists x(\pm x \wedge \ldots \wedge \bigcirc^n \pm x \wedge \bigcirc^{n+1}\square\neg x)$.

We will refer to the proof of all such tautologies as *finite marking* (FM). This class of tautologies can be used to dictate the interpretation of a quantified variable over the finite future. We note that finite marking is not strong enough to prove tautologies of the form $\vdash \exists x \square(x \leftrightarrow \neg\bigcirc x)$. Such proofs require the new rule QI.

Unfortunately we have no room to prove the following tautology, that states that if there is some assignment of propositions that makes a formula true in the future, there is some assignment of propositions that makes that formula true at the earliest possible moment.

(5) $\vdash \diamondsuit\alpha(X) \to \exists X \forall Y((\neg\alpha(Y))\mathcal{U}\alpha(X)))$.

4 Proof of completeness

In this section we will prove that **QF** is sound for QPTL and we will state that it is complete for QPTL. The full details of the completeness proof stretch over most of the rest of the paper. However, here we are able to easily reduce completeness to a question about the expressiveness of the fragment which consists just of PLTL formulas under existential quantification. In fact, we are left only to show that each QPTL formula has some existentially quantified equivalent and that our proof system can establish this equivalence.

LEMMA 1 *The proof system* **QF** *is sound for QPTL.*

Proof. Since the semantics of QPTL directly extend the semantics of PLTL, the soundness of the axioms C0-C5 and the rules MP and T-Gen can be assumed as they are known to be sound for PLTL. The proof is given by induction over the complexity of proofs so we only have to show that this induction holds for the remaining two axioms and two rules. As this is trivial for both axioms and ∀-Gen, we need only show that QI is sound.

Suppose β has temporal depth 1 in X and

(6) $\beta \to \exists Y(\bigcirc^{\leq 1}(Y \leftrightarrow X) \wedge \bigcirc\beta[Y\backslash X])$

is valid. Therefore if $\sigma \models \beta$ then σ has an X-variant σ' such that $\sigma_0 = \sigma_0'$, $\sigma_1 = \sigma_1'$ and $(\sigma', 1) \models \beta(X)$. Since (6) is a validity we can repeat the argument to get an X-variant of $(\sigma', 1)$, say $(\sigma'', 1)$. We can prefix $(\sigma'', 1)$ with any set of variables, so by repeating this process we can construct the sequence $\sigma^* = (\sigma_0, \sigma_1', \sigma_2'', \ldots)$. since β has temporal depth 1 in X its interpretation at $(\sigma^*, 0)$ depends only on σ_0 and σ_1'. By construction $(\sigma^*, i) \models \beta$ for all i and σ^* is an X-variant of σ. Therefore $\sigma \models \exists Y \Box \beta[Y \backslash X]$ and we are done. ∎

We define a Σ_1 formula (or an existentially quantified formula) to be of the form $\exists X \alpha$ where α is a PLTL formula, and X is a finite set of propositions

LEMMA 2 (Main Lemma) *Every QPTL formula α is provably equivalent in* **QF** *to an existentially quantified formula.*

This will be proven in Sections 6 and 7 and most of what follows leads to this result.

LEMMA 3 *A PLTL formula α is satisfiable if and only if $\exists X \alpha$ is satisfiable.*

Proof. If α has a PLTL model σ then σ is clearly also a model of $\exists X \alpha$. Conversely if $\exists X \alpha$ has some model σ then there is an X-variant of σ that is a model of α. ∎

THEOREM 4 **QF** *is complete for QPTL.*

Proof. We know that the satisfiable formulas of PLTL are exactly the consistent ones [4], and hence an existentially quantified formula of QPTL is satisfiable if and only if it is consistent (Lemma 3). Since every formula α of QPTL is provably equivalent to an existentially quantified formula in **QF** (Lemma 2), the satisfiability of α is equivalent to the consistency of α and thus **QF** is complete. ∎

5 ω-automata

Our completeness proof, like that in [5], will make serious use of automata. We will use an induction on the complexity of formulas to show the equivalence of QPTL formulas with existentially quantified ones. The case of negation is difficult and we resort to complementation of automata to carry this through. In this section we provide the background definitions and some results relating automata to QPTL and our proof system.

It has been shown [15] that the language QPTLP is equivalent to the language of ω-automaton. We will use this equivalence to show that **QF** is complete. More specifically, we will represent deterministic Rabin (DR) and non-deterministic Büchi (NB) automaton in the language by way of a *characteristic formula*. In our proof we are not only required to know every formula of QPTL is equivalent to an ω-automaton, we also need to show that such an equivalence is a theorem of **QF**.

We will first establish some notation. An automaton is given by the tuple $(\Sigma, Q, q_0, \rho, \mathcal{C})$. The automaton acts on the alphabet, $\Sigma = \wp(\mathcal{L})$ for some finite $\mathcal{L} \subset \mathcal{V}$. The automaton will consist of a finite set of states, Q, and an initial state $q_0 \in Q$ is specified. The transition function, ρ is given by $\rho = \delta : Q \times \wp(\mathcal{L}) \longrightarrow Q$ (a deterministic transition function) or $\rho = \lambda : Q \times \wp(\mathcal{L}) \longrightarrow \wp(Q) \setminus \emptyset$ (a non-deterministic transition function).

We consider a run, $r = (r_0, r_1, \ldots)$, of an automaton over a structure σ to be a sequence of states taken from Q where $r_0 = q_0$ and $r_{i+1} = \delta(r_i, (\sigma_i \cap \mathcal{L}))$, (or $r_{i+1} \in \lambda(r_i, (\sigma_i \cap \mathcal{L}))$, in the case of nondeterministic automaton). We denote the set of states that occur infinitely often in r as $inf(r)$.

In the case of Rabin automata the acceptance condition, $\mathcal{C}$, is given as a set of pairs $\{(L_i, U_i) | i = 1, \ldots, k\}$ where $L_i, U_i \subseteq Q$. A run r satisfies the acceptance condition if for some i, $inf(r) \cap L_i \neq \emptyset$ and $inf(r) \cap U_i = \emptyset$. The Büchi acceptance condition is given as $\mathcal{C} = F \subseteq Q$ and a run r satisfies the condition if $inf(r) \cap F \neq \emptyset$.

We say a formula α of QPTL is congruent to some ω-automaton A when A accepts some structure σ if and only if $\sigma \models \alpha$. We state without proof that every formula of QPTL is congruent to some DR-automaton and some NB-automaton, [18]. However we will show by construction that every ω-automaton A is congruent to some QPTL formula χ_A, the characteristic formula of A.

Suppose the automaton A is given as above. We let $Q = \{q_0, \ldots, q_{n-1}\} \subset \mathcal{V} \setminus \mathcal{L}$. While the states and atomic variables representing the states have the same names, the intended meaning is always clear from the context. We define the formula at_q_i to be $q_i \wedge \neg \bigvee_{j \neq i} q_j$ and given a set of atoms $R = \{r_0, \ldots, r_{n-1}\}$ we let $at\_q_i(R)$ be an abbreviation for $at\_q_i[R \setminus Q]$. This notation also extends to the following formulas.

For each subset $P \subseteq Q$ we define the formula in_P to be $\bigvee_{q_i \in P} at\_q_i$. Finally for each subset $a \subseteq \mathcal{L}$ we define the formula $\rho(a)$ to be $\bigwedge_{x \in a} x \wedge \bigwedge_{x \in \mathcal{L} \setminus a} \neg x$. We can now put these formulas together to simulate a run of the DR-automaton A. We define

$$run_A \;:\; \Box \bigvee_{r \in Q}^{a \subseteq \mathcal{L}} (at\_r \wedge \rho(a) \wedge \bigcirc at\_\delta(r,a)) \tag{7}$$

$$acc_A \;:\; \bigvee_{1 \leq i \leq k} (\Box \Diamond in\_L_i \wedge \Diamond \Box \neg in\_U_i) \tag{8}$$

$$\chi_A \;:\; \exists Q(at\_q_0 \wedge run_A \wedge acc_A). \tag{9}$$

If A were a NB-automaton it is easy to see that a variation of χ_A would suffice as a characteristic formula. The proof that the characteristic formula are congruent to the automaton A is left to the reader. We can derive the following theorem that is analogous with the rule AA of [13].

LEMMA 5 *For every DR-automaton A we can have* $\vdash \exists Q(at\_q_0 \wedge run_A)$.

6 The Σ_1 normal form

The proof of completeness will have two main steps. We will show that every formula of QPTL is equivalent to an existentially quantified formula of PLTL. We do so by induction on the construction of α showing that $O\exists X\alpha$ is provably equivalent to a Σ_1 formula for any operator $O \vee, \exists, \Diamond$ and $\neg$ and each α from PLTL. The only complicated case is the operator $\neg$, which we will look at now.

To show that Σ_1 is closed under negation, let $\exists Y\alpha$ be some Σ_1 formula. We must show that $\neg \exists Y\alpha$ is provably equivalent to some Σ_1 formula. To do this we will create a series of automata where

- A is a DR-automaton corresponding to α
- B is a NB-automaton accepting exactly the sequences that A does, hence also corresponding to α.
- C is a NB-automaton accepting every Y-variant of the sequences that B does, hence corresponding to $\exists Y\alpha$.
- D is a NB-automaton accepting exactly the sequences that C does not. Hence corresponding to $\neg \exists Y\alpha$.

We show $\vdash \chi_C \leftrightarrow \exists Y\alpha$ and $\vdash \chi_D \leftrightarrow \neg \exists Y\alpha$. Since χ_D is a Σ_1 formula the proof will follow. Let A be a DR-automaton corresponding to α. The first step is easy:

LEMMA 6 $\vdash \chi_A \leftrightarrow \alpha$.

We now create the NB-automaton B that accepts exactly the sequences that A does. We define the automaton B to have the set of states $Q^0 \cup (Q^1 \backslash U_1^1) \cup \ldots \cup (Q^k \backslash U_k^k)$ where the Q^i etc are disjoint copies of Q. The transition relation is given as $\lambda^B(q_i^0, a) = \{\delta(q_i, a)^j | 0 \le j \le k\}$ and $\lambda^B(q_i^\ell, a) = \{\delta(q_i, a)^\ell\}$. The accepting set is $\bigcup_{0 < i \le k} L_i^i$. Essentially the automaton B guesses which accepting pair to use and when to use it. This is straightforward:

LEMMA 7 $\vdash \chi_B \leftrightarrow \alpha$.

We now construct a NB-automaton C that accepts exactly all Y-variants of any sequence that B accepts. The states, initial state and acceptance condition of C are identical to those of B. The only change is in the alphabet it reads and the transition function. The alphabet is $\mathcal{L}' = \mathcal{L} \backslash Y$ and we define

$$\lambda^C(r, a') = \bigcup\{\lambda^B(r, a) | a \cap \mathcal{L}' = a'\}.$$

Essentially a run of C is a run of B that ignores the interpretations of $y_0, \ldots, y_m$ and it is easy to see that the construction is both sound and complete. We state the result for $Y = \{y\}$ and the general result can be found by recursively applying this lemma (proof omitted).

LEMMA 8 $\vdash \chi_C \leftrightarrow \exists y \alpha$.

To derive the automaton D we require a complementation procedure for Büchi automata. This is the most complicated part of the proof and has been left to subsection 7.2. We state the lemma here.

LEMMA 9 $\vdash \neg \exists Y \alpha \leftrightarrow \chi_D$.

Assuming the proof of Lemma 9 we can now give the proof of the main lemma, (that every formula of QPTL is provably equivalent to a Σ_1 formula).

Proof. (the proof of Lemma 2)

We have mentioned already that we can establish the substitution rule for **QF** using QX2 (amongst other rules and axioms). This is particularly useful for this lemma.

The basic idea is to work within the fragment generated by $\vee, \neg, \bigcirc, \diamondsuit$ and $\exists x$. It is clear that every formula has an equivalent in this fragment and the equivalence can be shown in **QF**.

We work within this fragment because the result can then be shown by induction on construction of the (rewritten) formula and all but one case is straightforward.

The cases of atoms and formulas $\exists x\alpha$ are trivial. The case of $\vee$ follows as $\vdash ((\exists x\alpha) \vee \beta) \leftrightarrow \exists x(\alpha \vee \beta)$ which itself is established by substitution from semantic validification. The case of $\bigcirc$ can be shown via C3 and QX1. The case of $\Diamond$ follows from the proof of (2).

The only difficult case is that for negation and that is exactly what we have provided in Lemma 9: χ_D is a Σ_1 formula. ∎

7 Complementation

The complementation problem for non-deterministic Büchi automata has been studied extensively, and several different constructions have been given [15, 14, 7]. To prove the Lemma 9 we give a complementation construction for Büchi automata and prove the correctness of the construction in the proof system **QF**. This construction is effectively a refinement of the construction given by Nils Klarlund [7] and has a slightly better complexity, though both are optimal up to a constant factor in the exponent [11]. The proof of the correctness is significantly different to the one given by Klarlund, and easier to express in terms of proof system, **QF**.

The construction will be given in two steps. First we define a game for which the existence of a winning strategy for one player is equivalent to the automata being unable to accept some model. We then can derive a NB-automata that accepts a model if and only if the said player has a winning strategy. Thus it is the complement of the original automata.

7.1 The game

Given the NB-automaton $A = (\Sigma, Q, q_0, \lambda, F)$ we define the game G by the tuple $(A, N, \mathbb{C}, \Gamma)$ where

- A is the automaton we are seeking to complement.

- N is the linearly ordered set of labels $\{0, \ldots, n\}$ where $|Q| = n$.

- $\mathbb{C}$ is the set of colours $\{\mathbb{R}, \mathbb{G}\}$ (respectively *red* and *green*).

- $\Gamma : Q \times \mathbb{C} \times N^2 \longrightarrow \mathbb{C}$ is the colouring function, where

  – $\Gamma(q, c, i, j) = \mathbb{G}$ if $j > i$,
  – $\Gamma(q, c, i, j) = \mathbb{R}$ if $j \leq i$ and $q \in F$,
  – $\Gamma(q, c, i, j) = c$ otherwise.

The game is played by two players: $\mathcal{A}$, the *automaton* who is trying find a sequence of the states Q that satisfies the accepting condition of A; and

$\mathcal{B}$, the *complement* who is trying to show that $\mathcal{A}$ will not be able to satisfy the acceptance condition. The game is played over a *board* $\sigma \in \Sigma^\omega$.

We let $S = Q \times N \times \mathbb{C}$ be the set of *game states*, and define a play of the game G over the board σ to be an infinite sequence $s \in S^\omega$ where:

1. $s_0 = (q_0, 0, \mathbb{R})$

2. If $s_i = (q_i, j_i, c_i)$ and $s_{i+1} = (q_{i+1}, j_{i+1}, c_{i+1})$, then
    (a) $q_{i+1} \in \lambda(q_i, \sigma_i)$,
    (b) $j_{i+1} \in \{k \geq j_i | k \in N\}$,
    (c) $c_{i+1} = \Gamma(q_i, c_i, j_i, j_{i+1})$.

Throughout a play of the game we assume player $\mathcal{A}$ chooses q_{i+1} from the set $\lambda(q_i, \sigma_i)$ and player $\mathcal{B}$ chooses j_{i+1} from the set $\{k \geq j_i | K \in N\}$. Player $\mathcal{B}$ wins if for infinitely many i, $c_i = \mathbb{G}$, and player $\mathcal{A}$ wins otherwise.

We now define the notion of a winning strategy. Given a play s of a game G over σ we write $s_i = (p_i, j_i, c_i)$.

DEFINITION 10 A *strategy* for player $\mathcal{A}$ in the game G over σ is a map $\delta : S \times N \to Q$ where $\delta(s_i, i) \in \lambda(p_i, \sigma_i)$ for every s_i belonging to some play s such that for all $k \leq i$, $p_k = \delta(s_{k-1}, k-1)$.

A *strategy*, ξ, for player $\mathcal{B}$ is a map $\xi : S \times N \to \{0, \ldots, n\}$ where $\xi(s_i, i) \geq j_i$, for every s_i belonging to some play s such that for all $k \leq i$, $j_k = \xi(s_{k-1}, k-1)$.

A play s is played according the strategies δ and ξ, if for all i, $s_{i+1} = (\delta(s_i, i), \xi(s_i, i), e)$ where e is the colour determined by the game G. A *winning strategy* for a player is a strategy that will ensure the player wins, regardless of the moves of the opponent.

We note that a strategy can depend on σ_k for all k. It is clear that for any σ, $\mathcal{A}$ and $\mathcal{B}$ cannot both have a winning strategy, and it is a consequence of the proofs below that either $\mathcal{A}$ or $\mathcal{B}$ has a winning strategy.

We now show that player $\mathcal{A}$ has a winning strategy in the game G over σ if and only if A accepts σ, and A accepts σ if and only if player $\mathcal{B}$ does not have a winning strategy. To this we require the following definitions and lemmas.

Given the run r of A over σ let δ^r be the strategy defined by $\delta(s_i, i) = r_{i+1}$. It is easy to verify that this is a valid strategy.

LEMMA 11 *If r is an accepting run for A over σ then the strategy δ^r is a winning strategy for $\mathcal{A}$ in the game G over σ.*

DEFINITION 12 A run r of A over σ is *non-accepting at i* if there does not exist some run t with some $k > j > i$ such that $t_i = r_i$, $t_j \in F$ and $t_k = r_k$.

LEMMA 13 A does not accept σ if and only if for all runs r of A over σ, for all j there is some $i > j$ such that r is non-accepting at i.

Proof. ($\longleftarrow$) Suppose A accepts σ. Then there is an accepting run r of A over σ, and hence $\forall i, \exists k > i$ such that $r_k \in F$. Therefore we can let the run $t = r$ and then for all i we have $t_i = r_i$ and $t_j = r_j \in F$ for some $j > i$. Thus r is not non-accepting for any i and the proof follows by the contrapositive.

($\longrightarrow$) Suppose there is some run r of A over σ and some j such that for all $i > j$, r is not non-accepting at i. Then for all $i > j$ there exists some run t^i and some $j^i > i$ such that $t^i_i = r_i$, $t^i_{j^i} = r_{j^i}$ and $t^i_k \in F$ for some k where $i < k < j^i$.

We define the sequence $I = (i_0, i_1, \ldots)$ inductively by $i_0 = j$, and $i_{m+1} = j^{i_m}$. We can now define an accepting run u, of A over σ that is accepting with respect to ϕ as:

$$u_k = r_k \quad \text{for } k < i_0$$
$$u_k = t^{i_n}_k \quad \text{for } i_n \leq k < j^{i_n}, \text{for all } n$$
$$u_k = r_k \quad \text{for } j^{i_n} \leq k < i_n, \text{for all } n$$

Since r and t^{i_n} are valid runs, and $t^i_i = r_i$, $t^i_{j^i} = r_{j^i}$, it follows that u is a valid run. Furthermore, from the construction of u it follows that for all m, there is some k such that $i_m < k < i_{m+1}$ and $u_k \in F$, and thus u is an accepting run. Therefore A accepts σ and we are done. ∎

We can now define the strategy ξ for player $\mathcal{B}$ as

1. $\xi(s, 0) = 0$ for all $s \in S$.

2. if $i > 0$ and $s = (p, j, c)$ then

$$\xi(s, i) = j + 1 \quad \text{if } c = \mathbb{R},\ j < n, \text{ and } p = r_i$$
$$ \text{for some run } r, \text{ non-accepting at } i,$$
$$\xi(s, i) = j \quad \text{otherwise.}$$

LEMMA 14 Given the strategy ξ, for any run r of A over σ if for all j, there exists $i > j$ such that r is non-accepting at i, then ξ will win against the strategy δ^r in the game G.

Proof. Let $s = (p_i, j_i, c_i)_{i \in \mathbb{N}}$ be a play of G according to the strategies ξ and δ^r, where r is non-accepting infinitely often. Suppose for some i, $c_i = \mathbb{R}$. Then there must be some least $k > i$ such that r_k is non-accepting at k. According to the strategy ξ, $c_k = \mathbb{R}$ and provided $j_k < n$, $c_{k+1} = \mathbb{G}$.

Thus, to show that player $\mathcal{B}$ wins we must simply show that for all i, $j_i < n$. Suppose, for contradiction, that $j_i = n$. Then there must be n indices $k_1 < k_2 < \ldots < k_n$ where for $a \leq n$, $r_{k_a} = r^a_{k_a}$ for some run r^a that is non-accepting at k_a and $c_{k_a} = \mathbb{R}$. For all $a < n$ we have $c_{k_a+1} = \mathbb{G}$ and $c_{k_a+1} = \mathbb{R}$, and thus there is some m where $k_a < m < k_{a+1}$ and $r_m \in F$. Since r^a is non-accepting at k_a then for all $\ell > m$, for all runs t with $t_m = r_m$ we have $r^a_\ell \neq t_\ell$, and $r^a_\ell \notin F$.

Thus if $j_i = n$ then there are $n-1$ runs $r^1, \ldots, r^{n-1}$, such that for all $i > k_{n-1}$, $r^a_i = r^b_i$ if and only if $a = b$. However since $|Q| = n$, $|F| > 0$ and $r^a_i \notin F$ it must be that $|F| = 1$ and for all $\ell > k_{n-1}$ we must have $r_\ell \in F$, contradicting the fact that r is non-accepting infinitely often. ∎

7.2 The automaton

Given the game $G = (A, N, \mathbb{C}, \Gamma)$, we now construct an automaton $\overline{A}$, where a run of $\overline{A}$ over σ corresponds to a set of plays of G over σ. We will show that there is an accepting run of $\overline{A}$ over σ if and only if player $\mathcal{B}$ has a winning strategy for G over σ, and hence $\overline{A}$ is a complement of A.

The states of the automaton $\overline{A}$ are maps from the set of states Q (of the automaton A) to a set of labels, $\{0, \ldots, n\} \times \mathbb{C}$ where $\mathbb{C}$ is the set of colours $\mathbb{C} = \{\mathbb{R}, \mathbb{A}, \mathbb{G}\}$ (*red, amber* and *green*). We define an ordering, $<$, over the labels where $(j, c) < (j', c')$ if

1. $j < j'$ or,
2. $j = j'$ and $c = \mathbb{G} \neq c'$ or
3. $j = j'$ and $c = \mathbb{A}$ and $c' = \mathbb{R}$.

We define the automata $\overline{A} = (\overline{Q}, \phi_0, \overline{\lambda}, \overline{F})$ where

- $\overline{Q} = \{\phi | \phi : Q \to [0, n] \times \mathbb{C}\}$ where $|Q| = n$.
- $\phi_0(q_0) = (0, \mathbb{R})$ and $\phi_0(q) = (0, \mathbb{G})$ for $q \neq q_0$.
- $\overline{F} = \{\phi | \phi(Q) \subseteq [0, n] \times \{\mathbb{A}, \mathbb{G}\}\}$

and for each $\phi \in \overline{Q}$ and $a \in \mathcal{L}$ we define the set $\overline{\lambda}(\phi, a) \subset \overline{Q}$ by $\psi \in \overline{\lambda}(\phi, a)$ if and only if for all $p, q \in Q$ with $q \in \lambda(p, a)$:

1. $\phi(p) \leq \psi(q)$;

2. if $\phi(p) \in [1,n] \times \{\mathbb{G}\}$ and $p \in F$ then $\phi(p) < \psi(q)$;

3. if $\phi(p) \in [1,n] \times \{\mathbb{A}\}$ and $\phi \in \overline{F}$ then $\phi(p) < \psi(q)$.

LEMMA 15 *If θ is an accepting run of $\overline{A}$ over σ then player $\mathcal{B}$ has a winning strategy over σ.*

Proof. We give this proof by constructing the winning strategy δ^* from the run θ. Let $s_i = (p_i, j_i, c_i)$. Then

$$\delta^*(s_i, i) = k, \text{ where } \theta_i(p_i) = (k, c^*) \text{ for some } c^* \in \mathbb{C}.$$

To show that δ^* is a valid strategy we have to show that $\delta^*(s_i, i) \geq j_i$ for all $i > 0$ in every play. By the definition of δ^* we have $\theta_{i-1}(p_{i-1}) = (j_i, c')$, for some c', and we also know that $p_i \in \lambda(p_{i-1}, \sigma_{i-1})$. It follows from the definition of $\overline{\lambda}$ that $\theta_i(p_i) \geq \theta_{i-1}(p_{i-1})$ and hence $\delta^*(s_i, i) \geq j_i$.

Now suppose for contradiction that δ^* were not a winning strategy. Player A could chose some run r that wins against the strategy δ^*. If s were the resulting play there would be some fixed j and some least k such that for all $i \geq k$, $s_i = (r_i, j, \mathbb{R})$. If $j = 0$, then it is clear that $\theta_i(r_i) = (0, \mathbb{R})$ for all $i \geq k$. If $j > 0$ then $s_k = (r_i, j, \mathbb{R})$ and, since k is least, $s_{k-1} = (r_{k-1}, j, \mathbb{G})$. Therefore $\theta_{k-1}(r_{k-1}) \geq (j, \mathbb{G})$ and $r_{k-1} \in F$, so by the definition of $\overline{\lambda}$ it follows that for all $i \geq k$, $\theta_i(r_i) \geq (j, \mathbb{A})$.

Using the definition of the strategy δ^*, we show that for all $i \geq k$, $\theta_i(r_i) \in \{j\} \times \{\mathbb{A}, \mathbb{R}\}$. To see that $\theta_i(r_i)$ cannot be green for any $i \geq k$ we must consider two possibilities. If $j = 0$ then $\theta_i(r_i) = (0, \mathbb{R})$ as r_i is the state of some run at i and thus greater than $\theta_0(q_0)$. If $j > 0$ then $s_k = (r_i, j, \mathbb{R})$ and, since k is least, $s_{k-1} = (r_{k-1}, j, \mathbb{G})$. Therefore $\theta_{k-1}(r_{k-1}) \geq (j, \mathbb{G})$ and $r_{k-1} \in F$. From the definition of $\overline{\lambda}$ it follows that for all $i \geq k$, $\theta_i(r_i) \geq (j, \mathbb{A})$.

If for any $m > k$, $\theta_m \in \overline{F}$ then for all $i > m$, $\theta_i(r_i) = (j, \mathbb{R})$ and thus $\theta_i \notin \overline{F}$. Therefore θ is not an accepting run giving us the desired contradiction. ∎

LEMMA 16 *If player $\mathcal{B}$ has a winning strategy over σ then there is an accepting run of $\overline{A}$ over σ.*

Proof. Given δ^*, a winning strategy for player $\mathcal{B}$, we construct the run θ in several steps. First we define a sequence $\eta = (\eta_0, \eta_1, \ldots)$ where for all i, $\eta_i : Q \longrightarrow \{0, \ldots, n\}$. For each $p \in Q$ and each i we define

$$\eta_i(p) = \max\{0, j | s_i = (p, j, c) \text{ for some } c \text{ and some play } s \text{ according to } \delta^*\}.$$

We then define a sequence $\gamma = (\gamma_0, \gamma_1, \ldots)$ where for all i, $\gamma_i : Q \longrightarrow \mathbb{C}$. To do this for all $p \in Q$ and for all i, we let $\mu(p, i) \subseteq Q$ be the set of *minimal successors* of p at i, defined as $\{q | q \in \lambda(p, \sigma_i)$ and $\eta_i(q) = \eta_i(p)\}$. We now define the sequence γ with the following induction:

- $\gamma_0(q_0) = \mathbb{R}$ and $\gamma_0(q) = \mathbb{G}$ for all $q \neq q_0$.

- for every $p \in Q$

  1. if $p \in \mu(q, i)$ for some q such that either
     (a) $\gamma_i(q) = \mathbb{R}$, or
     (b) $\gamma_i(q) = \mathbb{A}$ and $\gamma_i(Q) \subseteq \{\mathbb{G}, \mathbb{A}\}$
     then $\gamma_{i+1}(q) = \mathbb{R}$,

  2. otherwise if $p \in \mu(q, i)$ for some q such that either
     (a) $\gamma_i(q) = \mathbb{A}$, or
     (b) $\gamma_i(q) = \mathbb{G}$ and $q \in F$
     then $\gamma_{i+1}(q) = \mathbb{A}$,

  3. otherwise $\gamma_{i+1}(q) = \mathbb{G}$.

We define a run θ, of $\overline{A}$ over σ by $\theta_i(p) = (\eta_i(p), \gamma_i(p))$ for all $p \in Q$ and all i. To see that θ is a valid run of $\overline{A}$ over σ we observe that if $q \in \lambda(p, \sigma_i)$ then $\eta_{i+1}(q) \geq \eta_i(p)$ and the rest follows from the inductive definition of γ.

Now suppose, for contradiction, that θ is not an accepting run. Therefore there are only finitely many accepting states, so we let m be the index one greater than the last accepting state (i.e. for all $i \geq m$, $\theta_i \notin F$).

By the inductive definition of γ we can see that if $\gamma_i(p) = \mathbb{R}$ for some $i \geq m$ then there is some q with $p \in \lambda(q, \sigma_{i-1})$ with $\gamma_{i-1}(q) = \mathbb{R}$. Applying this recursively we can deduce that there is some run r of A over σ with $r_i = p$ and for all k with $m \leq k \leq i$, $\gamma_k(r_k) = \mathbb{R}$, and hence $\eta_k(r_k) = \eta_i(r_i)$. For all $k \geq m$ and for all $i \geq k$ let

$$P^j(k, i) = \{r_k \in Q | r \text{ is a run of } A \text{ with } \theta_k(r_k) = \theta_i(r_i) = (j, \mathbb{R})\}.$$

The set $P^j(k, i)$ is the set of states that belong to some run at i that has had the label $(j, \mathbb{R})$ since k. We choose $j \leq n$ to be such that for all $k \geq m$ and for all $i > k$, $P^j(k, i) \neq \emptyset$. Such a j exists since $P^j(k, i) = \emptyset$ implies either $P^j(k, i+1) = P^j(k+1, i) = \emptyset$ or $\theta_k \in \overline{F}$. Since $\theta_k \notin \overline{F}$ for $k > m$, if such a j did not exist there would be some value $\ell > m$ such that $P^j(\ell, \ell) = \emptyset$ for all j. Consequently we would have $\theta_\ell \in \overline{F}$, contradicting the fact that $\theta_k \notin \overline{F}$ for all $k > m$.

We can define a losing play of the game for $\overline{A}$ as follows. Since $|Q|$ is finite there must be some $q_m \in Q$ such that $q_m \in P^j(m,i)$ for infinitely many i. By the definition of θ we can assume that there is some play s, played according to δ^*, with $s_m = (q_m, j, \mathbb{R})$. Since $q_m \in P^j(m,i)$ for infinitely many i, there must be some $q \in \lambda(q_m, \sigma_m)$ such that $q \in P^j(m+1,i)$ for infinitely many i. Player A can choose $q_{m+1} = q$ and by the way that θ was constructed we can see that $\delta^*(s_m, i) = j$, so $s_{m+1} = (q_{m+1}, j, \mathbb{R})$. Since $q_{m+1} \in P^j(m+1,i)$ for infinitely many i, the same argument applies and player A can ensure that $s_{m+2} = (q_{m+2}, j, \mathbb{R})$. Continuing this way player A can ensure that for all $i > m$ $j_i = j$ and $c_i = \mathbb{R}$, and hence win the game, contradicting the fact that δ^* is a winning strategy. Therefore θ must be an accepting run and we are done. ∎

COROLLARY 17 *The automaton $\overline{A}$ accepts σ if and only if the automaton A does not.*

COROLLARY 18 *Given a non-deterministic Büchi automaton with n states, we can find a complementary non-deterministic Büchi automaton with $(3(n+1))^n$ states.*

This follows directly from the complementation construction. Furthermore, a close examination of the proof of correctness will show that the complementary automaton need only use the states

$$\{\phi | \phi : Q \longrightarrow \{1, \ldots, n-1\} \times \mathbb{C} \cup \{(0, \mathbb{G}), (0, \mathbb{R}), (n, \mathbb{R})\}\}.$$

We can therefore improve the complexity of this construction to $(3n)^n$. This is a slight improvement on Klarlund's construction [7] which has complexity $(4n)^n$.

7.3 Formal proof of correctness

Because of space limitations we omit the part of the proof which shows that the correctness of the complementation construction can be established in our proof system. Details will appear in the full paper.

8 Redundancy

We consider whether any parts of the axiom system are redundant. The two candidates are the new axiom QX0 and the rule QI.

To show that QX0 is necessary we consider an alternative "wrapped" semantics for QPTL formulas based on Kripke structures. Even consideration of runs on a one state Kripke structure will suffice. Suppose $(S, R) = (\{s_0\}, \{(s_0, s_0)\})$ a structure with one state $s_0 \subseteq \mathcal{V}$ and a transition relation R which allows a transition from s_0 to s_0.

Now consider sequences σ of states from S, i.e. $\sigma = (s_0, s_0, s_0, ...)$. We can redefine the semantics for QPTL, i.e. give a "wrapped" semantics, by using the same clauses for each operator as in the standard semantics apart from the case of $\forall y \alpha$. Instead define $(\sigma, i) \models \forall y \alpha$ iff for every structure (S', R) which differs from (S, R) only to the extent that $s_0 \setminus \{x\} = s'_0 \setminus \{x\}$, we have $(\sigma, i) \models \alpha$.

It is straightforward to show that QF without the QX0 rule is sound for these wrapped semantics: so anything derivable in this system is a validity of wrapped QPTL. However, $\exists x(x \wedge \Box \neg x)$ is clearly not sound itself and so not derivable in the rest of the system.

In order to show that QI is not redundant we again can change the semantics of the logic. Briefly, we only consider models in which atoms change truth value a finite number of times as we proceed through time. It can be shown that **QF** is sound for such a language, and the result follows. Full details will appear in the long version of the paper.

9 Past operators

As in [19], the traditional unquantified PLTL is often extended to a logic which we call PLTLP (although there is no standard name) by the addition of past-time temporal operators. The extension brings no extra expressivity (in a certain sense [4]) but there are technical reasons and advantages in practical convenience. We can also add these operators to quantified propositional temporal logic to form a logic QPTLP. The syntactic formation rules for QPTLP are simply those for QPTLalong with allowing $\ominus \alpha$ and $\boxminus \alpha$ if α and β are formulas.

The new operators are **weak previous** $\ominus$ and **past generally** $\boxminus$. Add the following semantic clauses:

$$(\sigma, j) \models \ominus \alpha \Leftrightarrow (j = 0) \text{ or } (\sigma, j-1) \models \alpha. \tag{10}$$

$$(\sigma, j) \models \boxminus \alpha \Leftrightarrow \forall k \leq j \quad (\sigma, k) \models \alpha. \tag{11}$$

Kesten and Pnueli's proof system [6] for QPTLP consists of axioms for PLTL, axioms for the past-time operators, extra PLTLP axioms describing the interaction of the past and future operators, QX1, QX2, MP, T-GEN and ∀-GEN. There is nothing like our QX0 or our QI rule. The complicated completeness proof, using automata as we do, is presented in that paper.

It is interesting to consider whether their proof system can be made into a complete system for QPTL by the simple removal of axioms which mention the past operators. The "wrapped" semantics which we introduced above shows that this is not possible: QX0 is not derivable. To derive QX0 one makes use of the Kesten and Pnueli axiom $\ominus \bot$. Furthermore, an examination of the completeness proof shows heavy reliance on the past-time operators for establishing various marking and inductive constructions. Our desire to avoid such essential and subtle use of the past-time operators led us to instead use the Σ_1 normal form and deterministic Rabin automata approach to our completeness proof.

We have our reasons for not wanting to use the past-time operators. It is worth considering whether there are any good reasons to add them to QPTL as they are not needed to obtain a complete axiom system. The answer, even more clearly than in the PLTL case, is that they are not useful. Not only do they add no extra expressiveness but they are easy to define in terms of the future operators and quantification. We will now be more precise with our claim here.

To proceed we define a translation from QPTLP into QPTL which preserves truth at the start of time but not necessarily at other times. Consider a QPTLP formula α. Recursively replace past operators with new atoms from the bottom up. For each sub-formula $\ominus \beta$ choose a new atom q_β say, replace $\ominus \beta$ by q_β in the formula and add a new conjunct $q_\beta \wedge \square(\beta \to \bigcirc q_\beta))$. Similarly choose a new atom p_β to replace each $\boxminus \beta$ and add a new conjunct $(\beta \leftrightarrow p_\beta) \wedge \square(p_\beta \to \bigcirc(\beta \to p_\beta))$. Finally existentially quantify over all the new atoms in the resulting past-free formula. Call the QPTL result α^+ say. A straightforward induction shows the following:

LEMMA 19 *For each $\alpha \in QPTLP$, for each structure σ, $(\sigma, 0) \models \alpha$ iff $(\sigma, 0) \models \alpha^+$.*

Thus it is easy to find an (initially) equivalent QPTL formula to any QPTLP formula. The translation is PTIME in contrast to the non-elementary process in the PLTL case. Initial equivalence is exactly what is needed for reasoning about specifications of systems.

In the long version of this paper, we also show that we obtain a sound and complete axiomatization for QPTLP by the simple addition to QF of the standard PLTLP axioms for the past time operators.

10 Conclusion

Deciding QPTL like QPTLP is non-elementarily complex [18]. However, it is worth noting, in the context of our reduction to Σ_1 formulas, that the satisfiability problem for existentially quantified QPTL is, like that for PLTL in PSPACE.

The main achievements here are: a sound and complete proof system for pure future quantified propositional temporal logic; a straightforward process for converting QPTL formulas to a Σ_1 normal form; and a new and interesting proof of correctness for an optimal complementation procedure for non-deterministic Büchi automata.

In future work we hope that alternative semantics for QPTL, such as the wrapped semantics, are considered and axiomatized and branching time extensions of QPTL are also axiomatized.

The extension of QPTL to the branching time case is complicated with a variety of possible semantics. See [8] and [2]. However, it is worth pointing out that our new proof system for QPTL allows us to side-step the complications caused by past operators in the move to branching time logic. The use of quantification in a branching time setting is complicated enough. The addition of the past time operators just adds even more subtle variations without any reward.

BIBLIOGRAPHY

[1] K. Engelhardt, R. van der Meyden, and Y. Moses. Knowledge and the logic of local propositions. In *Conf. on Theoret. Aspects of Rationality and Knowledge*. 1998.

[2] T. French. Decidability of quantified propositional branching time logics. In *Proc. of the 14th Aust. Joint Conf. on A. I. (AI'2001)*, pages 165–176, 2001.

[3] T. French. Quantified propositional branching time logics. Thesis, in prep.

[4] D. M. Gabbay, A. Pnueli, S. Shelah, and J. Stavi. On the temporal analysis of fairness. In *7th ACM Symp. on Princ. of Prog. Languages, Las Vegas*, pages 163–173, 1980.

[5] Y. Kesten and A. Pnueli. A complete proof systems for QPTL. In *Proc., 10th Annual IEEE Symp. on Logic in Computer Science*, pages 2–12, 1995.

[6] Y. Kesten, and A. Pnueli. Complete proof system for QPTL. *J. Logic and Computation*. To appear in 2003.

[7] N. Klarlund. Progress measures for complementation of ω-automata with applications to temporal logic. In *Proc. Foundations of Computer Science*, 1991.

[8] O. Kupferman. Augmenting branching temporal logics with existential quantification over atomic propositions. In *Computer Aided Verification, Proc. 7th Int. Conference*, volume 939 of *LNCS*, pages 325–338, Liege, July 1995. Springer-Verlag.

[9] L. Lamport. Specifying concurrent program modules. *ACM Transactions on Prog. Languages and Systems*, 5:190–222, 1983.

[10] L. Lamport. The temporal logic of actions. Tech. Report 79, DEC Systems Research Center, Palo Alto, Calif., USA, 1991.

[11] M. Michel. Complementation is more difficult with automata on infinite words. Manuscript, CNET, Paris, 1988.

[12] A. Pnueli. The temporal logic of programs. In *Proc. of the Eighteenth Symp. on Foundations of Computer Science*, pages 46–57, 1977. Providence, RI.

[13] M. Reynolds. An axiomatization of full computation tree logic. *J. Symbolic Logic*, 63(3):1011–1057, 2001.

[14] S. Safra. On complexity of ω-automata. In *Proc. Found. of Computer Science*. 1988

[15] A. Sistla, M. Vardi, and P. Wolper. The complementation problem for Buchi automata with applications to temporal logic. *Theo. Computer Science*, 49:217–237, 1987.

[16] A. P. Sistla. *Theoretical Issues in the Design and Verification of Distributed Systems*. PhD thesis, Harvard, 1983.

[17] P. Wolper. Temporal logic can be more expressive. *Inf.and Comp.*, 56(1–2):72–99, 1983.

[18] P. Wolper, M. Vardi, and A. Sistla. Reasoning about infinite computation paths. In *Proc. of 24th IEEE Symp. on the Found. of Computer Science*, 1983.

[19] L. Zuck, O. Lichtenstein, A. Pnueli. The glory of the past. *LNCS*, 193:196–218, 1985.

Tim French
Department of Computer Science and Software Engineering
University of Western Australia
35 Stirling Highway
Crawley, Western Australia, 6009
Australia
E-mail: tim@csse.uwa.edu.au

Mark Reynolds
School of Information Technology
Murdoch University
Murdoch, Perth, Western Australia 6150
Australia
E-mail: m.reynolds@murdoch.edu.au

8
Equational Logic of Polynomial Coalgebras[1]
ROBERT GOLDBLATT

ABSTRACT. Coalgebras of polynomial functors constructed from sets of observable elements have been found useful in modelling various kinds of data types and state-transition systems. This paper presents a calculus of terms for operations on such coalgebras, based on a simple type theory, and develops its semantics. The terms admit a single state-valued *parameter*, but may also have state-valued variables. In a "rigid" term all state-variables are bound.

Boolean combinations of equations between terms of observable type are shown to form a natural language for specifying properties of polynomial coalgebras, and for giving a Hennessy-Milner style logical characterisation of bisimilarity of states: two states are bisimilar when they satisfy the same rigid observable formulas. Also, our syntax of terms is expressive enough to show, alternatively, that two states are bisimilar when they assign the same values to all ground observable terms. The proof involves a characterisation of bisimulations as relations preserved by certain functions induced by "paths" between functors. An analysis of the definability of path actions shows that our language is capable of defining certain modalities. These can be used to express modal assertions to the effect that certain formulas will be true after the execution of a state transition.

1 Introduction and overview

If $T : \mathbf{Set} \to \mathbf{Set}$ is a functor on the category of sets, then a *T-coalgebra* is a pair (A, α) with A being a set and α a function of the form $A \to TA$. This notion has proven useful in modelling data structures such as lists, streams and trees; transition systems such as automata; and classes in object-oriented programming languages [29, 16, 31, 32]. Generically A is viewed as a set of *states*, and α as a *transition structure*.

In many of the examples just mentioned, the functor T is *polynomial*, i.e. is constructed from constant-valued functors and the identity functor by forming products, exponential functors with constant exponent (which

---
[1] Prepared using Paul Taylor's `diagrams` package.

we will call *power* functors), and coproducts. In that case (A, α) is a *polynomial coalgebra*. The aim of this paper is to set out an appropriate formal language for specifying properties of polynomial coalgebras and to develop its semantics. A first criterion for "appropriateness" is that the language should provide a logical characterisation, in the style of Hennessy and Milner [13], of the fundamental relation of *bisimilarity* between coalgebraic states. The form of this characterisation is that two states are bisimilar precisely when they satisfy the same formulas of the language under the given semantics. In addition the syntax should provide terms denoting the various operations on coalgebras that are naturally associated with the constructions defining polynomial functors; these operations including projections, pairings, injections, evaluations, lambda abstractions, and functional applications. A consequence of that requirement is that bisimilar states can be alternatively characterised as those assigning the same values to certain terms of "observable" type.

Now relational models of propositional modal logic can be viewed as coalgebras [31] and this has led to a number of proposals of languages with modalities for describing coalgebras [26, 21, 30, 18]. An alternative method used here is to develop a syntax of equations between terms for coalgebraic operations that is analogous to the standard logic of terms for abstract algebras, but subject to the principle that a coalgebraic term should have a single *state*-valued variable or parameter. The result is an equational language that is very similar to those of classical universal algebra and categorical logic, but at the same time is *implicitly modal* in nature. In particular it is able to express modal assertions to the effect that certain formulas will be true after the execution of state transitions (see Theorem 16, Corollary 17 and Section 8).

A previous article [9] developed such a calculus of terms and equations for polynomial coalgebras that are *monomial*, i.e. constructed without the use of coproducts. It was shown that Boolean combinations of equations between terms of "observable" type form a suitable language of formulas for specifying properties of monomial coalgebras and characterising bisimulation relations between them, both in the Hennessy-Milner style of equivalence of formula satisfaction and in terms of equality of term-evaluation. Our purpose now is to explain how that theory can be extended to include coproducts, whose presence introduces considerable complexity associated with the *partiality* of certain "path functions" expressing the dynamics of the transition structure α.

The approach we take is to use type theory [17] to describe the construction of sets-as-types from some base types by forming products, powers and coproducts, and to provide rules of syntax for terms that take values in these types. The base types denote fixed sets of observable elements. There is also the type St of states: this symbol St denotes the state set of a given coalgebra. The symbol s is reserved as the special state-valued parameter that appears in terms, and may be thought of as denoting the "current" state. The symbol tr denotes the transition structure, so that we are able to form the term tr(s), or more generally tr(M) for any state-valued term M. But the situation is more subtle than previously, because we now allow state variables distinct from the parameter s in coalgebraic specifications, provided that they are bound. In the syntax of [9] all variables of a term are free, but here we have variable-binding operations on terms (lambda-abstraction, case-formation). A given term M may contain free state variables. More generally it may have a number of free variables of various types that occur in state-valued subterms, and hence provide a number of ways of denoting states by varying the values of those variables. M is *rigid* if this does not hold, i.e. if any variable occurring in a state-valued subterm is bound in M itself (an example will be given shortly). Rigidity is imposed on M by requiring that the type of any free variable of M does not involve St.

Following established practice in categorical logic, the "case" operation is used to introduce terms associated with coproducts. The coproduct $A_1 + A_2$ of sets A_1, A_2 is their disjoint union, and comes equipped with injective *insertion* functions $\iota_j : A_j \to A_1 + A_2$ for $j = 1, 2$. Each element of $A_1 + A_2$ is equal to $\iota_j(a)$ for a unique j and a unique $a \in A_j$. Our syntax generates terms of the form

$$\text{case } N \text{ of } [\iota_1 v_1 \mapsto M_1 \mid \iota_2 v_2 \mapsto M_2],$$

where N is a term taking values in $A_1 + A_2$, M_1 and M_2 take values in some other set B, and the v_j's are variables that take values in A_j *and are bound in the overall* case *term*. The latter is evaluated by first obtaining the value d of N in $A_1 + A_2$ and then, if d is equal to $\iota_j(a)$, evaluating M_j with v_j assigned value a, giving an element of B as the desired value. Another notation for this term [17, Section 2.3] is

$$\text{unpack } N \text{ as } [\iota_1 v_1 \text{ in } M_1 , \iota_2 v_2 \text{ in } M_2].$$

EXAMPLE. To illustrate the use of rigid terms and case-formation in coalgebraic specification, here is an example adapted from [19, Section 4]. Let A be a set of (possibly infinite) binary trees. Each tree x either is a single node with no children, or has exactly two children obtained by deleting the top node of x. This gives an operation

$$children : A \longrightarrow 1 + (A \times A),$$

$$\begin{aligned}
size(\mathsf{s}) = \ & \mathsf{case}\ children(\mathsf{s})\ \mathsf{of} \\
& \iota_1 u \mapsto \iota_2 1 \\
& \iota_2 v \mapsto \mathsf{case}\ size(\pi_1 v)\ \mathsf{of} \\
& \qquad \iota_1 u \mapsto \iota_1 * \\
& \qquad \iota_2 n \mapsto \mathsf{case}\ size(\pi_2 v)\ \mathsf{of} \\
& \qquad \qquad \iota_1 u \mapsto \iota_1 * \\
& \qquad \qquad \iota_2 k \mapsto \iota_2(n+k+1) \\
& \qquad \qquad \mathsf{endcase} \\
& \qquad \mathsf{endcase} \\
& \mathsf{endcase}
\end{aligned}$$

Figure 1. case terms

where $1 = \{*\}$; $children(x) = \iota_1 *$ when x has no children, and $children(x) = \iota_2(x_1, x_2)$ when x_1 and x_2 are the left and right children of x. There is a size (number of nodes) operation

$$size : A \longrightarrow 1 + \mathbb{N},$$

where $\mathbb{N}$ is the set of positive integers and $size(x) = \iota_1 *$ when x is infinite. The two operations can be "tupled" into a single function

$$A \xrightarrow{\alpha} (1 + (A \times A)) \times (1 + \mathbb{N})$$

which is a coalgebra for the functor $T(X) = (1 + (X \times X)) \times (1 + \mathbb{N})$. The operations can be recovered from α as $children = \pi_1 \circ \alpha$ and $size = \pi_2 \circ \alpha$, where π_1 and π_2 are the left and right projections.

Now the size of a tree is 1 if it has no children, is infinite if at least one child is infinite, and otherwise is the sum of the sizes of the children plus 1. Thus our example *validates* the equation of Figure 1, in which the right-hand term M is obtained by iteration of case-formation. Validity means that the equation is satisfied no matter what member of A is denoted by the state parameter s. The variable v takes values in $A \times A$, so $\pi_1 v$ and $\pi_2 v$ take values in A. Although v is free in these subterms, and indeed in the subterms beginning case $size(\pi_j v)...$, v is bound in M itself. M is rigid.

The notion of a *bisimulation* first appeared in [27] as a relation of mutual simulation between states of two automata. Park showed that if two deterministic automata are related by a bisimulation, then they accept the same set of inputs. Hennessy and Milner [12, 13] introduced the idea of characterising observationally equivalent, or behaviourally indistinguishable, processes as those satisfying the same formulae of a logical language. They provided a simple propositional modal language that achieved this. Observational equivalence was defined as the relation which is in fact the largest bisimulation between processes. Equivalent processes were later dubbed *bisimilar*.

A category-theoretic definition of bisimulation relations between T-coalgebras (A, α) and (B, β) was given in [1]: a relation $R \subseteq A \times B$ is a bisimulation if there is a transition structure $\rho : R \to TR$ such that the projections $A \times B \to A$ and $A \times B \to B$ are coalgebraic morphisms from ρ to α and β (see Section 5). *Bisimilarity* is the largest such relation, which always exists because the union of any collection of bisimulations is a bisimulation. In [15] another approach was introduced by defining a "lifting" of the relation R to a relation $R^T \subseteq TA \times TB$ and taking R to be a bisimulation if it was preserved by the lifting, in the sense that xRy implies $\alpha(x) R^T \beta(y)$. Here we will show that when T is polynomial, the definition can be reformulated with the help of a notion from [18] of a *path* $T \xrightarrow{p} S$ from T to one of its component functors S. This p is a sequence of symbols that reflects the way S is structured within T. We will see that a path induces certain partial functions $p_X : TX \longrightarrow SX$ for any set X, and that R is a bisimulation when it is "preserved" by such functions (Theorem 9). Moreover, the action of a path function is definable by a term of our language for coalgebras. From this we obtain a logical characterisation of bisimilarity of states as meaning that they assign the same values to all ground terms of observable type, or that they satisfy the same equations between such terms, or indeed that they satisfy the same Boolean combinations of such equations.

To summarize, the main features of this paper are:

- The formulation of syntax and semantics of types and terms for operations in coalgebras of any polynomial functor (Sections 2, 3 and 4).

- The characterisation of bisimulation relations by their preservation under partial functions induced by "paths" between functors, and the term-definability of these path functions (Sections 5 and 6).

- The definition of *observable* formulas as Boolean combinations of equations between terms of observable type, and their use in logically characterising bisimilarity of states: two states are bisimilar when they assign the same values to all ground observable terms, or equivalently when they satisfy the same rigid observable formulas (Theorem 20).

This characterisation of bisimilarity is used in an associated article [10] to obtain a structural characterisation of classes of polynomial coalgebras definable by sets of rigid observable formulas. The result is an analogue for polynomial coalgebras of Birkhoff's famous characterisation in [2] of equational classes of abstract algebras as being those closed under direct products, homomorphic images and subalgebras. The coalgebraic analogue requires the development of a new notion of "observational ultrapower" of coalgebras. Yet another approach [7] involves a "Stone space" type of construction on coalgebras, similar to the ultrafilter enlargements used in modal model theory. A brief description of these further developments is provided here in the final Section 9.

Section 8 gives a comparison with other formalisms, including an explanation of why the language developed here subsumes the modal language for polynomial coalgebras of [30].

2 Polynomial functors

Standard notation for products, powers and coproducts of sets will be used. The *coproduct* $A_1 + A_2$ and associated *insertions* $\iota_j : A_j \to A_1 + A_2$ have already been described. $\pi_j : A_1 \times A_2 \to A_j$ is the *projection* function from the *product* set $A_1 \times A_2$ onto A_j. The *D-th power* of set A is the set A^D of all functions from set D to A. For each $d \in D$ there is the evaluation function $ev_d : A^D \to A$ having $ev_d(f) = f(d)$. The identity function on a set A is denoted $\text{id}_A : A \to A$.

The symbol $\circ\!\!\to$ will be used for partial functions. Thus $f : A \circ\!\!\to B$ means that f is a function with codomain B and domain $\text{Dom}\, f \subseteq A$. We may write $f(x){\downarrow}$ to mean that $f(x)$ is defined, i.e. $x \in \text{Dom}\, f$. Associated with each insertion $\iota_j : A_j \to A_1 + A_2$ is its partial inverse, the *extraction* function $\varepsilon_j : A_1 + A_2 \circ\!\!\to A_j$ having $\varepsilon_j(y) = x$ iff $\iota_j(x) = y$. Thus $\text{Dom}\, \varepsilon_j = \iota_j A_j$, i.e. $y \in \text{Dom}\, \varepsilon_j$ iff $y = \iota_j(x)$ for some $x \in A_j$. Extraction functions play a vital role in the analysis of coalgebras built out of coproducts, as will be seen below.

Consider the following constructions of endofunctors $T : \mathbf{Set} \to \mathbf{Set}$.

- For a fixed set $D \neq \emptyset$, the *constant functor* $\bar{D}$ has $\bar{D}(A) = D$ on sets A and $\bar{D}(f) = \text{id}_D$ on functions f.

- The *identity functor* Id has $\text{Id}A = A$ and $\text{Id}f = f$.

- The product $T_1 \times T_2$ of two functors has $T_1 \times T_2(A) = T_1 A \times T_2 A$, and, for a function $f : A \to B$, has $T_1 \times T_2(f)$ being the function

$$T_1(f) \times T_2(f) : T_1 A \times T_2 A \to T_1 B \times T_2 B$$

that acts by $(a_1, a_2) \mapsto (T_1(f)(a_1), T_2(f)(a_2))$.

- The coproduct $T_1 + T_2$ of two functors has $T_1 + T_2(A) = T_1 A + T_2 A$, and for $f : A \to B$, has $T_1 + T_2(f)$ being the function
$$T_1(f) + T_2(f) : T_1 A + T_2 A \to T_1 B + T_2 B$$
that acts by $\iota_j(a) \mapsto \iota_j(T_j(f)(a))$.

- The D-th power functor T^D of a functor T has $T^D A = (TA)^D$, and $T^D(f) : (TA)^D \to (TB)^D$ being the function $g \mapsto T(f) \circ g$.

A functor T is *polynomial* if it is constructed from constant functors and Id by finitely many applications of products, coproducts and powers. Note that any polynomial functor constructed without the use of Id is constant.

A *T-coalgebra* is a pair (A, α) comprising a set A and a function $A \xrightarrow{\alpha} TA$. A is the set of *states* and α is the *transition structure* of the coalgebra. Note that A is determined as the domain $\operatorname{Dom} \alpha$ of α, so we can identify the coalgebra with its transition structure, i.e. a T-coalgebra is any function of the form $\alpha : \operatorname{Dom} \alpha \to T(\operatorname{Dom} \alpha)$. A *morphism* from T-coalgebra α to T-coalgebra β is a function $f : \operatorname{Dom} \alpha \to \operatorname{Dom} \beta$ between their state sets which commutes with their transition structures in the sense that $\beta \circ f = Tf \circ \alpha$, i.e. the following diagram commutes:

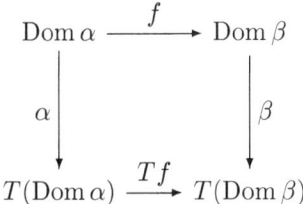

If $\operatorname{Dom} \alpha \subseteq \operatorname{Dom} \beta$, then α is a *subcoalgebra* of β iff the inclusion function $\operatorname{Dom} \alpha \hookrightarrow \operatorname{Dom} \beta$ is a morphism from α to β.

Every set $\{\alpha_i : i \in I\}$ of T-coalgebras has a *disjoint union* $\sum_I \alpha_i$, which is a T-coalgebra whose domain is the disjoint union of the $\operatorname{Dom} \alpha_i$'s and whose transition structure acts as α_j on the summand $\iota_j \operatorname{Dom} \alpha_j$ of $\operatorname{Dom} \sum_I \alpha_i$. More precisely, this transition is given by $\iota_j(a) \mapsto T(\iota_j)(\alpha_j(a))$, with the insertion $\iota_j : \operatorname{Dom} \alpha_j \to \operatorname{Dom} \sum_I \alpha_i$ being an injective morphism making α_j isomorphic to a subcoalgebra of the disjoint union (see [32, Section 4]).

3 Syntax of types, terms and formulas

Types

Fix a set $\mathbb{O}$ of symbols called *observable types*, and a collection $\{[\![o]\!] : o \in \mathbb{O}\}$ of non-empty sets indexed by $\mathbb{O}$. $[\![o]\!]$ is the *denotation* of o, and its members are called *observable elements*, or *constants* of type o.

EXAMPLE $\mathbb{O} = \{\mathsf{num}, \mathsf{bool}, 1\}$, with $[\![\mathsf{num}]\!] = \{0, 1, \ldots\}$, $[\![\mathsf{bool}]\!] = \{true, false\}$, $[\![1]\!] = \{*\}$.

The set of *types* over $\mathbb{O}$, or $\mathbb{O}$-*types*, is the smallest set $\mathbb{T}$ such that $\mathbb{O} \subseteq \mathbb{T}$, $\mathsf{St} \in \mathbb{T}$ and

(1). if $\sigma_1, \sigma_2 \in \mathbb{T}$ then $\sigma_1 \times \sigma_2$, $\sigma_1 + \sigma_2 \in \mathbb{T}$;

(2). if $\sigma \in \mathbb{T}$ and $o \in \mathbb{O}$, then $o \Rightarrow \sigma \in \mathbb{T}$.

A *subtype* of an $\mathbb{O}$-type τ is any type that occurs in the formation of τ. Thus the set $Sub\,\tau$ of subtypes of τ is just $\{\tau\}$ when $\tau \in \mathbb{O} \cup \{\mathsf{St}\}$; is equal to $\{\tau\} \cup Sub\,\sigma_1 \cup Sub\,\sigma_2$ when τ is $\sigma_1 \times \sigma_2$ or $\sigma_1 + \sigma_2$; and is $\{\tau\} \cup \{o\} \cup Sub\,\sigma$ when τ is $o \Rightarrow \sigma$.

St is a type symbol that will denote the state set of a given coalgebra. The symbol "o" will always be reserved for members of $\mathbb{O}$. $o \Rightarrow \sigma$ is a power type: such types will always have an observable exponent.

A type is *rigid* if it does not have St as a subtype. The set of rigid types is thus the smallest set that includes $\mathbb{O}$ and satisfies (1) and (2).

Each $\mathbb{O}$-type σ determines a polynomial functor $|\sigma| : \mathbf{Set} \to \mathbf{Set}$. For $o \in \mathbb{O}$, $|o|$ is the constant functor $\bar{D}$ where $D = [\![o]\!]$; $|\mathsf{St}|$ is the identity functor Id; and inductively

$$|\sigma_1 \times \sigma_2| = |\sigma_1| \times |\sigma_2|, \quad |\sigma_1 + \sigma_2| = |\sigma_1| + |\sigma_2|, \quad |o \Rightarrow \sigma| = |\sigma|^{[\![o]\!]}.$$

For denotations of types, we write $[\![\sigma]\!]_A$ for the set $|\sigma|A$. Thus we have $[\![o]\!]_A = [\![o]\!]$, $[\![\mathsf{St}]\!]_A = A$,

$$[\![\sigma_1 \times \sigma_2]\!]_A = [\![\sigma_1]\!]_A \times [\![\sigma_2]\!]_A$$
$$[\![\sigma_1 + \sigma_2]\!]_A = [\![\sigma_1]\!]_A + [\![\sigma_2]\!]_A$$
$$[\![o \Rightarrow \sigma]\!]_A = [\![\sigma]\!]_A^{[\![o]\!]}.$$

If σ is a rigid type then $|\sigma|$ is a constant functor, so $[\![\sigma]\!]_A$ is a fixed set whose definition does not depend on A and may be written $[\![\sigma]\!]$. A τ-*coalgebra* is a coalgebra (A, α) for the functor $|\tau|$, i.e. α is a function of the form $A \to [\![\tau]\!]_A$.

Terms

To define *terms* we fix a denumerable set *Var* of *variables* and define a *context* to be a finite (possible empty) list

$$\Gamma = (v_1 : \sigma_1, \ldots, v_n : \sigma_n)$$

of assignments of $\mathbb{O}$-types σ_i to variables v_i, with the proviso that $v_1,\ldots,v_n$ are all distinct. Γ is a *rigid* context if all of the σ_i's are rigid types. Concatenation of lists Γ and Γ' with disjoint sets of variables is written Γ,Γ'. A *term-in-context* is an expression of the form

$$\Gamma \triangleright M : \sigma,$$

which signifies that M is a "raw" term of type σ in context Γ. This may be abbreviated to $\Gamma \triangleright M$ if the type of the term is understood. If $\sigma \in \mathbb{O}$, then the term is *observable*.

Figure 2 gives axioms that legislate certain *base terms* into existence, and rules for generating new terms from given ones. Axiom (Con) states that an observable element is a constant term of its type, while the raw term s in axiom (St) is a special parameter which will be interpreted as the "current" state in a coalgebra. The rules for products, coproducts and powers are the standard ones for introduction and transformation of terms of those types.

Bindings of variables in raw terms occur in lambda-abstractions and case terms: the v in the consequent of rule (Abs) and the v_j's in the consequent of (Case) are bound in those terms. It is readily shown that in any term $\Gamma \triangleright M$, all free variables of M appear in the list Γ. A *ground* term is one of the form $\emptyset \triangleright M : \sigma$, which may be abbreviated to $M : \sigma$, or just to the raw term M. Thus a ground term has no free variables. Note that a ground term may contain the state parameter s, which behaves nonetheless as a variable in that it can denote any member of $\text{Dom}\,\alpha$, as will be seen in the semantics presented in Section 4.

LEMMA 1. *There exist ground terms of every type.*

Proof. (Con) provides ground terms of all observable types, since $[\![o]\!] \neq \emptyset$ when $o \in \mathbb{O}$, and (St) provides a term of type St. The rules (Pair), (In$_j$) and (Abs), with the help of (Weak), then allow the inductive formation of ground terms of all other types. ∎

A term is defined to be *rigid* if its context is rigid. This entails that any free variable of the term is assigned a rigid type by Γ, so its type is formed without use of St. Of course all ground terms are rigid.

τ-terms

For a given $\mathbb{O}$-type τ, a τ-*term* is any term that can be generated by the axioms and rules of Figure 2 together with the additional rule

$$(\tau\text{-Tr}) \quad \frac{\Gamma \triangleright M : \mathsf{St}}{\Gamma \triangleright \mathsf{tr}(M) : \tau}.$$

Axioms

(Var) $\dfrac{v \in Var}{v : \sigma \triangleright v : \sigma}$ \qquad (Con) $\dfrac{c \in [\![o]\!]}{\emptyset \triangleright c : o}$ \qquad (St) $\dfrac{}{\emptyset \triangleright \mathsf{s} : \mathsf{St}}$

Weakening

(Weak) $\dfrac{\Gamma, \Gamma' \triangleright M : \sigma}{\Gamma, v : \sigma', \Gamma' \triangleright M : \sigma}$ \qquad where v does not occur in Γ or Γ'.

Product Types

(Pair) $\dfrac{\Gamma \triangleright M_1 : \sigma_1 \quad \Gamma \triangleright M_2 : \sigma_2}{\Gamma \triangleright \langle M_1, M_2 \rangle : \sigma_1 \times \sigma_2}$

(Proj$_1$) $\dfrac{\Gamma \triangleright M : \sigma_1 \times \sigma_2}{\Gamma \triangleright \pi_1 M : \sigma_1}$ \qquad (Proj$_2$) $\dfrac{\Gamma \triangleright M : \sigma_1 \times \sigma_2}{\Gamma \triangleright \pi_2 M : \sigma_2}$

Coproduct Types

(In$_1$) $\dfrac{\Gamma \triangleright M : \sigma_1}{\Gamma \triangleright \iota_1 M : \sigma_1 + \sigma_2}$ \qquad (In$_2$) $\dfrac{\Gamma \triangleright M : \sigma_2}{\Gamma \triangleright \iota_2 M : \sigma_1 + \sigma_2}$

(Case) $\dfrac{\Gamma \triangleright N : \sigma_1 + \sigma_2 \quad \Gamma, v_1 : \sigma_1 \triangleright M_1 : \sigma \quad \Gamma, v_2 : \sigma_2 \triangleright M_2 : \sigma}{\Gamma \triangleright \mathsf{case}\ N\ \mathsf{of}\ [\iota_1 v_1 \mapsto M_1 \mid \iota_2 v_2 \mapsto M_2] : \sigma}$

Power Types

(Abs) $\dfrac{\Gamma, v : o \triangleright M : \sigma}{\Gamma \triangleright (\lambda v.M) : o \Rightarrow \sigma}$ \qquad (App) $\dfrac{\Gamma \triangleright M : o \Rightarrow \sigma \quad \Gamma \triangleright N : o}{\Gamma \triangleright M \cdot N : \sigma}$

Figure 2. Axioms and rules for generating terms

Equations

$$\text{(Eq)} \quad \frac{\Gamma \triangleright M_1 : \sigma \quad \Gamma \triangleright M_2 : \sigma}{\Gamma \triangleright M_1 \approx M_2}$$

Weakening

$$\text{(Weak)} \quad \frac{\Gamma, \Gamma' \triangleright \varphi}{\Gamma, v : \sigma', \Gamma' \triangleright \varphi} \quad \text{where } v \text{ does not occur in } \Gamma \text{ or } \Gamma'.$$

Connectives

$$\text{(Neg)} \quad \frac{\Gamma \triangleright \varphi}{\Gamma \triangleright \neg \varphi} \qquad \text{(Con)} \quad \frac{\Gamma \triangleright \varphi_1 \quad \Gamma \triangleright \varphi_2}{\Gamma \triangleright \varphi_1 \wedge \varphi_2}$$

Figure 3. Formation rules for formulas

Note that from this rule and the axiom (St) we can derive the ground τ-term

$$\emptyset \triangleright \mathsf{tr}(\mathsf{s}) : \tau.$$

The symbol tr denotes the transition structure of a τ-coalgebra $A \xrightarrow{\alpha} [\![\tau]\!]_A$. If M is interpreted as the state x of α, then $\mathsf{tr}(M)$ is interpreted as $\alpha(x)$.

τ-formulas

An *equation-in-context* has the form $\Gamma \triangleright M_1 \approx M_2$ where $\Gamma \triangleright M_1$ and $\Gamma \triangleright M_2$ are terms *of the same type*. A *formula-in-context* has the form $\Gamma \triangleright \varphi$, with the raw expression φ being constructed from equations $M_1 \approx M_2$ by propositional connectives. Formation rules for formulas are given in Figure 3, using the connectives $\neg$ and $\wedge$. The other standard connectives $\vee$, $\rightarrow$, and $\leftrightarrow$ can be introduced as definitional abbreviations in the usual way.

A formula $\emptyset \triangleright \varphi$ with empty context is *ground*, and may be abbreviated to φ. A *rigid* formula is one whose context is rigid.

A *τ-formula* is one that is generated by using only τ-terms as premises in the rule (Eq). An *observable* formula is one that uses only terms of observable type in forming its component equations.

4 Semantics of terms and formulas

Types σ and contexts $\Gamma = (v_1 : \sigma_1, \ldots, v_n : \sigma_n)$ are interpreted in a given τ-coalgebra $\alpha : A \to |\tau|A$ by putting $[\![\sigma]\!]_\alpha = |\sigma|(\mathrm{Dom}\,\alpha) = [\![\sigma]\!]_A$, and

$$[\![\Gamma]\!]_\alpha = [\![\sigma_1]\!]_\alpha \times \cdots \times [\![\sigma_n]\!]_\alpha$$

(so $[\![\emptyset]\!]_\alpha$ is the empty product 1). The *denotation* of each τ-term $\Gamma \triangleright M : \sigma$, relative to the coalgebra α, is a function

$$[\![\Gamma \triangleright M : \sigma]\!]_\alpha : A \times [\![\Gamma]\!]_\alpha \longrightarrow [\![\sigma]\!]_\alpha,$$

defined by induction on the formation of terms. For empty contexts,

$$A \times [\![\emptyset]\!]_\alpha = A \times 1 \cong A,$$

so we replace $A \times [\![\emptyset]\!]_\alpha$ by A itself and interpret a ground term $\emptyset \triangleright M : \sigma$ as a function $A \to [\![\sigma]\!]_\alpha$. The definition of denotations is as follows.

Var:

$[\![v : \sigma \triangleright v : \sigma]\!]_\alpha : A \times [\![\sigma]\!]_\alpha \to [\![\sigma]\!]_\alpha$ is the right projection function.

Con:

$[\![\emptyset \triangleright c : o]\!]_\alpha : A \to [\![o]\!]$ is the constant function with value c.

St:

$[\![\emptyset \triangleright \mathsf{s} : \mathsf{St}]\!]_\alpha : A \to [\![\mathsf{St}]\!]_\alpha$ is the identity function $A \to A$.

τ-Tr:

$[\![\Gamma \triangleright \mathsf{tr}(M) : \tau]\!]_\alpha : A \times [\![\Gamma]\!]_\alpha \to [\![\tau]\!]_\alpha$ is the composition of the functions

$$A \times [\![\Gamma]\!]_\alpha \xrightarrow{[\![\Gamma \triangleright M : \mathsf{St}]\!]_\alpha} A \xrightarrow{\alpha} [\![\tau]\!]_\alpha.$$

Weak:

$[\![\Gamma, v : \sigma', \Gamma' \triangleright M : \sigma]\!]_\alpha$ is the composition of $[\![\Gamma, \Gamma' \triangleright M : \sigma]\!]_\alpha$ with the projection

$$A \times [\![\Gamma]\!]_\alpha \times [\![\sigma']\!]_\alpha \times [\![\Gamma']\!]_\alpha \longrightarrow A \times [\![\Gamma]\!]_\alpha \times [\![\Gamma']\!]_\alpha.$$

Pair:

$[\![\Gamma \triangleright \langle M_1, M_2 \rangle : \sigma_1 \times \sigma_2]\!]_\alpha$ is the product map

$$A \times [\![\Gamma]\!]_\alpha \xrightarrow{\langle [\![\Gamma \triangleright M_1 : \sigma_1]\!]_\alpha, [\![\Gamma \triangleright M_2 : \sigma_2]\!]_\alpha \rangle} [\![\sigma_1]\!]_\alpha \times [\![\sigma_2]\!]_\alpha.$$

Proj$_j$:

$[\![\Gamma \triangleright \pi_j M : \sigma_j]\!]_\alpha$ is the composition of

$$A \times [\![\Gamma]\!]_\alpha \xrightarrow{[\![\Gamma \triangleright M : \sigma_1 \times \sigma_2]\!]_\alpha} [\![\sigma_1]\!]_\alpha \times [\![\sigma_2]\!]_\alpha \xrightarrow{\pi_j} [\![\sigma_j]\!]_\alpha.$$

Inj$_j$:

$[\![\Gamma \triangleright \iota_j M : \sigma_1 + \sigma_2]\!]_\alpha$ is the composition of

$$A \times [\![\Gamma]\!]_\alpha \xrightarrow{[\![\Gamma \triangleright M : \sigma_j]\!]_\alpha} [\![\sigma_j]\!]_\alpha \xrightarrow{\iota_j} [\![\sigma_1]\!]_\alpha + [\![\sigma_2]\!]_\alpha.$$

Case:

This is easier to describe at the function-value level. For $x \in A$ and $\gamma \in [\![\Gamma]\!]_\alpha$, let

$$[\![\Gamma \triangleright N : \sigma_1 + \sigma_2]\!]_\alpha(x, \gamma) = \iota_j(a) \in [\![\sigma_1]\!]_\alpha + [\![\sigma_2]\!]_\alpha$$

(which holds for a unique j and $a \in [\![\sigma_j]\!]_\alpha$). Then the element

$$[\![\Gamma \triangleright \text{case } N \text{ of } [\iota_1 v_1 \mapsto M_1 \mid \iota_2 v_2 \mapsto M_2] : \sigma]\!]_\alpha(x, \gamma)$$

of $[\![\sigma]\!]_\alpha$ is defined to be

$$[\![\Gamma, v_j : \sigma_j \triangleright M_j : \sigma]\!]_\alpha(x, \gamma, a).$$

Put more succinctly: if $[\![\Gamma \triangleright N]\!]_\alpha(x, \gamma) \in \text{Dom } \varepsilon_j$, then

$$[\![\Gamma \triangleright \text{case } N \text{ of}..M_1|..M_2]\!]_\alpha(x, \gamma) = [\![\Gamma, v_j : \sigma_j \triangleright M_j]\!]_\alpha(x, \gamma, \varepsilon_j [\![\Gamma \triangleright N]\!]_\alpha(x, \gamma)).$$

Abs:

$[\![\Gamma \triangleright (\lambda v.M) : o \Rightarrow \sigma]\!]_\alpha(x, \gamma)$ is the function $[\![o]\!] \to [\![\sigma]\!]_\alpha$ given by

$$a \mapsto [\![\Gamma, v : o \triangleright M : \sigma]\!]_\alpha(x, \gamma, a).$$

App:

$[\![\Gamma \triangleright M \cdot N : \sigma]\!]_\alpha(x, \gamma)$ is the element of $[\![\sigma]\!]_\alpha$ obtained by evaluating the function

$$[\![\Gamma \triangleright M : o \Rightarrow \sigma]\!]_\alpha(x, \gamma) : [\![o]\!] \longrightarrow [\![\sigma]\!]_\alpha$$

at $[\![\Gamma \triangleright N : o]\!]_\alpha(x, \gamma) \in [\![o]\!]$.

This completes the inductive definition of $[\![\Gamma \triangleright M : \sigma]\!]_\alpha$.

An important observation for what follows is that if $\Gamma \triangleright M : \sigma$ is "tr-free", meaning that the symbol **tr** does not occur in the raw term M, then the denotation $[\![\Gamma \triangleright M : \sigma]\!]_\alpha$ depends only on A, not on α, and may be written $[\![\Gamma \triangleright M : \sigma]\!]_A$. This is readily shown by induction on the formation of $\Gamma \triangleright M : \sigma$.

Substitution of terms

In working with this system it becomes essential to have available the operation $N[M/v]$ of substituting the raw term M for free occurrences of the variable v in N. The following rule is derivable:

$$\text{(Subst)} \quad \frac{\Gamma \triangleright M : \sigma \quad \Gamma, v : \sigma \triangleright N : \sigma'}{\Gamma \triangleright N[M/v] : \sigma'}$$

The semantics of terms obeys the basic principle that substitution is interpreted as *composition* of denotations [28, 2.2]. Because of the special role of the state set A, this takes the form

$$[\![\Gamma \triangleright N[M/v]]\!]_\alpha = [\![\Gamma, v : \sigma \triangleright N]\!]_\alpha \circ \langle \pi_1, \pi_2, [\![\Gamma \triangleright M]\!]_\alpha \rangle,$$

so that the following diagram commutes:

$$\begin{array}{ccc}
A \times [\![\Gamma]\!]_\alpha & \xrightarrow{\langle \pi_1, \pi_2, [\![M]\!]_\alpha \rangle} & A \times [\![\Gamma]\!]_\alpha \times [\![\sigma]\!]_\alpha \\
& \searrow{[\![N[M/v]]\!]_\alpha} & \downarrow{[\![N]\!]_\alpha} \\
& & [\![\sigma']\!]_\alpha
\end{array}$$

Substitution of a raw term can also take place in a *formula*, producing expressions $\varphi[M/v]$ where φ is a raw formula. Thus $(M_1 \approx M_2)[N/v]$ is $(M_1[N/v] \approx M_2[N/v])$ etc. With the help of (Subst), the following rule can be derived:

$$\frac{\Gamma \triangleright M : \sigma \quad \Gamma, v : \sigma \triangleright \varphi}{\Gamma \triangleright \varphi[M/v]}$$

Substitution for the state parameter

It is also possible to make substitutions $N[M/\mathsf{s}]$ for the state parameter s according to the derivable rule

$$\text{(s-Subst)} \quad \frac{\Gamma \triangleright M : \mathsf{St} \quad \Gamma \triangleright N : \sigma'}{\Gamma \triangleright N[M/\mathsf{s}] : \sigma'}$$

with the semantics $[\![\Gamma \triangleright N[M/\mathsf{s}]]\!]_\alpha = [\![\Gamma \triangleright N]\!]_\alpha \circ \langle [\![\Gamma \triangleright M]\!]_\alpha, \pi_2 \rangle$:

$$\begin{array}{ccc}
A \times [\![\Gamma]\!]_\alpha & \xrightarrow{\langle [\![M]\!]_\alpha, \pi_2 \rangle} & A \times [\![\Gamma]\!]_\alpha \\
& \searrow{[\![N[M/\mathsf{s}]]\!]_\alpha} & \downarrow{[\![N]\!]_\alpha} \\
& & [\![\sigma']\!]_\alpha
\end{array}$$

In the case of ground terms ($\Gamma = \emptyset$), this simplifies to

$$[\![N[M/s]]\!]_\alpha = [\![N]\!]_\alpha \circ [\![M]\!]_\alpha,$$

an equation that will play a significant role below.

Terms for transitions

In any τ-coalgebra (A, α), the term function $[\![\mathsf{tr}(\mathsf{s}) : \tau]\!]_\alpha$ is just α itself. If $\tau = \sigma_1 \times \sigma_2$, then $\pi_j \circ \alpha : A \to [\![\sigma_j]\!]_A$ is a σ_j-coalgebra, and a simple calculation shows that

$$[\![\mathsf{tr}(\mathsf{s}) : \sigma_j]\!]_{\pi_j \circ \alpha} = \pi_j \circ [\![\mathsf{tr}(\mathsf{s}) : \sigma_1 \times \sigma_2]\!]_\alpha = [\![\pi_j \mathsf{tr}(\mathsf{s}) : \sigma_j]\!]_\alpha. \qquad (4.1)$$

Similarly if τ is $o \Rightarrow \sigma$, then for each $d \in [\![o]\!]$, $\mathsf{ev}_d \circ \alpha : A \to [\![\sigma]\!]_A$ is a σ-coalgebra with

$$[\![\mathsf{tr}(\mathsf{s}) : \sigma]\!]_{\mathsf{ev}_d \circ \alpha} = [\![\mathsf{tr}(\mathsf{s}) \cdot d : \sigma]\!]_\alpha. \qquad (4.2)$$

Term-definability of extractions

To obtain a result for extractions similar to equations (4.1) and (4.2), we first show that the action of an extraction function ε_j is term-definable:

LEMMA 2. *For any term $\Gamma \triangleright M : \sigma_1 + \sigma_2$ of coproduct type there exist terms $\Gamma \triangleright \varepsilon_j M : \sigma_j$ for $j = 1, 2$ such that whenever $[\![\Gamma \triangleright M]\!]_\alpha(x, \gamma) \in \iota_j [\![\sigma_j]\!]_A$ then*

$$[\![\Gamma \triangleright \varepsilon_j M]\!]_\alpha(x, \gamma) = \varepsilon_j \circ [\![\Gamma \triangleright M]\!]_\alpha(x, \gamma) \in [\![\sigma_j]\!]_A.$$

Proof. Letting v_1, v_2 be variables not occurring in M, put

$$\varepsilon_1 M := \mathsf{case}\ M\ \mathsf{of}\ [\iota_1 v_1 \mapsto v_1 \mid \iota_2 v_2 \mapsto N_2],$$

where $\emptyset \triangleright N_2 : \sigma_1$ is any ground term of type σ_1 (Lemma 1). Using (Var) and (Weak) we derive terms $\Gamma, v_1 : \sigma_1 \triangleright v_1 : \sigma_1$ and $\Gamma \triangleright N_2 : \sigma_1$, from which the (Case) formation rule yields $\Gamma \triangleright \varepsilon_1 M : \sigma_1$. Then if $[\![\Gamma \triangleright M]\!]_\alpha(x, \gamma) = \iota_1(a)$, the definition of the denotation of case gives

$$[\![\Gamma \triangleright \varepsilon_1 M]\!]_\alpha(x, \gamma) = [\![\Gamma, v_1 : \sigma_1 \triangleright v_1 : \sigma_1]\!]_\alpha(x, \gamma, a) = a = \varepsilon_1([\![\Gamma \triangleright M]\!]_\alpha(x, \gamma))$$

as desired. Similarly, $\varepsilon_2 M := \mathsf{case}\ M\ \mathsf{of}\ [\iota_1 v_1 \mapsto N_1 \mid \iota_2 v_2 \mapsto v_2]$ with $\emptyset \triangleright N_1 : \sigma_2$. ∎

The "extraction terms" terms $\varepsilon_j M$ will play an important role later. Here we note that if (A, α) is a $\sigma_1 + \sigma_2$-coalgebra, and $\alpha_j : A \to [\![\sigma_j]\!]_A$ is an extension of the partial function $\varepsilon_j \circ \alpha$, then $\alpha(x) \in \iota_j [\![\sigma_j]\!]_A$ implies $\alpha_j(x) = \varepsilon_j(\alpha(x))$ and so by Lemma 2,

$$[\![\operatorname{tr}(s) : \sigma_j]\!]_{\alpha_j}(x) = \varepsilon_j([\![\operatorname{tr}(s) : \sigma_1 + \sigma_2]\!]_\alpha(x)) = [\![\varepsilon_j \operatorname{tr}(s) : \sigma_j]\!]_\alpha(x). \quad (4.3)$$

Also we will need a technical fact about the behaviour of $\varepsilon_j M$ under substitution in the case that M is a variable v: it is readily seen that

$$\varepsilon_j v[\operatorname{tr}(s)/v] = \varepsilon_j \operatorname{tr}(s). \quad (4.4)$$

Semantics of formulas

A τ-equation $\Gamma \triangleright M_1 \approx M_2$ is said to be *valid* in coalgebra α if the α-denotations $[\![\Gamma \triangleright M_1]\!]_\alpha$ and $[\![\Gamma \triangleright M_2]\!]_\alpha$ of the terms $\Gamma \triangleright M_j$ are identical. More generally we introduce a satisfaction relation

$$\alpha, x, \gamma \models \Gamma \triangleright \varphi,$$

for τ-formulas in τ-coalgebras, which expresses that $\Gamma \triangleright \varphi$ is *satisfied*, or *true*, in α at state x under the value-assigment $\gamma \in [\![\Gamma]\!]_\alpha$ to the variables of context Γ. This is defined inductively by

$$\begin{array}{lll} \alpha, x, \gamma \models \Gamma \triangleright M_1 \approx M_2 & \text{iff} & [\![\Gamma \triangleright M_1]\!]_\alpha(x, \gamma) = [\![\Gamma \triangleright M_2]\!]_\alpha(x, \gamma), \\ \alpha, x, \gamma \models \Gamma \triangleright \neg \varphi & \text{iff} & \text{not } \alpha, x, \gamma \models \Gamma \triangleright \varphi, \\ \alpha, x, \gamma \models \Gamma \triangleright \varphi_1 \wedge \varphi_2 & \text{iff} & \alpha, x, \gamma \models \Gamma \triangleright \varphi_1 \text{ and } \alpha, x, \gamma \models \Gamma \triangleright \varphi_2. \end{array}$$

$\Gamma \triangleright \varphi$ is *true at* x, written $\alpha, x \models \Gamma \triangleright \varphi$, if $\alpha, x, \gamma \models \Gamma \triangleright \varphi$ for all $\gamma \in \Gamma$. α is a *model of* $\Gamma \triangleright \varphi$, written $\alpha \models \Gamma \triangleright \varphi$, if $\alpha, x, \models \Gamma \triangleright \varphi$ for all states $x \in \operatorname{Dom} \alpha$. In that case we also say that $\Gamma \triangleright \varphi$ is *valid in* the coalgebra α.

Semantics under transitions

Results (4.1)–(4.3) give rise to corresponding results about satisfaction of certain formulas in a coalgebra α and in derivative coalgebras based on the same state set.

Suppose that a formula $(v : \tau \triangleright \varphi)$ is tr-free. Thus any raw term M in φ has no occurrence of tr, so in any τ-coalgebra (A, α) the term-denotation $[\![v : \tau \triangleright M]\!]_\alpha$ is independent of the transition α and therefore may be written $[\![v : \tau \triangleright M]\!]_A$. If $\tau = \sigma_1 \times \sigma_2$, then in the σ_j-coalgebra $\pi_j \circ \alpha : A \to [\![\sigma_j]\!]_A$ we find that in general

$$\pi_j \circ \alpha, x \models \varphi[\operatorname{tr}(s)/v] \quad \text{iff} \quad \alpha, x \models \varphi[\pi_j \operatorname{tr}(s)/v]. \quad (4.5)$$

To see this, suppose that φ is the equation $M_1 \approx M_2$, so $\varphi[\operatorname{tr}(s)/v]$ is the equation $M_1[\operatorname{tr}(s)/v] \approx M_2[\operatorname{tr}(s)/v]$, and likewise for $\varphi[\pi_j \operatorname{tr}(s)/v]$. But for $i = 1, 2$,

$$\begin{aligned}
& [\![M_i[\operatorname{tr}(s)/v]]\!]_{\pi_j \circ \alpha}(x) \\
&= [\![v : \tau \triangleright M_i]\!]_{\pi_j \circ \alpha}(x, [\![\operatorname{tr}(s) : \sigma_j]\!]_{\pi_j \circ \alpha}(x)) && \text{semantics of (Subst)} \\
&= [\![v : \tau \triangleright M_i]\!]_A(x, [\![\pi_j \operatorname{tr}(s) : \sigma_j]\!]_\alpha(x)) && \text{as } M_i \text{ is tr-free, and (4.1)} \\
&= [\![v : \tau \triangleright M_i]\!]_\alpha(x, [\![\pi_j \operatorname{tr}(s) : \sigma_j]\!]_\alpha(x)) && \text{as } M_i \text{ is tr-free} \\
&= [\![M_i[\pi_j \operatorname{tr}(s)/v]]\!]_\alpha(x),
\end{aligned}$$

from which (4.5) readily follows in this case. The inductive cases for formulas $\neg \varphi$ and $\varphi_1 \wedge \varphi_2$ are straightforward.

Similarly if τ is $o \Rightarrow \sigma$, then with the help of (4.2) we can show that for each $d \in [\![o]\!]$, the σ-coalgebra $ev_d \circ \alpha : A \to [\![\sigma]\!]_A$ has

$$ev_d \circ \alpha, x \models \varphi[\text{tr}(\text{s})/v] \quad \text{iff} \quad \alpha, x \models \varphi[\text{tr}(\text{s}) \cdot d/v]. \tag{4.6}$$

Finally there is the case $\tau = \sigma_1 + \sigma_2$. If $\alpha_j : A \to [\![\sigma_j]\!]_A$ is an extension of the partial function $\varepsilon_j \circ \alpha$, then using (4.3), if $\alpha(x) \in \iota_j[\![\sigma_j]\!]_A$,

$$\alpha_j, x \models \varphi[\text{tr}(\text{s})/v] \quad \text{iff} \quad \alpha, x \models \varphi[\varepsilon_j \text{tr}(\text{s})/v]. \tag{4.7}$$

5 Paths and bisimulations

If (A, α) and (B, β) are coalgebras for a functor T, then a relation $R \subseteq A \times B$ is a *T-bisimulation* from α to β if there exists a transition structure $\rho : R \to TR$ on R such that the projections from R to A and B are coalgebraic morphisms from ρ to α and β, i.e. the following diagram commutes:

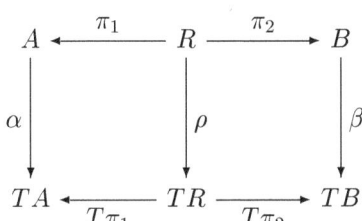

We may say that coalgebra β is the *image* of the bisimulation, or is the image of α under the bisimulation, if R is surjective, i.e. every member of B is in the image of R. Dually, α is the *domain* of the bisimulation if R is a total relation, i.e. $\text{Dom}\, R = A$.

A function $f : A \to B$ is a morphism from α to β iff its graph $\{(a, f(a)) : a \in A\}$ is a bisimulation from α to β [32, Theorem 2.5]: a morphism is essentially a functional bisimulation. When $\text{Dom}\, \alpha \subseteq \text{Dom}\, \beta$, α is a subcoalgebra of β iff the identity relation on $\text{Dom}\, \alpha$ is a bisimulation from α to β.

The above categorical definition of bisimulation appeared in [1]. It has a characterisation in terms of "liftings" of relations [14, 15]. For $R \subseteq A \times B$, define a relation $R^T \subseteq TA \times TB$ by induction on the formation of the polynomial functor T:

$$R^{\bar{D}} = \mathrm{id}_D$$
$$R^{\mathrm{Id}} = R$$
$$R^{T_1 \times T_2} = \{(x,y) : \pi_1 x R^{T_1} \pi_1 y \text{ and } \pi_2 x R^{T_2} \pi_2 y\}$$
$$R^{T_1 + T_2} = \{(\iota_1 x, \iota_1 y) : x R^{T_1} y\} \bigcup \{(\iota_2 x, \iota_2 y) : x R^{T_2} y\}$$
$$R^{T^D} = \{(f,g) : \forall d \in D,\ f(d) R^T g(d)\}.$$

These liftings preserve many basic properties of relations. Thus if R is total (Dom $R = A$) or surjective (onto B) or injective or functional, then R^T will also have the corresponding property.

THEOREM 3. *If $R \subseteq \mathrm{Dom}\,\alpha \times \mathrm{Dom}\,\beta$, where α and β are T-coalgebras, then R is a bisimulation from α to β if, and only if, for all states x in α and y in β,*

$$xRy \text{ implies } \alpha(x) R^T \beta(y). \blacksquare$$

The inverse of a bisimulation is a bisimulation, and the union of any collection of bisimulations from α to β is a bisimulation [32, Section 5]. Hence there is a largest bisimulation from α to β, which is a *symmetric* relation called *bisimilarity*. We denote this by $\sim$. States x and y are *bisimilar*, $x \sim y$, when xRy for some bisimulation R between α and β. This is intended to capture the notion that x and y are observationally indistinguishable.

Theorem 3 will now be used to show that bisimulation relations are stable under the action of term denotations, and in particular that related states assign the same values to observable terms. To explain this we need some more notation. Let (A,α) and (B,β) be τ-coalgebras, and $R \subseteq A \times B$. Then for each $\mathbb{O}$-type σ we have the lifted relation $R^{|\sigma|} \subseteq [\![\sigma]\!]_A \times [\![\sigma]\!]_B$, where $|\sigma|$ is the functor defined by σ. For observable σ, $R^{|\sigma|}$ is just the identity relation on $[\![\sigma]\!]$. The same is true whenever σ is a rigid type. For any context $\Gamma = (v_1 : \sigma_1, \ldots, v_n : \sigma_n)$ we define a relation $R^\Gamma \subseteq [\![\Gamma]\!]_A \times [\![\Gamma]\!]_B$ to be the direct product of the $R^{|\sigma_i|}$'s, i.e.

$$(\gamma_1, \ldots, \gamma_n)\, R^\Gamma\, (\gamma'_1, \ldots, \gamma'_n) \text{ iff } \gamma_i\, R^{|\sigma_i|}\, \gamma'_i \text{ for all } i \leq n.$$

For rigid Γ, R^Γ is just the identity relation on $[\![\Gamma]\!] = [\![\sigma_1]\!] \times \cdots \times [\![\sigma_n]\!]$.

THEOREM 4. *Suppose that R is a $|\tau|$-bisimulation from α to β, and let $\Gamma \triangleright M : \sigma$ be any τ-term.*

(1) *If $\gamma \in [\![\Gamma]\!]_\alpha$ and $\gamma' \in [\![\Gamma]\!]_\beta$ have $\gamma R^\Gamma \gamma'$, then*

xRy implies $[\![\,\Gamma \triangleright M\,]\!]_\alpha(x,\gamma)\, R^{|\sigma|}\, [\![\,\Gamma \triangleright M\,]\!]_\beta(y,\gamma')$.

(2) If $\Gamma \triangleright M$ is a term of observable type, and $\gamma R^\Gamma \gamma'$, then

$$xRy \text{ implies } [\![\,\Gamma \triangleright M\,]\!]_\alpha(x,\gamma) = [\![\,\Gamma \triangleright M\,]\!]_\beta(y,\gamma').$$

(3) If $\Gamma \triangleright M$ is a rigid term of observable type, and $\gamma \in [\![\,\Gamma\,]\!]$, then

$$xRy \text{ implies } [\![\,\Gamma \triangleright M\,]\!]_\alpha(x,\gamma) = [\![\,\Gamma \triangleright M\,]\!]_\beta(y,\gamma).$$

Proof. (3) is a special case of (2), which is a special case of (1). The latter is proven by induction on the formation of the term $\Gamma \triangleright M : \sigma$. To consider first the cases of base terms, suppose that $\gamma R^\Gamma \gamma'$ and xRy.

If $\Gamma \triangleright M : \sigma$ is $v : \sigma \triangleright v : \sigma$, then $[\![\,\Gamma \triangleright M\,]\!]_\alpha(x,\gamma) = \gamma$ and $[\![\,\Gamma \triangleright M\,]\!]_\beta(y,\gamma') = \gamma'$. But $\gamma R^{|\sigma|} \gamma'$ because $\gamma R^\Gamma \gamma'$ and $R^\Gamma = R^{|\sigma|}$ (so this case does not depend on whether xRy).

If $\Gamma \triangleright M : \sigma$ is $\emptyset \triangleright c : o$, then $[\![\,\Gamma \triangleright M\,]\!]_\alpha(x) = [\![\,\Gamma \triangleright M\,]\!]_\beta(y) = c$, and $cR^{|\sigma|}c$ as $R^{|\sigma|}$ is the identity relation on $[\![\,o\,]\!]$.

If $\Gamma \triangleright M : \sigma$ is $\emptyset \triangleright s : \mathsf{St}$, then $[\![\,\Gamma \triangleright M\,]\!]_\alpha(x) = x$ and $[\![\,\Gamma \triangleright M\,]\!]_\beta(y) = y$. But $xR^{|\sigma|}y$ because $R^{|\sigma|} = R^{|\mathsf{St}|} = R$ and xRy.

Now suppose $\Gamma \triangleright M : \sigma$ is $\Gamma \triangleright \mathsf{tr}(M) : \tau$, derived by rule ($\tau$-Tr), and assume that result (1) holds for $\Gamma \triangleright M : \mathsf{St}$. If $\gamma R^\Gamma \gamma'$ and xRy, then it follows from this assumption that $[\![\,\Gamma \triangleright M\,]\!]_\alpha(x,\gamma)\, R\, [\![\,\Gamma \triangleright M\,]\!]_\beta(y,\gamma')$. But here R is a $|\sigma|$-bisimulation, so Theorem 3 applies to yield $\alpha([\![\,\Gamma \triangleright M\,]\!]_\alpha(x,\gamma))\, R^{|\sigma|}\, \beta([\![\,\Gamma \triangleright M\,]\!]_\beta(y,\gamma'))$. Hence $[\![\,\Gamma \triangleright \mathsf{tr}(M)\,]\!]_\alpha(x,\gamma)\, R^{|\sigma|}\, [\![\,\Gamma \triangleright \mathsf{tr}(M)\,]\!]_\beta(y,\gamma')$, so result (1) holds for $\Gamma \triangleright \mathsf{tr}(M) : \tau$.

Next suppose $\sigma = \sigma_1 + \sigma_2$ and, for $j = 1,2$ make the assumption that (1) holds for the term $\Gamma \triangleright M : \sigma_j$. If $\gamma R^\Gamma \gamma'$ and xRy, let $a = [\![\,\Gamma \triangleright M\,]\!]_\alpha(x,\gamma)$ and $b = [\![\,\Gamma \triangleright M\,]\!]_\beta(y,\gamma')$. Then by hypothesis $aR^{|\sigma_j|}b$, and so $\iota_j(a)R^{|\sigma|}\iota_j(b)$. But $\iota_j(a) = [\![\,\Gamma \triangleright \iota_j M\,]\!]_\alpha(x,\gamma)$ and $\iota_j(b) = [\![\,\Gamma \triangleright \iota_j M\,]\!]_\beta(x,\gamma')$, so (1) holds for $\Gamma \triangleright \iota_j M : \sigma_1 + \sigma_2$.

Now let M be the raw term (case N of $[\iota_1 v_1 \mapsto M_1 \,|\, \iota_2 v_2 \mapsto M_2]$), and suppose $\Gamma \triangleright M : \sigma$ is derived by the rule (Case). Assume that (1) holds for the premises $\Gamma \triangleright N : \sigma_1 + \sigma_2$ and $\Gamma, v_j : \sigma_j \triangleright M_j : \sigma$ of that rule. If $\gamma R^\Gamma \gamma'$ and xRy, let $a = [\![\,\Gamma \triangleright N\,]\!]_\alpha(x,\gamma)$ and $b = [\![\,\Gamma \triangleright N\,]\!]_\beta(y,\gamma')$. Then by result (1) for $\Gamma \triangleright N$, $aR^{|\sigma_1 + \sigma_2|}b$, so for some j, $a \in \iota_j[\![\,\sigma_j\,]\!]_A$ and $b \in \iota_j[\![\,\sigma_j\,]\!]_B$ and $\varepsilon_j(a)R^{|\sigma_j|}\varepsilon_j(b)$. But then, by result (1) for $\Gamma, v_j : \sigma_j \triangleright M_j : \sigma$,

$$[\![\,\Gamma, v_j : \sigma_j \triangleright M_j\,]\!]_\alpha(x,\gamma,\varepsilon_j(a))\, R^{|\sigma|}\, [\![\,\Gamma, v_j : \sigma_j \triangleright M_j\,]\!]_\beta(y,\gamma',\varepsilon_j(b)).$$

Since the semantics of **case**-terms tells us that

$$[\![\Gamma \triangleright M]\!]_\alpha(x,\gamma) = [\![\Gamma, v_j : \sigma_j \triangleright M_j]\!]_\alpha(x,\gamma,\varepsilon_j(a)), \quad \text{and}$$
$$[\![\Gamma \triangleright M]\!]_\beta(y,\gamma') = [\![\Gamma, v_j : \sigma_j \triangleright M_j]\!]_\beta(y,\gamma',\varepsilon_j(b)),$$

we have $[\![\Gamma \triangleright M]\!]_\alpha(x,\gamma) \, R^{|\sigma|} \, [\![\Gamma \triangleright M]\!]_\beta(y,\gamma')$, and (1) holds for $\Gamma \triangleright M : \sigma$ in this case.

The inductive arguments for terms derived by the other rules of Figure 2 follow along similar lines, and are left to the reader. ∎

COROLLARY 5. *Suppose that R is a $|\tau|$-bisimulation from α to β, and let $\Gamma \triangleright M : \varphi$ be any observable τ-formula.*

(1) *If $\gamma R^\Gamma \gamma'$ and xRy, then*

$$\alpha, x, \gamma \models \Gamma \triangleright \varphi \text{ iff } \beta, y, \gamma' \models \Gamma \triangleright \varphi.$$

(2) *When $\Gamma \triangleright M$ is also rigid,*

$$\text{if } xRy, \text{ then} \quad \alpha, x \models \Gamma \triangleright \varphi \text{ iff } \beta, y \models \Gamma \triangleright \varphi.$$

Proof. If φ has the form $M_1 \approx M_2$, then (1) follows from part (2) of Theorem 4. For general φ, (1) then follows by induction on the formation of formulas. But (2) follows from (1) because $R^\Gamma = \text{id}$ when Γ is rigid. ∎

By part (1) of Corollary 5, if an observable formula $\Gamma \triangleright \varphi$ is valid in α and R is surjective, so that R^Γ is also surjective, $\Gamma \triangleright \varphi$ will be valid in β. On the other hand if $\beta \models \Gamma \triangleright \varphi$ and R is total, so that R^Γ is also total, then $\alpha \models \Gamma \triangleright \varphi$. In other words, validity is preserved in passing from α to β if β is the image of a bisimulation from α, and is preserved in passing from β to α if α is the domain of a bisimulation to β. If $\Gamma \triangleright \varphi$ is also rigid, then its validity is preserved by disjoint unions: given any element $\iota_j(a)$ of $\sum_I \alpha_i$ and any $\gamma \in [\![\Gamma]\!]$, if $\alpha_j \models \Gamma \triangleright \varphi$ we get $\sum_I \alpha_i, \iota_j(a), \gamma \models \Gamma \triangleright \varphi$, because $\alpha_j, a, \gamma \models \Gamma \triangleright \varphi$, $\gamma R^\Gamma \gamma$, and the insertion morphism ι_j is a bisimulation. To sum up:

THEOREM 6. *The class $\{\alpha : \alpha \models \Gamma \triangleright \varphi\}$ of all models of an observable formula is closed under domains and images of bisimulations, including domains and images of morphisms as well as subcoalgebras. If $\Gamma \triangleright \varphi$ is rigid and observable, then its class of models is also closed under disjoint unions.* ∎

The main objective of this article is to strengthen Theorem 4 to a logical characterisation of bisimilarity: states are bisimilar when they assign the same values to all ground terms of observable type, or equivalently when they satisfy the same rigid observable formulas (see Theorem 20). The key to this is the relation $\equiv_{\alpha\beta}$ defined by

$x \equiv_{\alpha\beta} y$ iff every *ground observable* term M has $[\![M]\!]_\alpha(x) = [\![M]\!]_\beta(y)$.

$\equiv_{\alpha\beta}$ is a bisimulation from α to β (Lemma 19), and turns out to be the largest one (Theorem 20). The proof of this requires the development of another characterisation of bisimulation, using the notion of "paths" between functors introduced in [18, Section 6].

A *path* is a finite list of symbols of the kinds π_j, ε_j, ev_d. Write $p.q$ for the concatenation of lists p and q. The notation $T \overset{p}{\rightsquigarrow} S$ means that p is a path from functor T to functor S, and is defined by the following conditions

- $T \overset{\langle\rangle}{\rightsquigarrow} T$, where $\langle\rangle$ is the empty path.

- $T_1 \times T_2 \overset{\pi_j.p}{\rightsquigarrow} S$ whenever $T_j \overset{p}{\rightsquigarrow} S$, for $j = 1, 2$.

- $T_1 + T_2 \overset{\varepsilon_j.p}{\rightsquigarrow} S$ whenever $T_j \overset{p}{\rightsquigarrow} S$, for $j = 1, 2$.

- $T^D \overset{ev_d.p}{\rightsquigarrow} S$ for all $d \in D$ whenever $T \overset{p}{\rightsquigarrow} S$.

It is evident that for any path $T \rightsquigarrow S$, S is one of the functors involved in the formation of T.

A path $T \overset{p}{\rightsquigarrow} S$ induces a partial function $p_A : TA \rightharpoonup SA$ for each set A, defined by induction on the length of p as follows.

- $\langle\rangle_A : TA \rightharpoonup TA$ is the identity function id_{TA}, so is totally defined.

- $(\pi_j.p)_A = p_A \circ \pi_j$, the composition of $T_1 A \times T_2 A \overset{\pi_j}{\longrightarrow} T_j A \overset{p_A}{\rightharpoonup} SA$. Thus $x \in \mathrm{Dom}\,(\pi_j.p)_A$ iff $\pi_j(x) \in \mathrm{Dom}\,p_A$.

- $(\varepsilon_j.p)_A = p_A \circ \varepsilon_j$, the composition of $T_1 A + T_2 A \overset{\varepsilon_j}{\rightharpoonup} T_j A \overset{p_A}{\rightharpoonup} SA$. Thus $x \in \mathrm{Dom}\,(\varepsilon_j.p)_A$ iff $x \in \mathrm{Dom}\,\varepsilon_j$ and $\varepsilon_j(x) \in \mathrm{Dom}\,p_A$.

- $(ev_d.p)_A = p_A \circ ev_d$, the composition of $(TA)^D \overset{ev_d}{\longrightarrow} TA \overset{p_A}{\rightharpoonup} SA$. Thus $f \in \mathrm{Dom}\,(ev_d.p)_A$ iff $f(d) \in \mathrm{Dom}\,p_A$.

A path $T \leadsto S$ is a *state path* if $S = \mathrm{Id}$, an *observation path* if $S = \bar{D}$ for some set D, and a *basic* path if it is either.

LEMMA 7. *Suppose T is a polynomial functor, A a set, and $x \in TA$. Then there exists a basic path $T \overset{p}{\leadsto} S$ with $x \in \mathrm{Dom}\, p_A$.*

Proof. By induction of the construction of T. If $T = \mathrm{Id}$ or $T = \bar{D}$, let p be the empty path $T \leadsto T$, so that $\mathrm{Dom}\, p_A = TA$. If $T = T_1 + T_2$, then $x \in \iota_j T_j A$ for some $j = 1, 2$. Then $\varepsilon_j(x) \in T_j A$ so by induction hypothesis there is a basic $T_j \overset{p}{\leadsto} S$ with $\varepsilon_j(x) \in \mathrm{Dom}\, p_A$. Then $T_1 + T_2 \overset{\varepsilon_j.p}{\leadsto} S$ is basic, with $x \in \mathrm{Dom}\, (\varepsilon_j.p)_A$.

The cases of $= T_1 \times T_2$ and $T = T_1^D$ are similarly straightforward. ■

A T-bisimulation can be characterised as a relation that is "preserved" by the partial functions induced by state and observation paths from T. To explain this we adopt the convention that whenever we write "$f(x)\, Q\, g(y)$" for some relation Q and some partial functions f and g we mean that $f(x)$ is defined iff $g(y)$ is defined, and $(f(x), g(y)) \in Q$ when they are both defined.

THEOREM 8. *Let $R \subseteq A \times B$, $x \in TA$, and $y \in TB$, where T is a polynomial functor. Then the following are equivalent.*

(1) $xR^T y$.

(2) *For all paths $T \overset{p}{\leadsto} S$, $p_A(x) R^S p_B(y)$.*

(3) *For all state paths $T \overset{p}{\leadsto} \mathrm{Id}$, $p_A(x)\, R\, p_B(y)$; and for all observation paths $T \overset{p}{\leadsto} \bar{D}$, $p_A(x) = p_B(y)$.*

Proof. (1) implies (2): this is proven by induction on the construction of T. If $T = \mathrm{Id}$ or $T = \bar{D}$, then the only path p from T is the empty path $T \leadsto T$, with $p_A = \mathrm{id}_{TA}$ and $p_B = \mathrm{id}_{TB}$. Thus if (1) holds, then $p_A(x) = x\, R^T\, y = p_B(y)$, giving (2). If $T = T_1 + T_2$, then from $xR^T y$ we conclude that for some $j = 1, 2$, $x \in \iota_j T_j A$, $y \in \iota_j T_j B$, and $\varepsilon_j(x) R^{T_j} \varepsilon_j(y)$. But now a path $T \overset{p}{\leadsto} S$ must have the form $T_1 + T_2 \overset{\varepsilon_{j'}.q}{\leadsto} S$ with $T_{j'} \overset{q}{\leadsto} S$ for some $j' = 1, 2$. If $j' \neq j$, then neither $\varepsilon_j(x) \in \mathrm{Dom}\, q_A$ nor $\varepsilon_j(y) \in \mathrm{Dom}\, q_B$, so neither $p_A(x)$ nor $p_B(y)$ is defined. If $j' = j$, then by induction hypothesis on T_j, since $\varepsilon_j(x) R^{T_j} \varepsilon_j(y)$, we have $q_A(\varepsilon_j(x))$ defined iff $q_B(\varepsilon_j(y))$ is defined, and when both are defined $p_A(x) = q_A(\varepsilon_j(x)) R^S q_B(\varepsilon_j(y)) = p_B(y)$ as desired. The cases of $= T_1 \times T_2$ and $T = T_1^D$ are similarly straightforward.

That (2) implies (3) is immediate from the definitions, so it remains to show (3) implies (1). If $T = \mathrm{Id}$ or $T = \bar{D}$, let p be the empty path

$T \rightsquigarrow T$. Then if (3) holds we have $x = p_A(x) \, R^T \, p_B(y) = y$, giving (1). If $T = T_1 + T_2$, then for some j, $x \in \iota_j T_j A$ and so $\varepsilon_j(x) \in T_j A$. Now by Lemma 7 there exists a basic path $T_j \overset{q}{\rightsquigarrow} S$ with $\varepsilon_j(x) \in \mathrm{Dom}\, q_A$, so $(\varepsilon_j.q)_A = q_A(\varepsilon_j(x))$ is defined. If (3) holds for T, then applying it to the path $p = \varepsilon_j.q$ we conclude that $(\varepsilon_j.q)_B(y)$ is defined, so $y \in \iota_j T_j B$ and $\varepsilon_j(y) \in \mathrm{Dom}\, q_B$. But if $T_j \overset{q'}{\rightsquigarrow} S'$ is *any* basic path out of T_j, the reasoning just given shows that $q'_A(\varepsilon_j(x))$ is defined iff $q'_B(\varepsilon_j(y))$ is defined, and when they are both defined, (3) for T ensures that $q'_A(\varepsilon_j(x)) R^S q'_B(\varepsilon_j(y))$. This proves that (3) holds for T_j, so by induction hypothesis on T_j, $\varepsilon_j(x) R^{T_j} \varepsilon_j(y)$, which implies $x R^T y$ as desired. Again the cases of $T = T_1 \times T_2$ and $T = T_1^D$ are left to the reader. ∎

Combining this result with Theorem 3 gives the desired "dynamic" characterisation of bisimulations:

THEOREM 9. *If $A \overset{\alpha}{\longrightarrow} TA$ and $B \overset{\beta}{\longrightarrow} TB$ are coalgebras for a polynomial functor T, then a relation $R \subseteq A \times B$ is a T-bisimulation if, and only if, xRy implies*

(1) *for all state paths $T \overset{p}{\rightsquigarrow} \mathrm{Id}$, $p_A(\alpha(x)) \, R \, p_B(\beta(y))$; and*

(2) *for all observation paths $T \overset{p}{\rightsquigarrow} \bar{D}$, $p_A(\alpha(x)) = p_B(\beta(y))$.*

Proof. By Theorem 8, (1) and (2) together are equivalent to $\alpha(x) R^T \beta(y)$. ∎

COROLLARY 10. *If $C \subseteq \mathrm{Dom}\, \alpha$, then C is a subcoalgebra of α iff $x \in C$ implies $p_A(\alpha(x)) \in C$ for all state paths $T \overset{p}{\rightsquigarrow} \mathrm{Id}$ such that $p_A(\alpha(x))\downarrow$.*

Proof. To say that C is a subcoalgebra of α means that there is some T-transition structure on C that is a subcoalgebra of α. Such a structure is unique, and exists iff the identity relation $\Delta_C = \{(x,x) : x \in C\}$ on C is a bisimulation relation on α [32, Proposition 6.2]. Now apply the Theorem with $R = \Delta_C$ and $\alpha = \beta$, and use the fact that $p_A(\alpha(x)) \Delta_C p_A(\alpha(x))$ iff $p_A(\alpha(x)) \in C$. ∎

This characterisation makes it easy to see that if R is a bisimulation from α to β, then $\mathrm{Dom}\, R$ is a subcoalgebra of α. For if $x \in \mathrm{Dom}\, R$ and $p_A(\alpha(x))\downarrow$, then xRy for some y, so $p_A(\alpha(x)) R p_B(\beta(y))$ by 8 and hence $p_A(\alpha(x)) \in \mathrm{Dom}\, R$. Similarly, the *image* of R is seen to be a subcoalgebra of β.

Theorem 9 also yields a characterisation of morphisms between polynomial coalgebras:

COROLLARY 11. *A function $f : A \to B$ is a T-morphism from (A, α) to (B, β) if, and only if,*

(1) *$f(p_A(\alpha(x))) = p_B(\beta(f(x)))$ for all state paths $T \xrightarrow{p} \mathrm{Id}$; and*

(2) *$p_A(\alpha(x)) = p_B(\beta(f(x)))$ for all observation paths $T \xrightarrow{p} \bar{D}$.*

Proof. Let R be the graph of f, i.e. xRy iff $y = f(x)$. Then f is a T-morphism iff R is a T-bisimulation. Now apply 9. ∎

6 Definability of path action

We have seen that path functions are an effective tool in the structural analysis of polynomial coalgebras. Their use in logical characterisations derives from the fact that the action of a path function is definable by a (ground) term, a fact that may be an intuitively expected feature of our formalism, but one which requires a delicate analysis to establish.

THEOREM 12. *For any path $|\tau| \xrightarrow{p} |\sigma|$ and variable v there exists a tr-free τ-term of the form*

$$v : \tau \triangleright \bar{p} : \sigma$$

such that for any τ-coalgebra (A, α) and any $x \in A$, if $\alpha(x) \in \mathrm{Dom}\, p_A$ then

$$p_A(\alpha(x)) = [\![\bar{p}[\mathsf{tr}(\mathsf{s})/v]]\!]_\alpha(x).$$

Proof. Note that since $\emptyset \triangleright \mathsf{tr}(\mathsf{s}) : \tau$ is a τ-term, so too is $\emptyset \triangleright \bar{p}[\mathsf{tr}(\mathsf{s})/v] : \sigma$ by the substitution rule (Subst). Hence $\bar{p}[\mathsf{tr}(\mathsf{s})/v]$ is a ground term of type σ.

Since the symbol tr does not occur in the raw term $\bar{p}$, the denotation $[\![v : \tau \triangleright \bar{p} : \sigma]\!]_\alpha$ depends only on A, not on α, and may be written $[\![v : \tau \triangleright \bar{p} : \sigma]\!]_A$.

The Theorem is proven by induction on the length of paths ending at $|\sigma|$. First, if $\tau = \sigma$ and p is the empty path $|\sigma| \rightsquigarrow |\sigma|$, let $\bar{p} = v$. It follows that $[\![\bar{p}[\mathsf{tr}(\mathsf{s})/v]]\!]_\alpha(x) = [\![\mathsf{tr}(\mathsf{s})]\!]_\alpha(x) = \alpha(x) = p_A(\alpha(x))$ since p_A is the identity function.

Now take the case $\tau = \tau_1 + \tau_2$. Then $p = \varepsilon_j.q$ for some j and some path $|\tau_j| \xrightarrow{q} |\sigma|$. From the base term $v : \tau \triangleright v : \tau$ we obtain a term $v : \tau \triangleright \varepsilon_j v : \tau_j$ by Lemma 2, and by induction hypothesis there is a term $v : \tau_j \triangleright \bar{q} : \sigma$ fulfilling the Theorem for q. By (Subst) these terms yield the term $v : \tau \triangleright \bar{p} : \sigma$ where $\bar{p} = \bar{q}[\varepsilon_j v/v]$. Then $\bar{p}[\mathsf{tr}(\mathsf{s})/v] = \bar{q}[(\varepsilon_j v[\mathsf{tr}(\mathsf{s})/v])/v] = \bar{q}[\varepsilon_j \mathsf{tr}(\mathsf{s})/v]$ by equation (4.4). If (A, α) is a τ-coalgebra and $\alpha(x) \in \mathrm{Dom}\, p_A$, then $\alpha(x) \in \iota_j[\![\tau_j]\!]_A$ and $\varepsilon_j(\alpha(x)) \in \mathrm{Dom}\, q_A$. Let $\alpha_j : A \to [\![\tau_j]\!]_A$ be any extension of the partial function $\varepsilon_j \circ \alpha$. Then

$[\![\bar{p}[\mathsf{tr}(\mathsf{s})/v]]\!]_\alpha(x)$
$= [\![\bar{q}[\varepsilon_j\mathsf{tr}(\mathsf{s})/v]]\!]_\alpha(x)$
$= [\![v : \tau_j \rhd \bar{q} : \sigma]\!]_\alpha(x, [\![\varepsilon_j\mathsf{tr}(\mathsf{s}) : \tau_j]\!]_\alpha(x))$ semantics of (Subst)
$= [\![v : \tau_j \rhd \bar{q} : \sigma]\!]_A(x, [\![\mathsf{tr}(\mathsf{s}) : \tau_j]\!]_{\alpha_j}(x))$ $\bar{q}$ is tr-free, and (4.3)
$= [\![\bar{q}[\mathsf{tr}(\mathsf{s})/v] : \tau_j]\!]_{\alpha_j}(x)$ semantics of (Subst)
$= q_A(\alpha_j(x))$ hypothesis on q
$= q_A(\varepsilon_j(\alpha(x))) = p_A(\alpha(x)).$

This establishes the Theorem for the case $\tau = \tau_1 + \tau_2$.

Next suppose $\tau = \tau_1 \times \tau_2$ and so $p = \pi_j.q$ for some j and some path $|\tau_j|\xrightarrow{q}|\sigma|$. Then $v : \tau \rhd \pi_j v : \tau_j$ (Proj$_j$) and $v : \tau_j \rhd \bar{q} : \sigma$ yield $v : \tau \rhd \bar{p} : \sigma$ with $\bar{p} = \bar{q}[\pi_j v/v]$ by (Subst). Moreover $\bar{p}[\mathsf{tr}(\mathsf{s})/v] = \bar{q}[\pi_j\mathsf{tr}(\mathsf{s})/v]$. This time if $\alpha(x) \in \mathrm{Dom}\, p_A$, then $\pi_j(\alpha(x)) \in \mathrm{Dom}\, q_A$. The proof of the Theorem for p now proceeds just as in the $\tau_1 + \tau_2$ case, but using the coalgebra $\pi_j \circ \alpha : A \to [\![\tau_j]\!]_A$ in place of α_j and (4.1) in place of (4.3), as well as the induction hypothesis on q.

Finally there is the case $\tau = (o \Rightarrow \tau_1)$ and $p = \mathsf{ev}_d.q$ with $d \in [\![o]\!]$ and $|\tau_1|\xrightarrow{q}|\sigma|$. From (Con) and (Weak) we get $v : \tau \rhd d : o$, and so by (App), $v : \tau \rhd v \cdot d : \tau_1$. Substitution and the hypothesis on q then allow us to put $\bar{p} = \bar{q}[v \cdot d/v]$. The rest of the proof is as in the $\tau_1 + \tau_2$ case, but using the coalgebra $\mathsf{ev}_d \circ \alpha : A \to [\![\tau_1]\!]_A$ in place of α_j and (4.2) in place of (4.3). ■

The term function $[\![\bar{p}[\mathsf{tr}(\mathsf{s})/v]]\!]_\alpha$ from Theorem 12 has domain A, and so may not be identical to $p_A \circ \alpha$ if p_A is partial. This is only an issue when the path p includes an extraction symbol ε_j (for otherwise p_A is total), but further use of **case** allows the construction of observable terms that "discriminate" between the two summands of a coproduct $[\![\tau_1]\!]_A + [\![\tau_2]\!]_A$ and determine whether $p_A(\alpha(x))$ is defined. For this to work we need the (plausible) assumption that we have available at least one observable type μ that is *non-trivial* in the sense that $[\![\mu]\!]$ has at least two distinct members, say c_1 and c_2. These can be used to form the term $v : \tau_1 + \tau_2 \rhd P : \mu$, where

$$P := \mathsf{case}\ v\ \mathsf{of}\ [\iota_1 v_1 \mapsto c_1 \mid \iota_2 v_2 \mapsto c_2]. \tag{6.1}$$

Then in any $\tau_1 + \tau_2$-coalgebra α, the ground term $P[\mathsf{tr}(\mathsf{s})/v] : \mu$ is a discriminator:

$$[\![P[\mathsf{tr}(\mathsf{s})/v]]\!]_\alpha(x) = c_j \quad \mathrm{iff} \quad \alpha(x) \in \iota_j[\![\tau_j]\!]_A = \mathrm{Dom}\,\varepsilon_j. \tag{6.2}$$

This observation will be applied to establish (see Corollary 14) that if states x in coalgebra α and y in coalgebra β assign the same values to all ground observable terms, then $p_A(\alpha(x))$ is defined iff $p_B(\beta(y))$ is defined. The key here is the following technical result, which uses similar ideas to those found in the proof of Theorem 12.

LEMMA 13. *Suppose that τ has at least one non-trivial observable subtype μ. For any path $|\tau| \overset{p}{\rightsquigarrow} |\sigma|$ there is a finite set X_p of tr-free τ-terms of the form $v : \tau \triangleright M : \mu$, and an associated set $X_p[\mathrm{tr}] = \{M[\mathrm{tr}(s)/v] : (v : \tau \triangleright M) \in X_p\}$ of ground observable terms of type μ, such that for any τ-coalgebras (A, α) and (B, β) and any $x \in A$ and $y \in B$, if $[\![N]\!]_\alpha(x) = [\![N]\!]_\beta(y)$ for all $N \in X_p[\mathrm{tr}]$, then $\alpha(x) \in \mathrm{Dom}\, p_A$ iff $\beta(y) \in \mathrm{Dom}\, p_B$.*

Proof. By induction on the length of p. If $\tau = \sigma$ and p is the empty path $|\sigma| \rightsquigarrow |\sigma|$, put $X_p = \emptyset$. Then $X_p[\mathrm{tr}] = \emptyset$ and the Lemma holds (vacuously) since p_A and p_B are total (identity) functions, so $\alpha(x) \in \mathrm{Dom}\, p_A$ and $\beta(y) \in \mathrm{Dom}\, p_B$ are both true.

For the inductive case of coproducts, suppose $\tau = \tau_1 + \tau_2$ and $p = \varepsilon_j.q$ for some j and some path $|\tau_j| \overset{q}{\rightsquigarrow} |\sigma|$. By induction hypothesis there is a finite set X_q of terms of the form $v : \tau_j \triangleright M : \mu$ that fulfils the Lemma for q. Apply substitution (Subst) to the term $v : \tau \triangleright \varepsilon_j v : \tau_j$ to generate the finite set

$$X_p = \{(v : \tau \triangleright M[\varepsilon_j v/v]) : (v : \tau_j \triangleright M) \in X_q\} \cup \{(v : \tau \triangleright P : \mu)\},$$

with P being the raw case-term defined in line (6.1) above. Note that if $(v : \tau_j \triangleright M) \in X_q$, then by equation (4.4),

$$M[\varepsilon_j \mathrm{tr}(s)/v] = M[(\varepsilon_j v[\mathrm{tr}(s)/v])/v] = (M[\varepsilon_j v/v])[\mathrm{tr}(s)/v] \in X_p[\mathrm{tr}].$$

Now suppose state x in τ-coalgebra α assigns the same values as y in β to all terms in $X_p[\mathrm{tr}]$. Then in particular $[\![P[\mathrm{tr}(s)/v]]\!]_\alpha(x) = [\![P[\mathrm{tr}(s)/v]]\!]_\beta(y)$, so for some $j' = 1, 2$, $\alpha(x) \in \iota_{j'}[\![\tau_{j'}]\!]_A$ and $\beta(y) \in \iota_{j'}[\![\tau_{j'}]\!]_B$. If $j' \neq j$, then $\alpha(x) \notin \iota_j[\![\tau_j]\!]_A$ and $\beta(y) \notin \iota_j[\![\tau_j]\!]_B$, so $\alpha(x) \notin \mathrm{Dom}\,(\varepsilon_j.q)_A$ and $\beta(y) \notin \mathrm{Dom}\,(\varepsilon_j.q)_B$, satisfying the Lemma. If however $j' = j$, let $\alpha_j : A \to [\![\tau_j]\!]_A$ be any extension of $\varepsilon_j \circ \alpha$ and $\beta_j : B \to [\![\tau_j]\!]_B$ be any extension of $\varepsilon_j \circ \beta$. For each term $v : \tau_j \triangleright M$ of X_q, reasoning as in the proof of Theorem 12 with M in place of $\bar{q}$ shows that $[\![M[\varepsilon_j \mathrm{tr}(s)/v]]\!]_\alpha(x) = [\![M[\mathrm{tr}(s)/v]]\!]_{\alpha_j}(x)$ and likewise $[\![M[\varepsilon_j \mathrm{tr}(s)/v]]\!]_\beta(y) = [\![M[\mathrm{tr}(s)/v]]\!]_{\beta_j}(y)$. But $M[\varepsilon_j \mathrm{tr}(s)/v] \in X_p[\mathrm{tr}]$ and x in α agrees with y in β on $X_p[\mathrm{tr}]$, so $[\![M[\varepsilon_j \mathrm{tr}(s)/v]]\!]_\alpha(x) = [\![M[\varepsilon_j \mathrm{tr}(s)/v]]\!]_\beta(y)$. Therefore $[\![M[\mathrm{tr}(s)/v]]\!]_{\alpha_j}(x) = [\![M[\mathrm{tr}(s)/v]]\!]_{\beta_j}(y)$. This shows that x in α_j agrees with y in β_j on all terms from $X_q[\mathrm{tr}]$, so by the induction hypothesis on q, $\alpha_j(x) \in \mathrm{Dom}\, q_A$ iff $\beta_j(y) \in \mathrm{Dom}\, q_B$. But $\alpha(x) \in \iota_j[\![\tau_j]\!]_A$ and $\beta(y) \in \iota_j[\![\tau_j]\!]_B$ (as j'=j), so $\alpha_j(x) = \varepsilon_j(\alpha(x))$ and $\beta_j(y) = \varepsilon_j(\beta(y))$. It follows that $\alpha(x) \in \mathrm{Dom}\,(\varepsilon_j.q)_A$

iff $\beta(y) \in \text{Dom}\,(\varepsilon_j.q)_B$, as desired, completing the proof for the case $\tau = \tau_1 + \tau_2$.

Next is the case $\tau = \tau_1 \times \tau_2$ with $p = \pi_j.q$ for some j and some path $|\tau_j| \overset{q}{\leadsto} |\sigma|$. With X_q given by the induction hypothesis on q, put $X_p = \{(v : \tau \triangleright M[\pi_j v/v]) : (v : \tau_j \triangleright M) \in X_q\}$. Then $(v : \tau_j \triangleright M) \in X_q$ implies $M[\pi_j \text{tr}(\text{s})/v] \in X_p[\text{tr}]$, and if x in α agrees with y in β on $X_p[\text{tr}]$, reasoning similar to the previous case leads to $[\![M[\text{tr}(\text{s})/v]]\!]_{\pi_j \circ \alpha}(x) = [\![M[\text{tr}(\text{s})/v]]\!]_{\pi_j \circ \beta}(y)$. Hence x in coalgebra $\pi_j \circ \alpha$ agrees with y in $\pi_j \circ \beta$ on $X_q[\text{tr}]$, so by hypothesis on q, $\pi_j \circ \alpha(x) \in \text{Dom}\,q_A$ iff $\pi_j \circ \beta(y) \in \text{Dom}\,q_B$, i.e. $\alpha(x) \in \text{Dom}\,(\pi_j.q)_A$ iff $\beta(y) \in \text{Dom}\,(\pi_j.q)_B$.

Finally there is the case $\tau = (o \Rightarrow \tau_1)$ and $p = ev_d.q$ with $d \in [\![o]\!]$ and $|\tau_1| \overset{q}{\leadsto} |\sigma|$. This time $X_p = \{(v : \tau \triangleright M[v \cdot d/v]) : (v : \tau_1 \triangleright M) \in X_q\}$. Similarly to the $\tau_1 + \tau_2$ case, we get that if x in α agrees with y in β on $X_p[\text{tr}]$, then x in coalgebra $ev_d \circ \alpha$ agrees with y in $ev_d \circ \beta$ on $X_q[\text{tr}]$, so $ev_d \circ \alpha(x) \in \text{Dom}\,q_A$ iff $ev_d \circ \beta(y) \in \text{Dom}\,q_B$, i.e. $\alpha(x) \in \text{Dom}\,(ev_d.q)_A$ iff $\beta(y) \in \text{Dom}\,(ev_d.q)_B$. ∎

COROLLARY 14. *Let (A, α) and (B, β) be τ-coalgebras, where τ has at least one non-trivial observable subtype μ. Let $x \in A$ and $y \in B$ have $[\![N]\!]_\alpha(x) = [\![N]\!]_\beta(y)$ for all ground τ-terms N of type μ. Then for any path $|\tau| \overset{p}{\leadsto} |\sigma|$, $\alpha(x) \in \text{Dom}\,p_A$ iff $\beta(y) \in \text{Dom}\,p_B$.* ∎

In a similar vein to Lemma 13, we can define *formulas* that characterise the property of belonging to the domain of a path function:

LEMMA 15. *Suppose that τ has at least one non-trivial observable subtype μ. For any path $|\tau| \overset{p}{\leadsto} |\sigma|$ and variable v there exists a tr-free τ-formula $v : \tau \triangleright \varphi_p$ such that for any τ-coalgebra (A, α) and any $x \in A$,*

$$\alpha, x \models \varphi_p[\text{tr}(\text{s})/v] \quad \text{iff} \quad \alpha(x) \in \text{Dom}\,p_A.$$

Proof. If p is the empty path, then $\alpha(x) \in \text{Dom}\,p_A$ for all $x \in A$. In that case let φ_p be the equation $c \approx c$ for some $c \in [\![\mu]\!]$. Then $\varphi_p[\text{tr}(\text{s})/v]$ is also $c \approx c$, which is satisfied in α at all $x \in A$.

For the inductive case of coproducts, suppose $\tau = \tau_1 + \tau_2$ and $p = \varepsilon_j.q$ for some j and some path $|\tau_j| \overset{q}{\leadsto} |\sigma|$. Let φ_p be $(P \approx c_j) \wedge \varphi_q[\varepsilon_j v/v]$, where P is the discriminating case-term of line (6.1), and φ_q is given by the induction hypothesis as fulfilling the Lemma for q. Now $\alpha(x) \in \text{Dom}\,p_A$ iff $\alpha(x) \in \iota_j[\![\tau_j]\!]_A$ and $\varepsilon_j \circ \alpha(x) \in \text{Dom}\,q_A$. But by line (6.2), $\alpha(x) \in \iota_j[\![\tau_j]\!]_A$ iff $\alpha, x \models P[\text{tr}(\text{s})/v] \approx c_j$. Also, if $\alpha_j : A \to [\![\tau_j]\!]_A$ is any extension of $\varepsilon_j \circ \alpha$, and $\varepsilon_j \circ \alpha(x)$ is defined, then by induction hypothesis on q, $\varepsilon_j \circ \alpha(x) \in \text{Dom}\,q_A$

iff $\alpha_j, x \models \varphi_q[\text{tr}(s)/v]$, which holds iff $\alpha, x \models \varphi_q[\varepsilon_j \text{tr}(s)/v]$ by (4.7). So $\alpha(x) \in \text{Dom}\, p_A$ iff the formula

$$(P[\text{tr}(s)/v] \approx c_j) \wedge \varphi_q[\varepsilon_j \text{tr}(s)/v]$$

is satisfied in α at x. But this formula is $\varphi_p[\text{tr}(s)/v]$ (see (4.4)), so the Lemma holds for the case $\tau = \tau_1 + \tau_2$.

If $\tau = \tau_1 \times \tau_2$ with $p = \pi_j.q$ for some j and some path $|\tau_j| \xrightarrow{q} |\sigma|$, let φ_p be $\varphi_q[\pi_j v/v]$. Now $\alpha(x) \in \text{Dom}\, p_A$ iff $\pi_j \circ \alpha(x) \in \text{Dom}\, q_A$ iff $\pi_j \circ \alpha, x \models \varphi_q[\text{tr}(s)/v]$ (by hypothesis on q) iff $\alpha, x \models \varphi_q[\pi_j \text{tr}(s)/v]$ by (4.5). But $\varphi_q[\pi_j \text{tr}(s)/v]$ is $\varphi_p[\text{tr}(s)/v]$ in this case.

If $\tau = (o \Rightarrow \tau_1)$ and $p = ev_d.q$ with $d \in [\![o]\!]$ and $|\tau_1| \xrightarrow{q} |\sigma|$, let φ_p be $\varphi_q[v \cdot d/v]$. Then analogously to the previous case, but using (4.6), we get $\alpha(x) \in \text{Dom}\, p_A$ iff $ev_d \circ \alpha(x) \in \text{Dom}\, q_A$ iff $ev_d \circ \alpha, x \models \varphi_q[\text{tr}(s)/v]$ iff $\alpha, x \models \varphi_q[\text{tr}(s) \cdot d/v]$ as desired. ∎

Definable modalities

The terms provided by Theorem 12 can be used to formalise modal assertions that certain formulas will be true after the execution of state transitions induced by path functions. This is a special case of the following result, which shows that we can express more general modal assertions associated with state transitions $x \mapsto [\![M]\!]_\alpha(x)$ defined by any term M of state type.

THEOREM 16. *If M is any ground term of type* St, *and φ any ground formula, then in any τ-coalgebra (A, α),*

$$\alpha, [\![M]\!]_\alpha(x) \models \varphi \quad \text{iff} \quad \alpha, x \models \varphi[M/s].$$

Proof. By induction on the formation of φ, with the inductive cases of the connectives $\neg$ and $\wedge$ being straightforward, so it suffices to consider the case that φ is an equation $N_1 \approx N_2$. For $i = 1, 2$, $[\![N_i]\!]_\alpha([\![M]\!]_\alpha(x))$ equals $[\![N_i[M/s]]\!]_\alpha(x)$ by the semantics of (s-Subst). Hence

$$[\![N_1]\!]_\alpha([\![M]\!]_\alpha(x)) = [\![N_2]\!]_\alpha([\![M]\!]_\alpha(x))) \quad \text{iff}$$
$$[\![N_1[M/s]]\!]_\alpha(x) = [\![N_2[M/s]]\!]_\alpha(x),$$

which proves the Theorem for this case of φ. ∎

Thus the formula $\varphi[M/s]$ is seen to expresses the modal assertion "after action M, φ". The above-mentioned case of state transitions induced by path functions is given by

COROLLARY 17. *Let $|\tau|\overset{p}{\leadsto}\mathsf{Id}$ be a state path and φ any ground formula. Then in any τ-coalgebra (A,α), if $\alpha(x) \in \operatorname{Dom} p_A$,*

$$\alpha, p_A(\alpha(x)) \models \varphi \quad \text{iff} \quad \alpha, x \models \varphi[M_p/\mathsf{s}],$$

where M_p is the ground term $\bar{p}[\mathsf{tr}(\mathsf{s})/v]$ of type St given by Theorem 12.

Proof. From Theorems 12 and 16, as $p_A(\alpha(x)) = [\![M_p]\!]_\alpha(x)$. ∎

Morphisms and term-value preservation

To round out this discussion of definability of path action, here is a characterisation of coalgebraic morphisms as functions preserving values of certain terms.

THEOREM 18. *A function $f : A \to B$ between τ-coalgebras (A,α) and (B,β) is a $|\tau|$-morphism if, and only if,*

(1) $f([\![M]\!]_\alpha(x)) = [\![M]\!]_\beta(f(x))$ *for all ground τ-terms M of type St; and*

(2) $[\![M]\!]_\alpha(x) = [\![M]\!]_\beta(f(x))$ *for all ground τ-terms M of observable type.*

Proof. If f is a morphism, then the relation xRy iff $y = f(x)$ is a bisimulation, so (1) and (2) follow from Theorem 4 parts (1) and (2).

For the converse it is enough to know that (1) and (2) hold whenever M is the ground term $\bar{p}[\mathsf{tr}(\mathsf{s})/v]$, given by Theorem 12, for the cases that p is a state path or an observation path, respectively. For then (1) gives $f(p_A(\alpha(x))) = p_B(\beta(f(x)))$ for any state path, while (2) gives $p_A(\alpha(x)) = p_B(\beta(f(x)))$ for any observation path. But by Corollary 11, this is enough to make f is a $|\tau|$-morphism. ∎

7 Logical characterisation of bisimilarity

Recall the relation $\equiv_{\alpha\beta}$ between the state sets of two τ-coalgebras, which has $x \equiv_{\alpha\beta} y$ iff $[\![M]\!]_\alpha(x) = [\![M]\!]_\beta(y)$ for every ground term M of observable type. We are now ready to show that $\equiv_{\alpha\beta}$ is identical to the bisimilarity relation between α and β, and that it has a characterisation in terms of satisfaction of observable formulas. The first step is

LEMMA 19. *If τ has at least one non-trivial observable subtype, then $\equiv_{\alpha\beta}$ is a bisimulation from α to β.*

Proof. We use the characterisation of bisimulations given by Theorem 9.

Let $x \equiv_{\alpha\beta} y$ and suppose $|\tau|\overset{p}{\leadsto}|\sigma|$ is a basic path from $|\tau|$. Since x and y assign the same values to all ground observable terms, it is immediate from

Corollary 14 that $p_A(\alpha(x))$ is defined iff $p_B(\beta(y))$ is defined. Suppose then that they are both defined.

(i) If p is a state path, i.e. $\sigma = \mathsf{St}$ and $|\sigma| = \mathsf{Id}$, we must show $p_A(\alpha(x)) \equiv_{\alpha\beta} p_B(\beta(y))$. Let $v : \tau \triangleright \bar{p} : \sigma$ be the term given by Theorem 12, and M the ground term $\bar{p}[\mathsf{tr}(\mathsf{s})/v]$ of type St. Then $p_A(\alpha(x)) = [\![M]\!]_\alpha(x) \in A$ and $p_B(\beta(y)) = [\![M]\!]_\beta(y) \in B$. Now if $N : o$ is any ground observable term, then the rule (s-Subst) gives the ground term $N[M/\mathsf{s}] : o$, and so as $x \equiv_{\alpha\beta} y$ we get $[\![N[M/\mathsf{s}]]\!]_\alpha(x) = [\![N[M/\mathsf{s}]]\!]_\beta(y)$. Hence by the semantics of (s-Subst), $[\![N]\!]_\alpha([\![M]\!]_\alpha(x)) = [\![N]\!]_\beta([\![M]\!]_\beta(y))$, i.e. $[\![N]\!]_\alpha(p_A(\alpha(x))) = [\![N]\!]_\beta(p_B(\beta(y)))$. Since this holds for all ground observable N, it means that $p_A(\alpha(x)) \equiv_{\alpha\beta} p_B(\beta(y))$ as required.

(ii) If p is an observation path, i.e. $\sigma \in \mathbb{O}$, then $\bar{p}[\mathsf{tr}(\mathsf{s})/v]$ is of observable type. Hence as $x \equiv_{\alpha\beta} y$, $[\![\bar{p}[\mathsf{tr}(\mathsf{s})/v]]\!]_\alpha(x) = [\![\bar{p}[\mathsf{tr}(\mathsf{s})/v]]\!]_\beta(y)$, so by Theorem 12 $p_A(\alpha(x)) = p_B(\beta(y))$ as required.

This completes the proof that $\equiv_{\alpha\beta}$ fulfills the criterion of Theorem 9 for being a bisimulation. ∎

Here now is our logical characterisation of bisimilarity for polynomial coalgebras:

THEOREM 20. *Let (A, α) and (B, β) be τ-coalgebras, where τ has at least one non-trivial observable subtype. Then for any $x \in A$ and $y \in B$, the following are equivalent:*

(1) *x and y are bisimilar: $x \sim y$.*

(2) *$\alpha, x \models \Gamma \triangleright \varphi$ iff $\beta, y \models \Gamma \triangleright \varphi$ for all rigid observable formulas $\Gamma \triangleright \varphi$.*

(3) *$\alpha, x \models \varphi$ iff $\beta, y \models \varphi$ for all ground observable formulas φ.*

(4) *$\alpha, x \models M \approx N$ iff $\beta, y \models M \approx N$ for all ground observable terms M and N.*

(5) *$\alpha, x \models M \approx N$ implies $\beta, y \models M \approx N$ for all ground observable terms M and N.*

(6) *$[\![M]\!]_\alpha(x) = [\![M]\!]_\beta(y)$ for all ground observable terms M, i.e. $x \equiv_{\alpha\beta} y$.*

Proof. (1) implies (2): let $x \sim y$. Thus xRy for some $|\tau|$-bisimulation R from α to β. Then (2) is given by Corollary 5(2). It is immediate that (2) implies (3), (3) implies (4), and (4) implies (5). If (5) holds and $c = [\![M]\!]_\alpha(x)$, where M has type $o \in \mathbb{O}$, then we can take N to be c in (5) to deduce $\beta, y \models M \approx c$, so $[\![M]\!]_\beta(y) = c = [\![M]\!]_\alpha(x)$, i.e. (6) holds. Finally, since $\equiv_{\alpha\beta}$ is a bisimulation and $\sim$ is the union of all bisimulations, $x \equiv_{\alpha\beta} y$ implies $x \sim y$, i.e. (6) implies (1). ∎

8 Comparison with other formalisms

For each functor $T : \mathbf{Set} \to \mathbf{Set}$ we can ask for a formalism of syntax and semantics that gives a logical characterisation of the T-bisimilarity relations between T-coalgebras: states are bisimilar precisely when they are logically indistinguishable in terms of satisfaction of formulas, or of valuation of terms. This programme, initiated by Hennessy and Milner [12, 13], has been carried out for many classes of coalgebras presented as labelled transition systems associated with theories of process algebra [23, 24, 25, 11], using some species of finitary or infinitary modal logic. Moss [26] showed that infinitary modal logic is also the appropriate logical formalism for characterising bisimilarity for very general functors acting on classes.

A characterisation of bisimilarity by term-valuation appears in the theory of hidden algebras in the work of Reichel and Malcolm (see [22]). Two elements of a hidden algebra are *behaviourally equivalent* if they assign the same values to all terms having a single variable of hidden sort. This relation proves to be equivalent to bisimilarity for a representation of hidden algebras as coalgebras of certain functors constructed out of products and powers [22, Example 2, Section 4.1].

Kurz [21] developed a finitary modal language for polynomial functors having the form $T(X) = \prod_{i=1}^{n}(B_i + C_i \times X)^{A_i}$. Rößiger [30] devised another kind of finitary language that is equivalent to Kurz's for the particular functors just mentioned, but is general enough to be able to characterise bisimilarity for all polynomial functors. It uses the notion of a *position*, which corresponds to what we have called a *basic path*. For each state path $T \xrightarrow{p} \mathrm{Id}$, Rößiger introduces a modal connective $[p]$, allowing formation of formulas $[p]\varphi$. For each observation path $T \xrightarrow{p} \bar{D}$, there is a formula $(p)c$ for each $c \in D$. Formulas in general are constructed from these atomic formulas $(p)c$ by applying the Boolean connectives and the modalities $[p]$. The semantics of [30, Definition 2.2] stipulates that

$$\alpha, x \models (p)c \quad \text{iff} \quad p_A(\alpha(x)) \text{ is defined and } p_A(\alpha(x)) = c,$$
$$\alpha, x \models [p]\varphi \quad \text{iff} \quad \text{either } p_A(\alpha(x)) \text{ is undefined or } \alpha, p_A(\alpha(x)) \models \varphi.$$

Now let M_p be the term $\bar{p}[\mathsf{tr}(\mathsf{s})/v]$ of Theorem 12 having $p_A(\alpha(x)) = [\![M_p]\!]_\alpha(x)$, and let $p\!\downarrow$ be the formula $\varphi_p[\mathsf{tr}(\mathsf{s})/v]$ of Lemma 15 having

$$\alpha, x \models p\!\downarrow \quad \text{iff} \quad \alpha(x) \in \mathrm{Dom}\, p_A.$$

Then it is evident that $(p)c$ has the same semantics as $p\!\downarrow \wedge (M_p \approx c)$, while by Corollary 17 $[p]\varphi$ has the same semantics as $(p\!\downarrow \to \varphi[M_p/\mathsf{s}])$. In this way the language of [30] can be identified with a subset of the language of the present paper. Rößiger's language does allow bisimilarity to be characterised by formula satisfaction, but it appears that the stronger syntax of terms developed here is required to give a charactersation in terms of equality of term-evaluation.

A complete deductive calculus of observational equations for monomial coalgebras is presented in [5, 6]. Another equational approach is that of [4, 3], which has "coterms" generated from symbols for operations of the form $X \to X_1 + \cdots + X_n$. Polynomial coalgebras can be represented as coalgebras built from such operations [3, Section 5], but an advantage of the present approach is that it provides a suitable syntax for coalgebras in the form in which they are given, rather than one for an equivalent representing coalgebra.

In summary, the language developed in this paper has all the expressive power of these other approaches and in addition provides a natural syntax for polynomial coalgebraic operations (projections, pairings, injections, case-formations, evaluations, lambda abstractions, functional applications). This makes it possible to characterise bisimilar states as those assigning the same values to terms of observable type. While the language is close in style to those of universal algebra and categorical logic, it has also been shown to be a species of modal logic. The idea of a term or formula having a single state-valued parameter is an inherently modal one: the standard translation of propositional modal logic into first-order logic associates with each modal formula a first-order formula that has a single free variable ranging over worlds or states [20, p. 234]. Moreover, our language is capable of defining certain modalities: as the above discussion and Theorem 16 and Corollary 17 show, the formula $\varphi[M_p/\mathsf{s}]$ expresses the modal assertion "after execution of the state transition $p_A \circ \alpha$, φ will be true", while more generally, $\varphi[M/\mathsf{s}]$ expresses "φ will be true after execution of the transition $x \mapsto [\![M]\!]_\alpha(x)$".

9 Towards Birkhoff's theorem

This paper confirms the fundamental role played by observable formulas in the theory of polynomial coalgebras, a role that could be said to be analogous to that played by equations in the theory of universal algebras. It is natural then to seek a characterisation of those classes of coalgebras that are defined by observational formulas. An approach to this has been developed in [10], and involves a new kind of ultrapower construction on coalgebras, which we now briefly describe.

Let U be an ultrafilter on a set I, i.e. U is a collection of subsets of I that is closed under finite intersections and supersets, and contains, for each $J \subseteq I$, exactly one of J and its complement $I - J$. Then for any set A the

relation
$$f =_U g \text{ iff } \{i \in I : f(i) = g(i)\} \in U$$
is an equivalence relation on the I-th power A^I of A. Each $f \in A^I$ has the equivalence class $f^U = \{g \in A^I : f =_U g\}$, and the quotient set $A^U = \{f^U : f \in A^I\}$ is called the *ultrapower of A modulo U*. Any function $\theta : A \to B$ has a *U-lifting* $\theta^U : A^U \to B^U$ given by $\theta^U(f^U) = (\theta \circ f)^U$.

Thus a τ-coalgebra $\alpha : A \to |\tau|A$ has the U-lifting $\alpha^U : A^U \to (|\tau|A)^U$, which is *not* a τ-coalgebra on A^U, since the latter would be a function of the form $A^U \to |\tau|(A^U)$. However a natural coalgebraic structure associated with A^U can be obtained by removing some of its points. Define f^U to be *observable* if for every ground observable τ-term $M : o$ there is an element $c \in [\![o]\!]$ such that the set $\{i \in I : [\![M]\!]_\alpha(f(i)) = c\}$ belongs to U. In essence this means that the U-lifting of the denotation $[\![M]\!]_\alpha : A \to [\![o]\!]$ gives f^U the value c.

If A^+ is the set of observable elements of A^U, then a coalgebra
$$\alpha^+ : A^+ \to [\![\tau]\!]_{A^+}$$
can be constructed such that if $\Gamma \triangleright \varphi$ is any observable formula, then
$$\alpha \models \Gamma \triangleright \varphi \text{ if, and only if, } \alpha^+ \models \Gamma \triangleright \varphi.$$

We call α^+ the *observational ultrapower* of α modulo U. Its definition, and the proof that it preserves validity, involve a lengthy and complex analysis of the U-liftings of the partial functions p_A induced by paths $|\tau| \overset{p}{\leadsto} |\sigma|$.

By choosing the right kind of ultrafilter U it can be arranged that α^+ is sufficiently "saturated" with states that the following holds:

> if every ground observable τ-formula valid in α is valid also in a τ-coalgebra β, then the bisimilarity relation from α^+ to β is surjective.

The proof makes use of our Theorem 20. The result itself is then used to give a structural characterisation of logically definable classes of coalgebras:

(†) *a class K of coalgebras is the class of all models of some set of rigid observable formulas if, and only if, K is closed under disjoint unions, images of bisimulations and observational ultrapowers.*

This may be viewed as an analogue for polynomial coalgebras of Birkhoff's famous characterisation of equational classes of abstract algebras as being those closed under direct products, homomorphic images and subalgebras. Details of the proof, and of the construction of α^+, are given in the paper [10].

Instead of working with ultrapowers, a notion of *ultrafilter enlargement* can be developed for a coalgebras, constructing from α a new coalgebra $E\alpha$ whose states are certain ultrafilters on the state set A of α. In modal model theory a construction like this is used that is based on the set of *all* ultrafilters on A, but for polynomial coalgebras we have to restrict to certain ultrafilters F that are *observationally rich*. This means that

for each ground observable term $M : o$ there exists some element $c \in [\![o]\!]$ such that $\{x \in A : [\![M]\!]_\alpha(x) = c\} \in F$.

If EA is the set of rich ultrafilters on A, then for any observational ultrapower (A^+, α^+) a map $\Phi : A^+ \to EA$ is given by

$$\Phi(f^U) = \{X \subseteq A : \{i : f(i) \in X\} \in U\}.$$

A transition structure $E\alpha : EA \to [\![\tau]\!]_{EA}$ can be defined such that Φ becomes a τ-morphism from α^+ to $E\alpha$. For saturated α^+, Φ is surjective. A fundamental feature of this construction is that for any $F \in EA$ and any ground observable formula φ,

$$E\alpha, F \models \varphi \quad \text{iff} \quad \varphi^\alpha \in F,$$

where $\varphi^\alpha = \{x \in A : \alpha, x \models \varphi\}$ is the subset of A *defined by* φ. This may be interpreted as saying that φ is true in $E\alpha$ at state F iff it is true in α at a set of states that is "large in the sense of F". The result is proved by transferring properties of α^+ to $E\alpha$ by the map Φ.

Yet another approach is to restrict attention to definable subsets of (A, α). The sets φ^α form a Boolean algebra, and we define ΔA to be the set of all observationally rich ultrafilters of this algebra. There is a natural map $\theta_\alpha : EA \to \Delta A$ that restricts each rich ultrafilter F on A to its members of the form φ^α. A transition $\Delta \alpha : \Delta A \to [\![\tau]\!]_{\Delta A}$ can be defined such that θ_α becomes a τ-morphism from $E\alpha$ to $\Delta \alpha$.

These constructions are related by a commuting diagram

$$\begin{array}{ccccc}
A^+ & \xrightarrow{\Phi} & EA & \xrightarrow{\theta_\alpha} & \Delta A \\
{\scriptstyle \alpha^+}\downarrow & & \downarrow{\scriptstyle E\alpha} & & \downarrow{\scriptstyle \Delta\alpha} \\
[\![\tau]\!]_{A^+} & \xrightarrow{|\tau|\Phi} & [\![\tau]\!]_{EA} & \xrightarrow{|\tau|\theta_\alpha} & [\![\tau]\!]_{\Delta A}
\end{array}$$

$(EA, E\alpha)$ is the *ultrafilter enlargement* of α, and $(\Delta A, \Delta \alpha)$ is the *definable enlargement*. It turns out that θ_α identifies two points of EA iff they are bisimilar, so in fact $\Delta\alpha$ is essentially the quotient of $E\alpha$ by bisimilarity. A version of the definable enlargement for monomial coalgebras was discussed in [8].

Now the above analogue (†) of Birkhoff's theorem remains true if "closed under observational ultrapowers" is replaced by "closed under ultrafilter enlargements", or "closed under definable enlargements". Details of the proof, and of the construction of $\Delta\alpha$, are given in the paper [7].

BIBLIOGRAPHY

[1] Peter Aczel and Nax Mendler. A final coalgebra theorem. In D. H. Pitt et al., editors, *Category Theory and Computer Science. Proceedings 1989*, volume 389 of *Lecture Notes in Computer Science*, pages 357–365. Springer-Verlag, 1989.

[2] Garret Birkhoff. On the structure of abstract algebras. *Proceedings of the Cambridge Philosophical Society*, 31:433–454, 1935.

[3] Corina Cîrstea. A coalgebraic equational approach to specifying observational structures. *Theoretical Computer Science*, to appear.

[4] Corina Cîrstea. A coequational approach to specifying behaviours. *Electronic Notes in Theoretical Computer Science*, 19, 1999. http://www.elsevier.nl/locate/entcs.

[5] Andrea Corradini. A complete calculus for equational deduction in coalgebraic specification. Technical Report SEN-R9723, Centrum voor Wiskunde en Informatica (CWI), Amsterdam, 1997.

[6] Andrea Corradini. A completeness result for equational deduction in coalgebraic specification. In Francesco Parisi Presicce, editor, *Recent Trends in Algebraic Development Techniques*, volume 1376 of *Lecture Notes in Computer Science*, pages 190–205. Springer-Verlag, 1998.

[7] Robert Goldblatt. Enlargements of polynomial coalgebras. In *Proceedings of the 8th Asian Logic Conference*, World Scientific (to appear). Manuscript available at http://www.mcs.vuw.ac.nz/~rob

[8] Robert Goldblatt. Duality for some categories of coalgebras. *Algebra Universalis*, 46(3):389–416, 2001.

[9] Robert Goldblatt. What is the coalgebraic analogue of Birkhoff's variety theorem? *Theoretical Computer Science*, 266:853–886, 2001.

[10] Robert Goldblatt. Observational ultraproducts of polynomial coalgebras. *Annals of Pure and Applied Logic*, to appear. Manuscript available at http://www.mcs.vuw.ac.nz/~rob

[11] M. Hennessy and X. Liu. A modal logic for message passing processes. *Acta Informatica*, 32:375–393, 1995.

[12] Matthew Hennessy and Robin Milner. On observing nondeterminism and concurrency. In J. W. de Bakker and J. van Leeuwen, editors, *Automata, Languages and Programming. Proceedings 1980*, volume 85 of *Lecture Notes in Computer Science*, pages 299–309. Springer-Verlag, 1980.

[13] Matthew Hennessy and Robin Milner. Algebraic laws for nondeterminism and concurrency. *Journal of the Association for Computing Machinery*, 32:137–161, 1985.

[14] Claudio Hermida. *Fibrations, Logical Predicates and Indeterminates*. PhD thesis, University of Edinburgh, 1993. Techn. rep. LFCS-93-277. Also available as Aarhus Univ. DAIMI Techn. rep. PB-462.

[15] Claudio Hermida and Bart Jacobs. Structural induction and coinduction in a fibrational setting. *Information and Computation*, 145:107–152, 1998.

[16] Bart Jacobs. Objects and classes, coalgebraically. In B. Freitag, C. B. Jones, C. Lengauer, and H.-J. Schek, editors, *Object-Orientation with Parallelism and Persistence*, pages 83–103. Kluwer Academic Publishers, 1996.

[17] Bart Jacobs. *Categorical Logic and Type Theory*. Elsevier, 1999.

[18] Bart Jacobs. Towards a duality result in coalgebraic modal logic. *Electronic Notes in Theoretical Computer Science*, 33, 2000. http://www.elsevier.nl/locate/entcs.
[19] Bart Jacobs. Exercises in coalgebraic specification. In R. Backhouse, R. Crole, and J. Gibbons, editors, *Algebraic and Coalgebraic Methods in the Mathematics of Program Construction*, volume 2297 of *Lecture Notes in Computer Science*, pages 237–280. Springer, 2002.
[20] Marcus Kracht. *Tools and Techniques in Modal Logic*, volume 142 of *Studies in Logic*. Elsevier, 1999.
[21] Alexander Kurz. Specifying coalgebras with modal logic. *Theoretical Computer Science*, 260:119–138, 2001.
[22] Grant Malcolm. Behavioural equivalence, bisimulation, and minimal realisation. In Magne Haveraaen, Olaf Owe, and Ole-Johan Dahl, editors, *Recent Trends in Data Type Specification*, volume 1130 of *Lecture Notes in Computer Science*, pages 359–378. Springer-Verlag, 1996.
[23] Robin Milner. Calculi for synchrony and asynchrony. *Theoretical Computer Science*, 25:267–310, 1983.
[24] Robin Milner. *Communication and Concurrency*. Prentice-Hall, 1989.
[25] Robin Milner, Joachim Parrow, and David Walker. Modal logics for mobile processes. *Theoretical Computer Science*, 114:149–171, 1993.
[26] Lawrence S. Moss. Coalgebraic logic. *Annals of Pure and Applied Logic*, 96:277–317, 1999.
[27] David Park. Concurrency and automata on infinite sequences. In P. Deussen, editor, *Theoretical Computer Science*, volume 104 of *Lecture Notes in Computer Science*, pages 167–183. Springer-Verlag, 1981.
[28] Andrew M. Pitts. Categorical logic. In S. Abramsky, D. M. Gabbay, and T. S. E. Maibaum, editors, *Handbook of Logic in Computer Science, Volume 5: Algebraic and Logical Structures*, chapter 2. Oxford University Press, 2000.
[29] Horst Reichel. An approach to object semantics based on terminal co-algebras. *Mathematical Structures in Computer Science*, 5:129–152, 1995.
[30] Martin Rößiger. From modal logic to terminal coalgebras. *Theoretical Computer Science*, 260:209–228, 2001.
[31] J.J.M.M. Rutten. A calculus of transition systems (towards universal coalgebra). In Alban Ponse, Maarten de Rijke, and Yde Venema, editors, *Modal Logic and Process Algebra*, CSLI Lecture Notes No. 53, pages 231–256. CSLI Publications, Stanford, California, 1995.
[32] J.J.M.M. Rutten. Universal coalgebra: a theory of systems. *Theoretical Computer Science*, 249(1):3–80, 2000.

Rob Goldblatt
Centre for Logic, Language and Computation
School of Mathematical and Computing Sciences
Victoria University of Wellington, New Zealand
E-mail: Rob.Goldblatt@vuw.ac.nz

9
Towards Uniform Reasoning via Structured Subset Spaces

BERNHARD HEINEMANN

ABSTRACT. This paper deals with subset spaces where points are structured as pairs of states. Our aim is to approach a formal model of uniform topological reasoning in this way. That is, in case a uniformity underlies the topological space under consideration, the refined reasoning model is to be sensitive to that. We show completeness and decidability of the basic modal logic arising from this setting, and discuss both possible extensions of the system and limitations.

1 Introduction

Approximately ten years ago, a bimodal logic for *topological reasoning* was proposed by Moss and Parikh [17]. The system developed in that paper connects the notion of knowledge, *cf* Fagin *et al.* [6], with fundamental concepts from general topology. In this way, a basic formal apparatus for reasoning tasks in the respective fields should be supplied.

Pursuing these ideas several classes of structures relevant to topology have been investigated since then; see for example Dabrowski *et al.* [5] (the journal version of Moss and Parikh [17], concerning general subset spaces and topological spaces respectively), Georgatos [9] (about treelike spaces), Heinemann [10] (spaces satisfying certain chain conditions), and Weiss and Parikh [19] (dealing with so-called *directed spaces,* which are closely related to bases of neighbourhood filters of topological spaces).

Why is topological reasoning significant for computer science and AI? First, because of its connection with the logic of knowledge. Summarizing the discussion in Dabrowski *et al.* [5], one can say suggestively that topological reasoning offers a natural way of reasoning about knowledge acquisition. Second, topology provides the solid mathematical framework for notions of location. Spatial objects and their mutual relationships like closeness, separation, or connectedness, are modelled adequately by topological means. Over and above this, there is yet another quite different idea inherent in topolgy: approximation. Approximating objects is topologically realized as a shrinking procedure applied to certain sets, viz *neighbourhoods* of the points of the considered topological space. Thus, at least from the point of view of a theorist, reasoning about topology should be the first objective in the development of spatial and (because of the procedural ingredient of approximation also) temporal reasoning formalisms. (We presently adopt a rather general point of view, dealing only with (generalizations of) the very basic structures of topology. Finer instruments of measurement like, *e.g.*, metrics would have to be incorporated in a more elaborate model.)

By means of the language underlying the system of Moss and Parikh one can speak about *subset frames*, *i.e.*, pairs $(X, \mathcal{O})$ where X is a non-empty set (of *points* or *states*) and $\mathcal{O}$ a set of subsets of X (the distinguished *system of opens* or *neighbourhoods*). Formulas are evaluated at *neighbourhood situations* consisting of a state and a neighbourhood thereof. The two modal connectives contained in the language, K and $\Box$, quantify across points *close* to the actual one and sets from $\mathcal{O}$ *refining* the current open, respectively. (One can think of K as a *knowledge operator* and $\Box$ as an *effort operator* assigned to some agent (being busy with approximating an object and acquiring knowledge, respectively).) Both modalities interact according to the topological structure being under discussion.

Now, it may happen that shrinking of opens proceeds in a *uniform* way. For example, computing a 0–1–stream $f : \mathbb{N} \longrightarrow \{0,1\}$ yields, step by step, a bisection of the actual open (consisting of all streams sharing the output obtained already). A similar phenomenon appears whenever the measurement precision has to be increased repeatedly during an experiment (say by one decimal place each time). Thus, *uniform* knowledge acquisition (about the function f and the measured object, respectively) turns up behind these examples.

The idea of uniformity has a formal counterpart in topology. Let be given a set X. Then, a *uniform structure* on X is a filter $\mathcal{U}$ in the powerset of $X \times X$ that satisfies certain closure properties. In particular, every *entourage* (that is, every element of $\mathcal{U}$) contains the diagonal of X. Furthermore, $\mathcal{U}$ is closed under converse and a kind of squaring; see Bourbaki [3, Definition II.1.1]. Note that relations of this type are subject to *two-dimensional modal logic;* cf Marx and Venema [16, Definition 2.1.2]. We will refer to them later on, in the concluding section of this paper.

Subsequently we extend the Moss–Parikh framework of topological reasoning to the effect that we take the first step towards a *modal logic of uniform subset spaces* (as a basic tool for *uniform reasoning*). Unlike the logic of subset frames mentioned above the *square* structure of sets must be taken into account now. Actually, we assume that we are given a system of distinguished entourages all of which are real squares. (Well-known metric spaces provide examples where one can find a base of the uniformity consisting of squares.) The language forming the basis of this approach is introduced in Section 2. A corresponding logical system **U** can be obtained as a synthesis of the logic of subset spaces and the two-dimensional logic of squares. We present the accompanying completeness result in Section 3. Afterwards, in Section 4, we give reasons for decidability of the set of **U**–theorems. Finally, we are heading for possible extensions towards a modal logic of uniform structures. In this connection a *hybrid* version of **U** is studied in Section 5. Concluding the paper we indicate other extensions and discuss the limitations of our approach (among other things).

2 Syntax, semantics, axioms

In this section we define precisely the syntax and semantics of the language of structured subset spaces. Moreover, we list the axioms of a logic corresponding to this class of models.

Let PROP be a denumerable set of symbols called *proposition variables*. We use $A, B, \ldots$ to denote typical elements of PROP. We define the set WFF of *well-formed formulas* of a trimodal language over PROP by the generative rule

$$\alpha ::= A \mid \neg\alpha \mid \alpha \wedge \beta \mid L_h\alpha \mid L_v\alpha \mid \Diamond\alpha.$$

That is, our language contains three unary diamonds L_h, L_v and $\Diamond$; the latter is intended to model *shrinking* of sets, whereas both modalities L_h and L_v correspond to the square structure of the subsets we consider. The missing boolean connectives $\top, \bot, \vee, \rightarrow, \leftrightarrow$ are treated as abbreviations, as needed. The duals of the modal operators are written K_h, K_v and $\Box$, respectively. (Our use of the letters L and K needs justification since it differs from that in traditional modal logic. Actually, the choice of names L and K, which is to emphasize the connection of these operators to possibility and knowledge, respectively, is the one that is usual for the common modal logic of subset spaces, *cf* Dabrowski *et al.* [5].)

We give next meaning to formulas. For this purpose we have to define the relevant domains first. We let $\mathcal{P}(X)$ designate the powerset of a given set X.

DEFINITION 1 (Structured subset spaces)   1. Given a non-empty set X, a subset $U \subseteq X \times X$ is called a *square (with respect to X)* iff $U = Y \times Y$ for some $Y \subseteq X$; in this case we write $Y = \mathrm{pr}(U)$ (the *projection* of U).

2. A *structured subset frame* $(X, \mathcal{U})$ consists of a non-empty set X and a set $\mathcal{U}$ of squares with respect to X.

3. Let $(X,\mathcal{U})$ be a structured subset frame. The set of *entourage situations* of $(X,\mathcal{U})$ is
$$\mathcal{N} := \{x,y,U \mid (x,y) \in U \text{ and } U \in \mathcal{U}\}.$$

4. A *structured subset space* or *structured model*[1] is a triple $(X,\mathcal{U},V)$, where $(X,\mathcal{U})$ is a structured subset frame and V a *(square) valuation*, i.e., a mapping
$$V : \text{PROP} \longrightarrow \mathcal{P}(X \times X).$$
We say that $\mathcal{M} := (X,\mathcal{U},V)$ is *based on* $(X,\mathcal{U})$.

For a given structured subset space $\mathcal{M}$ we define now the relation of satisfaction, $\models_{\mathcal{M}}$, between entourage situations of the underlying frame and formulas in WFF.

DEFINITION 2 (Satisfaction and validity) Let $\mathcal{M} := (X,\mathcal{U},V)$ be a structured subset space based on $\mathcal{S} := (X,\mathcal{U})$, and x,y,U an entourage situation of $\mathcal{S}$. Then

$$
\begin{array}{ll}
x,y,U \models_{\mathcal{M}} A & :\Longleftrightarrow (x,y) \in V(A) \\
x,y,U \models_{\mathcal{M}} \neg\alpha & :\Longleftrightarrow x,y,U \not\models_{\mathcal{M}} \alpha \\
x,y,U \models_{\mathcal{M}} \alpha \wedge \beta & :\Longleftrightarrow x,y,U \models_{\mathcal{M}} \alpha \text{ and } x,y,U \models_{\mathcal{M}} \beta \\
x,y,U \models_{\mathcal{M}} L_h\alpha & :\Longleftrightarrow z,y,U \models_{\mathcal{M}} \alpha \text{ for some } z \text{ such that } (z,y) \in U \\
x,y,U \models_{\mathcal{M}} L_v\alpha & :\Longleftrightarrow x,z,U \models_{\mathcal{M}} \alpha \text{ for some } z \text{ such that } (x,z) \in U \\
x,y,U \models_{\mathcal{M}} \Diamond\alpha & :\Longleftrightarrow \left\{ \begin{array}{l} x,y,U' \models_{\mathcal{M}} \alpha \text{ for some } U' \in \mathcal{U} \\ \text{such that } (x,y) \in U' \subseteq U, \end{array} \right.
\end{array}
$$

for all $A \in \text{PROP}$ and $\alpha, \beta \in \text{WFF}$. In case $x,y,U \models_{\mathcal{M}} \alpha$ is true we say that α *holds in* $\mathcal{M}$ *at* the entourage situation x,y,U. The formula α is called *valid in* $\mathcal{M}$, iff it holds in $\mathcal{M}$ at every entourage situation. (Manner of writing: $\mathcal{M} \models \alpha$.) (Moreover, the notion of validity can be extended as usual to structured subset frames $\mathcal{S}$ by quantifying over all structured subset spaces based on $\mathcal{S}$. However, frame validity will not be considered in the paper.)

Having defined the language we ask for an axiomatization of the validities of the class of structured subset spaces. Let $\mathbf{K}_3$ designate the minimal trimodal system over $\{L_h, L_v, \Diamond\}$. We propose the following list of axioms:

1. All $\mathbf{K}_3$–schemata

---

[1] Throughout this paper we will use the first term although the second one is closer to the manner of speaking in modal logic; however, we follow Dabrowski et al. [5] once again here.

2. $\alpha \to (L_i\alpha \wedge K_iL_i\alpha)$

3. $L_iL_i\alpha \to L_i\alpha$

4. $L_hL_v\alpha \leftrightarrow L_vL_h\alpha$

5. $(\Diamond A \to A) \wedge (\Diamond \neg A \to \neg A)$

6. $\alpha \to \Diamond\alpha$

7. $\Diamond\Diamond\alpha \to \Diamond\alpha$

8. $\Diamond L_i\alpha \to L_i\Diamond\alpha$,

where $i \in \{h, v\}$, $A \in \text{PROP}$, and $\alpha \in \text{WFF}$.

Note that the product logic **S5** × **S5** is axiomatized by the schemata 1 to 4, in particular; *cf* Marx and Venema [16, Section 2.2]. Axiom 5 is due to the special semantics of the proposition variables, and 1, 6 and 7 make the modality $\Diamond$ an **S4**–diamond. Finally, Axiom 8 describes the interaction of $\Diamond$ and the L–operators, corresponding to descending with respect to the set inclusion relation $\supseteq$.[2] Thus, for the basic system we consider in the present paper this axiom really reflects the underlying geometric ideas, as it forces shrinking of entourages.

Obviously, 'forgetting' the commutation axiom 4 and the subscripts to the L–modalities yields the well-known axiomatization of the usual modal logic of subset spaces; *cf* Dabrowski *et al.* [5].

3 Completeness

The topic of this section is to give a proof for the completeness part of Theorem 3 below.

Let **U** be the modal system obtained from the above axioms and the common derivation rules *modus ponens* and *necessitation with respect to each modality*:

$$\frac{\alpha \to \beta, \alpha}{\beta}$$

$$\frac{\alpha}{K_h\alpha} \quad \frac{\alpha}{K_v\alpha} \quad \frac{\alpha}{\Box\alpha},$$

for all $\alpha, \beta \in \text{WFF}$. Then, we have following theorem.

THEOREM 3 (Soundness and completeness of **U**) *Let $\alpha \in \text{WFF}$ be a formula. Then α is **U**–derivable, iff it is valid in every structured subset space.*

---

[2]It should be stressed that we do *not* have a commutation axiom like Schema 4 here. In particular, even disregarding Schema 5 the logic of structured subset spaces does not contain the Gabbay–Shehtman logic [**S4**, **S5**, **S5**], *cf* Gabbay and Shehtman [8, Section 3], but is 'two-dimensional' in essence.

While the soundness part of Theorem 3 is obvious, the main idea of proving completeness is to perform a particular step-by-step construction, similar as in Sec. 2.2 of the paper Dabrowski et al. [5]. Though step-by-step is brought out as a general method meanwhile, *cf* Blackburn et al. [1, 4.6], special arrangements have to be made here due to the interpretation of formulas in structured subset spaces.

Let $\mathcal{C}$ be the set of all maximal **U**–consistent set of formulas, and

$$\xrightarrow{L_h}, \xrightarrow{L_v}, \text{ and } \xrightarrow{\Diamond},$$

the distinguished accessibility relations on $\mathcal{C}$ induced by the modalities L_h, L_v and $\Diamond$, respectively. Suppose that $\Gamma_0 \in \mathcal{C}$ is to be realized. We choose a denumerably infinite set of points, Y, and fix an element $x_0 \in Y$. We will construct inductively a sequence of quadruples (X_n, P_n, d_n, s_n) such that, for every $n \in \mathbb{N}$,

- $X_n \subseteq Y$ is a finite set containing x_0,

- P_n is a finite partial order containing a least element p_0; moreover, for every $p \in P_n$ the set $\{q \in P_n \mid q \leq p\}$ is linearly ordered (that is, $(P_n, \leq)$ is a finite tree with root p_0),

- $d_n : P_n \longrightarrow \mathcal{P}(X_n \times X_n)$ is a mapping satisfying

    - $d_n(p)$ is a square for every $p \in P_n$,
    - $d_n(p_0) = X_n \times X_n$, and
    - $p \leq q \iff d_n(p) \supseteq d_n(q)$ for all $p, q \in P_n$, and

- $s_n : (X_n \times X_n) \times P_n \longrightarrow \mathcal{C}$ is a partial function such that, whenever $x, y, z \in X_n$ and $p, q \in P_n$,

    - $s_n(x, y, p)$ is defined iff $(x, y) \in d_n(p)$; in this case it holds that
        * if $(z, y) \in d_n(p)$, then $s_n(x, y, p) \xrightarrow{L_h} s_n(z, y, p)$,
        * if $(x, z) \in d_n(p)$, then $s_n(x, y, p) \xrightarrow{L_v} s_n(x, z, p)$, and
        * if $q \leq p$, then $s_n(x, y, q) \xrightarrow{\Diamond} s_n(x, y, p)$,
    - $s_n(x_0, x_0, p_0) = \Gamma_0$.

The intermediate structures (X_n, P_n, d_n, s_n) represent *approximations* to the desired model of a special kind. To be more precise, the construction will ensure, in particular, that

- $X_n \subseteq X_{n+1}$,

- $(P_{n+1}, \leq)$ is an *end extension* of $(P_n, \leq)$,
- $d_{n+1}(p) \cap X_n = d_n(p)$, for every $p \in P_{n+1}$, and
- $s_{n+1}|_{(X_n \times X_n) \times P_n} = s_n$,

for all $n \in \mathbb{N}$. Furthermore, the following requirements will be satisfied:

- if $L_\mathrm{h}\beta \in s_n(x,y,p)$, then there are $k > n$ and $z \in X_k$ such that $(z, y) \in d_k(p)$ and $\beta \in s_k(z, y, p)$,
- if $L_\mathrm{v}\beta \in s_n(x,y,p)$, then there are $k > n$ and $z \in X_k$ such that $(x, z) \in d_k(p)$ and $\beta \in s_k(x, z, p)$, and
- if $\Diamond\beta \in s_n(x,y,p)$, then for some $k > n$ there is a $q \in P_k$ such that $q \geq p$, $(x, y) \in d_k(q)$ and $\beta \in s_k(x, y, q)$.

Let us assume for the moment that the construction has been carried out successfully, meeting all these conditions. Let (X, P, d, s) be the *limit* of the structures (X_n, P_n, d_n, s_n), i.e.,

- $X = \bigcup\limits_{n \in \mathbb{N}} X_n$ and $P = \bigcup\limits_{n \in \mathbb{N}} P_n$,
- $d(p) = \bigcup\limits_{n \geq m} d_n(p)$, where m is the smallest number such that $p \in P_m$, and
- s is given by $s(x, y, p) := s_n(x, y, p)$, where n is the smallest number l such that $s_l(x, y, p)$ is defined $(x, y \in X$, and $p \in P)$.

We define a square valuation V by

$$V(A) := \{(x, y) \in X \mid A \in s(x, y, p_0)\},$$

for all $A \in \mathrm{PROP}$. Then, $\mathcal{M} := (X, d(P), V)$ is a structured subset space. In the same way as in the paper Dabrowski *et al.* [5], Proposition 2.6 and Lemma 2.5, one can prove the following *Truth Lemma* by induction on the structure of formulas.

LEMMA 4 (Truth lemma) *For all formulas $\beta \in \mathrm{WFF}$ and entourage situations $x, y, d(p)$ of the frame $(X, d(P))$ we have that*

$$x, y, d(p) \models_\mathcal{M} \beta \text{ iff } \beta \in s(x, y, p).$$

Letting $\beta := \neg\alpha$ be contained in $\Gamma_0 = s_0(x_0, x_0, p_0)$, Theorem 3 follows immediately from that.

It remains to define (X_n, P_n, d_n, s_n), for all $n \in \mathbb{N}$. At this point the differences in the technical details become apparent between the proof contained in Dabrowski *et al.* [5] and the present one.

The case $n = 0$ is more or less obvious, thus omitted. If $n \geq 1$, then an existential formula contained in some of the sets $s_m(x, y, p) \in \mathcal{C}$, where $m < n$, is to be realized by the actual procedure in a way that respects the above requirements. For this purpose the order of the construction steps must be arranged carefully. This is done in the following way. First, we choose an enumeration μ_n of the set $Q_{L_h}^n :=$

$$\left\{ (L_h\beta, x, y, p) \in \text{WFF} \times X_n \times X_n \times P_n \mid \left(\begin{array}{c} s_n(x, y, p) \text{ exists and} \\ L_h\beta \in s_n(x, y, p) \end{array} \right) \right\}$$

for every $n \in \mathbb{N}$, and enumerations ν_n, κ_n of the corresponding sets $Q_{L_v}^n$, $Q_\Diamond^n$ as well. Then, an appropriate 'global' enumeration λ is defined, along which the approximation of the desired model proceeds correctly. That is, λ schedules processing to quadruples $(L_h\beta, x, y, p)$, $(L_v\beta, x, y, p)$ and $(\Diamond\beta, x, y, p)$, respectively, in such a way that all of the sets $Q_{L_h}^n$, $Q_{L_v}^n$, and $Q_\Diamond^n$, are exhausted eventually. All this can be done in a fairly standard way; cf Dabrowski *et al.* [5, Section 2.2]. So, we will not regard these enumerations any more subsequently.

We treat the case '$L_h\beta$' of the inductive construction somewhat more detailedly now. So, let (X_n, P_n, d_n, s_n) be already defined and assume that $L_h\beta \in s_m(x, y, p)$ is going to be realized in step $n + 1$, where $m \leq n$. We choose $z \in Y \setminus X_n$ and let $X_{n+1} := X_n \cup \{z\}$. The partial order remains unaltered in this case, *i.e.*, $P_{n+1} := P_n$. Our next task is the definition of d_{n+1}. This mapping is given by

$$d_{n+1}(q) := \begin{cases} (\text{pr}\,(d_n(q)) \cup \{z\}) \times (\text{pr}\,(d_n(q)) \cup \{z\}) & \text{if } q \leq p \\ d_n(q) & \text{otherwise,} \end{cases}$$

for all $q \in P_{n+1}$.

The definition of s_{n+1} is a little more involved. Concerning the 'p-level', s_{n+1} has to remedy the defect caused by $L_h\beta$, in the sense of, *e.g.*, Marx and Venema [16, p. 16 ff]. This can be done with the aid of the **S5** × **S5**–axioms in a familiar way. But the new pairs having z as a component also occur at the 'q-level' for every $q \leq p$. Thus corresponding maximal **U**–consistent sets must be assigned to them in a faithful way, *i.e.*, reflecting the enlarged square structure at level q on $\mathcal{C}$. Here is the point where Axiom 8 plays its part, enabling us to 'lift' $\xrightarrow{L_h}$–arrows 'from right to left' with respect to $\leq$. All these facts are summed up in the following lemma.

LEMMA 5 (Crucial properties of the canonical model) *Let m, x, y, z, p as well as (X_n, P_n, d_n, s_n) and $L_\mathrm{h}\beta$ be as above. Then, for all $q \leq p$ and $u \in \mathrm{pr}\,(d_n(q)) \cup \{z\}$ there exist $\Gamma^q_{(u,z)}, \Gamma^q_{(z,u)} \in \mathcal{C}$ satisfying*

1. $\beta \in \Gamma^p_{(z,y)}$,

2. $s_m(x, y, q) \xrightarrow{L_\mathrm{h}} \Gamma^q_{(z,y)} \xrightarrow{L_\mathrm{v}} \Gamma^q_{(z,u)}$,

3. $s_m(x, y, q) \xrightarrow{L_\mathrm{v}} \Gamma^q_{(x,z)} \xrightarrow{L_\mathrm{h}} \Gamma^q_{(u,z)}$, *and*

4. *both* $\Gamma^q_{(z,u)} \xrightarrow{\diamond} \Gamma^{q'}_{(z,u)}$ *and* $\Gamma^q_{(u,z)} \xrightarrow{\diamond} \Gamma^{q'}_{(u,z)}$, *whenever $q \leq q' \leq p$.*

Proof. It suffices to prove the lemma in case $q = q'$ and q is the immediate $\leq$-predecessor of p. The general case is yielded by (backward) induction and transitivity of $\xrightarrow{\diamond}$.

Due to our construction we have that $s_m(x, y, q) \xrightarrow{\diamond} s_m(x, y, p)$. Because of $L_\mathrm{h}\beta \in s_m(x, y, p)$ there exists $\Gamma^p_{(z,y)} \in \mathcal{C}$ such that

$$\beta \in \Gamma^p_{(z,y)} \text{ and } s_m(x, y, p) \xrightarrow{L_\mathrm{h}} \Gamma^p_{(z,y)}.$$

We define $\Gamma^p_{(z,u)} := \Gamma^p_{(z,y)}$, for all $u \in \mathrm{pr}\,(d_n(q)) \cup \{z\}$. Then we get

$$s_m(x, y, p) \xrightarrow{L_\mathrm{h}} \Gamma^p_{(z,y)} \xrightarrow{L_\mathrm{v}} \Gamma^p_{(z,u)}.$$

Axiom 8 (for 'h') gives us now an element $\Gamma_1 \in \mathcal{C}$ such that

$$s_m(x, y, q) \xrightarrow{L_\mathrm{h}} \Gamma_1 \xrightarrow{\diamond} \Gamma^p_{(z,y)},$$

in the same way as the cross axiom has an effect in usual modal logic of subset spaces; cf Dabrowski et al. [5, Proposition 2.2]. Next, Axiom 8 (for 'v') delivers an element $\Gamma_2 \in \mathcal{C}$ such that

$$\Gamma_1 \xrightarrow{L_\mathrm{v}} \Gamma_2 \xrightarrow{\diamond} \Gamma^p_{(z,u)}.$$

We define, therefore, $\Gamma^q_{(z,y)} := \Gamma_1$ and $\Gamma^q_{(z,u)} := \Gamma_2$. This shows 5.2 and the first assertion of 5.4. The remaining parts of the proof are similar. ∎

Lemma 5 puts us in a position to define the mapping s_{n+1}:

$$s_{n+1}(u, v, q) := \begin{cases} s_n(u, v, q) & \text{if } u \neq z \text{ and } v \neq z \\ \Gamma^q_{(u,z)} & \text{if } u \in \mathrm{pr}\,(d_n(q)) \cup \{z\},\ v = z \text{ and } q \leq p \\ \Gamma^q_{(z,v)} & \text{if } v \in \mathrm{pr}\,(d_n(q)) \cup \{z\},\ u = z \text{ and } q \leq p \\ \text{undefined} & \text{otherwise,} \end{cases}$$

for all $u, v \in X_{n+1}$ and $q \in P_{n+1}$. It is easy to see that the above requirements on $(X_{n+1}, P_{n+1}, d_{n+1}, s_{n+1})$ are satisfied then.

The other modalities are treated similarly. The case of L_{V} is even completely analogous. In case $\Diamond\beta$ is to be processed, i.e., $\Diamond\beta \in s_m(x, y, p)$, the definitions are as follows. We let

$$X_{n+1} := X_n \cup \{z\} \text{ and } P_{n+1} := P_n \cup \{r\},$$

where z and r are 'new' each. Moreover, we let $p \leq r$ in P_{n+1}. Then, the interesting parts of the mappings d_{n+1} and s_{n+1} are given by

- $d_{n+1}(r) := \{(z, z)\}$ and
- $d_{n+1}(q) := (\mathrm{pr}\,(d_n(q)) \cup \{z\}) \times (\mathrm{pr}\,(d_n(q)) \cup \{z\})$ for all $q \leq p$,

and

- $s_{n+1}(r) :=$ any point Γ of the canonical model satisfying

$$\beta \in \Gamma \text{ and } s_m(x, y, p) \xrightarrow{\Diamond} \Gamma$$

- and, for all $q \leq p$,
  - $s_{n+1}(x', z, q) = s_n(x', y, q)$ for all $x' \in \mathrm{pr}\,(d_n(q))$,
  - $s_{n+1}(z, y', q) = s_n(x, y', q)$ for all $y' \in \mathrm{pr}\,(d_n(q))$,
  - $s_{n+1}(z, z, q) = s_n(x, y, q)$,

respectively. Again, the above requirements on $(X_{n+1}, P_{n+1}, d_{n+1}, s_{n+1})$ are satisfied. This finally yields the proof of completeness.

It should be mentioned that one of the referees of a preliminary version of this paper pointed a possible simplification of the above proof out to us, suggesting the use of game–theoretic arguments à la Gabbay et al. [7].

4 Decidability

We argue below that the set of **U**–theorems is decidable.[3] To this end we introduce a class of Kripke models with respect to which **U** is sound and complete, too, and satisfies the finite model property. We will, in particular, make use of the fact that subsets are structured as *two*-dimensional objects.

Let W be a non-empty set and R, R' two binary relations on W. We say that the pair (R, R') *satisfies the cross property,* iff

---

[3] This result should be compared with the well-known undecidability results for three-dimensional logics; see Gabbay et al. [7, Ch. 8] or Gabbay and Shehtman [8, Section 15].

for all $u, v, w \in W$ there exists $v' \in W$ such that, if $u\,R\,v\,R'\,w$, then $u\,R'\,v'\,R\,w$ (or, in other words, iff $R \circ R' \subseteq R' \circ R$).

Note that R and L_{h}, S and L_{v}, and T and $\diamond$, respectively, are intended to correspond to each other in the following.

DEFINITION 6 (Kripke–**U**–model) Let $\mathfrak{M} := (W, R, S, T, V)$ be a trimodal model, where $R, S, T \subseteq W \times W$ are binary relations and V a valuation. Then $\mathfrak{M}$ is called a *Kripke–**U**–model*, iff

1. R, S are equivalence relations, and T is reflexive and transitive,

2. the pairs of relations (R, S), (S, R), (T, R) and (T, S) satisfy the cross property, and

3. for all $w, w' \in W$ and $A \in \text{PROP} : w \in V(A) \iff w' \in V(A)$, whenever $w\,T\,w'$.

It turns out that the canonical model considered in the previous section is an example of a Kripke–**U**–model. Moreover, every structured subset space $\mathcal{M}$ induces a semantically equivalent Kripke–**U**–model $\mathfrak{M}$ as follows: the set W of points of $\mathfrak{M}$ equals the set of entourage situations of the subset frame underlying $\mathcal{M}$, and the accessibility relations on W are given by

$$
\begin{aligned}
(x, y, U)\,R\,(x', y', U') &: \iff y = y' \text{ and } U = U' \\
(x, y, U)\,S\,(x', y', U') &: \iff x = x' \text{ and } U = U' \\
(x, y, U)\,T\,(x', y', U') &: \iff x = x',\ y = y' \text{ and } U \supseteq U'.
\end{aligned}
$$

Finally, the valuation on $\mathfrak{M}$ is induced by the valuation on $\mathcal{M}$.

In this way we get with the aid of Theorem 3:

THEOREM 7 (Kripke completeness) *The system* **U** *is sound and complete with respect to the class of Kripke–**U**–models.*

We use *filtrations* for proving the finite model property of **U** with respect to Kripke–**U**–models. For a given $\alpha \in \text{WFF}$, we define a filter set $\Sigma \subseteq \text{WFF}$ as follows. We first join together the set $\text{sf}(\alpha)$ of subformulas of α and the set $\{\neg\beta \mid \beta \in \text{sf}(\alpha)\}$. Let $\Sigma^{\neg}$ denote the resulting set of formulas. Afterwards we form the boolean closure of $\Sigma^{\neg}$, in the following sense. We take the set Σ' of all finite disjunctions or conjunctions of pairwise distinct elements of $\Sigma^{\neg}$, and let then Σ^{bc} be the set of all finite conjunctions and disjunctions, respectively, of pairwise distinct elements of Σ'. Finally, we close under application of $O_i O'_j$, where $O, O' \in \{L, K\}$ and $i, j \in \{\mathrm{h}, \mathrm{v}\}$, and join together all these intermediate sets. Let Σ denote the resulting set of formulas. Note that Σ is finite and subformula closed. (Actually, 2^{2^n} is an upper bound for the cardinality of Σ (except for a constant), where n is the length of α.)

We consider now the *smallest* filtrations $\xrightarrow{L_h}{\Sigma}$ and $\xrightarrow{L_v}{\Sigma}$ of the relations $\xrightarrow{L_h}$ and $\xrightarrow{L_v}$, respectively, on the set $\mathcal{C}$ (introduced in the previous section); cf Blackburn *et al.* [1, p. 79]. The following property is crucial.

LEMMA 8 *Let* $[\Gamma], [\Gamma']$ *be the equivalence classes of* $\Gamma, \Gamma' \in \mathcal{C}$ *with respect to* Σ, *respectively (i.e.,* $[\Gamma] = \{\Delta \in \mathcal{C} \mid \Delta \cap \Sigma = \Gamma \cap \Sigma\}$, *and* $[\Gamma']$ *is given correspondingly). Moreover, let*

$$[\Gamma] \xrightarrow{L_i}{\Sigma} [\Gamma'], \text{ where } i \in \{h, v\}.$$

Then, for all $\Delta \in [\Gamma]$ *there is a* $\Delta' \in [\Gamma']$ *such that* $\Delta \xrightarrow{L_i} \Delta'$.

Proof. Let $\gamma_{\Gamma'} := \bigwedge_{\beta \in \Gamma' \cap \Sigma} \beta$. Regarding that

$$\Sigma = \Sigma^{bc} \cup \{O_i O'_j \beta \mid \beta \in \Sigma^{bc}, O, O' \in \{L, K\}, i, j \in \{h, v\}\},$$

one can show that the implication

$$L_i \gamma_{\Gamma'} \in \Delta \text{ for some } \Delta \in [\Gamma] \implies L_i \gamma_{\Gamma'} \in \Delta \text{ for all } \Delta \in [\Gamma]$$

is valid; to this end one has to utilize the definition of Σ and the **S5**-properties of the operator L_i (among other things). Since

$$L_i \gamma_{\Gamma'} \in \Delta \text{ implies } \exists \Theta : \left(\gamma_{\Gamma'} \in \Theta \text{ and } \Delta \xrightarrow{L_i} \Theta\right)$$

and

$$\gamma_{\Gamma'} \in \Theta \text{ implies } [\Theta] = [\Gamma'],$$

this suffices to prove the assertion of the lemma. ∎

A corresponding lemma for the one-dimensional case is due to Krommes [14]. It is decisive for the present (two-dimensional) case that every 'prefix' $\in \{L_h, L_v, K_h, K_v\}^*$ of length 3 has a provably equivalent prefix of length at most 2. (For example, $L_h K_v L_h$ and $K_v L_h$ as well as $L_v L_h L_v$ and $L_v L_h$ correspond to each other in this sense.) A 'prefix reduction' of the same kind is no longer possible in case of cubes of dimension > 2. (For example, $L_1 K_2 L_3 K_1$ cannot be reduced further.)[4]

---

[4] See Scherer [18] for an extensive study of knowledge prefixes.

It remains to choose a filtration of $\xrightarrow{\diamond}$. We take the *transitive closure of the smallest filtration* of $\xrightarrow{\diamond}$ for that; *cf* Chagrov and Zakharyaschev [4, p. 141 ff]. It is not hard to see that, in fact, a finite Kripke–U–model results. This gives us the desired finite model property.

THEOREM 9 (Finite model property of **U**) *For every formula α that is not derivable in* **U** *there is a finite Kripke–U–model falsifying α.*

Proof. The required properties of the filtration can be obtained as follows. Reflexivity of $\xrightarrow[\Sigma]{\diamond}$ (the filtration of $\xrightarrow{\diamond}$ chosen above), $\xrightarrow[\Sigma]{L_h}$ and $\xrightarrow[\Sigma]{L_v}$ is obvious. Symmetry of $\xrightarrow[\Sigma]{L_h}$ and $\xrightarrow[\Sigma]{L_v}$ is an easy consequence of the fact that we have taken the respective smallest filtration. Transitivity of $\xrightarrow[\Sigma]{\diamond}$ is obvious, too.

We show transitivity of $\xrightarrow[\Sigma]{L_h}$ with the aid of Lemma 8 now. Let $[\Gamma], [\Gamma']$ and $[\Gamma'']$ be equivalence classes with respect to Σ, and assume that

$$[\Gamma] \xrightarrow[\Sigma]{L_h} [\Gamma'] \text{ and } [\Gamma'] \xrightarrow[\Sigma]{L_h} [\Gamma'']$$

holds. Then, due to the definition of the smallest filtration there are $\Delta \in [\Gamma]$, $\Delta', \tilde{\Delta}' \in [\Gamma']$ and $\Delta'' \in [\Gamma']$ such that

$$\Delta \xrightarrow{L_h} \Delta' \text{ and } \tilde{\Delta}' \xrightarrow{L_h} \Delta''.$$

Since Δ' and $\tilde{\Delta}'$ belong to the same class some $\tilde{\Delta}'' \in [\Gamma'']$ exists such that $\Delta' \xrightarrow{L_h} \tilde{\Delta}''$, according to Lemma 8. We infer $\Delta \xrightarrow{L_h} \tilde{\Delta}''$ from this, for $\xrightarrow{L_h}$ is a transitive relation. Consequently, $[\Gamma] \xrightarrow[\Sigma]{L_h} [\Gamma'']$, as desired.

Furthermore, transitivity of $\xrightarrow[\Sigma]{L_v}$ as well as the various *cross properties*, can be proved similarly. Finally, Property 3 of Definition 6 can be established in a standard way. ∎

Note that the size of the refuting Kripke–U–model is doubly-exponential in the length of α, due to the cardinality of Σ mentioned above.

The claimed decidability result follows readily from Theorem 9.

COROLLARY 10 (Decidability) *The set of* **U**–*theorems is decidable.*

Some remarks on the complexity of the logic are contained in the final section of this paper.

5 Hybridization

As it stands, the system **U** is rather weakly expressive. In particular, only few of the necessary requirements for dealing with uniform structures are met. So, suitable extensions of the system should be considered in order to rectify that. And one should ask for generalizations besides.

For a start, the *hybrid* logic of a slightly more general class of structured subset spaces will be studied to some extent below. Since higher dimensions are involved and hybrid logic has not been developed into this direction up to now, this could be interesting *per se*. (As to a hybrid version of the usual logic of subset spaces and a discussion of its usefulness, *cf.* Heinemann [12]; see also Heinemann [13] for an application to reasoning about knowledge.)

We add a set NOM of *nominals* to the language of Section 2. The elements of NOM are denoted by $\mathfrak{e}, \mathfrak{f}, \mathfrak{g}, \ldots$ and $\mathfrak{e}', \mathfrak{f}', \mathfrak{g}', \ldots$, as needed.[5] Furthermore, a *satisfaction operator* $@_\mathfrak{e}$ is associated to each nominal $\mathfrak{e}$ (and added to the language as well).

The concept of structured subset spaces (see Definition 1) is modified as follows.

1. Let $X = X' \times X''$ be a non-empty product set. Then, a subset $U \subseteq X$ is called a *rectangle (with respect to X)* iff $U = Y \times Z$ for some $Y \subseteq X'$ and $Z \subseteq X''$.

2. A *structured subset frame* $(X, \mathcal{U})$ consists now of a non-empty product set X and a set $\mathcal{U}$ of rectangles with respect to X.

3. A *hybrid structured subset space* is yielded by adding a *hybridized valuation* V to a structured subset frame; *i.e.*, the denotation of a nominal with respect to V is a *single* entourage situation.

The following clauses have to be added with regard to satisfaction (see Definition 2):

$$x, y, U \models_\mathcal{M} \mathfrak{e} \quad :\iff\quad V(\mathfrak{e}) = x, y, U$$
$$x, y, U \models_\mathcal{M} @_\mathfrak{e} \alpha \quad :\iff\quad V(\mathfrak{e}) \models_\mathcal{M} \alpha,$$

where $\mathfrak{e} \in \text{NOM}$ and $\alpha \in \text{WFF}$.

An easy inspection shows that all the axioms from the list contained in Section 2 are sound with respect to the smaller class of hybrid structured subset spaces as well.

But what is missing for completeness? — There are several matters we have to take into account. First of all, hybrid logic has to be integrated

---

[5] This should remind one of the fact that nominals are intended to name *entourage situations*.

into our framework. To this end we adapt the corresponding Hilbert–style axiomatization given in Blackburn *et al.* [1, p. 438 ff] (see also Blackburn and Tzakova [2]), to the present case.

1. $@_\mathfrak{e}(\alpha \to \beta) \to (@_\mathfrak{e}\alpha \to @_\mathfrak{e}\beta)$
2. $@_\mathfrak{e}\alpha \leftrightarrow \neg @_\mathfrak{e} \neg \alpha$
3. $\mathfrak{e} \wedge \alpha \to @_\mathfrak{e}\alpha$
4. $@_\mathfrak{e}\mathfrak{e}$
5. $@_\mathfrak{e}\mathfrak{f} \wedge @_\mathfrak{f}\alpha \to @_\mathfrak{e}\alpha$
6. $@_\mathfrak{e}\mathfrak{f} \leftrightarrow @_\mathfrak{f}\mathfrak{e}$
7. $@_\mathfrak{f}@_\mathfrak{e}\alpha \leftrightarrow @_\mathfrak{e}\alpha$
8. $L_i@_\mathfrak{e}\alpha \to @_\mathfrak{e}\alpha$
9. $\Diamond@_\mathfrak{e}\alpha \to @_\mathfrak{e}\alpha$,

where $i \in \{\mathrm{h}, \mathrm{v}\}$, $\mathfrak{e}, \mathfrak{f} \in \mathrm{NOM}$ and $\alpha, \beta \in \mathrm{WFF}$. Note that the axiom schema called (*back*) in Blackburn *et al.* [1] occurs three times here (items 8 and 9 above) owing to the three modalities of the underlying system.

As to soundness, it can be seen easily again that Axioms 1 – 9 are valid in every hybrid structured subset space.

The hybrid rules we have to add are $@_\mathfrak{e}$-*necessitation* for every $\mathfrak{e} \in \mathrm{NOM}$, and the usual ones for *naming* and *pasting; cf.* Blackburn *et al.* [1, Section 7.3]. We do not write down these rules here, but mention that the latter also occurs in triplicate.

Giving this axiomatics of hybrid logic indicates already how we will try to obtain completeness, viz *canonically, i.e.,* à la Blackburn *et al.* [1], *loc cit.* (That is, with respect to a named and pasted model yielded by a distinguished maximal consistent set; we call this model 'canonical' below, for simplicity.) For this purpose we have to provide for *purity* of the axioms which go beyond $\mathbf{K}_3$. In particular, the schemata for structured subset spaces have to be re-formulated appropriately. Except for the axiom expressing persistency of the proposition variables (Schema 5 from the list contained in Section 2) this can obviously be done. That schema, 5, however, does not cause any problems on the canonical model at all.

Two points are left to be discussed. First, the general scenario of a hybrid subset space has to be arranged on the canonical model. *I.e.*, in particular, the denotation of every nominal must be unequivocally positioned there. The reader may consult Heinemann [12] in order to convince himself or herself that this can in fact be done with the aid of suitable hybrid axioms. (Clearly, some minor modifications caused by the additional dimension still must be incorporated.) We do not really go into detail concerning this.

We focus on the second point instead, viz establishing the required rectangle structure on the canonical model. Again we want suitable axioms to do this job. The following ones turn out to be right for that.

R1. $L_\mathrm{h}(\mathfrak{e} \wedge L_\mathrm{v}\mathfrak{g}) \wedge L_\mathrm{h}(\mathfrak{f} \wedge L_\mathrm{v}\mathfrak{g}) \to L_\mathrm{h}(\mathfrak{e} \wedge \mathfrak{f})$

R2. $L_\mathrm{v}(\mathfrak{e} \wedge L_\mathrm{h}\mathfrak{g}) \wedge L_\mathrm{v}(\mathfrak{f} \wedge L_\mathrm{h}\mathfrak{g}) \to L_\mathrm{v}(\mathfrak{e} \wedge \mathfrak{f})$

R3. $L_\mathrm{h}(\mathfrak{e} \wedge L_\mathrm{v}\mathfrak{g}) \wedge L_\mathrm{v}(\mathfrak{f} \wedge L_\mathrm{h}\mathfrak{g}) \wedge L_\mathrm{h}(\mathfrak{e}' \wedge L_\mathrm{v}\mathfrak{g}') \wedge L_\mathrm{v}(\mathfrak{f}' \wedge L_\mathrm{h}\mathfrak{g}') \wedge @_\mathfrak{g} \neg \mathfrak{g}' \to (@_\mathfrak{e} \neg \mathfrak{e}' \vee @_\mathfrak{f} \neg \mathfrak{f}')$,

where $\mathfrak{e}, \mathfrak{f}, \mathfrak{g}, \mathfrak{e}', \mathfrak{f}', \mathfrak{g}' \in \mathrm{NOM}$. Note that R1 – R3 are sound on hybrid structured subset spaces, too.

Let $\xrightarrow{L_\mathrm{h}}$ and $\xrightarrow{L_\mathrm{v}}$ designate the accessibility relations on the canonical model induced by the modalities L_h and L_v, respectively (as above). Furthermore, let x be an arbitrary point of that model, $[x]_\mathrm{h}$ the $\xrightarrow{L_\mathrm{h}}$-equivalence class and $[x]_\mathrm{v}$ the $\xrightarrow{L_\mathrm{v}}$-equivalence class of x. Finally, let

$$Q_x := \{y \mid (x, y) \in \xrightarrow{L_\mathrm{h}} \circ \xrightarrow{L_\mathrm{v}}\}.$$

Then it is claimed that $Q_x \cong [x]_\mathrm{h} \times [x]_\mathrm{v}$. A corresponding isomorphism can actually be defined as follows. According to Axiom R1 every element $y \in Q_x$ has a uniquely determined 'horizontal' component y_h (with respect to x), and according to Axiom R2 a uniquely determined 'vertical' component y_v as well. Thus the mapping

$$f : Q_x \longrightarrow [x]_\mathrm{h} \times [x]_\mathrm{v}, \quad \text{given by } y \mapsto (y_\mathrm{h}, y_\mathrm{v}) \text{ for all } y \in Q_x,$$

is well-defined. Injectivity of f is expressed by Axiom R3. Surjectivity follows from (the hybrid version of) the commutation axiom 4 from the list contained in Section 2, and from the fact that $\xrightarrow{L_\mathrm{h}}$ as well as $\xrightarrow{L_\mathrm{v}}$ are equivalence relations. Finally, it can easily be seen that f preserves relations in the sense that the property

$$\forall y, z \in Q_x : \left(y \xrightarrow{L_\mathrm{h}} z \iff y_\mathrm{h} \xrightarrow{L_\mathrm{h}} z_\mathrm{h} \text{ and } y_\mathrm{v} = z_\mathrm{v} \right)$$

holds for L_h, and an analogous property for L_v. — All in all we have shown that f is an isomorphism.

Summing up our exposition of hybridizing the logic of structured subset spaces we can now state:

THEOREM 11 (Hybrid completeness) *The hybrid system described above is sound and (canonically) complete with respect to the class of hybrid structured subset spaces.*

6 Concluding remarks

We showed above that the system **U** is sound and complete for the class of structured subset spaces, *i.e.*, spaces where the distinguished subsets are squares. Moreover, we proved that **U** satisfies the finite model property with respect to a suitable class of Kripke models and is decidable thus. (It should be mentioned here that we could not prove the finite model property with respect to structured subset spaces directly. Presumably this fmp does not hold; concerning a counter-example for the one-dimensional case see Dabrowski *et al.* [5, 1.3].)

Our decidability proof contained in Section 4 leads to a non-deterministic decision algorithm for satisfiability running in time $O\left(2^{2^n}\right)$ (because of the size of the refuting model estimated there). On the other hand, deciding **U**–satisfiability is clearly at least as hard as deciding satisfiability for the product logic $\mathbf{S5} \times \mathbf{S5}$. As is shown in Marx [15, Corollary 4.1] (see also Gabbay *et al.* [7, Theorem 5.24]), the latter problem is NEXPTIME–complete. Thus a gap appears between the upper and the lower bound. (Apart from few exceptions, *e.g.*, *topological nexttime logic, cf* Heinemann [11], this seems to be the normal case for modal logics of subset spaces up to now.)

> *Question 1. What is the exact complexity of the **U**–satisfiability problem?*

Maybe it is possible to use a filter set which is structured in the same way as the filter set applied to the basic modal logic of subset spaces, in order to obtain the fmp for **U**; *cf* Dabrowski *et al.* [7, Section 2.3]. This would show that the lower bound is matched.

In Section 5 we proposed a hybrid variant of **U**. Other extensions to be taken into account may concern, for example,

- further modalities typical of multi–dimensional modal logic, *e.g.*, the *diagonal modality* $\iota\delta$ and the *converse modality* $\otimes$,

- further modalities increasing the expressive power of the language like the *global modality, cf* Blackburn *et al.* [1, Section 7.1] or

- higher dimensions, *i.e.*, spaces of n–tuples of states where $n \geq 3$.

In the latter case the situation with respect to effectivity properties is hopeless; *cf* the negative results concerning this in Gabbay *et al.* [7, Ch.

8], or Gabbay and Shehtman [8, Section 15]. The obstacle found in Section 4 of the present paper (impossible prefix reduction) is, therefore, in fact unsurmountable.

In contrast to this we conjecture that at least a completeness results holds in the first case. If the connectives $\iota\delta$ and $\otimes$ are included in the language, then a proof of completeness should be possible along the lines of Marx and Venema [16, Section 2.3]. A system of this kind is very appropriate to the present framework since it is closer than **U** to uniform structures. However, it is not clear from the outset whether or not such a system is decidable.

Question 2. Is the logic of structured subset spaces in the language including $\iota\delta$ and $\otimes$ decidable?

Because of the squaring operation in uniform spaces it is also very desirable to have the *composition operator* to hand. Unfortunately this operator behaves badly from a computational point of view since we have neither completeness nor decidability then, actually; *cf* Marx and Venema [16, Section 2.4].

Question 3. How to modalize reasonably the squaring operation in uniform spaces?

An answer to this question would yield the decisive breakthrough towards a modal basis for uniform reasoning. Maybe the modalities mentioned in the second of the above items will prove to be helpful in that. But it is not clear so far if the addition of some of those so-called *logical modalities* has any substantial impact on the modal logic of subset spaces at all.

Question 4. To what extent can logical modalities be useful to the modal logic of subset spaces?

Each of the above questions indicates a future research duty.

Acknowledgements

I want to thank the referees very much for both their benevolent comments on the paper and their detailed suggestions how to improve it.

BIBLIOGRAPHY

[1] Patrick Blackburn, Maarten de Rijke, and Yde Venema. *Modal Logic*, volume 53 of *Cambridge Tracts in Theoretical Computer Science*. Cambridge University Press, Cambridge, 2001.
[2] Patrick Blackburn and Miroslava Tzakova. Hybrid Languages and Temporal Logic. *Logic Journal of the IGPL*, 7(1):27–54, 1999.
[3] Nicolas Bourbaki. *General Topology, Part 1*. Hermann, Paris, 1966.

[4] Alexander Chagrov and Michael Zakharyaschev. *Modal Logic*, volume 35 of *Oxford Logic Guides*. Clarendon Press, Oxford, 1997.
[5] Andrew Dabrowski, Lawrence S. Moss, and Rohit Parikh. Topological Reasoning and The Logic of Knowledge. *Annals of Pure and Applied Logic*, 78:73–110, 1996.
[6] Ronald Fagin, Joseph Y. Halpern, Yoram Moses, and Moshe Y. Vardi. *Reasoning about Knowledge*. MIT Press, Cambridge, MA, 1995.
[7] Dov M. Gabbay, Agnes Kurucz, Frank Wolter, and Michael Zakharyaschev. *Many-dimensional Modal Logics: Theory and Applications*. Studies in Logic, Elsevier, 2003.
[8] Dov M. Gabbay and Valentin B. Shehtman. Products of Modal Logics, Part 1. *Logic Journal of the IGPL*, 6:73–146, 1998.
[9] Konstantinos Georgatos. Knowledge on Treelike Spaces. *Studia Logica*, 59:271–301, 1997.
[10] Bernhard Heinemann. Topological Modal Logics Satisfying Finite Chain Conditions. *Notre Dame Journal of Formal Logic*, 39(3):406–421, 1998.
[11] Bernhard Heinemann. Topological Nexttime Logic. In M. Kracht, M. de Rijke, H. Wansing, and M. Zakharyaschev, editors, *Advances in Modal Logic 1*, volume 87 of *CSLI Publications*, pages 99–113, Stanford, CA, 1998. Kluwer.
[12] Bernhard Heinemann. Axiomatizing Modal Theories of Subset Spaces (An Example of the Power of Hybrid Logic). In *HyLo@LICS, 4th Workshop on Hybrid Logic, Proceedings*, pages 69–83, Copenhagen, Denmark, July 2002.
[13] Bernhard Heinemann. Knowledge over dense flows of time (from a hybrid point of view). In M. Agrawal and A. Seth, editors, *FST TCS 2002: Foundations of Software Technology and Theoretical Computer Science*, volume 2556 of *Lecture Notes in Computer Science*, pages 194–205, Berlin, 2002. Springer.
[14] Gisela Krommes. A new proof of decidability for the modal logic of subset spaces. In B. tenCate, editor *Proceedings of the Eighth ESSLLI Student Session*, pp. 137–147, Vienna, Austria, August, 2003.
[15] Maarten Marx. Complexity of Products of Modal Logics. *Journal of Logic and Computation*, 9(2):197–214, 1999.
[16] Maarten Marx and Yde Venema. *Multi-Dimensional Modal Logic*, volume 4 of *Applied Logic Series*. Kluwer Academic Publishers, Dordrecht, 1997.
[17] Lawrence S. Moss and Rohit Parikh. Topological Reasoning and The Logic of Knowledge. In Y. Moses, editor, *Proceedings of the 4th Conference on Theoretical Aspects of Reasoning about Knowledge (TARK 1992)*, pages 95–105, San Francisco, CA, 1992. Morgan Kaufmann.
[18] Beat G. Scherer. *Atome und Präfixe in der Wissenslogik*. PhD thesis, ETH Zürich, 1995.
[19] M. Angela Weiss and Rohit Parikh. Completeness of Certain Bimodal Logics for Subset Spaces. *Studia Logica*, 71:1–30, 2002.

Bernhard Heinemann
FernUniversität in Hagen
Fachbereich Informatik
PO Box 940,
D-58084 Hagen, Germany
E-mail: Bernhard.Heinemann@FernUni-Hagen.de

10
Controlled Model Exploration
GABRIEL G. INFANTE-LOPEZ, CARLOS ARECES, AND
MAARTEN DE RIJKE

ABSTRACT. We provide a detailed analysis of very weak fragments of modal logic. Our fragments lack connectives that introduce non-determinism and they feature restrictions on the modal operators, which may lead to substantial reductions in complexity. Our main result is a general game-based characterization of the expressive power of our fragments over the class of finite structures.

1 Introduction

The search for computationally well-behaved fragments of languages such as first-order and second-order logic has a long history. For instance, early in the twentieth century, Löwenheim already gave a decision procedure for the satisfiability of first-order sentences with only unary predicates. Some familiar fragments of first-order logic are defined by means of restrictions of the quantifier prefix of formulas in prenex normal forms. Finite-variable fragments of first-order logic are yet another family of fragments whose computational properties have been studied extensively, with decidability results going back to the early 1960s [20], while the late 1990s saw detailed complexity analyses of the two-variable fragment [10, 11, 16]. Despite the fact that the computational properties of prenex normal form and finite variable fragments have been (almost) completely investigated, these fragments leave something to be desired: their meta-logical properties are often poor, and, in particular, they usually do not enjoy a decent model theory that helps us to understand their computational properties. To overcome these drawbacks, there are ongoing research efforts to identify fragments of first-order or second-order logic that manage to combine good computational behavior with good logical properties.

One such effort takes modal logic as its starting point. Through the *standard* or *relational translation*, modal languages may be viewed as fragments of first-order languages [4]. Modal fragments are computationally very well-behaved; their satisfiability and model checking problems are of reasonably low complexity, and they are so in a robust way [21, 9]. The *guarded fragment* [1] was introduced as a generalization of the modal fragment, one that retains the good computational properties of modal fragments as much as possible. The good computational behavior of modal and guarded fragments has been explained in terms of the tree model property, and generalizations thereof.

In this paper we also search for well-behaved fragments of first-order logic by considering modal and modal-like languages, but we aim at a more fine-grained analysis. We start by taking a computationally well-behaved logic that can be translated into first-order logic, and try to generalize what we believe to be the main features responsible for the good computational behavior. Instead of modal logic, however, our starting point is taken from description logic. The description logic $\mathcal{FL}^-$ may be viewed as a restriction of the traditional modal language, where disjunctions are disallowed and the diamond operator is severely constrained. The restrictions built into $\mathcal{FL}^-$ yield significant reductions in computational complexity.

The aim of the paper is to provide a systematic exploration of the logical aspects of the restrictions built into $\mathcal{FL}^-$. We define a family of modal fragments inspired by $\mathcal{FL}^-$, briefly survey the computational complexity of their satisfiability problems, and spend most of the paper on providing a game-based characterization of their expressive power.

2 Description logics and $\mathcal{FL}^-$

Description logics have been proposed in the area of knowledge representation to specify systems in which structured knowledge can be expressed and reasoned with in a principled way [2]. They provide a logical basis to the well-known traditions of frame-based systems, semantic networks and KL-ONE-like languages, and now also for the semantic web. The main building blocks of languages of description logic are *concepts* and *roles*. The former are interpreted as subsets of a given domain, and the later as binary relations on the domain. Description logics differ in the constructions they admit for building complex concepts and roles.

Our starting point here is the logic $\mathcal{FL}^-$ [5]; its language has universal quantification, conjunction and unqualified existential quantification. That is, the legal concepts are generated by the following rule: $C ::= A \mid C \sqcap C \mid \forall R.C \mid \exists R.\top$, where A is an atomic concept, and R is an atomic role. In traditional modal logic notation, this production rule would be written as $\phi ::= p \mid \phi \wedge \phi \mid [R]\phi \mid \langle R \rangle \top$, or as $\phi ::= p \mid \phi \wedge \phi \mid \Box \phi \mid \Diamond \top$ when considering only one role.

Interpretations for description logics such as $\mathcal{FL}^-$ are pairs $\mathcal{I} = (\Delta, I)$, where Δ is a non-empty set, and I is a mapping that takes concepts to subsets of Δ and roles to subsets of $\Delta \times \Delta$. In (uni-)modal notation, a model is a tuple $\mathcal{A} = (W, R, V)$ where W is a non-empty set, R is a binary relation on W, and V is a function assigning subsets of W to proposition letters.

3 Taking a cue from $\mathcal{FL}^-$

The logic $\mathcal{FL}^-$ was carefully designed to control two important sources of computational complexity: non-determinism and deep model exploration. This aim shows up clearly in the syntactic constraints imposed on the language. The elimination of negation and disjunction restricts non-determinism (partial information cannot be expressed), while the restriction to unqualified existential quantification reduces model exploration to the bare minimum. As we will see in detail in Section 4, these design decisions have a significant impact on the computational complexity, making satisfiability checking trivial and subsumption checking polynomially tractable.

In contrast, standard modal logics (allowing full Boolean expressivity and qualified existential quantification) have PSPACE-complete satisfiability problems, as they allow one to code up models that are exponential in the size of the input formula [4]. The fact that restrictions on modal operators (the modal counterparts of description logic's quantifiers) produce computationally well behaved languages has also been studied in the modal logic community. Specifically, bounding the depth of nesting of modal operators may bring the complexity of the satisfiability problem down in dramatic ways, especially if one restricts the language even further by allowing only finitely many proposition letters (see [12]).

Despite the considerable computational impact of restricting non-determinism and existential quantification, a thorough analysis of its logical aspects, and especially of the expressive power, has been missing so far. The definitions below allow us to capture not just $\mathcal{FL}^-$ but a wide variety of additional fragments as well. In what follows we take the Boolean restrictions as they occur in $\mathcal{FL}^-$ mostly for granted (but we do include $\top$, $\bot$), and focus instead on its modal restrictions in a systematic way.

First, as we saw above, description and modal languages encode two kinds of information: *local* information depending only on the current node of evaluation, and *non-local* or *relational* information requiring model exploration (in controlled ways).

DEFINITION 1 (Local Formulas). A formula ϕ is a *local formula* if it is in the set of formulas LF generated by: $\phi ::= \top \mid \bot \mid p \mid \phi \wedge \phi \mid \Diamond\top$, where p is a proposition letter.

Second, we generalize the notion of unqualified existential quantification, by allowing complete control on which quantifiers are permitted at each level of nesting.

DEFINITION 2 (Fragment of $\mathcal{ML}$ modulo f). Let X be either $\mathbb{N}$ or an initial segment $\{1,\ldots,k\}$ of $\mathbb{N}$. Let $f : X \to \{\Diamond, \Box, \varheartsuit\}$. The *fragment of $\mathcal{ML}$ modulo f* (notation: $\mathcal{ML}^f$) is defined inductively as

$$\mathcal{ML}_0^f = \text{LF (the set of local formulas)}$$
$$\mathcal{ML}_{n+1}^f = \text{the closure under taking conjunctions of (LF}$$
$$\cup \{\Diamond \phi \mid \phi \in \mathcal{ML}_n^f \text{ and } f(n+1) = \Diamond\} \cup$$
$$\cup \{\Box \phi \mid \phi \in \mathcal{ML}_n^f \text{ and } f(n+1) = \Box\} \cup$$
$$\cup \{\Diamond \phi, \Box \phi \mid \phi \in \mathcal{ML}_n^f \text{ and } f(n+1) = \varheartsuit\}).$$

The *language* $\mathcal{ML}^f$ is defined as $\mathcal{ML}^f = \bigcup_{n \in X} \mathcal{ML}_n^f$.

A few comments are in order. First, the definition of our $\mathcal{ML}^f$-fragments depends on the choice of LF, the set of local formulas; in Section 7 we will vary this set.

Second, the function f used in the definition allows us to precisely control the legal arguments of the modalities at each node in the construction tree of a formula in $\mathcal{ML}^f$. In this manner we are able to cut up the full modal language in novel ways. However, the present definition does not yet allow us to define all of the standard modal language $\mathcal{ML}$; see Section 7 for more on this.

Third, let $f^\Box : \mathbb{N} \to \{\Diamond, \Box, \varheartsuit\}$ be such that $f(n) = \Box$ for all n. Obviously, if we were to allow $\top$ and $\bot$ in $\mathcal{FL}^-$, we would have $\mathcal{ML}^{f^\Box} = \mathcal{FL}^-$. Our definition captures $\mathcal{FL}^-$ in a very natural way: the function $f^\Box$ dictates that the modal box (and only the modal box) can have arguments of arbitrary complexity.

4 Computational aspects

In this section we provide a brief overview of some computational aspects of our $\mathcal{ML}^f$-fragments. First of all, recall that the satisfiability problem for the standard modal logic **K** is PSPACE-complete. By going down to $\mathcal{FL}^-$, that is, by disallowing disjunction (as well as negation, $\top$ and $\bot$) and by restricting ourselves to unqualified existential quantification, the satisfiability problem becomes trivial as all formulas in $\mathcal{FL}^-$ are satisfiable. More interesting is the fact that deciding subsumption (given two formulas $\phi, \psi \in \mathcal{FL}^-$ decide whether $\phi \to \psi$ is a theorem) is solvable in polynomial time [6]. We refer the reader to [8] for further discussion on the computational aspects of $\mathcal{FL}^-$ and its extensions.

In [12], Halpern shows that finiteness restrictions (both on the number of propositional symbols and on the nesting of operators) also lowers the complexity of the inference tasks. Satisfiability of the basic modal logic **K** becomes NP-complete when we only allow finite nesting of modalities, and it drops to linear time when we furthermore restrict the language to only a finite number of propositional symbols.

These results can immediately be extended to the appropriate $\mathcal{ML}^f$ fragments. For example, the results for $\mathcal{FL}^-$ directly imply similar results for the fragment $\mathcal{ML}^{f^\square}$ defined above. The following two results are more general, but also straightforward.

THEOREM 3. *Let $f : X \to \{\Diamond, \square, \mathbb{Q}\}$, where X is an initial segment or $X = \mathbb{N}$. The problem of deciding whether a formula in $\mathcal{ML}^f$ is satisfiable is in co-NP.*

Proof. For each $\mathcal{ML}^f$-fragment, we can reduce its satisfiability problem to the satisfiability problem for the description logic $\mathcal{ALE}$ [19]. $\mathcal{ALE}$ extends $\mathcal{FL}^-$ by allowing atomic negation, $\top$, $\bot$, and qualified existential quantification. That is, its set of legal concepts is given by $C ::= \top \mid \bot \mid \neg A \mid C \sqcap C \mid \forall R.C \mid \exists R.C$. The satisfiability problem for $\mathcal{ALE}$ is known to be co-NP-complete [8]. ∎

THEOREM 4. *Let X be $\mathbb{N}$ or an initial segment of $\mathbb{N}$, and let $f : X \to \{\Diamond, \square, \mathbb{Q}\}$ be such that $|\{n \mid f(n) = \Diamond \text{ or } f(n) = \mathbb{Q}\}|$ is finite. Assume that the set of local formulas LF is built using only finitely many proposition letters. Then deciding if a formula in $\mathcal{ML}^f$ is satisfiable can be done in linear time.*

Proof. The proof follows the lines of the similar proof in [12]. Let k be the maximal n such that $f(n) = \Diamond$ or $f(n) = \mathbb{Q}$. Given a formula ϕ in $\mathcal{ML}^f$, define ϕ^* by replacing every $\square$-subformula of ϕ that occurs at depth $k+1$ or deeper by $\square\bot$. It is easy to see that ϕ is satisfiable iff ϕ^* is. Hence, we only have to consider the fragment with formulas of modal depth at most $k+1$. A straightforward induction shows that there are only finitely many non-equivalent formulas in such fragments. Using this, one can find a fixed number of finite models such that a formula is satisfiable iff it is satisfiable on one of these models. This can be checked in time linear in the size of the formula being checked. ∎

5 A game-based characterization

Our next aim is to obtain an exact semantic characterization of the $\mathcal{ML}^f$-fragments. Games are a flexible and popular tool for obtaining results of this kind; see e.g., [7] for an introduction at the textbook level. Given an appropriate function f we define a game G^f that precisely characterizes $\mathcal{ML}^f$; in Section 6 below we build on this to capture the expressive power of our $\mathcal{ML}^f$-fragments.

DEFINITION 5. Let $\mathcal{A}$ be a model, and let X and X' be two subsets of its universe. We use $\mathcal{A}, X \models \phi$ to denote that $\mathcal{A}, w \models \phi$, for all $w \in X$.

The *children* of w in $\mathcal{A}$ are all v such that wRv. We say that $XR^\uparrow X'$ if for every x in X there exists x' in X' with xRx'. We say that $XR^\downarrow X'$ if for all x' of X' there exists x in X with xRx' (i.e., X' is a subset of the children of X).

Let $\mathcal{A}, \mathcal{B}$ be models with domains $W_\mathcal{A}$ and $W_\mathcal{B}$, respectively, and let $X_0 \subseteq W_\mathcal{A}$, $Y_0 \subseteq W_\mathcal{B}$. Let f be such that $\text{dom } f = \{1, \ldots, k\}$ or $\text{dom } f = \mathbb{N}$, and assume n in the domain of f. We write $G^f(\mathcal{A}, X_0, \mathcal{B}, Y_0, n)$ to denote the following game. The game is played by two players, called Di and Si, on relational structures $\mathcal{A}$ and $\mathcal{B}$ (intuitively, Di is trying to proof that $\mathcal{A}$ and $\mathcal{B}$ are *different*, while Si wants to show they are *similar*). A *position* in the game $G^f(\mathcal{A}, X_0, \mathcal{B}, Y_0, n)$ is given by a pair $\langle X, Y \rangle$ such that X is a set of elements in $\mathcal{A}$ and Y a set of elements in $\mathcal{B}$; $\langle X_0, Y_0 \rangle$ is the initial position. During the $(i+1)$-th round, the current position $\langle X_i, Y_i \rangle$ will change to the new position $\langle X_{i+1}, Y_{i+1} \rangle$ according to the following rules.

Rule 1 If $f(n-i) = \square$ then Di has to choose a set $Y_{i+1} \subseteq W_\mathcal{B}$ such that $Y_i R^\uparrow Y_{i+1}$, a counter-move of Si consists of choosing a set $X_{i+1} \subseteq W_\mathcal{A}$ such that $X_i R^\uparrow X_{i+1}$.

Rule 2 If $f(n-i) = \Diamond$ then Di has to choose a set $X_{i+1} \subseteq W_\mathcal{A}$ such that $X_i R^\downarrow X_{i+1}$. Si has to answer by choosing a set $Y_{i+1} \subseteq W_\mathcal{B}$ such that $Y_i R^\downarrow Y_{i+1}$.

Rule 3 If $f(n-i) = \varoast$ then Di can choose any of the previous rules to play by during this round.

The game *ends* on position $\langle X_i, Y_i \rangle$ when one of the following conditions fires:

Condition 1 There is a formula $\phi \in$ LF st. $\mathcal{A}, X_i \models \phi$ but $\mathcal{B}, Y_i \not\models \phi$.

Condition 2 $i < n$ and Si cannot move.

Condition 3 $i < n$ and Di cannot move.

Condition 4 Both players have made n moves ($i = n$) and none of the conditions above holds.

We say that Si *wins* the game if the game finishes because of conditions 3 or 4, otherwise Di wins. By $G^f(\mathcal{A}, X_0, \mathcal{B}, Y_0, \infty)$ (for f such that $\text{dom } f = \mathbb{N}$) we denote the game with an unbounded number of moves. In this case the game ends if Conditions 1, 2 or 3 are fired for some n. Di wins if the game ends under Condition 1 or 2. Si wins if Condition 3 holds, or if the game can continue indefinitely.

Two important characteristics of the definition of G^f are its directedness (in the rules and in Condition 1), and its use of sets, instead of elements, to represent positions. We will see that they are crucial in the following example.

EXAMPLE 6. To illustrate the definitions given so far we play $G^f(\mathcal{A}, \{a_1\}, \mathcal{B}, \{b_1\}, 1)$ with $\mathcal{A}$ and $\mathcal{B}$ as shown in Figure 1 for different values of f.

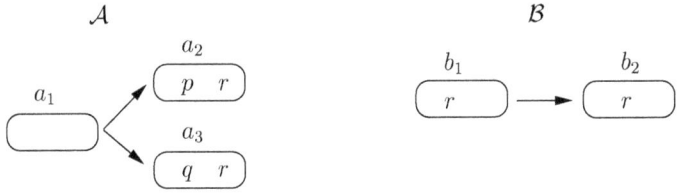

Figure 1. Playing a game

First take $f(1) = \Box$, then Si has a winning strategy: Di has to move in $\mathcal{B}$ with $Y' = \{b_2\}$. Si can choose either $\{a_1\}$, $\{a_2\}$ or $\{a_1, a_2\}$, and win the game. Note that all formulas $\Box\phi$, with ϕ local, that are satisfied in $(\mathcal{A}, a_1)$ are satisfied in $(\mathcal{B}, b_1)$. Now take $f(1) = \Diamond$; this time Di has a winning strategy: choose $\{a_2\}$ or $\{a_3\}$. In any of the two possibilities Si can only choose $Y' = \{b_2\}$ and in both cases there is a local formula, namely p or q such that $\mathcal{B}, \{b_2\} \not\models p$ or $\mathcal{B}, \{b_2\} \not\models q$, respectively.

The definition of G^f has been tailored to the restricted expressivity of the language $\mathcal{ML}^f$. We will now show that making the definition less tight would produce a mismatch in expressive power.

Suppose we weaken Condition 1 to make it symmetric, requiring that there is a formula $\phi \in$ LF such that either $\mathcal{A}, X_i \models \phi$ but $\mathcal{B}, Y_i \not\models \phi$; or $\mathcal{B}, Y_i \models \phi$ but $\mathcal{A}, X_i \not\models \phi$. Under this definition Di has a winning strategy for $G^f(\mathcal{A}, \{a_1\}, \mathcal{B}, \{b_1\}, k)$ for any f and k. This would be the case, in general, whenever two states disagree on the formulas in LF they satisfy. But this would be equivalent to allow atomic negation in LF! Under the new definition, the game G^f would be too discriminating for the expressive power of $\mathcal{ML}^f$. More generally, making Rules 1 to 3 symmetric would correspond to allowing full negation.

In a similar way, if we restrict positions to singleton sets (or equivalently, elements in the domain) the game would 'preserve' disjunctions, allowing Di to define a winning strategy on two models that only differ on disjunctive statements.

Games provide a mechanism for identifying differences between two models. Such differences may also be captured by logical formulas (in some language) that are true in one model but not in the other. The following theorem relates these two ideas for G^f-games and $\mathcal{ML}^f$-equivalence.

THEOREM 7. *Fix k and n such that $k \geq n > 0$, and let $f : \{1, \ldots, k\} \to \{\Diamond, \Box, \varnothing\}$ be given.*

1. *Si has a winning strategy for the game $G^f(\mathcal{A}, X, \mathcal{B}, Y, n)$ iff for every formula $\phi \in \mathcal{ML}_n^f$, if $\mathcal{A}, X \models \phi$ then $\mathcal{B}, Y \models \phi$.*

2. *Di has a winning strategy for the game $G^f(\mathcal{A}, X, \mathcal{B}, Y, n)$ iff there is a formula $\phi \in \mathcal{ML}_n^f$, such that $\mathcal{A}, X \models \phi$ and $\mathcal{B}, Y \not\models \phi$.*

Proof. 1. ($\Rightarrow$) We will prove this direction using induction on $n \leq k$.

Assume that Si has a winning strategy for the game $G^f(\mathcal{A}, X, \mathcal{B}, Y, 0)$. The theorem says that all the formulas in LF that are satisfied in all the elements of X have to be satisfied in all the elements of Y. As Si has a winning strategy, Condition 3 or 4 should hold. Condition 3 does not apply and hence Condition 4 ensures the needed condition. Assume that the result holds for $G^f(\mathcal{A}, X, \mathcal{B}, Y, n)$. Let ϕ in $\mathcal{ML}_{n+1}^f$ such that $\mathcal{A}, X \models \phi$. We first consider the case $f(n+1) = \Diamond$. If ϕ is a formula in LF, the truth of ϕ in $\mathcal{B}$ is given by Rule 1. If ϕ is a conjunction $\phi_1 \wedge \phi_2$, we can use a second inductive argument (on the number of $\wedge$-signs) to establish the claim. Next, ϕ may be of the form $\Diamond \phi_1$ with $\phi_1 \in \mathcal{ML}_n^f$. Since $\mathcal{A}, X \models \Diamond \phi_1$, we have that every element $x \in X$ has an R-child x' such that $\mathcal{A}, x' \models \phi_1$. Let us play with Di choosing $X' = \{x' \mid \mathcal{A}, x' \models \phi_1\}$. Since Si has a winning strategy for $G^f(\mathcal{A}, X, \mathcal{B}, Y, n+1)$, she will counter-play with a set Y' such that $YR^\uparrow Y'$ and will still have a winning strategy for $G^f(\mathcal{A}, X, \mathcal{B}, Y, n)$. Using the induction hypothesis we have that $\mathcal{B}, Y' \models \phi_1$ since $\mathcal{A}, X' \models \phi_1$. Hence, $\mathcal{B}, Y \models \phi$.

Suppose $f(n+1) = \Box$. Di has to move in $\mathcal{B}$. Suppose that $\phi = \Box \phi_1$ and $\mathcal{B}, Y \not\models \phi$. Then there is a set Y' such that $YR^\uparrow Y'$ and $\mathcal{B}, Y' \not\models \phi_1$. Let this Y' be the set chosen by Di. Since Si has a wining strategy she will choose a set X' such that $XR^\uparrow X'$. X' will be such that $\mathcal{B}, X' \models \phi_1$. As Si has a winning strategy for the game $G^f(\mathcal{A}, X', \mathcal{B}, Y', n)$, from this and the inductive hypothesis we can conclude that X' and Y' do indeed satisfy the same set of formulas, a contradiction.

The case where $f(n+1) = \varnothing$ reduces to one of the two cases above.

($\Leftarrow$) Assume that for every $\phi \in \mathcal{ML}_n^f$ we have that $\mathcal{A}, X \models \phi$ implies $\mathcal{B}, Y \models \phi$. We need to define a winning strategy for Si when she plays the game $G^f(\mathcal{A}, X, \mathcal{B}, Y, n)$. The proof is by induction on n. For $n = 0$ we need to check that $G^f(\mathcal{A}, X, \mathcal{B}, Y, 0)$ starts with a winning position for Si. This is true because by hypothesis we have that formulas in LF valid in X are also valid in Y.

Assume the result holds for n. Suppose that $f(n+1) = \diamond$ and that Di has chosen a set X' such that $XR^\downarrow X'$. Let $\Phi = \{\phi \in \mathcal{ML}_n^f \mid \mathcal{A}, X' \models \phi\}$. Let Si choose a set Y' such that $YR^\downarrow Y'$ and $\mathcal{B}, Y' \models \Phi$. The existence of such a set is given by hypothesis. After this move all the formulas in $\mathcal{ML}_n^f$ satisfied in X' are also satisfied in Y', and, by induction, Si can complete the strategy. For $f(n+1) = \square$, Di has to move in $\mathcal{B}$. Let us suppose that Di has chosen a set Y'. Define Φ as before. Let X' be a set of elements in $\mathcal{A}$ such that $\mathcal{A}, X' \models \phi$. X' will be the move of Si— by using the induction hypothesis again we have the complete strategy.

The $f(n+1) = \varobar$ case reduces to one of the two cases above.

2. ($\Rightarrow$) The left-to-right implication is similar to item 1, left-to-right. To prove the right-to-left implication one can build the required strategy for Di by induction on the size of the formula ϕ.

If the formula is an atomic proposition letter then Si wins immediately because of Rule 1. For the inductive case, we should decide the move for Di in the current position and the induction hypothesis will provide the rest of the winning strategy. Since $\mathcal{B}, Y \not\models \phi$, there is an element $w' \in Y$ such that $\mathcal{B}, w' \not\models \phi$. Suppose that $\phi = \square\phi_1$, since $\mathcal{B}, w' \not\models \square\phi_1$ implies that there is w_1' such that $w'Rw_1'$ and $\mathcal{B}, w_1' \not\models \phi_1$. Following the rules of the game, Di has to move in $\mathcal{B}$: Di has to choose any subset of the children of Y such that w_1 is included. Suppose that $\phi = \diamond\phi_1$, then there exists a set $YR^\downarrow Y'$ such that $\mathcal{B}, Y' \not\models \phi_1$. By inductive hypothesis, Di has a winning strategy for the game $G^f(\mathcal{A}, X', \mathcal{B}, Y', n)$. Y' as a first move together with the previous strategy, gives Di the complete strategy. ∎

COROLLARY 8. *For every game $G^f(\mathcal{A}, X, \mathcal{B}, Y, n)$ either Di or Si has a winning strategy, i.e., the game $G^f(\mathcal{A}, X, \mathcal{B}, Y, n)$ is deterministic.*

6 The expressive power of $\mathcal{ML}^f$

Van Benthem [3] proved the following preservation result: a class of models defined by a first-order sentence is closed under bisimulations iff it can be defined by a modal formula. Rosen [18] proved that this result remains true over the class of finite structures. Kurtonina and de Rijke [13] extended Van Benthem's result in a different direction, by proving analogous preservation results for broad classes of description logics, including both restrictions and extension of the basic modal language such as $\mathcal{FL}^-$; see also [14].

Below we prove a general preservation result, for each of the fragments defined in Definition 2, over the class of finite structures. Our proof, which is based on the games introduced in the previous section, follows the structure of Rosen's proof.

For the formulation of our results it is convenient to work with so-called *pointed models* $(\mathcal{A}, c^{\mathcal{A}})$; these are models with a distinguished element.

DEFINITION 9. Let $\mathcal{A}$ and $\mathcal{B}$ two models with distinguished elements $c^{\mathcal{A}}$ and $c^{\mathcal{B}}$ respectively. We write $\mathcal{A} \sim^n_{Gf} \mathcal{B}$ to denote that Si has winning strategies for both of the following games: $G^f(\mathcal{A}, \{c^{\mathcal{A}}\}, \mathcal{B}, \{c^{\mathcal{B}}\}, n)$ and $G^f(\mathcal{B}, \{c^{\mathcal{B}}\}, \mathcal{A}, \{c^{\mathcal{A}}\}, n)$. We write $\mathcal{A} \sim^\infty_{Gf} \mathcal{B}$ to denote that Si has winning strategies for the corresponding infinite games.

The first key theorem in Rosen's paper is the following.

THEOREM 10 (Rosen [18]). *Let $\mathcal{C}$ be any class of models (each model $\mathcal{A}$ with a distinguished node $c^{\mathcal{A}}$), closed under isomorphism. Let $\mathcal{C}'$ be any subclass of $\mathcal{C}$, also closed under isomorphism. Then for all n, the following conditions are equivalent:*

1. *For all $\mathcal{A} \in \mathcal{C}'$, $\mathcal{B} \in \mathcal{C} - \mathcal{C}'$, $\mathcal{A} \not\sim^n \mathcal{B}$ (where $\sim^n$ stands for n-bisimulation).*

2. *There is a modal formula of quantifier rank less or equal than n that defines $\mathcal{C}'$ over $\mathcal{C}$.*

We extend this result to be able to cope with our restricted fragments.

THEOREM 11. *Let $\mathcal{C}$ be any class of models (each model $\mathcal{A}$ with a distinguished node $c^{\mathcal{A}}$), closed under isomorphism. Let $\mathcal{C}'$ be a subclass of $\mathcal{C}$ also closed under isomorphism. For every n, let M_n be the biggest subset of $\mathcal{ML}^f_n$ such that all the formulas in M_n are satisfied in $c^{\mathcal{A}}$ by all the models $\mathcal{A}$ in $\mathcal{C}'$. If $M_n \neq \emptyset$ then the following conditions are equivalent:*

1. *For all $\mathcal{A} \in \mathcal{C}'$, $\mathcal{B} \in \mathcal{C} - \mathcal{C}'$, $\mathcal{A} \not\sim^n_{Gf} \mathcal{B}$.*

2. *There is a formula ϕ in M_n such that ϕ defines $\mathcal{C}'$ over $\mathcal{C}$.*

Observe the following relations between Theorems 10 and 11. The set $\mathcal{ML}^f_n$ in Rosen's theorem is given by all modal formulas in which the maximal number of nested modal operators is at most n. The set M_n (of formulas satisfied in all elements of $\mathcal{C}'$) is never empty: the set of modal formulas containing at most n nested modal operators is logically finite, and every model in $\mathcal{C}'$ will satisfy ϕ or $\neg\phi$ for ϕ with modal depth less than or equal to n. Form the disjunction of one such formula ϕ per model in $\mathcal{C}$, and this formula will be true in all the elements of $\mathcal{C}'$.

Proof of Theorem 11. ($1 \Rightarrow 2$) Suppose that $M_n \neq \emptyset$ and suppose that for all $\mathcal{A} \in \mathcal{C}'$, $\mathcal{B} \in \mathcal{C} - \mathcal{C}'$, $\mathcal{A} \not\sim^n_{Gf} \mathcal{B}$. By Theorem 7 this implies that for all $\mathcal{A} \in \mathcal{C}'$, $\mathcal{B} \in \mathcal{C} - \mathcal{C}'$, there is a formula ϕ in M_n such that $\mathcal{A}, c^{\mathcal{A}} \models \phi$ but $\mathcal{B}, c^{\mathcal{B}} \not\models \phi$. Let $\mathcal{A}$ be any model in $\mathcal{C}'$, we define $\Phi_{\mathcal{A}}$ by putting

$$\Phi_{\mathcal{A}} = \bigwedge \{\phi \in \mathcal{ML}^f_n \mid \mathcal{A} \models \phi \text{ and } \mathcal{B} \not\models \phi \text{ with } \mathcal{B} \in \mathcal{C} - \mathcal{C}'\}.$$

Note that since $\mathcal{ML}^f_n$ is finite, $\Phi_{\mathcal{A}}$ is a finite conjunction. Note also that $\Phi_{\mathcal{A}}$ belongs to $\mathcal{ML}^f_n$ and that it is satisfied by all the models in $\mathcal{C}$ (it belongs to M_n) but not by any model in $\mathcal{C}'$. Hence $\Phi_{\mathcal{A}}$ is the needed definition.

($2 \Rightarrow 1$) Suppose that $M_n \neq \emptyset$ and that there is a formula ϕ in M_n such that ϕ defines $\mathcal{C}'$ over $\mathcal{C}$. By Theorem 7, Di will have a winning strategy for the appropriate games and for all $\mathcal{A} \in \mathcal{C}'$, $\mathcal{B} \in \mathcal{C} - \mathcal{C}'$, it will be true that $\mathcal{A} \not\sim^n_{Gf} \mathcal{B}$. ∎

We need some further terminology. Given a model $\mathcal{A}$ and a node w in $\mathcal{A}$, we say that w is a *descendant* of v if wR^*v, where R^* is the transitive closure of R. The *family* of w in $\mathcal{A}$, written $\mathcal{F}^w_{\mathcal{A}}$, is the submodel of $\mathcal{A}$ with universe $\{w\} \cup \{v \mid v \text{ is a descendant of } w\}$. We say that w and v are *disjoint* iff $\mathcal{F}^w_{\mathcal{A}} \cap \mathcal{F}^v_{\mathcal{A}} = \emptyset$. The *r-neighborhood* of a node w, denoted $\mathcal{N}_r(w)$, is defined inductively. $\mathcal{N}_0(w)$, is the submodel of $\mathcal{A}$ with universe $\{w\}$, and for all $r+1$, $v \in \mathcal{N}_{r+1}(w)$ iff $v \in \mathcal{N}_r(w)$ or there is a $w' \in \mathcal{N}_r(w)$ such that $\mathcal{A} \models Rw'v \vee Rvw'$. An *r-tree* is a directed tree rooted at u of height $\leq r$. An *r-pseudotree* is a model such that $\mathcal{N}_r(u)$ is a tree with the property that all distinct pairs of its leaves are disjoint, as defined above. As is standard, $\cong$ denotes isomorphism.

PROPOSITION 12. *Assume that $(\mathcal{A}, c^{\mathcal{A}})$ and $(\mathcal{B}, c^{\mathcal{B}})$ are two models with $(\mathcal{A}, c^{\mathcal{A}}) \sim^n_{Gf} (\mathcal{B}, c^{\mathcal{B}})$. Then there are n-pseudotrees $(\mathcal{A}', c^{\mathcal{A}'})$ and $(\mathcal{B}', c^{\mathcal{B}'})$ with $(\mathcal{A}, c^{\mathcal{A}}) \sim^{\infty}_{Gf} (\mathcal{A}', c^{\mathcal{A}'})$, $(\mathcal{B}, c^{\mathcal{B}}) \sim^{\infty}_{Gf} (\mathcal{B}', c^{\mathcal{B}'})$ and $\mathcal{N}_n(c^{\mathcal{A}'}) \cong \mathcal{N}_n(c^{\mathcal{B}'})$.*

Proof. We specify an algorithm that transforms the two pointed models into models with isomorphic n-neighborhoods. After each step s ($s \leq n$) we have models $(\mathcal{A}_s, c^{\mathcal{A}}_s)$ and $(\mathcal{B}_s, c^{\mathcal{B}}_s)$ such that $(\mathcal{A}, c^{\mathcal{A}}) \sim^{\infty}_{Gf} (\mathcal{A}_s, c^{\mathcal{A}}_s)$ and $(\mathcal{B}, c^{\mathcal{B}}) \sim^{\infty}_{Gf} (\mathcal{B}_s, c^{\mathcal{B}}_s)$ while $c^{\mathcal{A}_s}$ and $c^{\mathcal{B}_s}$ have isomorphic s neighborhoods. At each step $s+1$, $\mathcal{A}_{s+1}$ (respectively $\mathcal{B}_{s+1}$) is obtained from $\mathcal{A}_s$ ($\mathcal{B}_s$) by adding or removing copies of families of nodes at distance $s+1$ from the root.

Let $\{a_1, \ldots, a_l, b_1, \ldots, b_m\}$ be the set of children of $c^{\mathcal{A}}$ and $c^{\mathcal{B}}$. We will build the models using the following two rules: If $f(n) = \square$ then for constructing $\mathcal{A}_1$ and $\mathcal{B}_1$ we just choose one a_i and b_j and drop all the remaining children. We will redefine the set of local formulas satisfied in a_i and b_j as the local formulas that are common to all states $\{a_1, \ldots, a_l, b_1, \ldots, b_m\}$.

All formulas in $\mathcal{ML}_n^f$ will either start with a box or will be local formulas. If a formula $\phi = \Box \phi_1$ is satisfied in $\mathcal{A}$ then ϕ_1 will be satisfied in all children of $c^{\mathcal{A}}$, and, in particular, in a_i, hence ϕ will be satisfied in $c^{\mathcal{A}_1}$.

If $f(n) = \Diamond$, the relation $\sim_{G^f}^{n-1}$ induces an equivalence classes on the set $\{a_1, \ldots, a_l, b_1, \ldots, b_m\}$. Note that not every equivalence class necessarily has a member in each $\mathcal{A}$ and $\mathcal{B}$. An example of such a configuration is as in (a) below:

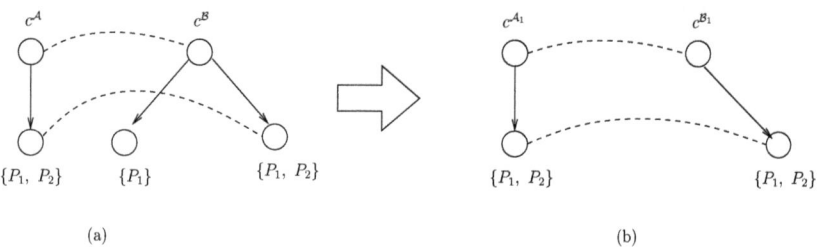

To obtain $\mathcal{A}_1$ and $\mathcal{B}_1$ with isomorphic 1-neighborhoods of c such that $\mathcal{A} \sim_{G^f}^1 \mathcal{A}_1$, we have to do two things. First we should add enough copies of families of the children a_i and b_j such that each equivalence class has an equal number of members in $\mathcal{A}_1$ and in $\mathcal{B}_1$. Second, if an element d is not related by $\sim_{G^f}^{n-1}$ to an element in the opposite model, we just drop its family.

We should now verify that we are not throwing away any states that provide the only way to satisfy a certain formula. Suppose for contradiction that this is the case: there is a state b_i and a formula ϕ such that $\mathcal{B}_1, b_j \models \phi$ (hence $\mathcal{B}, c^{\mathcal{B}} \models \Diamond \phi$) and b_j is the only child of $c^{\mathcal{B}}$ that satisfies this formula. Since $\mathcal{A}, c^{\mathcal{A}} \models \Diamond \phi$ there is a child a_i of $c^{\mathcal{A}}$ such that $\mathcal{A}, a_i \models \phi$. By hypothesis a_i is not $\sim_{G^f}^{n-1}$-related to b_j, meaning that there is a formula ϕ_1 such that $\mathcal{A}, a_i \models \phi_1$ but $\mathcal{B}, b_j \not\models \phi_1$. As $\phi \wedge \phi_1 \in \mathcal{ML}^f$ there is another child of $c^{\mathcal{B}}$ that satisfies ϕ, a contradiction. The case where $f(n+1) = \Box$ reduces to one of the two cases above. The next step on the algorithm is to move to each of the elements in the isomorphic neighborhoods and apply the same schema for each pair of nodes related by the isomorphism.

The models $\mathcal{A}_s$ and $\mathcal{B}_s$ constructed during the proof will be both isomorphic and $\sim_{G^f}^\infty$-related to $\mathcal{A}$ and $\mathcal{B}$, respectively, as needed. ∎

Before we can formulate our main expressiveness result, we need one more auxiliary result, due to Rosen. In formulating it, we write $\mathcal{A} \equiv^n \mathcal{B}$ to denote that $\mathcal{A}$ and $\mathcal{B}$ satisfy the same first-order sentences with at most n nested quantifiers.

THEOREM 13 (Rosen [18]). *Let $\mathcal{A}$ and $\mathcal{B}$ be two $(2H(n))$-pseudotrees for which $\mathcal{N}_{H(n)}(c^{\mathcal{A}}) \cong \mathcal{N}_{H(n)}(c^{\mathcal{B}})$ holds, where $H(x)$ is the Hanf function. Then there are $\mathcal{A}'$ and $\mathcal{B}'$ such that $\mathcal{A} \sim^{\infty}_{Gf} \mathcal{A}'$, $\mathcal{B} \sim^{\infty}_{Gf} \mathcal{B}'$, and $\mathcal{A}' \equiv^n \mathcal{B}'$.*

Actually, Rosen used $\sim$ (bisimulation) in his theorem instead of $\sim^{\infty}_{Gf}$, but if two models are related by $\sim$ they will be related by $\sim^{\infty}_{Gf}$.

THEOREM 14. *Let $\mathcal{C}$ be a class of finite pointed models and $\mathcal{C}'$ a subclass of $\mathcal{C}$ such that the set of formulas $\Phi \subseteq \mathcal{ML}^f$ that are satisfied in all the models in $\mathcal{C}'$ is non-empty. Let $\mathcal{C}'$ be defined by a first-order formula α. If $\mathcal{C}'$ is closed under $\sim^{\infty}_{Gf}$, then $\mathcal{C}'$ is definable by a formula in $\mathcal{ML}^f$.*

Proof. Suppose that $\mathcal{C}'$ is defined by a first-order sentence and closed under $\sim^{\infty}_{Gf}$ but not definable by a formula in $\mathcal{ML}^f$. We want to prove that for all n there are pointed models $(\mathcal{A}, c^{\mathcal{A}}) \in \mathcal{C}'$ and $(\mathcal{B}, c^{\mathcal{B}}) \in \mathcal{C} - \mathcal{C}'$ such that $(\mathcal{A}, c^{\mathcal{A}}) \equiv^n (\mathcal{B}, c^{\mathcal{B}})$, which would contradict the hypothesis. For any $n \in \mathbb{N}$, Theorem 11 says that there are models $(\mathcal{A}', c^{\mathcal{A}'})$ and $(\mathcal{B}', c^{\mathcal{A}'})$ such that $(\mathcal{A}', c^{\mathcal{A}'}) \sim^{H(n)}_{Gf} (\mathcal{B}', c^{\mathcal{A}'})$. Proposition 12 then lets us construct models such that $(\mathcal{A}'', c^{\mathcal{A}''}) \sim^{\infty}_{Gf} (\mathcal{B}'', c^{\mathcal{A}''})$, and $\mathcal{N}_{H(n)}(c^{\mathcal{A}''}) \cong \mathcal{N}_{H(n)}(c^{\mathcal{B}''})$. Finally, we apply Theorem 13 to obtain the needed $(\mathcal{A}, c^{\mathcal{A}})$ and $(\mathcal{B}, c^{\mathcal{B}})$ and the contradiction. ∎

7 Extensions

In this section we discuss some possible extensions of $\mathcal{ML}^f$ for which Theorem 14 still holds. Such extensions involve two main issues: modifying $\mathcal{ML}^f$ and finding the corresponding game.

First of all, we can easily cater for atomic negations, simply by expanding the definition of local formulas to also include negations of proposition letters. In this case the game definition is not affected.

Second, adding disjunctions is also straightforward. The new $\mathcal{ML}^f(\vee)$-fragments are closely related to the description logic $\mathcal{FLU}^-$ [5], just like the original $\mathcal{ML}^f$-fragments are closely related to $\mathcal{FL}^-$. Their definitions are given by:

$$\mathcal{ML}^f_0(\vee) = \text{LF}$$
$$\mathcal{ML}^f_{n+1}(\vee) = \text{the closure under } \wedge \text{ and } \vee \text{ of (LF}$$
$$\cup \{\Diamond \phi \mid \phi \in \mathcal{ML}^f_n(\vee) \text{ and } f(n+1) = \Diamond\}$$
$$\cup \{\Box \phi \mid \phi \in \mathcal{ML}^f_n(\vee) \text{ and } f(n+1) = \Box\}$$
$$\cup \{\Diamond \phi, \Box \phi \mid \phi \in \mathcal{ML}^f_n(\vee) \text{ and } f(n+1) = \varnothing\}).$$

Di and Si will have to play using singletons. In other words, the first designated position in each model will be a singleton and both Di and

Si have to choose singletons in following moves. Note that once we have added the usual connectives and modal operators, $\sim_{Gf}^{\infty}$ is equivalent to bisimulation and Theorem 10 is actually equivalent to Theorem 11.

Another natural extension is to go multi-modal; this can be done in many different ways. The obvious one is to replace $\Box$ with $[R_i]$ in each of the production rules in Definition 2. This method will not control the modal depth at which a particular relation is used. Alternatively, we can let our functions f choose which subset of modal operators is to be considered legal at each level in the definition of $\mathcal{ML}^f$. Our main complexity, characterization, and expressiveness results hold for both ways of going multi-modal.

Finally, we can go a step further and allow unqualified number restrictions, thus moving to modal counterparts of fragments of the description logic $\mathcal{FLN}^-$ [5]. Recall that unqualified number restrictions are formulas of the form $\Diamond^{\leq n}\top$ that are true in a state w iff there are $w_1, \ldots, w_k$ with $k \leq n$ such that $wRw_1, \ldots, wRw_k$. Unqualified number restrictions are still very local, and because of that it is easy to extend our setup to deal with them: we can simply add them to the set of local formulas, and our characterization and expressiveness results will continue to hold.

8 Conclusions and future work

We have introduced a novel mechanism for decomposing modal logic into fragments. Each of these fragments can be specified in a very fine-grained manner, and for each of them we have defined a notion of game that allows us to characterize the fragment's expressive power. We have also provided uniform upper bounds for the complexity of the satisfiability problem for each fragment. Our games provide a natural tool to understand the fragments and the constructors they admit.

The first natural next step is to extend our fragment so as to also capture more expressive modal logics, especially ones with qualified number restrictions. We also aim to further explore how these fragments behave computationally: not only by providing better upper bounds for the complexity of the satisfiability problem, but also by considering different reasoning tasks, for example model checking.

Finally, in our characterization of the expressive power of the $\mathcal{ML}^f$-fragments we adapted a proof due to Rosen. Otto [15] has recently given a alternative proof of Rosen's result, of a far less combinatorial nature than Rosen's. It would be instructive to see what additional insights this proof yields when adapted to our fragments.

Acknowledgmentss

We thank the anonymous referees for their comments. Gabriel Infante-Lopez is supported by the Netherlands Organization for Scientific Research (NWO) under project number 220-80-001. Carlos Areces is supported by a grant from NWO under project number 612.069.006. Maarten de Rijke is supported by NWO under project numbers 612.013.001, 612.069.006, 365-20-005, 220-80-001, 612.000.106, and 612.000.207.

BIBLIOGRAPHY

[1] H. Andréka, J. van Benthem, and I. Németi. Modal languages and bounded fragments of predicate logic. *Journal of Philosophical Logic*, 27:217–274, 1998.

[2] F. Baader, D. Calvanese, D. McGuinness, D. Nardi, and P. Patel-Schneider, editors. *The Description Logic Handbook: Theory, Implementation, and Applications*. Cambridge University Press, 2003.

[3] J. van Benthem. Correspondence theory. In D. Gabbay and F. Guenthner, editors, *Handbook of philosophical logic. Vol. II*, volume 165 of *Synthese Library*, pages 167–247. D. Reidel Publishing Co., Dordrecht, 1984. Extensions of classical logic.

[4] P. Blackburn, M. de Rijke, and Y. Venema. *Modal Logic*. Cambridge University Press, 2001.

[5] R. Brachman and H. Levesque. The tractability of subsumption in frame-based description languages. In *Proc. of the 4th Nat. Conf. on Artificial Intelligence (AAAI-84)*, pages 34–37, 1984.

[6] R. Brachman and H. Levesque. Expressiveness and tractability in knowledge representation and reasoning. *Computational Intelligence*, 3:78–93, 1987.

[7] K. Doets. *Basic Model Theory*. CSLI Publications, 1997.

[8] F. Donini. *Complexity of Reasoning*, chapter 3, pages 101–141. In Baader et al. [2], 2003.

[9] E. Grädel. Why are modal logics so robustly decidable? *Bulletin EATCS*, 68:90–103, 1999.

[10] E. Grädel, P. Kolaitis, and M. Vardi. On the decision problem for two-variable first-order logic. *Bulletin of Symbolc Logic*, 3:53–69, 1997.

[11] E. Grädel, M. Otto, and E. Rosen. Two-variable logic with counting is decidable. In *Proceedings of LICS 1997*, pages 306–317, 1997.

[12] J. Halpern. The effect of bounding the number of primitive propositions and the depth of nesting on the complexity of modal logic. *Artificial Intelligence*, 75:361–372, 1995.

[13] N. Kurtonina and M. de Rijke. Expressiveness of concept expressions in first-order description logics. *Artificial Intelligence*, 107:303–333, 1999.

[14] A. Nonnengart, H. Ohlbach, M. de Rijke, and D. Gabbay. *Encoding two-valued non-classical logics in classical logic*, pages 1403–1486. In Robinson and Voronkov [17], 2001.

[15] M. Otto. Modal and guarded characterisation theorems over finite transition systems. In *Proceedings LICS 2002*, 2002.

[16] L. Pacholsky, W. Szwast, and L. Tendera. Complexity of two-variable logic with counting. In *Proceedings of LICS 1997*, pages 318–327, 1997.

[17] A. Robinson and A. Voronkov, editors. *Handbook of Automated Reasoning*. Elsevier Science Publishers, 2001.

[18] E. Rosen. Modal logic over finite structures. *Journal of Logic, Language and Information*, 6(4):427–439, 1997.

[19] M. Schmidt-Schauss and G. Smolka. Attributive concept descriptions with complements. *Artificial Intelligence*, 48(1):1–26, 1991.

[20] D. Scott. A decision method for validity of sentences in two variables. *Journal of Symbolic Logic*, 27:377, 1962.
[21] M. Vardi. Why is modal logic so robustly decidable? In *DIMACS Ser. Disc. Math. Theoret. Comp. Sci. 31*, pages 149–184. AMS, 1997.

Carlos Areces
Langue et Dialogue Group, INRIA Lorraine
615, rue du Jardin Botanique
54602 Villers lès Nancy Cedex, France
E-mail: carlos.areces@loria.fr

Gabriel G. Infante-Lopez, Maarten de Rijke
Language & Inference Technology Group
ILLC, U. of Amsterdam
Nieuwe Achtergracht 166
1018 WV Amsterdam, The Netherlands
E-mail: {infante,mdr}@science.uva.nl

11
A Note on Relativised Products of Modal Logics

AGI KURUCZ AND MICHAEL ZAKHARYASCHEV

ABSTRACT. A natural idea of reducing the strong interaction between modal operators of product logics in hope of obtaining decidable but still expressive and useful many-dimensional formalisms is to consider (not necessarily generated) subframes of product frames. Worlds are still tuples, the relations still act coordinate-wise, but not all tuples of the Cartesian product are present, and so the commutativity and Church–Rosser properties do not necessarily hold. This kind of restriction on the domains of modal operators is similar to relativisations of the quantifiers in first-order logic and algebraic logic, where it indeed results in improving the bad algorithmic behaviour. Relativised product logics are often located between the fusions and the products of their components. In this paper, we show that arbitrary and locally cubic relativised products of many standard modal logics in fact coincide with their fusions. This result can also be considered as a new many-dimensional semantical characterisation of such fusions. We give some examples of arbitrarily relativised products sitting properly between fusions and products. We also provide some observations on expanding and decreasing relativised products, and their connections with first-order modal logics.

1 Introduction

One may think of many ways of combining normal modal logics representing various aspects of an application domain. Two 'canonical' constructions, supported by a well-developed mathematical theory, are fusions [17, 6, 7] and products [8, 7].

The *fusion* $L_1 \otimes \ldots \otimes L_n$ of $n \geq 2$ propositional normal unimodal logics L_i with the boxes $\Box_i$ is the smallest multimodal logic in the language with n boxes $\Box_1, \ldots, \Box_n$ (and their duals $\Diamond_1, \ldots, \Diamond_n$) that contains all the L_i. This means that if the L_i are axiomatised by sets Ax_i of axioms, then $L_1 \otimes \ldots \otimes L_n$ is axiomatised by the union $Ax_1 \cup \ldots \cup Ax_n$. Thus, fusions are useful when the modal operators of the combined logics are not supposed to interact (see e.g. [1] which provides numerous examples of applications of fusions in description logic). It is this absence of interaction axioms that ensures the transfer of good algorithmic properties from the components to their fusion [17, 6, 33]. In particular, it is possible to reduce reasoning in the fusion to reasoning in the components. On the semantic level, the fusion of Kripke complete modal logics $L_1, \ldots, L_n$ can be characterised by the class of all *n-frames* $\langle W, R_1, \ldots, R_n \rangle$ such that each $\langle W, R_i \rangle$ is a frame for L_i [17, 6]. Note that although frames for fusions have n different accessibility relations, they cannot be regarded as 'genuinely many-dimensional' in the geometric sense.

Products of modal logics do have real many-dimensional frames. Given n Kripke frames $\mathfrak{F}_1 = \langle U_1, R_1 \rangle, \ldots, \mathfrak{F}_n = \langle U_n, R_n \rangle$, their *product* $\mathfrak{F}_1 \times \cdots \times \mathfrak{F}_n$ is the n-frame

$$\langle U_1 \times \cdots \times U_n, \overline{R}_1, \ldots, \overline{R}_n \rangle,$$

where each $\overline{R}_i$ is a binary relation on $U_1 \times \ldots \times U_n$ defined by taking

$$\langle u_1, \ldots, u_n \rangle \, \overline{R}_i \, \langle u'_1, \ldots, u'_n \rangle \quad \text{iff} \quad u_i R_i u'_i \text{ and } u_k = u'_k, \text{ for } k \neq i.$$

The *product* $L_1 \times \ldots \times L_n$ of Kripke complete unimodal logics $L_1, \ldots, L_n$ is the logic of the class of all product frames $\mathfrak{F}_1 \times \cdots \times \mathfrak{F}_n$, where each $\mathfrak{F}_i$ is a frame for L_i. (Here by 'the logic of a class' we mean the set of those modal formulas with n boxes $\square_1, \ldots, \square_n$ that are valid in each frame of the class.) For example, $\mathbf{K}^n$ is the logic of all n-ary product frames. It is not hard to see that $\mathbf{S5}^n$ is the logic of all n-ary products of *universal* frames having the same worlds, that is, frames $\langle U, R_i \rangle$ with $R_i = U \times U$. We refer to product frames of this kind as *cubic universal product $\mathbf{S5}^n$-frames*. Note that the 'i-reduct' $\mathfrak{F}^{(i)} = \langle U_1 \times \cdots \times U_n, \overline{R}_i \rangle$ of $\mathfrak{F}_1 \times \cdots \times \mathfrak{F}_n$ is a union of n disjoint copies of $\mathfrak{F}_i$. Thus, $\mathfrak{F}^{(i)}$ and $\mathfrak{F}_i$ validate the same formulas, and so

$$L_1 \otimes \cdots \otimes L_n \subseteq L_1 \times \cdots \times L_n.$$

There is a strong interaction between the modal operators of product logics. Every n-ary product frame satisfies the following two properties, for each pair $i \neq j$, $i, j = 1, \ldots, n$:

Commutativity:

$$\forall x \forall y \forall z \Big(\big(x\overline{R}_i y \wedge y\overline{R}_j z \to \exists u \, (x\overline{R}_j u \wedge u\overline{R}_i z) \big) \\ \wedge \big(x\overline{R}_j y \wedge y\overline{R}_i z \to \exists u \, (x\overline{R}_i u \wedge u\overline{R}_j z) \big) \Big).$$

Church–Rosser property:

$$\forall x \forall y \forall z \, \big(x\overline{R}_i y \wedge x\overline{R}_j z \to \exists u \, (y\overline{R}_j u \wedge z\overline{R}_i u) \big).$$

This means that the corresponding modal interaction formulas

$$\Box_i\Box_j p \leftrightarrow \Box_j\Box_i p \quad \text{and} \quad \Diamond_i\Box_j p \to \Box_j\Diamond_i p$$

belong to every n-dimensional product logic. The geometrically intuitive many-dimensional structure of product frames makes them a perfect tool for constructing formalisms suitable for, say, spatio-temporal representation and reasoning (see e.g. [34, 35]) or reasoning about the behaviour of multi-agent systems (see e.g. [4]). However, the price we have to pay for the use of products is an extremely high computational complexity—even the product of two NP-complete logics can be non-recursively enumerable (see e.g. [30, 28]). In higher dimensions practically all products of 'standard' modal logics are undecidable and non-finitely axiomatisable [16].

A natural idea of reducing the strong interaction between modal operators of product logics in hope of obtaining more 'user-friendly' but still expressive and useful many-dimensional formalisms is to consider (not necessarily generated) *subframes*[1] of product frames. Worlds are still tuples, the relations still act coordinate-wise, but not all tuples of the Cartesian product are present, and so the commutativity and Church–Rosser properties do not necessarily hold. This kind of restriction on the 'domains' of modal operators is similar to 'relativisations' of the quantifiers in first-order logic and algebraic logic, where it indeed results in improving the bad algorithmic behaviour, cf. [25, 20]. As a modification of the product construction, 'relativisation' was first suggested in [22].

This idea gives rise to the following 'product-like' combinations of logics. First, we choose a class of 'desirable' subframes of product frames. This can be any class: the class of all such subframes, the so-called 'locally cubic' frames, frames that 'expand' along one of the coordinates (see below for definitions), a class of frames satisfying some (modal or first-order) formulas, etc. Having chosen such a class $\mathcal{K}$, we then take the logic determined by those subframes of the appropriate product frames that belong to $\mathcal{K}$. Thus, each choice of $\mathcal{K}$ defines a new product-like operator on logics. More precisely, the $\mathcal{K}$-*relativised product* $(L_1 \times \ldots \times L_n)^{\mathcal{K}}$ of Kripke complete unimodal logics $L_1, \ldots, L_n$ is the logic of the class of those frames in $\mathcal{K}$ that are subframes of product frames $\mathfrak{F}_1 \times \ldots \times \mathfrak{F}_n$ such that each $\mathfrak{F}_i$ is a frame for L_i. Observe that if we choose $\mathcal{K}$ to be the class of all product frames $\mathfrak{F}_1 \times \cdots \times \mathfrak{F}_n$, where $\mathfrak{F}_i$ is a frame for L_i, then the $\mathcal{K}$-relativised product of the L_i is just their usual product.

The aim of this note may appear to be rather modest. Instead of investigating the decision and complexity problems for relativised product logics

---

[1] Recall that an n-frame $\langle U, S_1, \ldots, S_n \rangle$ is a *subframe* of $\langle W, R_1, \ldots, R_n \rangle$ if $U \subseteq W$ and $S_i = R_i \cap (V \times V)$, for $i = 1, \ldots, n$.

right away, we prefer to (cautiously) begin by trying to find out what kind of 'creatures' these relativised products are and how they are related to standard products and fusions. The results of investigation are somewhat surprising. As we shall see, relativised product logics are often located between the fusions and the products of their components. However, our main statements in Sections 2 and 3 show that 'arbitrary' and 'locally cubic' relativised products of many standard modal logics in fact *coincide* with their fusions (which justifies our cautious approach and gives 'automatic' solutions to the decision and complexity problems). This result can also be considered as a nice new many-dimensional semantical characterisation of such fusions. We also give some interesting (and natural) examples of 'arbitrarily' relativised products sitting properly between fusions and products. Finally, in Section 4, we provide some observations on 'expanding' and 'decreasing' relativised products, and their connections with first-order modal logics.

2 Arbitrary relativisations

We begin by considering the product operator determined by the class SF_n of *all* subframes of n-ary product frames. SF_n-relativised products of logics will be called *arbitrarily relativised products*. Since SF_n contains frames that do not satisfy commutativity and/or Church–Rosser properties, clearly we have

$$(L_1 \times \cdots \times L_n)^{\mathsf{SF}_n} \subsetneq L_1 \times \cdots \times L_n.$$

On the other hand, unlike product logics, arbitrarily relativised products do not necessarily contain the fusion of their components. For example, consider the minimal deontic logic **D**, which is known to be characterised by the class of *serial* frames. Now, the formula $\Diamond_2 \top$ clearly belongs to the fusion $\mathbf{K} \otimes \mathbf{D}$, but is refuted in any *finite* subframe of, say, $\langle \omega, < \rangle \times \langle \omega, < \rangle$, and so $\Diamond_2 \top \notin (\mathbf{K} \times \mathbf{D})^{\mathsf{SF}_2}$. However, as we shall see below, for a large class of natural logics, arbitrarily relativised products do contain the fusions.

A Kripke complete modal logic L is called a *subframe logic* if the class of Kripke frames for L is closed under taking (not necessarily generated) subframes. (For a general theory of subframe logics consult [5, 2, 32] and references therein.) Typical examples of subframe logics are modal logics whose classes of Kripke frames are definable by universal first-order formulas, such as **K**, **Alt**, **T**, **K4**, **S4**, **S5**, **K5**, **K45**, **S4.3**, and **K4.3**. Note, however, that subframe logics like **GL**, **GL.3**, **Grz** are not first-order definable.

PROPOSITION 1. *If $L_1, \ldots, L_n$ are all subframe logics, then*

$$L_1 \otimes \ldots \otimes L_n \subseteq (L_1 \times \ldots \times L_n)^{\mathsf{SF}_n}.$$

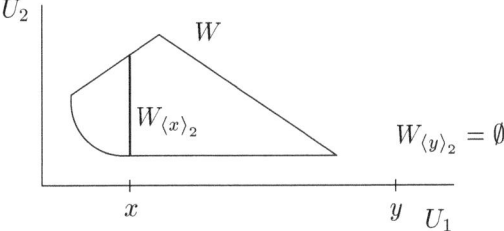

Figure 1. 'Coordinate-wise' subframes.

Proof. Suppose that an n-frame $\mathfrak{G} = \langle W, \overline{S}_1, \ldots, \overline{S}_n \rangle$ is a subframe of some product frame

$$\langle U_1, R_1 \rangle \times \cdots \times \langle U_n, R_n \rangle,$$

where $\langle U_i, R_i \rangle$ is a frame for L_i, $i = 1, \ldots, n$. Fix some i, $1 \le i \le n$. For every $n-1$-tuple $\overline{u}_i = \langle u_1, \ldots, u_{i-1}, u_{i+1}, \ldots, u_n \rangle$ with $u_j \in U_j$, for $j \ne i$, we take the set

$$W_{\overline{u}_i} = \{ \langle u_1, \ldots, u_n \rangle \in W \mid u_i \in U_i, \langle u_1, \ldots, u_{i-1}, u_{i+1}, \ldots, u_n \rangle = \overline{u}_i \},$$

and let $S_{\overline{u}_i}$ be the restriction of $\overline{S}_i$ to $W_{\overline{u}_i}$, i.e., $S_{\overline{u}_i} = \overline{S}_i \cap (W_{\overline{u}_i} \times W_{\overline{u}_i})$; see Fig. 1. Then clearly we have the following:

- if $W_{\overline{u}_i}$ is not empty then $\langle W_{\overline{u}_i}, S_{\overline{u}_i} \rangle$ is isomorphic to a subframe of $\langle U_i, R_i \rangle$;

- $\langle W, \overline{S}_i \rangle$ is the disjoint union of the frames $\langle W_{\overline{u}_i}, S_{\overline{u}_i} \rangle$, for all possible $n-1$-tuples $\overline{u}_i$ with non-empty $W_{\overline{u}_i}$.

Therefore, since L_i is a subframe logic, $\langle W, \overline{S}_i \rangle$ is a frame for L_i. ∎

As we shall see below, the converse of the inclusion in Proposition 1 does not always hold. However, as the following theorem shows, for many standard subframe logics, their arbitrarily relativised product coincides with their fusion. Thus, 'arbitrary relativisation' can be regarded as a 'many-dimensional' semantical characterisation of fusions of these logics.

THEOREM 2. *Let $L_i \in \{\mathbf{K}, \mathbf{T}, \mathbf{K4}, \mathbf{S4}, \mathbf{S5}, \mathbf{S4.3}\}$, for $i = 1, \ldots, n$. Then*

$$(L_1 \times \cdots \times L_n)^{\mathsf{SF}_n} = L_1 \otimes \cdots \otimes L_n.$$

Proof. It is well-known (see [6, 17]) that all fusions $L_1 \otimes \cdots \otimes L_n$ mentioned in the formulation of the theorem are characterised by countable (in fact, finite) rooted n-frames $\mathfrak{G} = \langle W, S_1, \ldots, S_n \rangle$, where $\langle W, S_i \rangle$ is a frame for L_i, $i = 1, \ldots, n$. We prove now the following:

LEMMA 3. *Let $L_i \in \{\mathbf{K, T, K4, S4, S5, S4.3}\}$, $i = 1, \ldots, n$, and let $\mathfrak{G} = \langle W, S_1, \ldots, S_n \rangle$ be a countable rooted n-frame such that each $\langle W, S_i \rangle$ is a frame for L_i. Then $\mathfrak{G}$ is a p-morphic image of a subframe of some product frame for $L_1 \times \cdots \times L_n$.*

Proof. First we show that every countable rooted n-frame

$$\mathfrak{G} = \langle W, S_1, \ldots, S_n \rangle$$

is a p-morphic image of a subframe of some product frame. We will construct, step-by-step, frames $\mathfrak{F}_i = \langle U_i, R_i \rangle$ ($i = 1, \ldots, n$), a subframe $\mathfrak{H}$ of $\mathfrak{F}_1 \times \cdots \times \mathfrak{F}_n$, and a p-morphism f from $\mathfrak{H}$ onto $\mathfrak{G}$. One way of formalising this straightforward step-by-step argument is by defining a *game* $G(\mathfrak{G})$ between two players ∀ (male) and ∃ (female) over $\mathfrak{G}$. (Our game and its properties are similar to those of [14], where games are played over relation algebras. For a detailed treatment of games over many-dimensional structures consult [15].)

Define a $\mathfrak{G}$-*network* to be a tuple

$$N = \langle U_1^N, \ldots, U_n^N, V^N, R_1^N, \ldots, R_n^N, f^N \rangle$$

such that $\mathfrak{F}_i^N = \langle U_i^N, R_i^N \rangle$ are finite intransitive trees for all $i = 1, \ldots, n$, $V^N \subseteq U_1^N \times \cdots \times U_n^N$, and f^N is a homomorphism from the subframe $\mathfrak{H}^N$ of $\mathfrak{F}_1^N \times \ldots \times \mathfrak{F}_n^N$, having V^N as its set of worlds, to $\mathfrak{G}$. In other words, for all $u_1 \in U_1, \ldots, u_n \in U_n$, $i = 1, \ldots, n$, and $u_i' \in U_i$,

if $\langle u_1, \ldots, u_n \rangle \in V^N$, $\langle u_1, \ldots, u_{i-1}, u_i', u_{i+1}, \ldots, u_n \rangle \in V^N$ and $u_i R_i^N u_i'$
then $f^N(u_1, \ldots, u_i, \ldots, u_n) S_i f^N(u_1, \ldots, u_i', \ldots, u_n)$.

The players ∀ and ∃ build a countable 'expanding' sequence of finite $\mathfrak{G}$-networks as follows.

In round 0, ∀ picks the root r of $\mathfrak{G}$. ∃ responds with a $\mathfrak{G}$-network N_0 such that all the $U_i^{N_0}$ are singleton sets, $V^{N_0} = U_1^{N_0} \times \cdots \times U_n^{N_0}$, the relations $R_i^{N_0}$ are all empty, and f^{N_0} maps the only n-tuple in V^{N_0} to r.

Suppose now that in round j, $0 < j < \omega$, the players have already built a finite $\mathfrak{G}$-network N_{j-1}. Now player ∀ challenges player ∃ with a possible *defect* of N_{j-1} which indicates that the homomorphism $f^{N_{j-1}}$ is not a p-morphism onto $\mathfrak{G}$ yet. ∀ picks such a defect which consists of

- an n-tuple $\langle u_1, \ldots, u_n \rangle \in V^{N_{j-1}}$,
- a coordinate $i \in \{1, \ldots, n\}$, and
- a world w in $\mathfrak{G}$ such that $f^{N_{j-1}}(u_1, \ldots, u_n) R_i w$.

Player $\exists$ can respond in two ways. If there is some u'_i such that

$$\langle u_1, \ldots, u'_i, \ldots, u_n \rangle \in V^{N_{j-1}}, \; u_i R_i^{N_{j-1}} u'_i$$
$$\text{and } f^{N_{j-1}}(u_1, \ldots, u'_i, \ldots, u_n) = w,$$

then she responds with $N_j = N_{j-1}$. Otherwise, she responds with the following $\mathfrak{G}$-network N_j extending N_{j-1}:

- $U_i^{N_j} = U_i^{N_{j-1}} \cup \{u^+\}$, where u^+ is a fresh point,
- $R_i^{N_j} = R_i^{N_{j-1}} \cup \{\langle u_i, u^+ \rangle\}$,
- $V^{N_j} = V^{N_{j-1}} \cup \{\langle u_1, \ldots, u_{i-1}, u^+, u_{i+1}, \ldots, u_n \rangle\}$,
- $\mathfrak{F}_k^{N_j} = \mathfrak{F}_k^{N_{j-1}}$ for all $k \neq i$, and
- $f^{N_j}(u_1, \ldots, u^+, \ldots, u_n) = w$.

Observe that $\exists$ can *always* respond this way. In other words, she always has a winning strategy in the ω-long game $G(\mathfrak{G})$. It is straightforward to see that the union (in the natural sense) of the constructed $\mathfrak{G}$-networks gives the required p-morphism f from a subframe $\mathfrak{H} = \langle V, \ldots \rangle$ of a product frame $\mathfrak{F}_1 \times \cdots \times \mathfrak{F}_n$ onto $\mathfrak{G}$. This proves the lemma for $L_i = \mathbf{K}$, $i = 1, \ldots, n$.

However, in the other cases nothing guarantees that the 'coordinate' frames $\mathfrak{F}_i = \langle U_i, R_i \rangle$ are actually frames for L_i. In what follows we fix some i, $1 \leq i \leq n$, and try to transform $\mathfrak{F}_i$ into a frame for L_i keeping all other frames $\mathfrak{F}_j$ for $j \neq i$ and the set V intact. Without loss of generality we may assume that $i = 1$.

To begin with, we show that the frames $\mathfrak{F}_i$ and the subframe $\mathfrak{H} = \langle V, \ldots \rangle$ have some useful properties. First, it should be clear from the construction that

(1) for each $i = 1, \ldots, n$, the frame $\mathfrak{F}_i$ is an intransitive tree.

To formulate another property, we require an auxiliary definition. Given an odd natural number k, a sequence $\langle v^0, \ldots, v^k \rangle$ of distinct n-tuples $v^\ell = \langle v_1^\ell, \ldots, v_n^\ell \rangle$, $\ell \leq k$, from V is called a *path in V between v^0 and v^k* if the following two conditions hold:

- for each even number $\ell < k$, $v_j^\ell = v_j^{\ell+1}$ whenever $j \neq 1$, and

- for each odd number $\ell < k$, $v_1^\ell = v_1^{\ell+1}$

(see Fig. 2). We call k the *length* of this path. If in addition $v_1^0 = v_1^k$ also

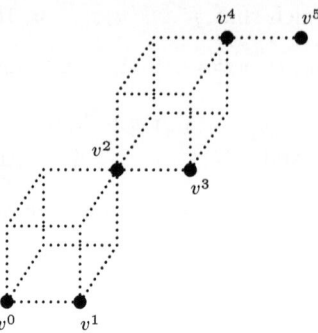

Figure 2. A 3-dimensional path of length 5.

holds then we call $\langle v^0, \ldots, v^k \rangle$ a *circle in* V (since all the n-tuples are distinct in a path, this can happen only if $k \geq 3$; see Fig. 3). Observe that

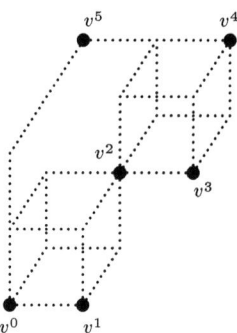

Figure 3. A 3-dimensional circle.

if $\langle v^0, \ldots, v^k \rangle$ is a circle then, for every $\ell \leq k$ and every i, $1 \leq i \leq n$, there exists an $\ell' \leq k$, $\ell' \neq \ell$, such that $v_i^\ell = v_i^{\ell'}$.

The second important property is that

(2) there are no circles in V.

Indeed, suppose otherwise. Take a circle $\langle v^0, \ldots, v^k \rangle$ in V and enumerate all of its n-tuples according to their 'creation time' in the game. Let v^ℓ be the last one in this list. By the rules of the game, one of the coordinates of v^ℓ must be fresh, contrary to the observation above.

Note that, as a special case of (2), we obtain that there are no *squares* in V, i.e., four distinct n-tuples of the form $\langle x, w_2, \ldots, w_n\rangle$, $\langle x, w'_2, \ldots, w'_n\rangle$, $\langle y, w_2, \ldots, w_n\rangle$, and $\langle y, w'_2, \ldots, w'_n\rangle$.

Now, in order to transform $\mathfrak{F}_1 = \langle U_1, R_1\rangle$ into a frame for L_1, we will extend, step-by-step, the accessibility relation R_1 (but always leave the sets U_1, V and the frames $\mathfrak{F}_j$ for $j \ne 1$ unchanged).

First, let $L_1 = \mathbf{K4}$. Define an infinite ascending chain

$$R_1^0 \subseteq R_1^1 \subseteq \ldots \subseteq R_1^m \subseteq \ldots$$

of binary relations on U_1 by taking $R_1^0 = R_1$ and, for $m < \omega$,

$$R_1^{m+1} = R_1^m \cup$$
$$\{\langle x_1, y_1\rangle \in U_1 \times U_1 \mid x_1 R_1^m z_1 \text{ and } z_1 R_1^m y_1 \text{ for some } z_1 \in U_1\}.$$

For every $m < \omega$, let $\mathfrak{F}_1^m = \langle U_1, R_1^m\rangle$ and let $\mathfrak{H}^m$ be the subframe of

$$\mathfrak{F}_1^m \times \mathfrak{F}_2 \times \cdots \times \mathfrak{F}_n$$

with V as its set of worlds. Finally, let

$$R_1^\infty = \bigcup_{m<\omega} R_1^m, \qquad \mathfrak{F}_1^\infty = \langle U_1, R_1^\infty\rangle,$$

and let $\mathfrak{H}^\infty$ be the corresponding subframe of $\mathfrak{F}_1^\infty \times \mathfrak{F}_2 \times \cdots \times \mathfrak{F}_n$.

Clearly, $\mathfrak{F}_1^\infty$ is a frame for $\mathbf{K4}$. We are about to show that f is still a p-morphism from $\mathfrak{H}^\infty$ onto $\mathfrak{G}$. Since the 'backward' p-morphism condition always holds after extending the accessibility relation of the pre-image, it is enough to show that f is a homomorphism from $\mathfrak{H}^\infty$ onto $\mathfrak{G}$. We will prove by parallel induction on m that the following two statements hold, for all $m < \omega$ and for $x_1, y_1 \in U_1$:

(a) If $x_1 R_1^m y_1$ then there exist $x_2, \ldots, x_n, y_2, \ldots, y_n$ such that there is a path in V between $x = \langle x_1, x_2, \ldots, x_n\rangle$ and $y = \langle y_1, y_2, \ldots, y_n\rangle$.

(b) If $x_1 R_1^m y_1$ and both $w^x = \langle x_1, w_2, \ldots, w_n\rangle$ and $w^y = \langle y_1, w_2, \ldots, w_n\rangle$ are in V for some $w_j \in U_j$, $j = 2, \ldots, n$, then $f(w^x) S_1 f(w^y)$. In other words, f is a homomorphism from $\mathfrak{H}^m$ onto $\mathfrak{G}$.

Suppose first that $m = 0$. Then, by the definition of $\mathfrak{H}$, (b) holds and there exist $w_2, \ldots, w_n$ such that both $w^x = \langle x_1, w_2, \ldots, w_n\rangle$ and $w^y = \langle y_1, w_2, \ldots, w_n\rangle$ are in V. By (1), $x_1 \ne y_1$ and so the sequence $\langle w^x, w^y\rangle$ is a path in V as required.

Let us assume inductively that (a) and (b) hold for some $m < \omega$, and let $x_1, y_1 \in U_1$ be such that $x_1 R_1^{m+1} y_1$, but $x_1 R_1^m y_1$ does not hold. Then there is a $z_1 \in U_1$ such that $x_1 R_1^m z_1$ and $z_1 R_1^m y_1$. It is not hard to see that, by (1), x_1, y_1 and z_1 should be all distinct. By item (a) of the induction hypothesis, there are x_j, z_j, z'_j, y_j, for $j = 2, \ldots, n$, such that

- there is a path in V between the n-tuples $x = \langle x_1, x_2, \ldots, x_n \rangle$ and $z = \langle z_1, z_2, \ldots, z_n \rangle$;

- there is a path in V between the n-tuples $z' = \langle z_1, z'_2, \ldots, z'_n \rangle$ and $y = \langle y_1, y_2, \ldots, y_n \rangle$.

If $z \ne z'$ then the concatenation of these two paths gives a path between x and y. If $z = z'$ then we leave out z from the concatenated sequence, and the rest gives a path as required in (a).

For (b), suppose that $w^x = \langle x_1, w_2, \ldots, w_n \rangle$ and $w^y = \langle y_1, w_2, \ldots, w_n \rangle$ are in V for some $w_j \in U_j$, $j = 2, \ldots, n$. Let $w^z = \langle z_1, w_2, \ldots, w_n \rangle$. Consider the n-tuples x, y, z, z' given above. We claim that

(3) $\quad x = w^x$, $y = w^y$ and $z = z' = w^z$.

For suppose otherwise. Then several cases are possible. We will show that each of them means that there is a circle in V, contrary to (2). Let $\rho = \langle x, v^1, \ldots, v^k, y \rangle$ denote the path in V between x and y given by (a).

- Suppose first that $x \ne w^x$ and $y \ne w^y$. Then the concatenation of $\langle w^y, w^x \rangle$ and ρ is a circle in V.

- Suppose that $x = w^x$ and $y \ne w^y$. Then the length of ρ is ≥ 3, and so $\langle w^y, v^1, \ldots, v^k, y \rangle$ is a circle in V. The case when $x \ne w^x$ and $y = w^y$ is similar.

- Finally, suppose $x = w^x$ and $y = w^y$. If $z \ne w^z$ and $z' \ne w^z$ then the length of ρ should be ≥ 5, and $\langle v^k, v^1, \ldots, v^{k-1} \rangle$ is a circle in V. The cases when one of z and z' coincides with w^z, but the other does not, are similar.

As a consequence of (3), we obtain that w^z is in V. So by item (b) of the induction hypothesis,

$$f(w^x) S_1 f(w^z) \quad \text{and} \quad f(w^z) S_1 f(w^y).$$

Since S_1 is transitive, we have $f(w^x) S_1 f(w^y)$, which completes the proof of Lemma 3 for $L_1 = \mathbf{K4}$.

If $L_1 = \mathbf{S4}$ or $L_1 = \mathbf{T}$, we simply make all worlds of $\mathfrak{F}_1$ reflexive. Clearly, f will still be a p-morphism. In the case of $L_1 = \mathbf{S5}$, we have to 'close' $\mathfrak{F}_1$ under both transitivity and symmetry. It is not hard to see that this causes no problem, since there are no squares in V.

For $L_1 = \mathbf{S4.3}$ we need a slight modification of the above proof for $\mathbf{K4}$. We have to turn $\mathfrak{F}_1$ to a reflexive, transitive and weakly connected frame. To this end, we modify the definition of the accessibility relation R_1^{m+1} ($m < \omega$). First, we make all the points in U_1 reflexive. Then, for all distinct $x_1, y_1 \in U_1$, we define $\langle x_1, y_1 \rangle$ to be in R_1^{m+1} iff one of the following three conditions hold:

- $x_1 R_1^m y_1$;

- there is a $z_1 \in U_1$ such that $x_1 R_1^m z_1$ and $z_1 R_1^m y_1$;

- there is a $z_1 \in U_1$ such that $z_1 R_1^m x_1$, $z_1 R_1^m y_1$, and
  - either there are no $w_2, \ldots, w_n$ such that both n-tuples $w^x = \langle x_1, w_2, \ldots, w_n \rangle$ and $w^y = \langle y_1, w_2, \ldots, w_n \rangle$ are in V,
  - or there exist $w_2, \ldots, w_n$ such that both w^x and w^y are in V, and $f(w^x) S_1 f(w^y)$ holds. (Note that although $w^x \neq w^y$, it can happen that $f(w^x) = f(w^y)$.)

Since there are no squares in V, R_1^{m+1} is well-defined. The very same inductive proof as above shows that the frame $\mathfrak{F}_1^\infty$ obtained this way is reflexive, transitive and weakly connected, and f is still a p-morphism from $\mathfrak{H}^\infty$ onto $\mathfrak{G}$. ∎

Now we can complete the proof of Theorem 2. Let $\varphi \notin L_1 \otimes \cdots \otimes L_n$. Take a countable rooted n-frame $\mathfrak{G} = \langle W, S_1, \ldots, S_n \rangle$ refuting φ and such that, for every $i = 1, \ldots, n$, $\langle W, S_i \rangle$ is a frame for L_i. Using Lemma 3, we can find a subframe $\mathfrak{H}$ of a product frame for $L_1 \times \cdots \times L_n$ having $\mathfrak{G}$ as its p-morphic image. It follows that $\mathfrak{H} \not\models \varphi$, and so $\varphi \notin (L_1 \times \cdots \times L_n)^{\mathsf{SF}_n}$. Thus, $(L_1 \times \cdots \times L_n)^{\mathsf{SF}_n} \subseteq L_1 \otimes \cdots \otimes L_n$. Proposition 1 gives the converse inclusion. ∎

It is not clear how far Theorem 2 can be generalised. On the one hand, we conjecture that it holds for $L_i \in \{\mathbf{K4.3}, \mathbf{Grz}, \mathbf{GL}, \mathbf{GL.3}\}$ as well. For $\mathbf{K4.3}$ even Lemma 3 may hold, although a somewhat different, 'more careful' proof would be needed. However, it is not true that every countable (even finite) frame for, say, $\mathbf{Grz} \otimes \mathbf{Grz}$ is a p-morphic image of a subframe of a product of two $\mathbf{Grz}$-frames. Consider, for instance, the 2-frame $\langle \{x, y, z, w\}, R_1, R_2 \rangle$ with $x R_1 y R_2 z R_1 w R_2 x$. It is not hard to see that if this frame is a p-morphic image of a subframe of $\mathfrak{F}_1 \times \mathfrak{F}_2$ then both $\mathfrak{F}_1$ and $\mathfrak{F}_2$ must contain infinite ascending chains of distinct points, and so cannot be frames for **Grz**.

On the other hand, Theorem 2 does not hold for all subframe logics, not even for those of them that (unlike **Grz**) are characterised by universally first-order definable classes of frames. Take, for instance, the logic

$$\mathbf{K5} = \mathbf{K} \oplus \Diamond\Box p \to \Box p.$$

It is well-known (see e.g. [2]) that **K5** is Kripke complete and characterised by the class of Euclidean frames, i.e., frames $\langle W, R \rangle$ satisfying the universal (Horn) sentence

$$\forall x \forall y \forall u \, \big(R(u,x) \wedge R(u,y) \to R(x,y)\big).$$

In particular, frames for **K5** have the property

$$\forall x \forall u \, \big(R(u,x) \to R(x,x)\big).$$

Now consider the formula

$$\varphi = \Diamond_1\big(p \wedge \Diamond_2(q \wedge \neg p)\big) \wedge \Box_1\Box_2(q \to \neg\Diamond_1 q).$$

It is clearly satisfiable in the following frame for $\mathbf{K5} \otimes \mathbf{K}$:

On the other hand, it is not hard to see that φ is not satisfiable in any subframe of a product frame for $\mathbf{K5} \times \mathbf{K}$. Therefore,

$$\mathbf{K5} \otimes \mathbf{K} \subsetneq (\mathbf{K5} \times \mathbf{K})^{\mathsf{SF}_2} \subsetneq \mathbf{K5} \times \mathbf{K}.$$

In fact, a similar statement holds for any logic $\mathbf{K} \oplus \Diamond^i \Box p \to \Box^i p$ ($i \geq 1$) in place of **K5**. Further, the same argument shows that

$$\mathbf{K45} \otimes \mathbf{K4} \subsetneq (\mathbf{K45} \times \mathbf{K4})^{\mathsf{SF}_2} \subsetneq \mathbf{K45} \times \mathbf{K4},$$

where $\mathbf{K45} = \mathbf{K4} \oplus \Diamond\Box p \to \Box p$.

Another kind of logics for which Theorem 2 does not hold are those having frames with a finite bound on their branching, e.g. **Alt**. Recall that $\langle W, R \rangle$ is a frame for **Alt** iff every point in W has at most one R-successor. Now consider the formula

$$\psi = p \wedge \Diamond_1\big(\neg p \wedge \Diamond_2 q\big) \wedge \Diamond_2\big(\neg p \wedge \Diamond_1 r\big) \wedge \Box_1\Box_2(q \to \neg r).$$

It is clearly satisfiable in the $\mathbf{Alt} \otimes \mathbf{Alt}$-frame

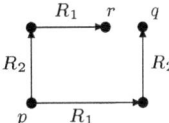

On the other hand, it should be clear that ψ is not satisfiable in any subframe of a frame for $\mathbf{Alt} \times \mathbf{Alt}$. Thus,

$$\mathbf{Alt} \otimes \mathbf{Alt} \subsetneq (\mathbf{Alt} \times \mathbf{Alt})^{\mathsf{SF}_2} \subsetneq \mathbf{Alt} \times \mathbf{Alt}.$$

However, in general the behaviour of arbitrarily relativised products remains unexplored. It would be of interest, for instance, to find solutions to the following problems.

Question 1. *Are arbitrarily relativised products of finitely axiomatisable logics also finitely axiomatisable (in those cases when they differ from the fusions)?*

Question 2. *Are arbitrarily relativised products of decidable logics also decidable?*

Question 3. *Find a general characterisation of those arbitrarily relativised products of logics that coincide with their fusions.*

3 Cubic and locally cubic relativisations

To motivate another kind of relativisations, let us briefly discuss a possible way of creating new, more expressive logics from products. Given the product of n unimodal logics, one may want to add new operations to $\Box_1, \ldots, \Box_n$ that 'connect' the different dimensions. Perhaps the simplest and most natural operations of this sort are the *diagonal constants* d_{ij}. Given two natural numbers i and j, $1 \leq i, j \leq n$, the truth-relation for the constant d_{ij} in models over subframes of n-ary product frames is defined as follows:

$$(\mathfrak{M}, \langle u_1, \ldots, u_n \rangle) \models \mathsf{d}_{ij} \quad \text{iff} \quad u_i = u_j.$$

The set of n-tuples satisfying d_{ij} is usually called the (i,j)-*diagonal element*.

Actually, the main reason for introducing such constants is to give a 'modal treatment' of equality of classical first-order logic. Let $(\mathbf{S5}^n)^=$ denote the logic (in the language with n boxes and the diagonal constants) determined by the class of cubic universal product $\mathbf{S5}^n$-frames extended with the diagonal elements (interpreting the d_{ij}). Modal algebras for this logic are called *representable cylindric algebras* and are extensively studied in the algebraic logic literature; see e.g. [13, 12, 15] and references therein. By the algebraic results of [23] and [18], $(\mathbf{S5}^n)^=$ is neither finitely axiomatisable nor decidable. Note also that $(\mathbf{S5}^n)^=$ is not a conservative extension of $\mathbf{S5}^n$ [12].

Another natural way of connecting dimensions is via so-called 'jump' modalities. Given a function $\pi : \{1, \ldots, n\} \to \{1, \ldots, n\}$ (such a map can be called a *jump*), define the truth-relation for the unary modal operator s_π in models over subframes of product frames as follows:

$$(\mathfrak{M}, \langle u_1, \ldots, u_n \rangle) \models s_\pi \varphi \quad \text{iff} \quad (\mathfrak{M}, \langle u_{\pi(1)}, \ldots, u_{\pi(n)} \rangle) \models \varphi.$$

These modal operators are often called (*generalised*) *substitutions*, since (when added to $\mathbf{S5}^n$) they are the 'modal counterparts' of variable substitutions in classical first-order logic; see [20]. Note that in cubic universal product $\mathbf{S5}^n$-frames certain substitutions are expressible with the help of the boxes and the diagonal constants [12]. Various versions of the modal algebras corresponding to products of $\mathbf{S5}$ logics with substitutions and with or without diagonal constants (e.g., *polyadic* and *substitution algebras*) are studied in [9, 10, 26, 27]; see also [3, 24, 29]. Again, the algebraic results show that most of these logics are non-finitely axiomatisable and undecidable.

Arbitrary relativisations of these extensions of $\mathbf{S5}$-products do result in new, decidable many-dimensional logics; see [25, 31]. Moreover, both the diagonal constants and the substitutions can 'detect' some properties of the set of worlds, so it makes sense to consider, for example, those frames whose sets of worlds are closed under jumps. A non-empty set W of n-tuples is called a *local n-cube* if for all maps

$$\pi : \{1, \ldots, n\} \to \{1, \ldots, n\}$$

and all $\langle u_1, \ldots, u_n \rangle \in W$, we have $\langle u_{\pi(1)}, \ldots, u_{\pi(n)} \rangle \in W$. It is easy to see that W is a local n-cube iff for every $\langle u_1, \ldots, u_n \rangle \in W$, the Cartesian power $\{u_1, \ldots, u_n\}^n$ is a subset of W, that is, W is the union of 'n-dimensional cubes.' In particular, local 2-cubes are just the reflexive and symmetric binary relations.

A set W such that $W = U^n$, for some non-empty set U, will be called an *n-cube*. Clearly, n-cubes are special cases of local n-cubes. Let

$$\mathsf{LC}_n = \{\langle W, S_1, \ldots, S_n\rangle \in \mathsf{SF}_n \mid W \text{ is a local } n\text{-cube}\},$$
$$\mathsf{C}_n = \{\langle W, S_1, \ldots, S_n\rangle \in \mathsf{SF}_n \mid W \text{ is an } n\text{-cube}\}.$$

Note that cubic universal product frames (as introduced in Section 1) belong to C_n. In general, we will refer to frames whose sets of worlds are n-cubes as *cubic*.

Locally cubic relativisations of the above extensions of $\mathbf{S5}$-products again give new logics that are also different from the arbitrarily relativised versions. Moreover, all these 'extended relativised $\mathbf{S5}$-products' turn out to be decidable and often finitely axiomatisable. A comprehensive treatment of relativised versions of $(\mathbf{S5}^n)^=$ and products of $\mathbf{S5}$ logics extended with substitutions can be found in [20] under the respective names of *cylindric modal logics* and *modal logics of relations*.

Note that one can also establish connections between different dimensions by introducing polyadic modal operators on product frames. This is the road taken by *arrow logics*, where a binary modal operator is considered. Relativised versions of arrow logics are among the main topics of [20]; see also references therein.

Question 4. *What can be said about extensions with diagonals and/or substitutions of arbitrarily and locally cubic relativised products of modal logics different from* $\mathbf{S5}$?

As mentioned in [22], decidability of these extensions of relativised $\mathbf{K}^n$ can be proved by a reduction to the $n+1$-variable packed fragment of first-order logic. According to [21], the mosaic method (which has been so successful for extensions of relativised $\mathbf{S5}^n$) can also be used to show decidability of extensions of $(L_1 \times \cdots \times L_n)^{\mathsf{LC}_n}$ and $(L_1 \times \cdots \times L_n)^{\mathsf{SF}_n}$, whenever $L_i \in \{\mathbf{K}, \mathbf{T}, \mathbf{K4}, \mathbf{S4}, \mathbf{S5}\}$.

The following two propositions show that if we do not enrich the language of n boxes, then locally cubic and cubic relativisations do not yield anything new.

PROPOSITION 4. *For all Kripke complete unimodal logics* $L_1, \ldots, L_n$ *and all classes* $\mathcal{K}$ *such that* $\mathsf{LC}_n \subseteq \mathcal{K} \subseteq \mathsf{SF}_n$,

$$(L_1 \times \cdots \times L_n)^{\mathsf{LC}_n} = (L_1 \times \cdots \times L_n)^{\mathcal{K}} = (L_1 \times \cdots \times L_n)^{\mathsf{SF}_n}.$$

Proof. The inclusions

$$(L_1 \times \cdots \times L_n)^{\mathsf{LC}_n} \supseteq (L_1 \times \cdots \times L_n)^{\mathcal{K}} \supseteq (L_1 \times \cdots \times L_n)^{\mathsf{SF}_n}$$

are obvious. To prove the converse ones, take any rooted subframe $\mathfrak{H}$ of a product frame $\mathfrak{F} = \mathfrak{F}_1 \times \cdots \times \mathfrak{F}_n$, where $\mathfrak{F}_i = \langle U_i, R_i \rangle$ is a frame for L_i for each $i = 1, \ldots, n$. We show that $\mathfrak{H}$ is isomorphic to a generated subframe of some $\mathfrak{G} \in \mathsf{LC}_n$, where $\mathfrak{G}$ is a subframe of some product frame $\mathfrak{F}_1^+ \times \cdots \times \mathfrak{F}_n^+$, with each $\mathfrak{F}_i^+$ being a frame for L_i. Indeed, take an isomorphic copy of $\mathfrak{F}$ such that the U_i are pairwise disjoint. By Makinson's theorem [19], for each Kripke complete unimodal logic L, either the one-element reflexive frame ($\circ$) or the one-element irreflexive frame ($\bullet$) is a frame for L. For all $i, j \in \{1, \ldots, n\}$, we define binary relations R_i^j on U_j by taking

$$R_i^j = \begin{cases} R_i & \text{if } i = j, \\ \emptyset & \text{if } \bullet \text{ is a frame for } L_i, \\ \{\langle u, u \rangle \mid u \in U_j\} & \text{if } \circ \text{ is a frame for } L_i. \end{cases}$$

Now let $U = \bigcup_{1 \leq i \leq n} U_i$. For every $i \in \{1, \ldots, n\}$, set $R_i^+ = \bigcup_{1 \leq j \leq n} R_i^j$, and take $\mathfrak{F}_i^+ = \langle U, R_i^+ \rangle$. Since each $\mathfrak{F}_i^+$ is a disjoint union of L_i-frames, the product frame

$$\mathfrak{F}^+ = \mathfrak{F}_1^+ \times \cdots \times \mathfrak{F}_n^+$$

is then a frame for $L_1 \times \cdots \times L_n$. Let W denote the set of worlds of $\mathfrak{H}$. Define W^+ to be the smallest local n-cube containing W, that is,

$$W^+ = \bigcup \{\{u_1, \ldots, u_n\}^n \mid \langle u_1, \ldots, u_n \rangle \in W\},$$

and let $\mathfrak{G}$ be the subframe of $\mathfrak{F}^+$ with W^+ as its set of worlds (see Fig. 4 for the case $n = 2$). Then clearly $\mathfrak{G} \in \mathsf{LC}_n$ holds, and $\mathfrak{H}$ is a subframe of $\mathfrak{G}$.

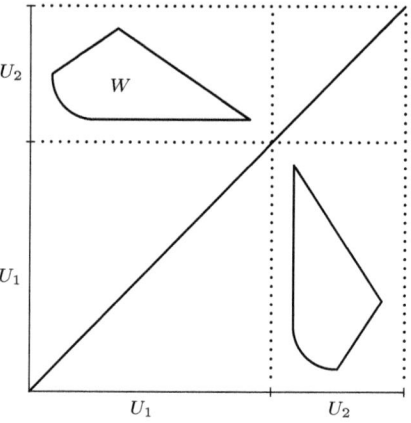

Figure 4. The smallest local 2-cube containing W.

It is not hard to see that $\mathfrak{H}$ is in fact a *generated* subframe of $\mathfrak{G}$, because $W^+ \cap (U_1 \times \cdots \times U_n) = W$. ∎

PROPOSITION 5. *For all subframe logics $L_1, \ldots, L_n$,*

$$(L_1 \times \cdots \times L_n)^{C_n} = L_1 \times \cdots \times L_n.$$

Proof. The inclusion $L_1 \times \cdots \times L_n \subseteq (L_1 \times \cdots \times L_n)^{C_n}$ is easy, since the L_i are subframe logics and every cubic subframe of a product frame is in fact a product of some subframes of the components.

To prove the converse, we show that every frame $\mathfrak{F} = \mathfrak{F}_1 \times \cdots \times \mathfrak{F}_n$, with each $\mathfrak{F}_i$ being a frame for L_i, is a p-morphic image of a cubic product frame, that is, a frame $\mathfrak{G} = \mathfrak{G}_1 \times \cdots \times \mathfrak{G}_n$ such that every $\mathfrak{G}_i$ has the same set of worlds and each $\mathfrak{G}_i$ is a frame for L_i. Indeed, take a cardinal $\kappa \geq \max_{1 \leq i \leq n} |\mathfrak{F}_i|$ and let $\mathfrak{G}_i$ be the disjoint union of κ-many copies of $\mathfrak{F}_i$. Then $|\mathfrak{G}_i| = \kappa$ and $\mathfrak{F}_i$ is a p-morphic image of $\mathfrak{G}_i$, whenever $1 \leq i \leq n$. Thus, it is easy to see (cf. e.g. [8]) that the 'product' of these p-morphisms gives a p-morphism from $\mathfrak{G}$ onto $\mathfrak{F}$. Since all the $\mathfrak{G}_i$ are of the same cardinality, we may assume that they are frames over the same set of worlds. ■

4 Expanding and decreasing relativisations

First-order modal and intuitionistic logics as well as modal description logics motivate our third group of relativisations. Fix a subset N of $\{1, \ldots, n\}$. An n-frame $\mathfrak{G} = \langle W, S_1, \ldots, S_n \rangle$ is called an N-*expanding* (or N-*decreasing*) *relativised product frame* if there are frames $\mathfrak{F}_1 = \langle U_1, R_1 \rangle, \ldots, \mathfrak{F}_n = \langle U_n, R_n \rangle$ such that

- $\mathfrak{G}$ is a subframe of $\mathfrak{F}_1 \times \cdots \times \mathfrak{F}_n$;
- for all $\langle w_1, \ldots, w_n \rangle \in W$, $j \in N$, and $u \in U_j$,

    if $w_j R_j u$ (or $u R_j w_j$) then $\langle w_1, \ldots, w_{j-1}, u, w_{j+1}, \ldots, w_n \rangle \in W$.

If $N = \{1\}$ then we call $\mathfrak{G}$ an (n-ary) *expanding* (*decreasing*) *relativised product frame*. Examples of decreasing relativised product frames are the two-dimensional frames for the Halpern–Shoham interval temporal logic (they are also $\{2\}$-expanding); see [11, 20]. In what follows we consider only expanding relativisations. The reader should have no problem in reformulating all notions and results for the case of decreasing ones.

Define EX_n to be the class of all n-ary expanding relativised product frames. In case $n = 2$, we omit the subscript and write EX.

It is easy to see that every expanding relativised product frame satisfies 'left' commutativity

$$\forall x \forall y \forall z \, (x R_i y \wedge y R_1 z \to \exists u \, (x R_1 u \wedge u R_i z)),$$

and Church–Rosser properties between coordinates 1 and i, for all $i = 2, \ldots, n$. Therefore, the formulas $\Box_1 \Box_i p \to \Box_i \Box_1 p$ and $\Diamond_1 \Box_i p \to \Box_i \Diamond_1 p$ are valid in expanding relativised product frames for all $i = 2, \ldots, n$.

Let us consider first the axiomatisation problem for two-dimensional expanding relativisations. Given logics L_1 and L_2, define

$$[L_1, L_2]^{\mathsf{EX}} = (L_1 \otimes L_2) \oplus \Box_1 \Box_2 p \to \Box_2 \Box_1 p \oplus \Diamond_1 \Box_2 p \to \Box_2 \Diamond_1 p.$$

A formula

$$\forall x \forall y \forall \bar{z} \, (\psi(x, y, \bar{z}) \to R(x, y))$$

of the first-order language with a binary predicate R is called a *universal Horn sentence* if $\psi(x,y,\bar{z})$ is built up from atoms using only $\wedge$ and $\vee$. We call a unimodal formula φ a *Horn formula*, if there is a universal Horn sentence χ_φ such that for all frames $\mathfrak{F}$,

$$\mathfrak{F} \models \varphi \quad \text{iff} \quad \mathfrak{F} \models \chi_\varphi.$$

A unimodal formula is called *variable free* if it contains no propositional variables, i.e., all of its atomic subformulas are constants $\bot$ or $\top$. We call a unimodal logic *Horn axiomatisable* if it is axiomatisable by only Horn and variable-free formulas. Examples of Horn axiomatisable logics are **K**, **D**, **K4**, **K5**, **S4**, **KD45**, **T**, **B**, **S5**.

THEOREM 6. *Let L_1 and L_2 be Kripke complete logics such that $L_1 \in \{\mathbf{K}, \mathbf{T}, \mathbf{K4}, \mathbf{S4}, \mathbf{S5}\}$ and L_2 is Horn axiomatisable. Then*

$$(L_1 \times L_2)^{\mathsf{EX}} = [L_1, L_2]^{\mathsf{EX}}.$$

Proof. It is easy to see that if $\square_1\square_2 p \to \square_2\square_1 p$ and $\Diamond_1\square_2 p \to \square_2\Diamond_1 p$ are valid in a frame $\mathfrak{F} = \langle W, R_1, R_2\rangle$ with symmetric R_1, then $\square_1\square_2 p \leftrightarrow \square_2\square_1 p$ is also valid in $\mathfrak{F}$. By a result of [8], we then have

$$(\mathbf{S5} \times L_2)^{\mathsf{EX}} = \mathbf{S5} \times L_2 = [\mathbf{S5}, L_2] = [\mathbf{S5}, L_2]^{\mathsf{EX}}$$

(here $[\mathbf{S5}, L_2] = (\mathbf{S5} \otimes L_2) \oplus \square_1\square_2 p \leftrightarrow \square_2\square_1 p \oplus \Diamond_1\square_2 p \to \square_2\Diamond_1 p$). In the other cases one can show, by a step-by-step construction similar to the one in the proof of Lemma 3, that every countable rooted 2-frame validating $\square_1\square_2 p \to \square_2\square_1 p$ and $\Diamond_1\square_2 p \to \square_2\Diamond_1 p$ is a p-morphic image of an expanding relativised product frame. Then, as L_2 is Horn definable, we can add the missing pairs to R_1 and R_2, if needed (see [8] for a similar proof). By adding new pairs to R_1 we are not forced to extend the set of worlds, since $L_1 \in \{\mathbf{T}, \mathbf{K4}, \mathbf{S4}\}$. ∎

Question 5. *What can be said about axiomatisations of higher-dimensional expanding relativised products?*

As to decidability, expanding relativisations can be reduced to products. Let φ be a multimodal formula of the language with n boxes and let e be a propositional variable which does not occur in φ. Define by induction on the construction of φ a multimodal formula φ^e as follows:

$$\begin{aligned}
p^e &= p \quad (p \text{ a propositional variable}),\\
(\psi \wedge \chi)^e &= \psi^e \wedge \chi^e,\\
(\neg\psi)^e &= \neg\psi^e,\\
(\square_1\psi)^e &= \square_1\psi^e,\\
(\square_i\psi)^e &= \square_i(e \to \psi^e) \quad (i = 2, \ldots, n).
\end{aligned}$$

A straightforward induction on the construction of φ proves the following:

THEOREM 7. *For all Kripke complete unimodal logics $L_1, \ldots, L_n$ and all multimodal formulas φ, the following conditions are equivalent:*

- $\varphi \in (L_1 \times \cdots \times L_n)^{\mathsf{EX}_n}$;
- $\left(e \wedge \Box_1^{\leq md(\varphi)} \mathsf{M}_{(2,n)}^{\leq md(\varphi)} (e \to \Box_1 e) \right) \to \varphi^e \in L_1 \times \cdots \times L_n$,

where $md(\varphi)$ is the modal depth of φ, $\mathsf{M}_{(2,n)}^{\leq 0} \psi = \psi$ and

$$\mathsf{M}_{(2,n)}^{\leq k+1} \psi = \mathsf{M}_{(2,n)}^{\leq k} \psi \wedge \bigwedge_{i=2}^{n} \Box_i \mathsf{M}_{(2,n)}^{\leq k}.$$

In particular, for $n = 2$,

$$\varphi \in (L_1 \times L_2)^{\mathsf{EX}} \text{ iff } \left(e \wedge \Box_1^{\leq md(\varphi)} \Box_2^{\leq md(\varphi)} (e \to \Box_1 e) \right) \to \varphi^e \in L_1 \times L_2.$$

As a consequence of this theorem we obtain that expanding relativised products are decidable in all those cases when the corresponding products are decidable. Unfortunately, for $n \geq 3$ the only known cases of this kind are products of tabular and **Alt** logics [8, 16].

Question 6. *Does the decidability of an expanding relativised product logic imply that the corresponding product logic is decidable as well?*

Let us conclude this section by observing the (lack of) connections between expanding relativised products and finite variable fragments of first-order modal logics with expanding domains. To begin with, we reduce product logics of the form

$$L \times \overbrace{\mathbf{S5} \times \cdots \times \mathbf{S5}}^{n}$$

to n-variable fragments of first-order modal logics $\mathbf{Q}L$ with *constant domains*. Define a translation † from the multimodal language with boxes $\Box_1, \ldots, \Box_{n+1}$ to unimodal first-order formulas with variables $x_1, \ldots, x_n$ by taking

$$\begin{aligned}
p_i^\dagger &= P_i(x_1, \ldots, x_n) \\
(\varphi \wedge \psi)^\dagger &= \varphi^\dagger \wedge \psi^\dagger, \\
(\neg \varphi)^\dagger &= \neg \varphi^\dagger, \\
(\Box_1 \psi)^\dagger &= \Box \psi^\dagger, \\
(\Box_j \psi)^\dagger &= \forall x_{j-1} \psi^\dagger, \quad \text{for } j = 2, \ldots, n+1.
\end{aligned}$$

It is not hard to see that the product logic $L \times \mathbf{S5} \times \cdots \times \mathbf{S5}$ is determined by product frames of the form $\mathfrak{F} \times \mathfrak{G} \times \cdots \times \mathfrak{G}$, where $\mathfrak{F}$ is a frame for L and $\mathfrak{G} = \langle V, V \times V \rangle$ for some non-empty set V. Now, there is a one-to-one correspondence between (propositional) modal models $\mathfrak{M}$ based on such product frames and first-order Kripke models of the form $\mathfrak{N} = \langle \mathfrak{F}, V, I \rangle$ where, for all $w \in W$,

$$I(w) = \left\langle V, P_0^{I(w)}, \ldots \right\rangle$$

is a first-order structure such that, for all $a_1, \ldots, a_n \in V$,

$$\langle a_1, \ldots, a_n \rangle \in P_i^{I(w)} \quad \text{iff} \quad (\mathfrak{M}, \langle w, a_1, \ldots, a_n \rangle) \models p_i.$$

It should be clear that in fact for *all* multimodal formulas φ of the language with $n + 1$ boxes, we have

$$I(w) \models \varphi^\dagger[a_1, \ldots, a_n] \quad \text{iff} \quad (\mathfrak{M}, \langle w, a_1, \ldots, a_n \rangle) \models \varphi,$$

for all $a_1, \ldots, a_n \in V$ and $w \in W$. As a consequence we obtain the following:

PROPOSITION 8. *Let L be a Kripke complete unimodal logic. Then for every formula φ,*

$$\varphi \in L \times \overbrace{\mathbf{S5} \times \cdots \times \mathbf{S5}}^{n} \quad \text{iff} \quad \varphi^\dagger \in \mathbf{Q}L.$$

On the one hand, it is readily checked that if $n = 1$ then the translation † reduces $(L \times \mathbf{S5})^{\mathsf{EX}}$ to the 1-variable fragment of the first-order modal logic $\mathbf{Q}L_e$ having models with *expanding domains*.

On the other hand, as far as we see, for $n \geq 3$ there is no such reduction of expanding relativised products of the form

(4) $\qquad (L \times \overbrace{\mathbf{S5} \times \cdots \times \mathbf{S5}}^{n})^{\mathsf{EX}_{n+1}}$

to $\mathbf{Q}L_e$, since quantifiers $\forall x_i$ and $\forall x_j$ of the latter always commute, while there is no interaction between the boxes $\Box_i$ and $\Box_j$ of the former whenever $i \neq j$ and $i, j > 1$. An alternative approach could be to consider the (2-dimensional) product of L and the multimodal[2] logic $\mathbf{S5}^n$ and then take instead of (4) the 2-dimensional expanding relativised product

$$(L \times \mathbf{S5}^n)^{\mathsf{EX}}.$$

Note that in general it is not known whether the product operator is associative. In particular, for $n \geq 3$ it is not known whether 2-dimensional product

---

[2] Products of multimodal logics can be defined similarly to the unimodal case.

logic $L \times \mathbf{S5}^n$ and the $n+1$-dimensional product logic $L \times \mathbf{S5} \times \cdots \times \mathbf{S5}$ are the same. Moreover, since we do not know how frames for $\mathbf{S5}^n$ look like when $n \geq 3$ (it is not even decidable whether a finite n-frame is frame for $\mathbf{S5}^n$ [16]), it is not clear how to turn a model for $(L \times \mathbf{S5}^n)^{\mathsf{EX}}$ into a model for $\mathbf{Q}L_e$.

For $n = 2$ we do have a characterisation of $\mathbf{S5} \times \mathbf{S5}$-frames. As is shown in [8], each countable rooted frame for $\mathbf{S5} \times \mathbf{S5}$ is in fact a p-morphic image of a product of two $\mathbf{S5}$-frames. Therefore, it is not hard to see that we have the required reduction: for every formula φ,

$$\varphi \in (L \times (\mathbf{S5} \times \mathbf{S5}))^{\mathsf{EX}} \quad \text{iff} \quad \varphi^\dagger \in \mathbf{Q}L_e.$$

Acknowledgements

Research was partially supported by U.K. EPSRC grants GR/R42474/01, GR/R45369/01 and by Hungarian National Foundation for Scientific Research grants T30314 and 035192. The authors are grateful to Valentin Shehtman and Frank Wolter for stimulating discussions and comments on the first version of this paper.

BIBLIOGRAPHY

[1] F. Baader, C. Lutz, H. Sturm, and F. Wolter. Fusions of description logics and abstract description systems. *Journal of Artificial Intelligence Research*, 16:1–58, 2002.
[2] A. Chagrov and M. Zakharyaschev. *Modal Logic*, volume 35 of *Oxford Logic Guides*. Clarendon Press, Oxford, 1997.
[3] A. Daigneault and J.D. Monk. Representation theory for polyadic algebras. *Fundamenta Mathematicae*, 52:151–176, 1963.
[4] R. Fagin, J. Halpern, Y. Moses, and M. Vardi. *Reasoning about Knowledge*. MIT Press, 1995.
[5] K. Fine. Logics containing $K4$, part II. *Journal of Symbolic Logic*, 50:619–651, 1985.
[6] K. Fine and G. Schurz. Transfer theorems for stratified modal logics. In J. Copeland, editor, *Logic and Reality, Essays in Pure and Applied Logic. In memory of Arthur Prior*, pages 169–213. Oxford University Press, 1996.
[7] D. Gabbay, A. Kurucz, F. Wolter, and M. Zakharyaschev. *Many-Dimensional Modal Logics: Theory and Applications*. Studies in Logic. Elsevier, 2003.
[8] D. Gabbay and V. Shehtman. Products of modal logics. Part I. *Journal of the IGPL*, 6:73–146, 1998.
[9] P. Halmos. Algebraic logic, IV. *Transactions of the AMS*, 86:1–27, 1957.
[10] P.R. Halmos. *Algebraic logic*. Chelsea Publishing Company, New York, 1962.
[11] J. Halpern and Y. Shoham. A propositional modal logic of time intervals. In *Proceedings of the First Symposium of Logic in Computer Science*, Washington, 1986. IEEE Computer Society Press.
[12] H. Henkin, J.D. Monk, and A. Tarski. *Cylindric Algebras, Part II*. North Holland, 1985.
[13] L. Henkin, J.D. Monk, A. Tarski, H. Andréka, and I. Németi. *Cylindric set algebras*, volume 883 of *Lecture Notes in Maths*. Springer, 1981.
[14] R. Hirsch and I. Hodkinson. Step by step — building representations in algebraic logic. *Journal of Symbolic Logic*, 62:225–279, 1997.

[15] R. Hirsch and I. Hodkinson. *Relation algebras by games.* Studies in Logic. Elsevier, 2002.
[16] R. Hirsch, I. Hodkinson, and A. Kurucz. On modal logics between $\mathbf{K} \times \mathbf{K} \times \mathbf{K}$ and $\mathbf{S5} \times \mathbf{S5} \times \mathbf{S5}$. *Journal of Symbolic Logic,* 67:221–234, 2002.
[17] M. Kracht and F. Wolter. Properties of independently axiomatizable bimodal logics. *Journal of Symbolic Logic,* 56:1469–1485, 1991.
[18] R. Maddux. The equational theory of CA_3 is undecidable. *Journal of Symbolic Logic,* 45:311–315, 1980.
[19] D.C. Makinson. Some embedding theorems for modal logic. *Notre Dame Journal of Formal Logic,* 12:252–254, 1971.
[20] M. Marx and Y. Venema. *Multi-dimensional Modal Logic.* Kluwer Academic Publishers, 1997.
[21] Sz. Mikulás. Personal communication, 2000.
[22] Sz. Mikulás and M. Marx. Relativized products of modal logics. In Procs. *Advances in Modal Logic,* Leipzig, Germany, October 2000.
[23] J.D. Monk. Nonfinitizability of classes of representable cylindric algebras. *Journal of Symbolic Logic,* 34:331–343, 1969.
[24] I. Németi. Algebraization of quantifier logics: an overview. *Studia Logica,* 3,4:485–569, 1991.
[25] I. Németi. Decidable versions of first-order predicate logic and cylindric relativized set algebras. In L. Csirmaz, D. Gabbay, and M. De Rijke, editors, *Logic Colloquium'92.* CSLI Publications, 1995.
[26] C. Pinter. A simple algebra of first order logic. *Notre Dame Journal of Formal Logic,* 1:361–366, 1973.
[27] C. Pinter. Algebraic logic with generalized quantifiers. *Notre Dame Journal of Formal Logic,* 16:511–516, 1975.
[28] M. Reynolds and M. Zakharyaschev. On the products of linear modal logics. *Journal of Logic and Computation,* 11:909–931, 2001.
[29] I. Sain and R. Thompson. Strictly finite schema axiomatization of quasipolyadic algebras. In H. Andréka, J.D. Monk, and I. Németi, editors, *Algebraic logic,* pages 539–571. North-Holland, 1991.
[30] E. Spaan. *Complexity of Modal Logics.* PhD thesis, Department of Mathematics and Computer Science, University of Amsterdam, 1993.
[31] Y. Venema and M. Marx. A modal logic of relations. In E. Orlowska, editor, *Logic at Work,* pages 124–167. Springer-Verlag, 1999.
[32] F. Wolter. The structure of lattices of subframe logics. *Annals of Pure and Applied Logic,* 86:47–100, 1997.
[33] F. Wolter. Fusions of modal logics revisited. In M. Kracht, M. de Rijke, H. Wansing, and M. Zakharyaschev, editors, *Advances in Modal Logic,* pages 361–379. CSLI, Stanford, 1998.
[34] F. Wolter and M. Zakharyaschev. Spatio-temporal representation and reasoning based on RCC-8. In *Proceedings of the Seventh Conference on Principles of Knowledge Representation and Reasoning, KR2000, Breckenridge, USA,* pages 3–14, Morgan Kaufman, 2000.
[35] F. Wolter and M. Zakharyaschev. Qualitative spatiotemporal representation and reasoning: a computational perspective. In G. Lakemeyer and B. Nebel, editors, *Exploring Artificial Intelligence in the New Millenium,* pages 175–216, Morgan Kaufmann, 2002.

Agi Kurucz and Michael Zakharyaschev
Department of Computer Science
King's College London, Strand, London WC2R 2LS, UK
E-mail: {kuag,mz}@dcs.kcl.ac.uk

Notes on the Space Requirements for Checking Satisfiability in Modal Logics

MARCUS KRACHT

ABSTRACT. Recently, there has been growing attention to the space requirements of tableau methods (see for example [7], [1], [12]). We have proposed in [10] a method of reducing modal consequence relations to the global and local consequence relation of (polymodal) K. The reductions used there did however not establish good time complexity bounds. In this note we shall use reduction functions to obtain rather sharp space bounds. These bounds can also be applied to show completeness of ordinary tableau systems, which in turn yield space bounds that are slightly different from the ones derived by applying the reduction functions alone.

It has been shown by Hudelmaier ([7]) that satisfiability in S4 is $O(n^2 \log n)$–space computable, while satisfiability in K and KT are $O(n \log n)$–space computable. An $O(n \log n)$–space bound for KD has been obtained by Basin, Matthews and Viganò ([1]). Viganò ([15]) has shown that satisfiability in K4, KD4 and S4 is in $O(n^2 \log n)$–space. Nguyen has reduced these bounds to $O(n \log n)$ for K4, K4D and S4 in [12]. The bound for K has been improved to $O(n)$ by Hemaspaandra in [5].

We shall deal here with an abstract method for obtaining space bounds, using reduction functions. Reduction functions have been introduced in [9] and further developed in [10]. These functions were used to obtain a number of folklore results on standard modal systems in a uniform way. Examples were the finite model property and interpolation, but also complexity. However, the bounds for local satisfiability obtained there were not good enough. Here we shall improve these results rather drastically. In some cases we shall obtain linear bounds (namely for K.D, K.alt$_1$), while in other cases we shall obtain a bound which is a product of the number of subformulae and the modal depth. In some cases this is somewhat less than the best known result (it can be quadratic in the length), in others it is better. Typically, with the help of tableau methods we can also establish the $O(n \log n)$ space bound for most systems.

Advances in Modal Logic, Volume 4, 243–264.
© 2003, the author.

1 Preliminaries

We refer to [9] for details concerning modal logic. The language consists of the set $\{p_i : i \in \omega\}$ of variables and the symbols $\bot$, $\neg$, $\wedge$ and, finally, the modal operators $\Box_i$, $i < q$. (The letter 'q' is reserved throughout this paper for the number of basic operators.) The symbols $\vee$ and $\rightarrow$ are used here, but they are only abbreviations. For sets of formulae we use the following shorthand notation. For a formula φ, $\mathrm{dg}(\varphi)$ is the maximum nesting of modal operators. It is called the **modal degree of** φ. For a finite set Δ, $\mathrm{dg}(\Delta)$ is the maximum of the modal degrees of its members.

As usual, $\varphi;\psi$ denotes $\{\varphi,\psi\}$, $\Delta;\varphi$ denotes $\Delta \cup \{\varphi\}$ and $\Delta;\Theta$ denotes $\Delta \cup \Theta$. Thus $\Delta;\varphi;\varphi$ is the same set as $\Delta;\varphi$. For a formula φ, $\mathrm{sf}(\varphi)$ denotes the set of subformulae of φ, $\mathrm{sf}^{\neg}(\varphi) := \mathrm{sf}(\varphi) \cup \{\neg\delta : \delta \in \mathrm{sf}(\varphi)\}$, $\mathrm{var}(\varphi)$ the set of variables occurring in φ. For a set Δ of formulae we put

$$\mathrm{sf}(\Delta) := \bigcup\{\mathrm{sf}(\delta) : \delta \in \Delta\} \tag{1}$$

$$\mathrm{sf}^{\neg}(\Delta) := \bigcup\{\mathrm{sf}^{\neg}(\delta) : \delta \in \Delta\} \tag{2}$$

$$\mathrm{var}(\Delta) := \bigcup\{\mathrm{var}(\delta) : \delta \in \Delta\} \tag{3}$$

A modal logic is identified here with its set of theorems. The least monomodal logic is called K. With q the number of basic operators, the least q–modal logic is denoted by K_q. (So, K_0 is PC, and $\mathsf{K}_1 = \mathsf{K}$.) With a modal logic L, two consequence relations are associated: $\vdash_L$ and $\Vdash_L$. The first is called the **local consequence relation** and is defined as follows. $\Delta \vdash_L \varphi$ iff φ can be proved from $\Delta \cup L$ using only Modus Ponens. The second is called the **global consequence relation** and defined as follows. $\Delta \Vdash_L \varphi$ iff φ can be derived from $\Delta \cup L$ using Modus Ponens and Necessitation.

Problems are functions $f : A^* \rightarrow \{0,1\}$, where A is a finite set of so–called **letters**. We shall study the problems '$\Delta \vdash_L \varphi$' and '$\Delta \Vdash_L \varphi$' for various logics L, which are the following. Given a finite set Δ and a formula φ, determine whether $\Delta \vdash_L \varphi$ ($\Delta \Vdash_L \varphi$) and if so, output 1, and if not, output 0. Given a complexity class $\mathcal{C}$, a problem f is said to be **in** $\mathcal{C}$ if f is computable in $\mathcal{C}$; f is $\mathcal{C}$–**hard**, if for any problem g in $\mathcal{C}$ there is a function h in $\mathcal{C}$ such that $f \circ h = g$; and f is $\mathcal{C}$–**complete** if it is both in $\mathcal{C}$ and $\mathcal{C}$–hard. We assume that $\mathcal{C}$ is closed under composition.

DEFINITION 1. Let L be a modal logic and $\mathcal{C}$ a complexity class. We say that L is **locally** in $\mathcal{C}$ (is **locally** $\mathcal{C}$–**hard**, is **locally** $\mathcal{C}$–**complete**) if the problem '$\Delta \vdash_L \varphi$' is in $\mathcal{C}$ (is $\mathcal{C}$–hard, is $\mathcal{C}$–complete). Likewise, L is **globally** in $\mathcal{C}$ (is **globally** $\mathcal{C}$–**hard**, is **globally** $\mathcal{C}$–**complete**) if the problem '$\Delta \Vdash_L \varphi$' is in $\mathcal{C}$ (is $\mathcal{C}$–hard, is $\mathcal{C}$–complete).

Recall the following. First, let $\Box^k \varphi$ be defined by induction as usual.

$$\Box^{<1}\varphi := \varphi \tag{4}$$
$$\Box^{<k+1}\varphi := \varphi \wedge \Box\Box^{<k}\varphi \tag{5}$$

Also,

$$\Box^k \Delta := \{\Box^k \delta : \delta \in \Delta\} \tag{6}$$
$$\Box^{<k}\Delta := \{\Box^{<k}\delta : \delta \in \Delta\} \tag{7}$$

LEMMA 2. *(a) $\Delta \Vdash_L \varphi$ iff for some k: $\Box^{<k}\Delta \vdash_L \varphi$. (b) $\Vdash_L \varphi$ iff $\vdash_L \varphi$.*

Given a finite set Δ and a formula φ, we have $\Delta \vdash_L \varphi$ iff $\vdash_L \bigwedge \Delta \to \varphi$ iff $\Vdash_L \bigwedge \Delta \to \varphi$. Thus, local derivability problems are equivalent to local satisfiability problems, which are subproblems of global derivability problems. (Moreover, the transformation is log–space computable.)

PROPOSITION 3. *Let C be a complexity class and L a modal logic. If L is locally C–hard, it is also globally C–hard. If it is globally in C, it is also locally in C.*

Since satisfiability in consistent modal logics is at least NP–hard, the complexity of satisfiability in modal logics does not change under linear reductions. (For the inconsistent logic the satisfiability and derivability problems are of course trivial.)

2 Reduction functions

DEFINITION 4 (Local reduction function). Let $L \subseteq M$ be two logics. Let Y be a function defined on the set of finite sets of formulae, with range the set of finite sets of formulae. Y is called a **local reduction function** from M to L if $\Delta \vdash_M \varphi$ iff (1) $\Delta; Y(\Delta; \varphi) \vdash_L \varphi$, and (2) for all Δ $\text{var}(Y(\Delta)) \subseteq \text{var}(\Delta)$.

The condition on variables will not be needed here but is essential in proofs concerning interpolation (see [9]).

DEFINITION 5 (Global reduction function). Let $L \subseteq M$ be two logics. Let X be a function defined on the set of finite sets of formulae, with range the set of finite sets of formulae. X is called a **global reduction function** from M to L if $\Delta \Vdash_M \varphi$ iff (1) $\Delta; X(\Delta; \varphi) \Vdash_L \varphi$ and (2) for all Δ $\text{var}(X(\Delta)) \subseteq \text{var}(\Delta)$.

These definitions are as given in [10], where the following were established to be global reduction functions:

$$X_4(\Delta) := \{\Box\chi \to \Box\Box\chi : \Box\chi \in \mathrm{sf}(\Delta)\} \tag{8}$$
$$X_T(\Delta) := \{\Box\chi \to \chi : \Box\chi \in \mathrm{sf}(\Delta)\} \tag{9}$$
$$X_B(\Delta) := \{\neg\chi \to \Box\neg\Box\chi : \Box\chi \in \mathrm{sf}(\Delta)\} \tag{10}$$
$$X_G(\Delta) := \{\neg\Box\chi \to \neg\Box(\chi \vee \neg\Box\chi) : \Box\chi \in \mathrm{sf}(\Delta)\} \tag{11}$$
$$X_{\mathrm{Grz}}(\Delta) := \{\neg\Box\chi \to \neg\Box(\chi \vee \neg\Box(\chi \to \Box\chi)) : \Box\chi \in \mathrm{sf}(\Delta)\} \tag{12}$$
$$X_{\mathrm{alt}_1}(\Delta) := \{\neg\Box\chi \to \Box\neg\chi : \Box\chi \in \mathrm{sf}(\Delta)\} \tag{13}$$

Additional reduction functions for K.D and K.5 are

$$X_D(\Delta) := \{\neg\Box\bot\} \tag{14}$$
$$X_5(\Delta) := \{\neg\Box\chi \to \Box\neg\Box\chi : \Box\chi \in \mathrm{sf}(\Delta)\} \tag{15}$$

The functions X_G, X_5, are reduction functions to K4, X_{Grz} is a reduction to S4. All others are to K.

The results of this paper could be slightly improved if we define a function on the basis of the pair $\langle \Delta, \varphi \rangle$ rather than the set $\Delta; \varphi$. However, this gives no improvement on the theoretical complexity of the algorithms. Hence we have ignored this finesse here.

3 Local and global tableaux

We shall first sketch a method for obtaining space bounds for (polymodal) K. The first calculus, basically folklore (see [13]), shall be called the **local tableau calculus**, to distinguish it from the second, the so-called **global tableau calculus**.

The calculi as presented below operate on *sets* of formulae, while they should be thought of as operating on strings. This is just a matter of presentation. The space estimates will be always given in terms of the input string, which is an ordered sequence of formulae. For the purpose of the calculus, let $\Delta_{\Box_j} := \{\chi : \Box_j \chi \in \Delta\}$. The rules are as follows.

$$(\neg E) \frac{\Delta; \neg\neg\varphi}{\Delta; \varphi} \qquad (\wedge E) \frac{\Delta; \varphi \wedge \psi}{\Delta; \varphi; \psi}$$

$$(\vee E) \frac{\Delta; \neg(\varphi \wedge \psi)}{\Delta; \neg\varphi \mid \Delta; \neg\psi} \qquad (\Box_j E) \frac{\Delta; \neg\Box_j \varphi}{\Delta_{\Box_j}; \neg\varphi}$$

A **tableau** for Δ is a maximally 2–branching tree with nodes labeled by finite sets of formulae and a name of a rule (except for leaves) such that the root is labeled Δ, and the following holds.

1. If x has label $(\neg E)$, $(\wedge E)$ or $(\Box_j E)$, then it has a unique daughter and the daughter carries the appropriate label.

2. If x has label $(\vee E)$ then it has exactly two daughters, one for each set mentioned by the rule.

A tableau is **closed** if all leaves carry a Δ such that $\bot \in \Delta$ or both $\varphi \in \Delta$ and $\neg\varphi \in \Delta$ for some φ. Δ is **rejected** by the tableau calculus if it has a closed tableau. We remark without proof that this calculus is sound and complete for K in the sense that a set of formulae is rejected iff it is inconsistent. For the purpose of computing with tableaux we shall think of the sets of formulae as sequences possibly containing multiple occurrences of formulae. The semicolon is then to be thought of as sequence concatenation. It is possible to remove double occurrences of formulae (to save space). In this case we agree to keep the first occurrence of a formula in the sequence.

The local calculus is just part of the global calculus, which is as follows (see also [3]). It runs on pairs of sets of formulae, separated by $\dagger$. We interpret $\Theta \dagger \Delta$ as a set of formulae of which the members of Θ hold globally and the members of Δ hold only locally. The rules are as follows.

$$(\neg E) \frac{\Theta \dagger \Delta; \neg\neg\varphi}{\Theta \dagger \Delta; \varphi} \qquad (\wedge E) \frac{\Theta \dagger \Delta; \varphi \wedge \psi}{\Theta \dagger \Delta; \varphi; \psi}$$

$$(\vee E) \frac{\Theta \dagger \Delta; \neg(\varphi \wedge \psi)}{\Theta \dagger \Delta; \neg\varphi | \Theta \dagger \Delta; \neg\psi} \qquad (\Box_j E) \frac{\Theta \dagger \Delta; \neg\Box_j\varphi}{\Theta \dagger \Theta; \Delta_\Box; \neg\varphi}$$

This calculus is also sound and complete in the following sense. $\Theta \dagger \Theta; \Delta$ has a closed tableau iff $\neg \bigwedge \Delta$ globally follows in K from Θ. This is easily seen. If we put $\Theta := \varnothing$ we obtain the local calculus.

Notice that both tableau calculi use a somewhat larger set than the set of subformulae. Namely, if the starting set is Δ (or $\Theta \dagger \Delta$), then the sets in the tableau remain within the set $\text{sf}^\neg(\Delta)$ ($\text{sf}^\neg(\Theta; \Delta)$).

4 Space requirements of the tableaux

We shall deal now with the space requirement for these tableau calculi.

DEFINITION 6 (Syntax). Let q be the number of basic modal operators. $A := \{p, 0, 1, \wedge, \neg, \bot, \Box\}$, $S := \{;, \dagger\}$. Members of A are called **logical symbols**. The set of q–modal fomulae is a subset of A^*, which is defined as follows.

1. $\bot$ is a formula.

2. If $\alpha \in \{0,1\}^*$, then $p\alpha$ is a variable. A variable is a formula.

3. If φ is a formula, so is $\neg\varphi$.

4. If φ is a formula, and α the binary code of a number $< q$ then $\square\alpha\varphi$ is a formula.

5. If φ and χ are formulae, so is $\wedge\varphi\chi$.

If $q = 1$, we use $\square$ in place of $\square 0$ in Item 4. Sequences of formulae (and pairs thereof) are elements of $(A \cup S)^*$, which are defined as follows.

1. Every formula is a sequence of formulae.

2. If Δ and Δ' are sequences of formulae, so is $\Delta; \Delta'$.

3. If Θ and Δ are sequences of formulae, $\Theta \dagger \Delta$ is a pair of sequences of formulae.

Unique readability can be shown for this notation. Notice that formulae are coded in Polish Notation, but quoted in running text in the usual way. The *length* of a formula, sequence or pair of sequences of formulae is its **length** as a string, denoted by $|\Delta|$. To eliminate dependency on variable names, one defines the **modified length**, $||\Delta||$, as follows.

1. $||\bot|| := ||p\alpha|| := 1$.

2. $||\neg\varphi|| := ||\square\varphi|| := ||\square\alpha\varphi|| := ||\varphi|| + 1$.

3. $||\varphi \wedge \psi|| := ||\varphi|| + ||\psi|| + 1$.

4. $||\Delta; \Delta'|| := ||\Delta|| + ||\Delta'||$.

In general, $||\Delta|| \leq |\Delta|$. The most common length count is actually $||\Delta||$, which works however only if the number of operators is finite. (Otherwise, the length of α must be counted as well in the second clause above, but not in the first.) For details see [14].

It is worthwhile to explain here a measure of length that is used in [10]. Suppose that Δ is given. Then let $\sharp\Delta := \text{card}(\text{sf}(\Delta))$ be the number of subformulae of Δ. We remark here that $||\square^{<k}\varphi|| = k(||\varphi|| + 1) - 1$ and $\sharp(\square^{<k}\varphi) \leq (\sharp\varphi) + 2(k-1)$. For sets we have

$$||\square^{<k}\Delta|| \leq k(||\Delta||) - \text{card}(\Delta) \tag{16}$$

$$\sharp(\square^{<k}\Delta) \leq 2(k-1)\text{card}(\Delta) + \sharp\Delta \tag{17}$$

Most complexity results could be stated alternatively with this length function, though it is somewhat different from $|\Delta|$. Notice that the space needed to code Δ is not predictable from $\sharp\Delta$ alone, since Δ contains variable names of unpredictable length.

PROPOSITION 7. $\log_2 ||\Delta|| \leq \sharp\Delta \leq ||\Delta||$.

None of the bounds can be improved. Let Δ be a set of formulae without $\bot$ in which every variable occurs at most once. Then $\sharp\Delta = ||\Delta||$. For the other inequality, define

$$\chi_0 := \mathrm{p}0, \qquad \chi_{n+1} := \wedge \chi_n \chi_n \qquad (18)$$

χ_n has $n+1$ distinct subformulae and is of length $2^{n+1} + 2^{n-1} - 1$. Thus, $\log_2 ||\chi_n||$ is asymptotically equal to $\sharp\chi_n$. (This shows that $\log_2 ||\Delta|| \leq \sharp\Delta$ can in general not be improved.)

Computations considered here are implemented as computations over strings of formulae, which however represent sets of formulae. Since in sets we do not count repeated items, this leads to a slight improvement of the space complexity bounds.

LEMMA 8. *A member of* $\mathrm{sf}(\Delta)$ *needs* $O(\log \sharp\Delta)$ *space to code.*

To that effect, think of Δ as written onto a tape. A set of subformulae of Δ can be coded by marking on Δ all positions where a subformula belonging to that set begins. This explains the bound $\log_2 |\Delta|$. However, notice that (a) there are positions where no subformula begins (so the bound may be $\log_2 ||\Delta||$), and (b) Δ may contain the same subformula several times (which reduces the bound to $\log_2 \sharp\Delta$). So we do the following. We use a second tape, where cell number j contains $\circ$ iff cell number j on the input tape begins a subformula, and moreover, no cell number $i < j$ on the input tape begins the same subformula. All other cells are marked by $\bullet$. Evidently, since the rules manipulate effectively only sets of subformulae, we need to deal exclusively with those cells marked $\circ$ in place of all cells. The auxiliary tape has length $|\Delta|$ but only $\sharp\Delta$ many cells marked $\circ$. Hence, the binary code of these cells (counted ignoring the $\bullet$ cells) consumes only $\log_2 \sharp\Delta$ space. To generate that tape, a few read heads are needed to scan the input. Hence we can eliminate the auxiliary tape by introducing several read heads on the input tape (equivalently, consuming additional $O(\log_2 |\Delta|)$ space). Since the initial sequence, Δ, is on the input tape, its space requirement is not counted for the space complexity.

A **binary tree domain** is a subset T of $\{0,1\}^*$ such that (a) if $\vec{s}\,\smallfrown j \in T$ then $\vec{s} \in T$, (b) if $\vec{s}\,\smallfrown 1 \in T$ then $\vec{s}\,\smallfrown 0 \in T$. A binary branching tree can be coded by a binary tree domain. A tableau can be coded as a set $\mathcal{T}$ of quadruples of the form $w = \langle \vec{s}, \rho, \vec{u}, \Delta \rangle$, where $\vec{s}$ is a binary sequence, ρ the name of a rule, $\vec{u}$ a binary sequence, and Δ a sequence of formulae. Moreover, the set of $\vec{s}$ such that there is $\langle \vec{s}, \rho, \vec{u}, \Delta \rangle$ shall form a tree domain, such that the following holds: for each $w = \langle \vec{s}^\frown j, \rho, \vec{u}, \Delta\rangle$ and $w' = \langle \vec{s}, \sigma, \vec{v}, \Delta'\rangle$ in $\mathcal{T}$, Δ is obtained from Δ' by applying rule ρ to the nth formula of Δ', where $\vec{u}$ is the binary code of n. Moreover, if $\rho = (\vee\text{E})$, the nth formula is $\neg(\varphi \wedge \psi)$ and $j = 0$, then $\neg\varphi$ is chosen to form Δ, and if $j = 1$, then $\neg\psi$ is chosen. (Notice that if $\vec{s} = \varepsilon$, the empty word, ρ and $\vec{u}$ can be anything, for example 0.) $\vec{s}$ is called the **name** of w, and Δ its **label**. The pair $\rho, \vec{u}$ is called the **transition code**. Notice immediately that ρ is redundant in presence of $\vec{u}$. If v has label $\vec{s}^\frown j$, $j \in \{0, 1\}$, then v is called a **daughter of** w.

We shall first deal with the local tableau calculus.

LEMMA 9. *Let Δ' be the label of a daughter of w in a local tableau, and Δ the label of w. Then $|\Delta'| \leq |\Delta|$. Moreover, $||\Delta'|| < ||\Delta||$.*

It follows that the length of a branch is linearly bounded in the number of logical symbols of Δ_0, the starting sequence. Now notice the following. Each daughter node is uniquely determined from its mother node by naming: (a) which formula the rule has been applied on, (b) whether we choose the left hand branch in case the formula is of the form $\neg(\varphi \wedge \psi)$, or the right hand branch. The information (b) is actually present in the name of the node and need not be given, since for the leftmost formula the node named $\vec{s}0$ is chosen and for the rightmost formula the node named $\vec{s}1$. All other information can be recomputed. So we may store the tableau as follows. Instead of $\langle \vec{s}, \rho, \vec{u}, \Delta\rangle$ for $\vec{s} \neq \varepsilon$ we only store $\langle \vec{s}, \vec{u}\rangle$. We have $|\vec{s}| \leq \log_2 ||\Delta||$, as we need to perform tableaux rules only once per occurrence of a subformula. Also, $|\vec{u}| \leq \log_2 |\Delta|$. By Lemma 8, $\vec{u}$ has length $\leq \log_2 \sharp\Delta$. Finally, if we code only branches, $\vec{s}$ can be reduced to its last member, so is in effect constant. So, a branch for Δ in a local tableau needs $O(||\Delta||\log_2 \sharp\Delta)$ space to code.

Here is an algorithm that computes only a single branch at a time. Start with the tableau consisting of just one node and the opening sequence, and call x *unfinished*.

1. If x is unfinished and $(\neg\text{E})$, $(\wedge\text{E})$ or $(\vee\text{E})$ is applicable on a formula then: x remains unfinished. Apply the rules once to the leftmost possible formula. In case of $(\vee\text{E})$, choose the leftmost disjunct. This creates y. y is unfinished. Continue with y.

2. If x is unfinished, and only $(\square\text{E})$ is applicable at x then: apply $(\square\text{E})$ to the leftmost possible formula, create a successor y. y is unfinished. Continue with y.

3. If x is unfinished and no rule is applicable then: if it contains a formula and its negation, x becomes closed. If not, x becomes open. Continue

with x.

4. If x is closed and has no parent node: exit. 'There is a closing tableau.'

5. If x is closed and has parent y:

    (a) If y is a ($\vee$E)–node and x is the left hand daughter of y then: create right hand daughter z. z is unfinished. Continue with z.

    (b) Else: y becomes closed. Continue with y.

6. If x is open and has no parent: exit. 'There is no closing tableau.'

7. If x is open and has parent y:

    (a) If y is a ($\vee$E)–node, a ($\wedge$E)–node or a ($\neg$E)–node then: y becomes open. Continue with y.

    (b) If y is a ($\Box$E)–node then: take the next suitable formula to apply ($\Box$E). If it does not exist, y becomes open. If it exists, y becomes unfinished. Continue with y.

Since branches are bounded in length by $||\Delta||$ we get the following.

THEOREM 10. *It can be checked in $O(||\Delta|| \log_2 \sharp\Delta)$ space whether a given set Δ of fomulae is satisfiable in (polymodal)* K.

Since $\sharp\Delta \leq ||\Delta|| \leq |\Delta|$ this trivially implies that also $O(||\Delta|| \log_2 ||\Delta||)$ is also sufficient. Since complexity is typically measured in terms of $||\Delta||$, Theorem 10 is slightly better than the standard ones.

The tableau method can be applied also to other logics. For example, [13] has shown that if one adds the following rule

$$(\Box\text{T}) \quad \frac{\Delta; \Box\varphi}{\Delta; \varphi} \qquad (19)$$

the resulting calculus is sound and complete for K.T. (This can be proved also using reduction functions on the basis of the completeness of the calculus for K.) Evidently, no branch needs to be longer than $\sharp\Delta$, so we can restrict the space to $O(||\Delta|| \log \sharp\Delta)$.

Recently, Hemaspaandra [5] has observed that the space bounds can be improved even further. First, observe the following.

LEMMA 11. *A subset of* sf(Δ) *needs $O(\sharp\Delta)$ space to code.*

Namely, just note that a subset can be coded as a set of cells on the auxiliary tape (see the remarks following Lemma 8). Hemaspaandra steps directly from a downward saturated set to another downward saturated set using the following single rule:

$$(\Box H) \quad \frac{\Delta; \neg\Box_j \chi}{(\Delta_{\Box_j}; \neg\chi)^*}$$

Here, Θ^* denotes a saturated closure of Θ. This rule eliminates the rules $(\neg E)$, $(\wedge E)$ and $(\vee E)$. A downward saturated set is simply a subset of $\sharp\Delta$. However, Hemaspaandra makes the following crucial observation. Let $\mathrm{sf}(d, \Delta)$ be the set of occurrences of subformulae that are exactly inside the scope of d many $\Box$. It is clear that $\mathrm{sf}(d, \Delta) \cap \mathrm{sf}(d', \Delta) = \emptyset$ whenever $d \neq d'$. (Think of occurrences as cells of the input string.) Now, if the set of occurrences above the line of ($\Box H$) was within $\mathrm{sf}^\frown(d, \Delta)$, then the set of occurrences below the line is within $\mathrm{sf}^\frown(d+1, \Delta)$. Therefore the entire tableau can be coded inside $\mathrm{sf}^\frown(\Delta)$. Namely, a tableau is a sequence of the form $\langle\langle \Delta_i, \chi_i \rangle : i < n\rangle$, where Δ_i is a downward saturated subset of $\mathrm{sf}^\frown(i, \Delta)$ and $\chi_i \in \Delta_i$ the formula on which the rule is operated next. It is checked in linear space whether a set is downward saturated. The nondeterminism in choosing a successor set does not affect the space complexity. We can backtrack on the saturation (the saturated closures can be effectively enumerated), and we can backtrack on the formulae on which the rules have operated.

Now put

$$\sharp_\delta(\Delta) := \mathrm{card}(\bigcup_{n \in \omega} \mathrm{sf}(n, \Delta)) \qquad (20)$$

(So, in $\sharp_\delta \Delta$ we count two occurrences of a subformula as different just in case their degree of embedding is different.) Hemaspaandra's result can be improved to $O(\sharp_\delta \Delta)$. It can be seen that $\sharp_\delta(\Delta) \leq \mathrm{dg}(\Delta)\sharp\Delta \leq (\sharp\Delta)^2$. Namely, the degree of embedding is bounded from above by the number of subformulae. Hence there are at most $\sharp\Delta$ occurrences of a subformula that have pairwise different degree of modal embedding. This bound can, however, not be improved. Define

$$\sigma_0 := p0, \qquad \sigma_{n+1} := \wedge \sigma_n \Box \sigma_n \qquad (21)$$

Then $\sharp\{\sigma_n\} = 2n+1$, $\mathrm{dg}(\sigma_n) = n$, $\sharp_\delta(\{\sigma_n\}) = (n+1)^2$. With badly designed formulae, therefore, Hemaspaandra's result gives us only $O((\sharp\Delta)^2)$. It is such type of formulae that occur a lot with the reduction functions.

In order to boost this up for the minimal q-modal logic K_q we just have to replace the number d by a sequence of numbers $< q$. Thus, we define the following sets. Let Δ be given. Then for each occurrence of a subformula we define the following.

1. For all $\delta \in \Delta$: $\delta \in \text{sf}(\varepsilon, \Delta)$.
2. If $\neg \chi \in \text{sf}(\vec{\alpha}, \Delta)$, then $\chi \in \text{sf}(\vec{\alpha}, \Delta)$.
3. If $\varphi \wedge \chi \in \text{sf}(\vec{\alpha}, \Delta)$, then $\varphi, \chi \in \text{sf}(\vec{\alpha}, \Delta)$.
4. If $\Box_j \chi \in \text{sf}(\vec{\alpha}, \Delta)$, then $\chi \in \text{sf}(\vec{\alpha}\smallfrown j, \Delta)$.

Thus for each occurrence of a subformula χ there exists a unique sequence $\vec{\alpha}$ such that $\chi \in \text{sf}(\vec{\alpha}, \Delta)$. (The reader should not be mislead by the notation 'χ' to think of a formula; rather, one should think of it as an occurrence of a subformula. Otherwise, $\vec{\alpha}$ is not unique.)

THEOREM 12 (Hemaspaandra). *It can be checked in $O(\|\Delta\|)$ space whether a given set Δ of formulae is satisfiable in (polymodal) K.*

Now we shall deal with global tableaux. In a global tableau for $\Theta \dagger \Delta$, a node contains only members from $\text{sf}(\Delta; \Theta)$. Let n be the cardinality of this set. As remarked earlier, we can design the calculus in such a way that no formula occurs twice in a sequence. Now, a formula may or may not occur in a set, and it may occur negated or unnegated. Hence there are at most 3^n different labels. If this is so, then we immediately see that if a branch is of length $\geq 3^n$ and closed, then there is a branch of length $< 3^n$ that is also closed. This means that a branch of length $\geq 3^n$ may simply be regarded as open. This bound can be sharpened. We let $p := \sharp \Theta$ and $q := \sharp \Delta$. Then branches need only be $3^p + q$ deep. This is immediately lowered to $2^p + q$ by noting that if a branch closes with Δ, it closes with any superset $\Delta' \supseteq \Delta$. In other words, partiality may help to keep branches short, but it does not increase the length. This is cognate to the following theorem, which is a refinement of Lemma 3.1.9 of [9] (for which almost literally the same proof can be used).

THEOREM 13. *Let Δ be a set of formulae and φ a formula. Then put $p := \sharp \Delta$ and $q := \sharp \varphi$. Then $\Delta \Vdash_K \varphi \Leftrightarrow \Box^{<2^p+q+1} \Delta \vdash_K \varphi$.*

COROLLARY 14. *Let $q < \omega$ and $K := K_q$. It can be checked in $O(2^n \log n)$–space whether or not $\Delta \Vdash_K \varphi$, where $n := \sharp(\Delta; \varphi)$.*

This follows from the fact that the length of branches has an exponential upper bound. Notice that this length bound cannot be significantly reduced. In [10] we have shown that the number of ($\Box$E)–steps is in some cases $O(2^{c\sqrt{n}})$ for some $c > 0$, and this means that it is unlikely that one can do with less than exponential space. It can be shown on the other hand that there is an exponential time algorithm checking global satisfiability. Moreover, the following is also known (see [14]).

THEOREM 15 (Spaan). *K is globally EXPTIME–complete.*

Notice that this result holds also if a different measure of length is taken, namely the number of subformulae. This is of some importance later on. From this we shall obtain upper bounds for the global time complexity for many modal logics.

5 Space bounds via reduction functions

LEMMA 16. *Let $L \supseteq K$. Then the following holds with respect to the number of subformulae, with respect to the length, and the modified length. Let X be a global reduction function from L to K. If X is polynomial, L is globally EXPTIME.*

The global reduction functions defined above are quadratic in the (modified) length. Hence it is established that the logics discussed here are generally globally in EXPTIME. This follows from the following observation. Suppose that $L = K \oplus \Xi$. It is easy to see that $X(\Delta)$ consists of substitution instances of some members of Ξ. Thus, all we need to know is what to substitute for the variables. The functions given earlier have the property that the formulae that are substituted for the variables are from $\mathrm{sf}^-(\Delta)$. For example, $X_\mathsf{B}(\Delta) = \{\neg\Box\chi \to \Box\neg\Box\chi : \Box\chi \in \mathrm{sf}(\Delta)\}$ is obtained by substituting $\neg\Box\chi$ for p in $p \to \Box\neg\Box\neg p$, and $\mathsf{B} = \mathsf{K} \oplus p \to \Box\Diamond p$. Of course, $\Diamond$ abbreviates $\neg\Box\neg$; and we have replaced $\neg\neg\Box\chi$ by $\Box\chi$, but this is harmless cosmetics. This motivates the following.

DEFINITION 17. *Let X be a reduction function. Suppose that there is a finite set Θ such that $X(\Delta)$ consists of some or all substitution instances of Θ by members of $\mathrm{sf}^-(\Delta)$ for its variables. Further, let n be the number of variables in Θ. Then we call X an n-**analytic global reduction function** with **skeletal set** Θ.*

It is checked that all reduction functions defined above are 1–analytic.

LEMMA 18. *Let X be n–analytic with skeletal set Θ. Then $\sharp X(\Delta) \leq \sharp\Theta \cdot (\sharp\Delta)^n$.*

THEOREM 19. *Let L and M be modal logic. Suppose that X is an n–analytic skeletal global reduction function from M to L. Then if L is globally in EXPTIME with respect to any of the measures, then so is M.*

Clearly, this theorem can be generalized in the following way. If X is a global reduction function from L to M satisfying the analogous conditions, and if M is in $\mathcal{C}$, then so is L given that X is $\mathcal{C}$–computable. Notice that if X reduces L to M and Y reduces M to N, then $Y \circ X$ reduces L to N. Moreover, the property required in the previous theorem is also preserved. So, the reductions can be cascaded. This is needed when we want to reduce G to K, for example. The reduction functions given are only from G to K4.

However, we also have reduction functions from K4 to K. We shall refrain in the sequel from noting these obvious generalizations.

The local satisfiability problem still needs discussion. It will be treated by factoring the local reduction functions into a global reduction function plus a function bounding the depth of the tableau. The reduction functions given above are 1–analytic. This is not always so, for example with K4.3. The following is a global reduction function for K4.3 to K4:

$$X_3(\Delta) := \{\neg\Box\chi \wedge \neg\Box\chi' \to \neg\Box(\chi \vee \Box\chi') \vee \neg\Box(\chi' \vee \Box\chi)$$
$$\vee \neg\Box(\chi \vee \chi') : \Box\chi, \Box\chi' \in \mathrm{sf}(\Delta)\} \quad (22)$$

Of course it requires proof that this is a reduction function. Notice however that these formulae are instances of the characteristic axiom: replace p by $\neg\chi$ and q by $\neg\chi'$ in

$$\Diamond p \wedge \Diamond q \to \Diamond(p \wedge \Diamond q) \vee \Diamond(p \wedge \Diamond p) \vee \Diamond(p \wedge q) \quad (23)$$

X_3 is skeletal but only 2–analytic. (In fact, there does not exist a 1–analytic reduction function; otherwise S4.3 would have interpolation, contrary to fact.)

LEMMA 20. *Let X be a global reduction function from M to L. Then there exists a function ρ from sets of formulae to natural numbers such that*

$$Y(\Delta) := \{\Box^{<\rho(\Delta)+1}\vartheta : \vartheta \in X(\bigwedge \Delta \to \varphi)\} \quad (24)$$

is a local reduction function from M to L.

Proof. $\Delta \vdash_M \varphi$ implies $\Vdash_M \bigwedge \Delta \to \varphi$, which in turn implies $X(\bigwedge \Delta \to \varphi) \Vdash_L \bigwedge \Delta \to \varphi$ and so $\Box^{<\mu}X(\Delta;\varphi) \vdash_L \bigwedge \Delta \to \varphi$ for some number μ depending only on $\bigwedge \Delta \to \varphi$. Put $\rho(\bigwedge \Delta \to \varphi) := \mu$. Then $\Delta; \Box^{<\mu}X(\bigwedge \Delta \to \varphi) \vdash_L \varphi$, as promised. ∎

There is a slight problem in the definition of ρ. We have defined ρ such that $\rho(\bigwedge \Delta \to \varphi) := \mu$, while technically ρ should depend only on $\Delta; \varphi$ (that is, not knowing what is premiss and what is conclusion). This can be dealt with in two ways: (a) We make the reduction functions sensitive to this distinction (which gives more subtle bounds), (b) we take the maximum over all numbers for partitions of $\Delta; \varphi$ into a conclusion and a premiss set. (b) is less optimal but asymptotically the difference can usually (and in the present cases definitely) be ignored.

DEFINITION 21. Let M and L be modal logics, X, Y and ρ as in the previous lemma. If X is n–analytic with skeletal set Θ Y is also called **n–analytic** and **depth reduction function** ρ.

Notice that the local reduction need not be skeletal even if X is. However, there is an important case in which Y is once again skeletal.

DEFINITION 22. A reduction function X **splits** if

$$X(\Delta; \varphi \wedge \psi) = X(\Delta; \varphi \to \psi) = X(\Delta; \varphi; \psi) \qquad (25)$$

The following is shown in [9] and establishes the relevance of this concept.

THEOREM 23. *Suppose that X is a splitting global reduction function from M to L. Then if L has interpolation, so does M.*

If X is splitting, $X(\bigwedge \Delta \to \varphi) = X(\Delta; \varphi)$. All reduction functions defined above are in fact splitting and skeletal.

PROPOSITION 24. *Suppose that X is a splitting n–analytic global reduction function with skeletal set Θ. Further assume that ρ is a function from sets of formulae to numbers such that*

$$Y(\Delta) := \{\Box^{<\rho(\Delta)+1}\vartheta : \vartheta \in X(\Delta)\} \qquad (26)$$

is a local reduction function from M to L. Then Y is n–analytic with skeletal set Θ, and splitting. Moreover,

$$\sharp Y(\Delta) \leq 2\rho(\Delta)\,\mathrm{card}(X(\Delta)) + \sharp X(\Delta) \qquad (27)$$
$$\leq 2\rho(\Delta)(\sharp\Delta)^n + (\sharp\Delta)^n \sharp\Theta$$

Notice that for the logics discussed in this paper, $\rho(\Delta) \leq \sharp\Delta$. So, if $n > 0$, the second term wins for large $\sharp\Delta$. This is still suboptimal. Define

$$Y'(\Delta) := \Box^{<\rho(\Delta)+1} \bigwedge X(\Delta) \qquad (28)$$

Then

$$\sharp Y'(\Delta) \leq 2\rho(\Delta) + \sharp(\bigwedge X(\Delta)) \leq 2\rho(\Delta) + 2\sharp\Theta(\sharp\Delta)^n \qquad (29)$$

In order to use Hemaspaandra's results we need to bound the number $\sharp_\delta(Y(\Delta))$. However, notice that for any set Γ of formulae $\sharp_\delta(\Gamma) \leq \sharp\Gamma \cdot \mathrm{dg}(\Gamma)$.

THEOREM 25. *Let $n > 0$. Suppose that Y is an n–analytic local reduction function from M to K with skeletal set Θ and depth reduction function ρ. Assume that $\rho(\Delta) \leq \sharp\Delta$. Then for given Δ a tableau can be computed using $O((\sharp\Delta)^n \cdot \mathrm{dg}(\Delta))$ space.*

Proof. For checking satisfiability, we check whether $\Delta \vdash_L \bot$. We construct a tableau for $\Sigma := \Delta; Y'(\Delta; \bot)$ in place of Δ. $\sharp_\delta(\Sigma) \leq \text{dg}(\Sigma) \cdot \sharp\Sigma$, which is asymptotically of the magnitude $O(\text{dg}(\Delta)(\sharp\Delta)^n)$. This is also the space bound. ∎

In certain cases we can eliminate the additional factor $\text{dg}(\Delta)$. X_D for example is 0-analytic. Here, $Y(\Delta) := \Box^{<\text{dg}(\Delta)+1}\neg\Box\bot$, which has length $O(\text{dg}(\Delta))$. Hence the reduction function adds sublinear material.

COROLLARY 26. *Satisfiability of Δ in K.D is in $O(||\Delta||)$ space.*

This can be sharpened to $O(\sharp_\delta \Delta)$. For other logics Hemaspaandra's method gives $O(\sharp\Delta \cdot \text{dg}(\Delta))$. A second case is reduction functions which are degree homogeneous, such as X_{alt_1}. Here the following is a local reduction function:

$$Y_{\text{alt}_1}(\Delta) := \{\Box^d(\neg\Box\chi \to \Box\neg\chi) : \Box\chi \in \text{sf}(d, \Delta), d \leq \text{dg}(\Delta)\} \quad (30)$$

Also here we can apply a little bit of cosmetics. Put

$$Y^*_{\text{alt}_1}(\Delta) := \bigwedge X_{\text{alt}_1}(\text{sf}(0, \Delta)) \quad (31)$$
$$\wedge (\Box \bigwedge X_{\text{alt}_1}(\text{sf}(1, \Delta)))$$
$$\wedge (\Box \bigwedge X_{\text{alt}_1}(\text{sf}(2, \Delta)))$$
$$\wedge \ldots))$$

Here, every member of $\text{sf}(d, \Delta)$ is multiplied by two occurrences, and both have the same depth of embedding. In addition, there are $O(\sharp\Delta)$ new occurrences of subformulae. Thus, $\sharp_\delta Y^*_{\text{alt}_1}(\Delta)$ is linear in $\sharp_\delta \Delta$. Thus we have shown the following result (which can also easily be shown by adapting Hemaspaandra's calculus).

THEOREM 27. *K.alt$_1$ is locally in $O(||\Delta||)$-space.*

COROLLARY 28. *Let M be any union of the following logics: K, K.T, K.B, K.alt$_1$, K.D. Then M is locally in $O(\sharp\Delta \, \text{dg}(\Delta))$-space.*

For a proof we only need to show that the depth reduction function is linear in $\sharp\Delta$. This in turn means that the length of a maximal path in a minimal model for a consistent formula set must be linear in $\sharp\Delta$. For the model is basically created in all these cases from bundling open paths in open K–tableaux into a model (and defining the relation suitably). Clearly, for the postulates T, B, D, and alt$_1$ this is satisfied. A model needs to be only as deep as the modal depth of the formula set. (See the model construction procedure of [9].) By way of example, we give a proof of the fact that X_B is a skeletal reduction function with linear depth reduction function.

THEOREM 29. *Let L be a subframe logic whose class of frames is closed under passing from the relation to its symmetric closure. Then $X_\mathbf{B}$ is a 1-analytic skeletal global reduction function of $L.\mathbf{B}$ to L. Moreover, $Y_\mathbf{B}(\Delta) := \Box^{<\mathrm{dg}(\Delta)+1} X_\mathbf{B}(\Delta)$ is a local reduction function.*

Proof. The skeletality of $X_\mathbf{B}$ is easy to see. We prove that $Y_\mathbf{B}$ is a local reduction function; it follows directly that $X_\mathbf{B}$ is a global reduction function. Assume that $\Delta; \Box^{\leq q} X_\mathbf{B}(\Delta; \varphi)$ is L-consistent, where $q := \sharp\Delta$. We will show that Δ is $L.\mathbf{B}$-consistent, the converse being easy. Pick an L-frame $\mathfrak{F} = \langle F, \lhd \rangle$, a valuation β and a world x such that $\langle \mathfrak{F}, \beta, x \rangle \vDash \Delta; \Box^{\leq q} X_\mathbf{B}(\Delta; \varphi)$.

Let $\mathfrak{F}^B$ be obtained by replacing $\lhd$ by its symmetric closure, $\blacktriangleleft$. $\mathfrak{F}^B \vDash L.\mathbf{B}$, by assumption on L. By induction on χ we show that for all w reachable in at most $q - \mathrm{dg}(\chi)$ steps from x:

$$\langle \mathfrak{F}^B, \beta, w \rangle \vDash \chi \quad \Leftrightarrow \quad \langle \mathfrak{F}, \beta, w \rangle \vDash \chi \tag{32}$$

The only critical step is $\chi = \Box\tau$. From left to right this follows from the fact that if $x \lhd y$ then also $x \blacktriangleleft y$. For the other direction, assume we have $\langle \mathfrak{F}^B, \beta, w \rangle \nvDash \Box\tau$. Then there is a v such that $w \blacktriangleleft v$ and $\langle \mathfrak{F}^B, \beta, v \rangle \vDash \neg\tau$. If $w \lhd v$, we are done. Otherwise, $v \lhd w$. However, we have $\langle \mathfrak{F}, \beta, v \rangle \vDash \neg\tau \to \Box\neg\Box\neg\tau$. (For $x \vDash \Box^{\leq q}(\neg\tau \to \Box\neg\Box\neg\tau)$, and v is reachable in at most q steps from x.) So, if $\langle \mathfrak{F}, \beta, v \rangle \vDash \neg\tau$, we have $\langle \mathfrak{F}, \beta, w \rangle \vDash \neg\Box\tau$. This means that $\langle \mathfrak{F}, \beta, w \rangle \nvDash \Box\tau$, as promised.

Now let G be the set of all points reachable from x in at most q steps; let $\mathfrak{G}$ be the induced subframe of $\mathfrak{F}^B$ and let $\gamma(p) := \beta(p) \cap G$. Since L is a subframe logic, $\mathfrak{G}$ is a frame for L. It is symmetric, therefore a frame for $L.\mathbf{B}$. We claim $\langle \mathfrak{G}, \gamma, x \rangle \vDash \Delta$. Namely, the following is established by an easy induction: for all points y reachable in at most p steps from x and all subformulae χ of Δ of depth $\leq q - p$, $\langle \mathfrak{G}, \gamma, y \rangle \vDash \chi$ iff $\langle \mathfrak{F}^B, \gamma, y \rangle \vDash \chi$. The claim now follows since Δ has degree q and therefore every member of it is true at w. Hence Δ is $L.\mathbf{B}$-consistent. ∎

We remark that in the definition of $Y_\mathbf{B}$ we could have dropped all formulae of degree $> \mathrm{dg}(\Delta)$. This gives an improvement by a factor $1/2$, which is however ignored in the O-notation.

6 Transitive logics

A logic L is **transitive** if it contains the axiom 4: $\Box p \to \Box\Box p$. It is immediate that for transitive logics $\Delta \Vdash_L \varphi$ iff $\Delta; \Box\Delta \vdash_L \varphi$. This transformation of problems is linear. Hence, a transitive logic is globally $\mathcal{C}$-hard (globally in $\mathcal{C}$, globally $\mathcal{C}$-complete) if and only if it is locally $\mathcal{C}$-hard (locally in $\mathcal{C}$, locally $\mathcal{C}$-complete). Also, if X is a global reduction function from M to

$L \supseteq \mathsf{K4}$, $Y(\Delta) := \Box^{<2} X(\Delta)$ is a local reduction function from M to L. One immediate consequence is the following.

COROLLARY 30. *Let $\mathcal{C}$ be closed under linear transforms. Then if $\mathsf{K4}$ is locally in $\mathcal{C}$, so is G, $\mathsf{S4}$, $\mathsf{K4.D}$, and Grz.*

Thus, we only need to establish the local complexity of $\mathsf{K4}$. We can try Hemaspaandra's methods again. In place of the rule ($\Box$H) we take the following rule:

$$(\Box \mathrm{H}4) \quad \frac{\Gamma; \Box\Delta; \neg\Box\Sigma; \neg\Box\chi}{(\Delta; \Box\Delta; \neg\chi)^*} \tag{33}$$

where Γ is a set of formulae of degree 0. Once again the calculus steps from downward saturated sets to downward saturated sets. A downward saturated set takes $O(\sharp\Delta)$ space to code. The formula on which the rule operates takes $O(\log\|\Delta\|)$ space. Now observe the following. (a) If $\Box\delta$ appears above the line, it also appears below. (b) If $\neg\Box\delta$ appears below the line, it also appears above. So, the set below the line is characterized uniquely by the set of $\Box\delta$ that occur in it in addition to those that are above the line. Further, we we shall show it is not necessary to use the rule twice on the same formula $\neg\Box\chi$. Thus, for backtracking, we need a record only of (a) the variables occurring in the set, (b) the formulae $\Box\delta$ that are being added, (c) the formulae $\neg\Box\delta$ that are being retracted, (d) the formula $\neg\Box\chi$. So the tableau is stored as a sequence $\langle\langle A_i, B_i, C_i, \delta_i\rangle : i < n\rangle$, where A_i is a set of variables, B_i a set of subformulae of the form $\Box\delta$, C_i a set of subformulae or their negations of the form $\neg\Box\delta$, and δ_i a subformula. Then a variable is true at node i iff it is in A_i; a subformula $\Box\delta$ is true at i iff it is in some B_j, $j < i$; finally, $\neg\Box\delta$ is true iff it is not in any C_j for $j < i$, and not equal to any of the δ_j, $j < i$. Clearly, since the B_i and the C_i are pairwise disjoint but otherwise of any size, they can only be globally estimated. The bound is $\sharp\Delta \cdot \log \sharp\Delta$. Given that the length of a branch is bounded by $\sharp\Delta$, the overall bound is $\sharp\Delta \cdot \mathrm{card}(\mathrm{var}(\Delta)) + 3\sharp\Delta \log \sharp\Delta$. Finally, look at the transition

$$\frac{\Gamma; \Box\Delta; \neg\Box\Sigma; \neg\Box\chi}{(\Delta; \Box\Delta; \neg\chi)^*} \tag{34}$$

The set Γ of nonmodal formulae does not influence the formula set below the line. So, A_j, $j < i$, is not needed in backtracking. In other words, we do not backtrack to another choice of the values for the variables. This means that all we need to do is to keep a record of $\langle\langle B_j, C_j, \delta_j\rangle : j < i\rangle$, which is of size $O(\sharp\Delta \log \sharp\Delta)$.

THEOREM 31 (Nguyen). *Satisfiability in $\mathsf{K4}$ can be checked using only $O(\sharp\Delta \cdot \log \sharp\Delta)$ space.*

COROLLARY 32. *Satisfiability in G, S4, Grz is in $O(\sharp\Delta \cdot \log \sharp\Delta)$–space.*

There is also a proof using global tableaux. This proceeds by bounding the number of applications of the rule ($\Box$E). We shall start with a global K–tableau for

$$X_4(\Delta); \{p \vee \neg p : p \in \text{var}(\Delta)\} \dagger \Delta \qquad (35)$$

Suppose that no closed tableau exists. (This is the same as to say that Δ does not globally follow from $X_4(\Delta)$ and $\{p \vee \neg p : p \in \text{var}(\Delta)\}$.) Then we construct a model as follows. The nodes are all downward saturated sets occurring in any open branch of a tableau that has the form $\Sigma \dagger \Theta$ such that Σ did not appear in a previous node of that branch. By the fact that we have a tableau in which $p \vee \neg p$ is contained everywhere in Θ for each variable occurring in Δ, downward saturated sets will either contain p or $\neg p$ for these variables.

We put $\Theta \dagger \Sigma_1 \lhd \Theta \dagger \Sigma_2$ iff $\Theta \dagger \Sigma_1$ appeared above $\Theta \dagger \Sigma_2$ in that branch. Then we put $\blacktriangleleft := \lhd^+$, the transitive closure. This defines a frame. We define $\beta(p) := \{\Theta \dagger \Sigma : p \in \Sigma\}$. It can be checked that in this model, χ holds at $\Theta \dagger \Sigma$ iff it is in Σ. The model is transitive, and therefore we have a K4–model for Δ.

LEMMA 33. *Suppose that b is a branch in a global K–tableau for*

$$X_4(\Delta); \{p \vee \neg p : p \in \text{var}(\Delta)\} \dagger \Delta \qquad (36)$$

Suppose further that ($\Box$E) has least priority and that there exist no two downward saturated nodes $\Theta \dagger \Sigma_1$ and $\Theta \dagger \Sigma_2$ such that $\Sigma_1 = \Sigma_2$. Then ($\Box$E) has been applied at most $3 \cdot 2^n \sharp \Delta$ times, where n is the number of variables occurring in Δ.

Proof. In a local tableau, the number of applications of ($\Box$E) within a branch is bounded by the modal depth of the formula. More precisely, it is bounded by the number of nested negative occurrences of $\Box$. In a global tableau, the left hand set feeds the right hand side each time ($\Box$E) is executed. However, the set that is being added is always the same. We have

$$X_4(\Delta) = \{\neg(\Box\chi \wedge \neg\Box\Box\chi) : \Box\chi \in \text{sf}(\Delta)\} \qquad (37)$$

A downward saturated set is of the form $O; P; N$, where $O \subseteq \{p, \neg p : p \in \text{var}(\Delta)\}$, $P \subseteq \{\Box\chi, \Box\Box\chi : \Box\chi \in \text{sf}(\Delta)\}$ and $N \subseteq \{\neg\Box\chi : \Box\chi \in \text{sf}(\Delta)\}$. Suppose that some set contains $\Box\chi$ for some $\Box\chi \in \text{sf}(\Delta)$. Then, since it is the saturation of a set containing $\neg(\Box\chi \wedge \neg\Box\Box\chi)$, it also contains $\Box\Box\chi$. Thus, after one application of ($\Box$E) we have that $\Box\chi$ is contained in the set below the line as well, and after saturation also $\Box\Box\chi$. Hence, the set P

cannot shrink, while the set N cannot grow. So, in passing from $O; P; N$ to $O'; P'; N'$ we must have either $N \supsetneq N'$, or $P \subsetneq P'$, or $O' \neq O$. Moreover, as long as $N = N'$ and $P = P'$, O may never occur again. This yields the bound. ∎

The bound obtained in this way is too rough for our purposes. However, notice that downward saturated sets correspond to nodes in a model. Moreover, if $N' = N$ and $P' = P$, this means that we move inside a cluster. The cluster size is at most 2^n. However, it is not necessary to form big clusters all the time. We can avoid forming nontrivial clusters in most cases. In effect, one can show that if there exists a closed tableau then there exists a closed tableau with the following property. If at some step $N = N'$ and $P = P'$, then there is a formula $\neg \Box \chi$ for which ($\Box$E) is applied, and this formula will never appear in a successor cluster again. In terms of models this means that if there exists a model then there exists a model such that if we have a nontrivial cluster where $\neg \Box \chi$ is true, then this formula will be false in all clusters succeeding that cluster. This is what we shall show now.

Let $\mathfrak{F} = \langle F, \triangleleft \rangle$ be a frame. A **path of length** n is a sequence $\Pi = \langle w_i : i < n + 1 \rangle$ such that $w_i \triangleleft w_{i+1}$ for all $i < n$. Π is **nonrepeating** if for no $i < j$, $w_i = w_j$. We shall show that the size of nonrepeating paths in a minimal model is linear in $\sharp \Delta$. The following proof is an adaptation of the proof methods in [10].

LEMMA 34. *Let $M \supseteq \mathsf{K4}$ be a cofinal subframe logic. Then a consistent formula set Δ has a model in which every nonrepeating path has length $\leq \sharp \Delta$.*

Proof. Take a model $\langle \mathfrak{F}, \beta, x \rangle \vDash \Delta$. Let $\psi \in \mathrm{sf}(\Delta)$. Call a point y χ-**critical**, if (a) $\langle \mathfrak{F}, \beta, y \rangle \vDash \chi$ and (b) from $\langle \mathfrak{F}, \beta, z \rangle \vDash \chi$ and $y \triangleleft z$ follows $z \triangleleft y$. Evidently, if y and y' are χ-critical and $y \triangleleft y'$ then $y' \triangleleft y$ as well. Moreover, if y satisfies $\Diamond \chi$, then there is a critical z such that $y \triangleleft z$ and z satisfies χ. Take a cluster $C(u) = \{v : u \triangleleft v \triangleleft u\}$. If $|C(u)| = 1$, we remove $C(u)$ if it is not critical. If $|C(u)| > 1$, and it contains critical points, then we keep for each $\chi \in \mathrm{sf}(\Delta)$ exactly one χ-critical point, if the cluster contains one. If $|C(u)| > 1$ and $C(u)$ contains no critical points, we remove all points. Let G be the set of remaining points. Call the induced subframe $\mathfrak{G}$. Denote by γ the valuation induced by β on $\mathfrak{G}$. We claim that $\langle \mathfrak{G}, \gamma, x \rangle \vDash \Delta$. In fact, we show that for each formula $\chi \in \mathrm{sf}(\Delta)$ and each $y \in G$:

$$\langle \mathfrak{G}, \gamma, y \rangle \vDash \chi \quad \Leftrightarrow \quad \langle \mathfrak{F}, \beta, y \rangle \vDash \chi \tag{38}$$

Indeed, if χ is a variable, this holds by definition of γ. The only nonobvious step is $\chi = \Diamond \chi'$. Assume that $\langle \mathfrak{G}, \gamma, y \rangle \vDash \Diamond \chi'$. Then there exists a $z \in G$

such that $y \triangleleft z$ and $\langle \mathfrak{G}, \gamma, z \rangle \vDash \chi'$. By induction hypothesis, $\langle \mathfrak{F}, \beta, z \rangle \vDash \chi'$ and so $\langle \mathfrak{F}, \beta, y \rangle \vDash \Diamond \chi' = \chi$. Now assume that the latter holds. Then there is a χ'-critical successor z of y. Therefore, $\langle \mathfrak{F}, \beta, z \rangle \vDash \chi'$ and since $z \in G$, we have $\langle \mathfrak{G}, \gamma, z \rangle \vDash \chi'$ by inductive hypothesis. Therefore $\langle \mathfrak{G}, \gamma, y \rangle \vDash \Diamond \chi' = \chi$.

It is clear that a point is χ-critical in $\langle \mathfrak{F}, \beta \rangle$ iff it is χ-critical in $\langle \mathfrak{G}, \gamma \rangle$. Finally, we count the number of points in a nonrepeating path. Let $y, y' \in G$, and $y \triangleleft y'$. Since y and y' are critical they must be critical for different formulae. Hence we have at most $\sharp \Delta$ points in a nonrepeating path. ∎

This gives the following.

THEOREM 35.
$$\Delta \vdash_{\mathsf{K4}} \varphi \quad \Leftrightarrow \quad \Delta; \Box^{<\sharp(\Delta;\varphi)+1} X_4(\Delta;\varphi) \vdash_{\mathsf{K}} \varphi \tag{39}$$

Using Theorem 25 we get

COROLLARY 36. *K4 is in $O(\sharp \Delta \operatorname{dg}(\Delta))$–space.*

It follows with the help of the previous results that also G, Grz, and S4 are in $O(\sharp \Delta \operatorname{dg}(\Delta))$–space. This is somewhat different from the bound $O(\|\Delta\| \log \|\Delta\|)$.

Using the formulae of [10] polynomial space bounds can be obtained for all subframe logics, the degree of the polynomial equal to the number of variables needed to axiomatize that logic. This is not entirely straightforward as they result from substituting into the skeletal set some complex formulae built essentially from maximal consistent subsets $\operatorname{sf}^-(\Delta)$. However, with a set Δ given, a subset is linearly codable, and so the substitution instances need only $O((\sharp \Delta)^n)$–space to code.

It is known that extensions of S4.3 are cofinal subframe logics (see [2]). For an extension of S4.3 only a single tableau needs to be computed. Hence, such an extension is in NP and therefore NP–complete if consistent (see [14]). The space bounds established here are $O((\sharp \Delta)^k \log \sharp \Delta)$, where k is the number of variables needed to axiomatize the logic. Hence $k = 2$ for S4.3 itself. This may in many cases be suboptimal.

In a similar fashion, a lot more results can be shown. We can prove that satisfiability in tense logic is in $O(\|\Delta\| \log \sharp \Delta)$, and that PDL with converse is EXPTIME–complete (see [4]). Furthermore, the technique of [8] shows that many splitting axioms preserve complexity bounds above K4 and S4. This applies in particular (above K4) to .1, .2 and (above S4) .Dum, to name a few. Using the standard Gödel–translation, which is linear, we get from Corollary 32 the space bound $O(\|\Delta\| \log \|\Delta\|)$ not only for intuitionistic propositional logic, obtained in [6], but also for extensions of Int (for example, KC = Int $+ \neg p \vee \neg \neg p$, the logic corresponding to S4.2, or LC = Int $+ p \rightarrow q \vee q \rightarrow p$, the logic corresponding to S4.3).

7 Conclusion

Although there exist quite sophisticated tableau calculi specially adapted for logics extending **K**, the known (worst case) complexity bounds can typically be established using a purely combinatorial method based on reduction functions. The method may not be as useful in actual applications; however, it has the advantage of being uniform, and it can be generalized in various ways. A straightforward application of the method typically yields only $O(\sharp\Delta \cdot dg(\Delta))$, but this latter bound is not necessarily worse than the standardly proved $O(||\Delta|| \log ||\Delta||)$ (which typically can be improved to $O(\sharp\Delta \log \sharp\Delta)$ at no cost). Also, based on tableau methods, the latter can typically easily be established. Moreover, in the transitive case we obtain the latter bound directly as well, using a modification of Hemaspaandra's calculus adapted for **K4**.

Acknowledgements

The research for this paper has been made possible by a Heisenberg–grant by the Deutsche Forschungsgemeinschaft. I wish to express my thanks to Rajeev Goré, Edith Hemaspaandra, Fabio Massaci and two anonymous referees for useful discussions. Special thanks go to Frank Wolter for his help. The responsibility for errors and omissions is solely mine.

BIBLIOGRAPHY

[1] D. Basin, S. Matthews, and L. Viganò. A new method for bounding the complexity of modal logic. In G. Gottlob, A. Leitsch, and D. Mundici, editors, *Proceedings of the 5th Kurt Gödel Colloquium on Computational Logic and Proof Theory (KGC'97)*, number 1289 in Lecture Notes in Computer Science, pages 89 – 102, Heidelberg, 1997. Springer.

[2] Alexander Chagrov and Michael Zakharyaschev. *Modal Logic*. Oxford University Press, Oxford, 1997.

[3] Melvin Fitting. *Proof Methods for Modal and Intuitionistic Logic*. Number 169 in Synthese Library. Reidel, Dordrecht, 1983.

[4] Giuseppe de Giacomo. Eliminating "Converse" from Converse PDL. *Journal of Logic, Language and Information*, 5:193 – 208, 1996.

[5] Edith Hemaspandra. Modal Satisfiability is in Deterministic Linear Space. In *Computer Science Logic*, number 1786 in Lecture Notes in Computer Science, pages 332 – 343, Heidelberg, 2000. Springer Verlag.

[6] Jörg Hudelmaier. An $O(n \log n)$–Space Decision Procedure for Intuitionistic Propositional Logic. *Journal of Logic and Computation*, 3:63 – 75, 1993.

[7] Jörg Hudelmaier. Improved Decision Procedures for the Modal Logics K, T and S4. In H. Kleine Büning, editor, *Proceedings of CSL '95*, number 1092 in Lecture Notes in Computer Science, pages 320 – 334, 1996.

[8] Marcus Kracht. Splittings and the finite model property. *Journal of Symbolic Logic*, 58:139 – 157, 1993.

[9] Marcus Kracht. *Tools and Techniques in Modal Logic*. Number 142 in Studies in Logic. Elsevier, Amsterdam, 1999.

[10] Marcus Kracht. Reducing Modal Consequence Relations. *Journal of Logic and Computation*, 11:879 – 907, 2001.

[11] R. E. Ladner. The computational complexity of provability in systems of modal logic. *SIAM Journal of Computing*, 6:467 – 480, 1977.
[12] Linh Anh Nguyen. A New Space Bound for the Logics K4, KD4, and S4. In M. Kutyłowski, L. Pacholski, and T. Wierzbicky, editors, *Proceedings of MFCS'99*, LNCS 1675, pages 321 – 331, 1999.
[13] Wolfgang Rautenberg. Modal tableau calculi and interpolation. *Journal of Philosophical Logic*, 12:403 – 423, 1983.
[14] Edith Spaan. *Complexity of Modal Logics*. PhD thesis, Department of Mathematics and Computer Science, University of Amsterdam, 1993.
[15] L. Viganò. *A framework for non-classical logics*. PhD thesis, Universität des Saarlandes, Saarbrücken, Germany, 1997.

Marcus Kracht
Fachbereich Mathematik und Informatik
Institut fr Mathematik II
Freie Universitt Berlin
Arnimallee 3 (Villa, Raum 010)
D-14195 Berlin, Germany
E-mail: kracht@math.fu-berlin.de

13
Description Logics with Concrete Domains—A Survey
CARSTEN LUTZ

ABSTRACT. Description logics (DLs) are a family of logical formalisms that have initially been designed for the representation of conceptual knowledge in artificial intelligence and are closely related to modal logics. In the last two decades, DLs have been successfully applied in a wide range of interesting application areas. In most of these applications, it is important to equip DLs with expressive means that allow to describe "concrete qualities" of real-world objects such as their weight, temperature, and spatial extension. The standard approach is to augment description logics with so-called concrete domains, which consist of a set (say, the rational numbers), and a set of n-ary predicates with a fixed extension over this set. The "interface" between the DL and the concrete domain is then provided by a new logical constructor that has, to the best of our knowledge, no counterpart in modal logics. In this paper, we give an overview over description logics with concrete domains and summarize decidability and complexity results from the literature.

1 Introduction

Description logics (DLs) are a family of logical formalisms that originated in the field of artificial intelligence as a tool for the representation of conceptual knowledge. Since then, DLs have been successfully used in a wide range of application areas such as knowledge representation, reasoning about class-based formalisms (e.g. conceptual database models and UML diagrams), and ontology engineering in the context of the semantic web [10; 19; 8]. The basic syntactic entity of description logics are *concepts*, which are constructed from concept names (unary predicates) and role names (binary relations) using the set of concept and role constructors provided by a particular DL. For example, the following concept is formulated in the basic propositionally closed description logic $\mathcal{ALC}$ and could be used, e.g., in a knowledge-based process engineering application (as in [54; 49]) to describe a production process that has an expensive (specially trained) operator:

Process ⊓ ∀subproc.Process ⊓ ∃operator.(Human ⊓ Expensive)

In this example, Process, Human, and Expensive are concept names while subproc and operator are role names.

Viewed from a logical perspective, description logics are closely related to modal logics [55; 25]. For example, the DL $\mathcal{ALC}$ can be viewed as a notational variant of the modal logic K_ω, i.e., multimodal K with infinitely many accessibility relations: concept names correspond to propositional variables, role names correspond to (names for) accessibility relations, the $\forall$ constructor of $\mathcal{ALC}$ can be read as a modal box operator, and $\exists$ can be read as a diamond. However, there also exist several means of expressivity that are frequently used in description logics, but usually not considered in modal logics.

An important example are so-called *concrete domains*, which allow the integration of "concrete qualities" such as numbers, time intervals, and strings into description logic concepts. Suppose, for example, that we want to refine the description of a process given above by replacing the concept name Expensive with a concept expressing that the process operator earns at least 20 euro per hour. Then we need a proper way to talk about numbers such as "20" and comparisons between numbers such as "at least 20 euro". As another example, we may want to express that the time interval describing the working time of the operator should contain the time interval describing the execution time of the process. Here we obviously need to represent time intervals and relations between them.

The need for extending the expressive power of DLs in the described direction arises in almost all relevant application areas, let us review two (more) examples:

1. Semantic web. In this application, DLs are used to describe the contents of web pages in order to facilitate the development of more powerful web services such as advanced search engines [8; 15]. It is obvious and has always been emphasized that the representation of "concrete datatypes" such as numbers and strings is an important issue [23; 33]: if, for example, we want to describe the web page of a wine seller, then we need numbers to represent vintages and prices, and strings to represent the names of regions and wine producers. It should be clear that such concrete datatypes are precisely what we have described as "concrete qualities".

2. Conceptual database models. Entity Relationship (ER) diagrams are the predominant formalism for constructing conceptual models of relational databases [21; 60]. For example, an ER diagram could describe two entities Employee and Company related by a relationship employs such that each Employee is employed by exactly one Company, and each Company employs at least one Employee. DLs can be used to encode and reason about ER

diagrams, which allows to detect inconsistencies and implications that are only implicitly represented in the diagram [19; 24]. However, the standard translation of ER diagrams into DLs does not take into account so-called "numerical attribute dependencies", which can e.g. be used to express that no Employee was hired prior to his Company's founding. As argued in [46], it is important to include these dependencies when using DLs for reasoning about ER diagrams since they can be an (additional) source for inconsistencies and unnoticed ramifications. In order to do this, the target DL must be able to represent "concrete" objects such as numbers and comparisons between numbers.

The necessity of representing concrete qualities in description logics has been realized almost since the beginnings of the field, and, indeed, many early description logic reasoners such as MESON [22] and CLASSIC [18] provided for "ad hoc" solutions of this problem. The first formal treatment of the issue was presented by Baader and Hanschke in [11], who proposed to extend the description logic $\mathcal{ALC}$ with concrete domains. Formally, a concrete domain consists of a set such as the natural numbers and a set of predicates such as the binary "<" and the ternary "+" with a *fixed* extension over this set. Enriching $\mathcal{ALC}$ with such a concrete domain $\mathcal{D}$, we obtain the basic DL with concrete domains $\mathcal{ALC}(\mathcal{D})$. More precisely, $\mathcal{ALC}(\mathcal{D})$ is obtained from $\mathcal{ALC}$ by augmenting it with

- *abstract features*, i.e. roles interpreted as functional relations;

- *concrete features*: a new syntactic type that is interpreted as a partial function from the logical domain into the concrete domain;

- a new concept constructor that allows to describe constraints on concrete values using predicates from the concrete domain.

Let us view two example $\mathcal{ALC}(\mathcal{D})$-concepts: the concept

Process ⊓ ∀subproc.Process ⊓ ∃operator.(Human ⊓ ∃wage.$\geq_{20}$)

refines the process description from above by replacing the concept name Expensive with a concrete domain-based description of the operator's wage— which is at least 20 euro per hour. In this example, operator is an abstract feature while wage is a concrete feature. We use a concrete domain based on the natural numbers and assume that $\geq_{20}$ is a unary predicate with the obvious extension. The (sub)concept ∃wage.$\geq_{20}$ is an instantiation of the concrete domain constructor and must not be confused with the existential restriction as used in ∃operator.Human. Observe that concrete features such as wage are the "link" between the description logic and the concrete domain: they allow to associate concrete values such as numbers with logical objects such as the one representing the operator in the above example.

In the second concept, we use a concrete domain based on time intervals to describe a constraint on the execution time of processes as proposed above:

Process ⊓ ∀subproc.Process ⊓ ∃(operator worktime), (exectime).contains

Here, worktime and exectime (for "execution time") are concrete features, and contains is a binary concrete domain predicate. The last conjunct is an instantiation of the concrete domain constructor expressing that the time interval describing the working time of the operator contains the time interval describing the execution time of the process.

Since their first appearance in 1991, description logics with concrete domains have been extensively studied. The purpose of this paper is to survey the proposed logics and available results, focusing on decidability and computational complexity. It is organized as follows: in Section 2, we formally introduce concrete domains and the description logic $\mathcal{ALC}(\mathcal{D})$. Section 3 discusses results that have been obtained for $\mathcal{ALC}(\mathcal{D})$ and several of its extensions: in Section 3.1, we treat $\mathcal{ALC}(\mathcal{D})$ itself, Section 3.2 is concerned with extensions considered "standard" in the area of description logics, and Section 3.3 focuses on specifically concrete-domain related extensions. Most of the discussed results do not consider a particular concrete domain, but are of a general nature. Finally, Section 4 gives a brief overview over concrete domains that have been proposed in the literature.

2 The description logic $\mathcal{ALC}(\mathcal{D})$

In this section, we formally introduce Baader and Hanschke's basic description logic with concrete domains $\mathcal{ALC}(\mathcal{D})$ [11]. To do this, we must first define the underlying notion of concrete domains.

DEFINITION 1 (Concrete domain) A *concrete domain* $\mathcal{D}$ is a pair $(\Delta_\mathcal{D}, \Phi_\mathcal{D})$, where $\Delta_\mathcal{D}$ is a set and $\Phi_\mathcal{D}$ a set of predicate names. Each predicate name $P \in \Phi_\mathcal{D}$ is associated with an arity n and an n-ary predicate $P^\mathcal{D} \subseteq \Delta_\mathcal{D}^n$.

For many application areas, the most interesting concrete domains are *numerical* ones. Hence, let us introduce a typical numerical concrete domain Q to illustrate Definition 1: as the set Δ_Q, we use the rational numbers ℚ. The following predicates are available:

- unary predicates P_q for each $P \in \{<, \leq, =, \neq, \geq, >\}$ and each $q \in \mathbb{Q}$ with $(P_q)^Q = \{q' \in \mathbb{Q} \mid q' \, P \, q\}$;
- binary predicates $<, \leq, =, \neq, \geq, >$ with the obvious extension;
- ternary predicates $+$ and $\overline{+}$ with $(+)^Q = \{(q, q', q'') \in \mathbb{Q}^3 \mid q+q' = q''\}$ and $(\overline{+})^Q = \mathbb{Q}^3 \setminus (+)^Q$;
- a unary predicate $\top_Q$ with $(\top_Q)^Q = \mathbb{Q}$ and a unary predicate $\bot_Q$ with $(\bot_Q)^Q = \emptyset$.

The presence of the predicates $\top_Q$ and $\bot_Q$ and of the negation of the "+" predicate is related to the *admissibility* of concrete domains and will be discussed in Section 3.1. We will further discuss the concrete domain Q and its relatives in Section 4.

DEFINITION 2 ($\mathcal{ALC}(\mathcal{D})$ syntax) Let N_C, N_R, and N_{cF} be pairwise disjoint and countably infinite sets of *concept names*, *role names*, and *concrete features*. Furthermore, let N_{aF} be a countably infinite subset of N_R. The elements of N_{aF} are called *abstract features*. A *path* u is a composition $f_1 \cdots f_n g$ of n abstract features $f_1, \ldots, f_n$ ($n \geq 0$) and a concrete feature g. For $\mathcal{D}$ a concrete domain, the set of $\mathcal{ALC}(\mathcal{D})$-concepts is the smallest set such that

- every concept name is a concept, and
- if C and D are concepts, R is a role name, g is a concrete feature, $u_1, \ldots, u_n$ are paths, and $P \in \Phi_\mathcal{D}$ is a predicate of arity n, then the following expressions are also concepts: $\neg C$, $C \sqcap D$, $C \sqcup D$, $\exists R.C$, $\forall R.C$, $\exists u_1, \ldots, u_n.P$, and $g\uparrow$.

As usual, we use $\top$ as abbreviation for an arbitrary propositional tautology and $\bot$ as abbreviation for $\neg \top$. Additionally, if $u = f_1 \cdots f_k g$ is a path then $u\uparrow$ is used as abbreviation for $\forall f_1. \cdots .\forall f_k.g\uparrow$. As an example $\mathcal{ALC}(\mathsf{Q})$-concept, consider the process description

Process $\sqcap$ ∀subproc.Process $\sqcap$ ∃operator.(Human $\sqcap$ ∃wage.$\geq_{20}$)

$\sqcap$ ∃(operator wage), (cost).$\leq$,

where the second line states that the hourly cost of the process is at least as high as the hourly wage of its operator. We now introduce the semantics of $\mathcal{ALC}(\mathcal{D})$-concepts and the relevant reasoning problems.

DEFINITION 3 ($\mathcal{ALC}(\mathcal{D})$ semantics) An *interpretation* $\mathcal{I}$ is a pair $(\Delta_\mathcal{I}, \cdot^\mathcal{I})$, where $\Delta_\mathcal{I}$ is a set called the *domain* and $\cdot^\mathcal{I}$ is the *interpretation function*. The interpretation function maps

- each concept name C to a subset $C^\mathcal{I}$ of $\Delta_\mathcal{I}$,
- each role name R to a subset $R^\mathcal{I}$ of $\Delta_\mathcal{I} \times \Delta_\mathcal{I}$,
- each abstract feature f to a partial function $f^\mathcal{I}$ from $\Delta_\mathcal{I}$ to $\Delta_\mathcal{I}$, and
- each concrete feature g to a partial function $g^\mathcal{I}$ from $\Delta_\mathcal{I}$ to $\Delta_\mathcal{D}$.

If $u = f_1 \cdots f_n g$ is a path, then $u^\mathcal{I}(d)$ is defined as $g^\mathcal{I}(f_n^\mathcal{I} \cdots (f_1^\mathcal{I}(d)) \cdots)$. The interpretation function is extended to arbitrary concepts as follows:

$$(\neg C)^\mathcal{I} := \Delta_\mathcal{I} \setminus C^\mathcal{I}$$
$$(C \sqcap D)^\mathcal{I} := C^\mathcal{I} \cap D^\mathcal{I}$$
$$(C \sqcup D)^\mathcal{I} := C^\mathcal{I} \cup D^\mathcal{I}$$
$$(\exists R.C)^\mathcal{I} := \{d \in \Delta_\mathcal{I} \mid \{e \mid (d,e) \in R^\mathcal{I}\} \cap C^\mathcal{I} \neq \emptyset\}$$
$$(\forall R.C)^\mathcal{I} := \{d \in \Delta_\mathcal{I} \mid \{e \mid (d,e) \in R^\mathcal{I}\} \subseteq C^\mathcal{I}\}$$
$$(\exists u_1, \ldots, u_n.P)^\mathcal{I} := \{d \in \Delta_\mathcal{I} \mid \exists x_1, \ldots, x_n \in \Delta_\mathcal{D} : u_i^\mathcal{I}(d) = x_i \text{ for } 1 \leq i \leq n$$
$$\text{and } (x_1, \ldots, x_n) \in P^\mathcal{D}\}$$
$$(g\uparrow)^\mathcal{I} := \{d \in \Delta_\mathcal{I} \mid g^\mathcal{I}(d) \text{ undefined}\}$$

Let $\mathcal{I}$ be an interpretation. Then $\mathcal{I}$ is a *model* of a concept C iff $C^\mathcal{I} \neq \emptyset$. A concept C is *satisfiable* iff C has a model. A concept C is *subsumed by* a concept D (written $C \sqsubseteq D$) iff $C^\mathcal{I} \subseteq D^\mathcal{I}$ for all interpretations $\mathcal{I}$.

While satisfiability is familiar from modal and classical logics, subsumption deserves a brief comment: this reasoning task is rather important in description logics since DLs are frequently used to capture the terminological knowledge of an application domain, and subsumption can then be used to arrange the defined notions (represented by concepts) in a taxonomy. Logically, subsumption can obviously be understood as the validity of implications. It should thus be clear that, in $\mathcal{ALC}(\mathcal{D})$, concept subsumption can be reduced to concept (un)satisfiability and vice versa: $C \sqsubseteq D$ iff $C \sqcap \neg D$ is unsatisfiable and C is satisfiable iff $C \not\sqsubseteq \bot$.

It is not hard to see that "the $\mathcal{ALC}$ part" of $\mathcal{ALC}(\mathcal{D})$ is a syntactical variant of multimodal K (see Section 1). However, to the best of our knowledge, the concrete domain constructor has no counterpart in modal logic. Moreover, even for very simple concrete domains $\mathcal{D}$ there does not exist a translation from $\mathcal{ALC}(\mathcal{D})$-concepts into formulas of the two-variable fragment of first-order logic or of the guarded fragment—a property enjoyed by most modal and description logics [61; 16]. The reason for this is that we admit paths of length greater one inside the concrete domain constructor.

For most application areas, the reasoning tasks "concept satisfiability" and "subsumption" have to take into account so-called TBoxes. Such TBoxes are sets of concept equations, which are used to store terminological knowledge and background knowledge about the application domain. For example, we could use a TBox to define the notion "expensive process" by writing

$$\text{ExpensiveProcess} \doteq \text{Process} \sqcap \exists \text{cost.} \geq_{20}$$

Moreover, we could capture the "background knowledge" that every process controlled by an expensive operator is an expensive process:

$$\top \doteq (\text{Process} \sqcap \exists \text{operator.} \exists \text{wage.} \geq_{20}) \rightarrow \text{ExpensiveProcess}$$

In the DL literature, there exist various TBox formalisms with vast differences in expressive power. In this paper, we will only consider the two TBox formalisms that are used most frequently.

DEFINITION 4 (TBox) A *concept equation* is an expression $C \doteq D$, where C and D are concepts. A *general TBox* is a finite set of concept equations.

A concept equation $C \doteq D$ is called a *concept definition* if C is a concept name. A finite set of concept definitions $\mathcal{T}$ is called an *acyclic TBox* if the following conditions are satisfied:

1. the left-hand sides of concept definitions are unique, i.e., if $\{A \doteq C, A' \doteq C'\} \subseteq \mathcal{T}$, then $C \neq C'$ implies $A \neq A'$;

2. $\mathcal{T}$ is acyclic: there are no concept definitions $\{A_0 \doteq C_0, \ldots, A_{k-1} \doteq C_{k-1}\} \subseteq \mathcal{T}$ such that the concept name A_i occurs in $C_{i+1 \bmod k}$ for $i < k$.

An interpretation $\mathcal{I}$ is a *model* of a (general or acyclic) TBox $\mathcal{T}$ if $C^{\mathcal{I}} = D^{\mathcal{I}}$ for all $C \doteq D \in \mathcal{T}$. A concept C is *satisfiable w.r.t. a TBox* $\mathcal{T}$ iff C and $\mathcal{T}$ have a common model. A concept C is *subsumed by* a concept D w.r.t. a *TBox* $\mathcal{T}$ (written $C \sqsubseteq_{\mathcal{T}} D$) iff $C^{\mathcal{I}} \subseteq D^{\mathcal{I}}$ for all models $\mathcal{I}$ of $\mathcal{T}$.

From a modal logic perspective, the expressive power provided by general TBoxes is closely related to the expressiveness of the universal modality—see e.g. Section 2.2.1 of [43] for a thorough discussion. While general TBoxes are a rather powerful tool, the expressive power provided by acyclic TBoxes is relatively weak: due to acyclicity, they can be viewed as macro definitions, i.e., as providing a set of abbreviations for concepts. As we will see in Section 3.2, acyclic TBoxes can also be expanded like macros. Note, however, that acyclic TBoxes are still powerful enough to define terminologies as in the first example presented above.

To distinguish concept satisfiability without TBoxes from concept satisfiability with TBoxes, we will in the following sometimes use the term "pure concept satisfiability" to refer to the former.

3 Description logics with concrete domains

In this section, we consider the basic description logic with concrete domains $\mathcal{ALC}(\mathcal{D})$ and several of its extensions. We start with $\mathcal{ALC}(\mathcal{D})$ itself and then discuss "standard extensions" that are frequently considered in description logics and, in principle, independent of concrete domains. Finally, we give an overview over extensions of $\mathcal{ALC}(\mathcal{D})$ that concern the "concrete domain part" of this logic.

3.1 The basic formalism

In their original 1991 paper, Baader and Hanschke present a tableau algorithm that is capable of deciding (pure) $\mathcal{ALC}(\mathcal{D})$-concept satisfiability. Using the reduction from the previous section, this algorithm also yields a decision procedure for concept subsumption. Baader and Hanschke's decidability result is a rather general one since it does not concern a particular concrete domain, but applies to any concrete domain that satisfies some weak conditions. These conditions are derived from the fact that any satisfiability algorithm *not* committing itself to a particular concrete domain must call some concrete domain reasoner as a subprocedure via a well-defined "interface". In the case of Baader and Hanschke's tableau algorithm, such a concrete domain reasoner is required to decide the satisfiability of finite conjunctions of concrete domain predicates. This leads to the notion of *admissibility*.

DEFINITION 5 (Admissible) Let $\mathcal{D}$ be a concrete domain and V a set of variables. A $\mathcal{D}$-*conjunction* is a predicate conjunction of the form

$$c = \bigwedge_{i<k}(x_0^{(i)}, \ldots, x_{n_i}^{(i)}) : P_i,$$

where P_i is an n_i-ary predicate for $i < k$ and the $x_j^{(i)}$ are variables from V. A $\mathcal{D}$-conjunction c is *satisfiable* iff there exists a function δ mapping the variables in c to elements of $\Delta_\mathcal{D}$ such that $(\delta(x_0^{(i)}), \ldots, \delta(x_{n_i}^{(i)})) \in P_i^\mathcal{D}$ for each $i < k$. Such a function is called a *solution* for c. We say that the concrete domain $\mathcal{D}$ is *admissible* iff

1. its set of predicate names is closed under negation and contains a name $\top_\mathcal{D}$ for $\Delta_\mathcal{D}$ and

2. the satisfiability of $\mathcal{D}$-conjunctions is decidable.

We refer to the satisfiability of $\mathcal{D}$-conjunctions as *$\mathcal{D}$-satisfiability*.

Property 1 of admissibility has to be satisfied since the description logic $\mathcal{ALC}(\mathcal{D})$ provides for negation. For example, the concept

$$\neg(g_1\uparrow) \sqcap \neg(g_2\uparrow) \sqcap \neg(\exists g_1,g_2.<)$$

expresses that $g_1^{\mathcal{I}} \geq g_2^{\mathcal{I}}$ without explicitly using a "$\geq$" predicate, and such information must be passed to the concrete domain reasoner. Note that the concrete domain Q presented in Section 2 satisfies Property 1 of admissibility—in Section 4, we will see that Property 2 is also satisfied. The result obtained in [11] can now be formulated as follows:

THEOREM 6 (Baader, Hanschke) *Pure $\mathcal{ALC}(\mathcal{D})$-concept satisfiability and subsumption are decidable if $\mathcal{D}$ is admissible.*

We should briefly comment on a minor difference between the logic $\mathcal{ALC}(\mathcal{D})$ as defined in [11] and in Section 2: Baader and Hanschke's variant of $\mathcal{ALC}(\mathcal{D})$ uses only a single type of feature that is interpreted in partial functions from $\Delta_{\mathcal{I}}$ to $\Delta_{\mathcal{I}} \times \Delta_{\mathcal{D}}$ and thus combines our abstract and concrete features. It is not very hard to see that the difference in expressivity is negligible. However, the separation of abstract and concrete features necessitates the presence of the $g\uparrow$ constructor: without this constructor, we would not be able to remove negations in front of the concrete domain constructor when converting $\mathcal{ALC}(\mathcal{D})$-concepts into equivalent ones in negation normal form (NNF), for details see Section 3.3.[1]

The complexity of reasoning with $\mathcal{ALC}(\mathcal{D})$ has been analyzed in [45]. There, the tableau algorithm of Baader and Hanschke is refined by using the so-called *tracing technique*: instead of keeping entire tableaux in memory (which may become exponentially large), a tree-shaped tableau is constructed in a depth-first manner keeping only paths of the tree in memory. Since such paths are of at most polynomial length, this allows to devise a PSPACE algorithm. However, the complexity of reasoning with $\mathcal{ALC}(\mathcal{D})$ clearly depends on the complexity of $\mathcal{D}$-satisfiability:

THEOREM 7 *Pure $\mathcal{ALC}(\mathcal{D})$-concept satisfiability and subsumption are* PSPACE-*complete if $\mathcal{D}$ is admissible and $\mathcal{D}$-satisfiability is in* PSPACE.

Thus, reasoning with $\mathcal{ALC}(\mathcal{D})$ is not harder than reasoning with $\mathcal{ALC}$. As we will see in Section 4, Q-satisfiability can be decided in PTIME, and thus Theorem 7 yields a tight complexity bound for (pure) reasoning with $\mathcal{ALC}(\mathsf{Q})$.

---

[1] A concept is in NNF if negation does only occur in front of concept names. This normal form is frequently used to devise decision procedures for DLs.

3.2 Standard extensions

We now discuss the extension of $\mathcal{ALC}(\mathcal{D})$ with several means of expressivity that can be considered "standard" in the area of description logics. Let us start with general TBoxes.

General TBoxes

In [12], it is proved that $\mathcal{ALC}(\mathsf{R})$ extended with a transitive closure constructor on roles (similar to the star-operator of propositional dynamic logic) is undecidable, where R is a concrete domain based on Tarski algebra. The undecidability proof, which uses a reduction of the Post Correspondence Problem (PCP), can easily be adapted to $\mathcal{ALC}(\mathsf{R})$ extended with general TBoxes, which is thus also undecidable. This adaption is performed in [41; 44], where not only R is considered, but a more general result is obtained that applies to a large class of concrete domains.

THEOREM 8 *For concrete domains $\mathcal{D}$ such that (i) $\mathbb{N} \subseteq \Delta_\mathcal{D}$ and (ii) $\Phi_\mathcal{D}$ provides a unary predicate for equality with 0, a binary equality predicate, and a binary predicate for incrementation, $\mathcal{ALC}(\mathcal{D})$-concept satisfiability and subsumption w.r.t. general TBoxes are undecidable.*

Note that there exist rather simple (and admissible) concrete domains satisfying the conditions listed in the theorem, an example being the concrete domain Q.[2] Since Q-satisfiability can be decided in PTIME (Section 4), it should be clear that the reason for undecidability is an interaction between general TBoxes and concrete domains and *not* reasoning with arithmetic concrete domains themselves.

Since general TBoxes play a very important role in most application areas and are provided by almost all state-of-the-art description logics, the above result is rather discouraging. There are two ways for regaining decidability: either use a less powerful concrete domain constructor or very carefully choose the concrete domains used.

The first approach was adopted in [28] and [35]. In the former article, Haarslev et al. propose to allow only concrete features inside the concrete domain constructor instead of paths of arbitrary length. In the following, we call concepts satisfying this condition *path-free*. More precisely, Haarslev et al. introduce the rather powerful description logic $\mathcal{SHN}(\mathcal{D})$[3], which extends $\mathcal{ALC}(\mathcal{D})$ with expressive means such as unqualified number restrictions (a weak form of graded modalities) and role hierarchies (TBox-like assertions

---

[2] Strictly speaking, Q does not contain a predicate for addition with 1, but this is compensated by the predicates "$=_1$" and $+$.
[3] This logic is also called $\mathcal{ALCNH}_{R+}(\mathcal{D})$.

that allow to state inclusions between roles). If path-freeness is not assumed, then $\mathcal{SHN}(\mathcal{D})$-concept satisfiability and subsumption is undecidable since reasoning with general TBoxes can be reduced to reasoning without general TBoxes—the so-called "internalization of TBoxes", c.f. [55; 34]. However, using a tableau algorithm Haarslev et al. [28] were able to show the following:

THEOREM 9 (Haarslev et al.) *If the concrete domain $\mathcal{D}$ is admissible, then path-free $\mathcal{SHN}(\mathcal{D})$-concept satisfiability and subsumption w.r.t. general TBoxes are decidable.*

Horrocks and Sattler [35] propose to admit only unary concrete domain predicates to overcome undecidability. Under this restriction, they prove decidability of reasoning with the very expressive description logic $\mathcal{SHOQ}(\mathcal{D})$ and general TBoxes by devising an appropriate tableau algorithm. However, allowing only unary predicates is strictly less expressive than requiring path-freeness: the concept $\exists f_1 \cdots f_k g.P$ (with P unary predicate) can clearly be replaced with the equivalent one $\exists f_1.\exists f_2.\cdots.\exists f_k.\exists g.P$ that does not use paths of length greater than one. In [51], the initial result is strengthened by admitting concrete domain predicates of arbitrary arity, adopting path-freeness, and adding some additional means of expressivity (see Section 3.3). The resulting DL is called $\mathcal{SHOQ}(\mathcal{D}_n)$.

THEOREM 10 (Horrocks, Pan, Sattler) *If the concrete domain $\mathcal{D}$ is admissible, then path-free $\mathcal{SHOQ}(\mathcal{D}_n)$-concept satisfiability and subsumption w.r.t. general TBoxes are decidable.*

A more general result has been obtained in Section 5.3 of [9], where it is shown that any description logic $\mathcal{L}$, such that (i) $\mathcal{L}$-concept satisfiability w.r.t. general TBoxes is decidable and (ii) $\mathcal{L}$ is "closed under disjoint unions" (see [9] for details), can be extended with the path-free variant of the concrete domain constructor without losing decidability of reasoning with general TBoxes. This result generalizes Theorem 9 but not Theorem 10 since $\mathcal{SHOQ}$ does not satisfy Property (ii). Indeed, the "harmlessness" of the path-free concrete domain constructor is not very surprising since dropping paths deprives concrete domains of most of their expressive power: in Section 2.4.1 of [43], it is shown that the path-free variant of the concrete domain constructor can be "simulated" by concept names, which is not possible for the variant admitting paths of arbitrary length. In the same section, it is proved that path-free $\mathcal{ALC}(\mathcal{D})$-concept satisfiability and subsumption w.r.t. general TBoxes are ExpTime-complete if $\mathcal{D}$ is admissible and $\mathcal{D}$-satisfiability is in ExpTime.

We now discuss the second approach to overcome undecidability of $\mathcal{ALC}(\mathcal{D})$ with general TBoxes, namely to keep the original version of the concrete domain constructor and look for concrete domains that are both interesting and do not destroy decidability of reasoning with general TBoxes. The first positive result following this route was established in [40], where a concrete domain C is considered that is based on the rational numbers $\mathbb{Q} = \Delta_\mathsf{C}$, and provides for the binary predicates $<, \leq, =, \neq, \geq, >$ with the obvious extension. Using an automata-based approach, the following result is obtained:

THEOREM 11 $\mathcal{ALC}(\mathsf{C})$-*concept satisfiability and subsumption w.r.t. general TBoxes are* EXPTIME-*complete.*

It is then shown that this result can be extended to an interval-based, temporal concrete domain. Theorem 11 has subsequently been generalized in [42]: first, the concrete domain C has been extended to C^+ which, additionally, admits unary predicates $=_q$ for each $q \in \mathbb{Q}$ (with the obvious extension). Second, the "description logic part" is extended from $\mathcal{ALC}$ to the very expressive DL $\mathcal{SHIQ}$ that plays an important role in many application areas [31]. The following theorem is proved in [42], also using an automata-theoretic approach:

THEOREM 12 $\mathcal{SHIQ}(\mathsf{C}^+)$-*concept satisfiability and subsumption w.r.t. general TBoxes are* EXPTIME-*complete.*

Note that this logic is called $\mathbb{Q}$-$\mathcal{SHIQ}$ in [42]. It is very unlikely that C^+ can be extended with any form of arithmetics without losing decidability. For example, if $\mathcal{ALC}(\mathsf{C}^+)$ is extended with a binary predicate for incrementation with one, we obtain undecidability of reasoning w.r.t. general TBoxes from Theorem 8. An interesting open question is whether a unary predicate int can be added whose extension are the integers. Such a predicate would be very useful for many applications.

Acyclic TBoxes

If we restrict ourselves to acyclic TBoxes rather than admitting general ones, the situation becomes much simpler: it is well-known that concept satisfiability w.r.t. acyclic TBoxes can be reduced to concept satisfiability without TBoxes by using *unfolding* [50]: given an input concept C and an acyclic TBox $\mathcal{T}$, we can exhaustively replace concept names in C that appear on the left-hand side of a concept definition in $\mathcal{T}$ with the corresponding right-hand side. This process terminates since $\mathcal{T}$ is acyclic. Moreover, it is not hard to see that the resulting concept is satisfiable iff C is satisfiable w.r.t.

$\mathcal{T}$. Thus, Theorem 6 implies that $\mathcal{ALC}(\mathcal{D})$-concept satisfiability and subsumption w.r.t. acyclic TBoxes are decidable if $\mathcal{D}$ is admissible—although unfolding involves an exponential blow-up in size.

Concerning complexity, the results obtained for reasoning with $\mathcal{ALC}(\mathcal{D})$ and acyclic TBoxes are much more surprising: it is well-known that, for almost all description logics considered in the literature, adding acyclic TBoxes does not increase the complexity of reasoning. For example, $\mathcal{ALC}$-concept satisfiability and subsumption are PSPACE-complete, both with and without acyclic TBoxes [57; 39]. Interestingly, this is not the case for $\mathcal{ALC}(\mathcal{D})$: although pure $\mathcal{ALC}(\mathcal{D})$-concept satisfiability is PSPACE-complete, in [41; 44] a large class of so-called *arithmetic* concrete domains $\mathcal{D}$ is identified for which $\mathcal{ALC}(\mathcal{D})$-concept satisfiability w.r.t. acyclic TBoxes is considerably harder, namely NEXPTIME-complete.

DEFINITION 13 (Arithmetic) A concrete domain $\mathcal{D}$ is called *arithmetic* iff $\Delta_\mathcal{D}$ contains the natural numbers and $\Phi_\mathcal{D}$ contains

- unary predicates for equality with zero and with one,
- a binary equality predicate, and
- ternary predicates expressing addition and multiplication.

A NEXPTIME-complete variant of the Post Correspondence Problem is used to show the following result:

THEOREM 14 *For any arithmetic concrete domain $\mathcal{D}$, $\mathcal{ALC}(\mathcal{D})$-concept satisfiability w.r.t. acyclic TBoxes is* NEXPTIME-*hard.*

Since concept satisfiability can be reduced to *non*-subsumption, this implies a co-NEXPTIME lower bound for $\mathcal{ALC}(\mathcal{D})$-concept subsumption if $\mathcal{D}$ is arithmetic. A corresponding upper bound is established using a tableau algorithm:

THEOREM 15 *$\mathcal{ALC}(\mathcal{D})$-concept satisfiability w.r.t. acyclic TBoxes is in* NEXPTIME *if $\mathcal{D}$ is admissible and $\mathcal{D}$-satisfiability is in* NP.

Again, we have to consider the complementary complexity class for subsumption. It is interesting that the addition of the seemingly harmless acyclic TBoxes results in a leap of complexity from PSPACE-completeness to NEXPTIME-completeness.

Concept- and role-constructors

Interestingly, acyclic TBoxes are not the only means of expressivity that considerably increases the complexity of reasoning if added to $\mathcal{ALC}(\mathcal{D})$. In [41; 44; 4], analogues of Theorems 14 and 15 have been proved for the following extensions of $\mathcal{ALC}(\mathcal{D})$:

- Inverse roles. We can now additionally use expressions R^- inside the $\exists R.C$ and $\forall R.C$ constructors, where R may also be an abstract feature. The interpretation $(R^-)^\mathcal{I}$ of R^- is obtained by taking the converse of the relation $R^\mathcal{I}$. Inverses of abstract (or even concrete) features inside the concrete domain constructor are *not* allowed since the inverse of a feature is not necessarily functional.

- Role conjunction. We admit roles like $R_1 \sqcap \cdots \sqcap R_n$ inside the $\exists R.C$ and $\forall R.C$ constructors, where the R_i may also be abstract features. The interpretation $(R_1 \sqcap \cdots \sqcap R_n)^\mathcal{I}$ of $R^\mathcal{I}$ is obtained by taking the intersection of the relations $R_1^\mathcal{I}, \ldots, R_n^\mathcal{I}$. Conjunctions of abstract (or even concrete) features inside the concrete domain constructor are *not* allowed.

- Nominals. Nominals (known, e.g., from hybrid logic [3]) are a new syntactic type that is used in the same way as concept names, but interpreted in *singleton* sets.

All these means of expressivity (with the possible exception of nominals) are usually considered "harmless" w.r.t. complexity, i.e., in most cases they do not increase the complexity of reasoning when added to a description logic. The above results thus show that the PSPACE upper complexity bound for reasoning with $\mathcal{ALC}(\mathcal{D})$ is *not robust*, but rather quite unstable w.r.t. extensions of the language.

Although formal proofs are missing, most other standard means of expressivity are very likely to preserve decidability and the PSPACE upper bound when added to $\mathcal{ALC}(\mathcal{D})$. Such means of expressivity are, e.g., qualified number restrictions (the DL counterpart of graded modalities) and transitive roles (i.e., a new sort of role names interpreted in transitive relations—not to be confused with the transitive closure role constructor). Concerning decidability, some results can be obtained by using the transfer results for fusions of description logics presented in [9]. This is to some extent discussed in Section 5.6 of [43], where the following result is obtained:

THEOREM 16 *If $\mathcal{D}$ is admissible, then pure $\mathcal{ALCQ}_{R+}^-(\mathcal{D})$-concept satisfiability is decidable.*

Here, $\mathcal{ALCQ}^-_{R+}(\mathcal{D})$ is $\mathcal{ALC}(\mathcal{D})$ extended with inverse roles, qualifying number restrictions, and transitive roles. It should also be noted that concrete domains can be combined with so-called feature agreements and disagreements without spoiling the PSPACE upper complexity bound [45].

3.3 Concrete domain-related extensions

We now review various proposals for enhancing the expressive power of $\mathcal{ALC}(\mathcal{D})$ by extending the "concrete domain part" of this logic.

Generalized concrete domain constructor

In the original version of $\mathcal{ALC}(\mathcal{D})$ as defined in Section 2, we only allow abstract features to be used in the concrete domain concept constructor instead of admitting arbitrary role names. This observation leads to a natural generalization of the concrete domain constructor that has first been proposed by Hanschke [30].

DEFINITION 17 ($\mathcal{ALCP}(\mathcal{D})$) A sequence $U = R_1 \cdots R_k g$ where $R_1, \ldots, R_k \in \mathsf{N_R}$ ($k \geq 0$) and $g \in \mathsf{N_{cF}}$ is called a *role path*. For an interpretation $\mathcal{I}$, $U^{\mathcal{I}}$ is defined as

$$\{(d, x) \subseteq \Delta_{\mathcal{I}} \times \Delta_{\mathcal{D}} \mid \exists d_1, \ldots, d_{k+1} : d = d_1,$$
$$(d_i, d_{i+1}) \in R_i^{\mathcal{I}} \text{ for } 1 \leq i \leq k, \text{ and } g^{\mathcal{I}}(d_{k+1}) = x\}.$$

$\mathcal{ALCP}(\mathcal{D})$ is obtained from $\mathcal{ALC}(\mathcal{D})$ by allowing the use of concepts of the form $\forall U_1, \ldots, U_n.P$ and $\exists U_1, \ldots, U_n.P$ in place of concept names, where $P \in \Phi_{\mathcal{D}}$ is of arity n and $U_1, \ldots, U_n$ are role paths. The semantics of the generalized concrete domain constructors is defined as follows:

$$(\forall U_1, \ldots, U_n.P)^{\mathcal{I}} := \{d \in \Delta_{\mathcal{I}} \mid \text{ For all } x_1, \ldots, x_n \text{ with } (d, x_i) \in U_i^{\mathcal{I}},$$
$$\text{we have } (x_1, \ldots, x_n) \in P^{\mathcal{D}}\}$$

$$(\exists U_1, \ldots, U_n.P)^{\mathcal{I}} := \{d \in \Delta_{\mathcal{I}} \mid \text{ There exist } x_1, \ldots, x_n \text{ with } (d, x_i) \in U_i^{\mathcal{I}}$$
$$\text{and } (x_1, \ldots, x_n) \in P^{\mathcal{D}}\}$$

Obviously, every path is also a role path. Hence, the $\exists U_1, \ldots, U_n.P$ constructor of $\mathcal{ALCP}(\mathcal{D})$ is a generalization of the $\exists u_1, \ldots, u_n.P$ constructor of $\mathcal{ALC}(\mathcal{D})$. For paths $u_1, \ldots, u_n$, the $\mathcal{ALCP}(\mathcal{D})$-concept $\forall u_1, \ldots, u_n.P$ is equivalent to the $\mathcal{ALC}(\mathcal{D})$-concept $u_1\uparrow \sqcup \cdots \sqcup u_n\uparrow \sqcup \exists u_1, \ldots, u_n.P$. This is the reason why $\mathcal{ALC}(\mathcal{D})$ does not provide for a counterpart of the $\forall U_1, \ldots, U_n.P$ constructor.

Using the generalized constructors, we can, for example, express that the duration of subprocesses is bounded by the duration of the mother process *without committing to a particular number of subprocesses*:

Process $\sqcap \forall$(duration), (subproc duration).$\leq$, where **duration** is a concrete feature. The existential version of the generalized concrete domain constructor can then be used to express that there exists a subprocess whose duration is strictly shorter than the duration of the mother process:

$$\text{Process} \sqcap \exists(\text{duration}), (\text{subproc duration}).{<}. \qquad (*)$$

Note, however, that it is now impossible to state that the subprocess with the shorter duration is a **DangerousProcess**. This observation suggests that role hierarchies are a useful extension of $\mathcal{ALCP(D)}$: in the resulting DL, we can modify $(*)$ by replacing the role **subproc** with an abstract feature **specialSubprocess**, adding the conjunct $\exists \text{specialSubprocess}.\text{DangerousProcess}$, and finally using a role hierarchy to state that **specialSubprocess** is a subrole of **subproc**, i.e. that we have $\text{specialSubProcess}^{\mathcal{I}} \subseteq \text{subproc}^{\mathcal{I}}$. In the following, however, we will stick with the original variant of $\mathcal{ALCP(D)}$ that does not admit role hierarchies.

As shown in [30], satisfiability and subsumption of $\mathcal{ALCP(D)}$-concepts are decidable if $\mathcal{D}$ is admissible. However, when investigating the complexity of $\mathcal{ALCP(D)}$, it becomes clear that initially restricting ourselves to abstract features inside the concrete domain constructor is a sensible idea since it allows a more fine-grained complexity analysis: it is shown in [43] that, while reasoning with $\mathcal{ALC(D)}$ is PSPACE-complete, reasoning with $\mathcal{ALCP(D)}$ is much harder. Indeed, the complexity of (pure) reasoning with $\mathcal{ALCP(D)}$ parallels the complexity of reasoning with $\mathcal{ALC(D)}$ extended with acyclic TBoxes. The lower bound is determined by reduction of a NExpTime-complete variant of the PCP:

THEOREM 18 *For any arithmetic concrete domain $\mathcal{D}$, pure $\mathcal{ALCP(D)}$-concept satisfiability is NExpTime-hard.*

It is interesting to note that this lower bound does even hold if abstract features are dropped from the language. As in the case of acyclic TBoxes, there exists a corresponding upper bound which is established using a tableau algorithm:

THEOREM 19 *Pure $\mathcal{ALCP(D)}$-concept satisfiability is in NExpTime if $\mathcal{D}$ is admissible and $\mathcal{D}$-satisfiability is in NP.*

We obtain corresponding co-NExpTime complexity bounds for concept subsumption. Another generalization of the concrete domain constructor has been proposed in [51]: the authors replace concrete features by concrete roles, which are not required to be functional. Additionally, they allow the application of number restrictions to concrete roles. This allows, for example, to state that each person has exactly one age (attached via a concrete role age) while being allowed to have many telephone numbers (attached via a concrete role tel).

A concrete domain role constructor

Another natural extension of the original variant of $\mathcal{ALC}(\mathcal{D})$ is obtained by using the concrete domain not only to define concepts, but by additionally allowing the definition of complex roles with reference to concrete domain predicates. Such an extension has first been proposed in [27].

DEFINITION 20 ($\mathcal{ALCRP}(\mathcal{D})$) A *concrete domain role* is an expression of the form
$$\exists (u_1, \ldots, u_n), (v_1, \ldots, v_m).P$$
where $u_1, \ldots, u_n$ and $v_1, \ldots, v_m$ are paths and P is an $n+m$-ary predicate. The semantics is given as follows:

$(\exists(u_1, \ldots, u_n), (v_1, \ldots, v_m).P)^{\mathcal{I}} :=$
    $\{(d, e) \in \Delta_{\mathcal{I}} \times \Delta_{\mathcal{I}} \mid$ There exist $x_1, \ldots, x_n$ and $y_1, \ldots, y_m$
    such that $u_i^{\mathcal{I}}(d) = x_i$ for $1 \leq i \leq n$, $v_i^{\mathcal{I}}(e) = y_i$ for $1 \leq i \leq m$, and
    $(x_1, \ldots, x_n, y_1, \ldots, y_m) \in P^{\mathcal{D}}\}$

$\mathcal{ALCRP}(\mathcal{D})$ is obtained from $\mathcal{ALC}(\mathcal{D})$ by allowing the use of concrete domain roles inside the $\exists R.C$ and $\forall R.C$ constructors.

Note that concrete domain roles are *not* allowed inside the concrete domain concept constructor. Let us view an example $\mathcal{ALCRP}(\mathcal{D})$-concept. Assume that we use a concrete domain based on temporal intervals and binary predicates describing the possible relationships between such intervals. Then the concept

Process $\sqcap \forall (\exists(\text{exectime}), (\text{exectime}).\text{overlaps}).\neg\text{DangerousProcess}$,

describes processes that are not temporally overlapping with dangerous processes. Note that $\exists(\text{exectime}), (\text{exectime}).\text{overlaps}$ is a concrete domain role defined in terms of the binary predicate overlaps and the concrete feature exectime which associates processes with the time interval in which they are executed.

In [47], a reduction of the Post Correspondence Problem is used to prove that there exist concrete domains $\mathcal{D}$ such that the satisfiability of $\mathcal{ALCRP}(\mathcal{D})$-concepts is undecidable. It is straightforward to generalize this result to the class of concrete domains identified in Theorem 8:

THEOREM 21 (Lutz, Möller) *For concrete domains $\mathcal{D}$ such that (i) $\mathbb{N} \subseteq \Delta_\mathcal{D}$ and (ii) $\Phi_\mathcal{D}$ provides a unary predicate for equality with 0, a binary equality predicate, and a binary predicate for incrementation, pure $\mathcal{ALCRP}(\mathcal{D})$-concept satisfiability and subsumption are undecidable.*

In [27] a fragment of $\mathcal{ALCRP}(\mathcal{D})$ is identified that is closed under negation, strictly extends $\mathcal{ALC}(\mathcal{D})$, and is decidable for all admissible concrete domains. To introduce this fragment, we need a way to convert $\mathcal{ALCRP}(\mathcal{D})$-concepts into equivalent ones in NNF. Assuming that $\mathcal{D}$ is admissible, this conversion can be done by eliminating double negation and using de Morgan's rules, the duality between $\exists R.C$ and $\forall R.C$, and the equivalences

$$\neg(\exists u_1, \ldots, u_n.P) \equiv \exists u_1, \ldots, u_n.\overline{P} \sqcup u_1\uparrow \sqcup \cdots \sqcup u_n\uparrow$$
$$\neg(g\uparrow) \equiv \exists g.\top_\mathcal{D}$$

where, for P an n-ary predicate, $\overline{P}$ denotes the predicate satisfying $\overline{P}^\mathcal{D} = \Delta_\mathcal{D}^n \setminus P^\mathcal{D}$, which exists since $\mathcal{D}$ is admissible. In the following, $\mathsf{sub}(C)$ refers to the set of subconcepts of the concept C (including C itself).

DEFINITION 22 (Restricted $\mathcal{ALCRP}(\mathcal{D})$-concept) An $\mathcal{ALCRP}(\mathcal{D})$-concept C is called *restricted* iff the result C' of converting C to NNF satisfies the following conditions:

1. For any $\forall R.D \in \mathsf{sub}(C')$, where R is a concrete domain role,

   (a) $\mathsf{sub}(D)$ does not contain any concepts $\exists S.E$ with S a concrete domain role, and

   (b) if $\mathsf{sub}(D)$ contains a concept $\exists u_1, \ldots, u_n.P$, then $u_1, \ldots, u_n \in \mathsf{N_{cF}}$.

2. For any $\exists R.D \in \mathsf{sub}(C')$, where R is a concrete domain role,

   (a) $\mathsf{sub}(D)$ does not contain any concepts $\forall S.E$ with S a concrete domain role, and

   (b) if $\mathsf{sub}(D)$ contains a concept $\exists u_1, \ldots, u_n.P$, then $u_1, \ldots, u_n \in \mathsf{N_{cF}}$.

It is easily seen that the set of restricted $\mathcal{ALCRP}(\mathcal{D})$-concepts is closed under negation. Hence, subsumption of restricted $\mathcal{ALCRP}(\mathcal{D})$-concepts can still be reduced to satisfiability of restricted $\mathcal{ALCRP}(\mathcal{D})$-concepts (and vice versa). Decidability of restricted $\mathcal{ALCRP}(\mathcal{D})$-concept satisfiability and subsumption has been shown in [27], where it is also illustrated that this fragment of $\mathcal{ALCRP}(\mathcal{D})$ is still useful for reasoning about spatio-terminological knowledge. The complexity of reasoning has been investigated in [43], where it is shown that, once more, we can use a NExpTime-complete variant of the PCP and a tableau algorithm to prove the following:

THEOREM 23 *Let $\mathcal{D}$ be a concrete domain. If $\mathcal{D}$ is arithmetic, then (pure) satisfiability of restricted $\mathcal{ALCP}(\mathcal{D})$-concepts is NExpTime-hard. If $\mathcal{D}$ is admissible and $\mathcal{D}$-satisfiability is in NP, then (pure) satisfiability of restricted $\mathcal{ALCRP}(\mathcal{D})$-concepts can be decided in NExpTime.*

Again, we obtain corresponding co-NExpTime bounds for concept subsumption.

Aggregation functions

Aggregation is a useful mechanism available in many expressive representation formalisms such as database schema and query languages. It is thus a natural idea to extend description logics providing for concrete domains with aggregation as proposed in [13]. Consider, for example, a process description

Process ⊓ ∃duration.$>_0$ ⊓ ∀subproc.(Process ⊓ ∃duration.$>_0$).

The aggregation function "sum" is needed if we want to express that the duration of the mother process is identical to the sum of the durations of all its subprocesses (of which there may be arbitrarily many).

DEFINITION 24 (Aggregation) A *concrete domain with aggregation* is a concrete domain that, additionally, provides for a set of aggregation functions agg($\mathcal{D}$), where each $\Gamma \in$ agg($\mathcal{D}$) is associated with a partial function $\Gamma^{\mathcal{D}}$ from the set of finite multisets of dom($\mathcal{D}$) into dom($\mathcal{D}$).[4]

To distinguish concrete domains with aggregation from those without, we denote the former with Σ. Typical aggregation functions are min, max, sum, count, and average (with the obvious extensions).

$\mathcal{ALC}(\Sigma)$-concepts are now defined in the same way as $\mathcal{ALC}(\mathcal{D})$-concepts except that *aggregated features* may be substituted for concrete features, where an aggregated feature is an expression $\Gamma(R \circ g)$ with R role, g concrete feature, and Γ an aggregation function from Σ. The semantics of aggregated features is defined via multisets:

DEFINITION 25 (Semantics of $\mathcal{ALC}(\Sigma)$) Let $\mathcal{I}$ be an interpretation. For each $d \in \Delta_{\mathcal{I}}$ such that the set $\{e \mid (d,e) \in R^{\mathcal{I}}\}$ is finite, we use $M_d^{R \circ g}$ to

---

[4]Intuitively, a multiset is a set that may contain the same element multiple (but only finitely many) times.

denote the multiset that, for each $z \in \Delta_\mathcal{D}$, contains z exactly $|\{e \mid (d,e) \in R^\mathcal{I}$ and $g^\mathcal{I}(e) = z\}|$ times. The semantics of aggregated features is now defined as follows:

$$\Gamma(R \circ g))^\mathcal{I}(d) := \begin{cases} \Gamma^\Sigma(M_d^{R \circ g}) & \text{if } \{e \mid (d,e) \in R^\mathcal{I}\} \text{ is finite} \\ \text{undefined} & \text{otherwise.} \end{cases}$$

Returning to the initial example, we can now express the fact that the duration of the mother process is identical to the sum of the durations of all its subprocesses by writing

$$\exists \mathsf{duration}, \mathsf{sum}(\mathsf{subproc} \circ \mathsf{duration}).=.$$

The investigations performed by Baader and Sattler [13] reveal that the expressive power provided by aggregation functions is hard to tame. The following result is proved by a reduction of Hilbert's 10-th problem.

THEOREM 26 (Baader, Sattler) *For concrete domains with aggregation Σ where (i) $\mathrm{dom}(\Sigma)$ includes the non-negative integers, (ii) Φ_Σ contains a (unary) predicate for equality with 1 and a (binary) equality predicate, and (iii) $\mathrm{agg}(\Sigma)$ contains* min, max, *and* sum, *pure $\mathcal{ALC}(\Sigma)$-concept satisfiability and subsumption are undecidable.*

This lower bound does even apply if we admit only conjunction, the $\forall R.C$ constructor, and the concrete domain constructor, but drop all other concept constructors. Rather strong measures have to be taken in order to regain decidability: either, we have to drop the $\forall R.C$ constructor from the language or we have to confine ourselves with "well-behaved" aggregation functions. Following the first approach, one may replace the logic $\mathcal{ALC}(\Sigma)$ with the DL $\mathcal{EL}(\Sigma)$ that only provides for the following concept constructors: atomic negation (i.e. restricted to concept names), conjunction, disjunction, the $\exists R.C$ constructor, and the concrete domain constructor. When devising decision procedures for $\mathcal{EL}(\Sigma)$, requiring concrete domains to be admissible is no longer sufficient since the multisets underlying aggregation functions need to be dealt with: Σ-conjunctions may, additionally, contain multiset variables and inclusion statements between multisets (for a precise definition see [13]). If the satisfiability of such extended Σ-conjunctions is decidable, we call Σ *aggregation-admissible*. Baader and Sattler [13] prove the following result by devising a tableau algorithm:

THEOREM 27 (Baader, Sattler) *For concrete domains with aggregation Σ that are aggregation-admissible, pure $\mathcal{EL}(\Sigma)$-concept satisfiability is decidable.*

However, subsumption of $\mathcal{EL}(\Sigma)$-concepts is, in general, still undecidable. Following the second approach, Baader and Sattler found out that only min and max can be considered well-behaved, obtaining the following result also by construction of a tableau algorithm:

THEOREM 28 (Baader, Sattler) *For concrete domains with aggregation Σ such that (i) Σ is admissible, (ii) Φ_Σ contains a binary equality predicate and a binary predicate for a linear ordering on Δ_Σ, and (iii) $\mathsf{agg}(\Sigma) = \{\min, \max\}$, pure $\mathcal{ALC}(\Sigma)$-concept satisfiability and subsumption are decidable.*

Keys

In several applications, it is useful to identify a set of concrete features whose values *uniquely* determine logical objects. Say, for example, that there exists a concrete feature socnum associating humans with their social security number. Then, if a human is American, she should be *uniquely* identified by this number. In other words, there should be no two distinct domain elements that are both in the extension of American and share the same value of the concrete feature socnum. This idea leads to the definition of key boxes, which have been proposed in [2].

DEFINITION 29 (Key box) A *key box* is a finite set of *key definitions*

$$(u_1, \ldots, u_n \text{ keyfor } C),$$

where $u_1, \ldots, u_n$ are paths and C is a concept. An interpretation $\mathcal{I}$ *satisfies* a key definition $(u_1, \ldots, u_n \text{ keyfor } C)$ iff, for any $a, b \in C^\mathcal{I}$,

$$u_i^\mathcal{I}(a) = u_i^\mathcal{I}(b) \text{ for } 1 \leq i \leq n \text{ implies } a = b.$$

$\mathcal{I}$ is a *model* of a key box $\mathcal{K}$ iff $\mathcal{I}$ satisfies all key definitions in $\mathcal{K}$.

Clearly, key boxes are a natural choice in database applications such as the one described in Section 1: they correspond to so-called functional dependencies which are the most popular type of constraint for relational databases. For this reason, keys for description logics have also been considered in a non-concrete domain context [17; 20; 36].

From a logical perspective, there exists a close relationship between nominals and key boxes. For example, if used together with the key definition $(g \text{ keyfor } \top)$, then the $\mathcal{ALC}(\mathsf{Q})$-concept $\exists g.=_q$ "behaves" like a nominal for each $q \in \mathbb{Q}$: it is interpreted either in the empty set or in a singleton set.

Indeed, key boxes are a quite powerful expressive means. This is reflected by the computational complexity of $\mathcal{ALCK}(\mathcal{D})$, the extension of $\mathcal{ALC}(\mathcal{D})$

with key boxes, which is investigated in [2]. The following undecidability result is proved by a reduction of the PCP:

THEOREM 30 (Areces et al.) *For any arithmetic concrete domain $\mathcal{D}$, pure $\mathcal{ALCK}(\mathcal{D})$-concept satisfiability and subsumption w.r.t. key boxes are undecidable.*

Decidability can be regained by allowing only Boolean combinations of concept names on the right-hand side of key definitions. Key boxes satisfying this property are called *Boolean*. Pure $\mathcal{ALCK}(\mathcal{D})$-concept satisfiability and subsumption w.r.t. Boolean key boxes are NExpTime-hard for arithmetic concrete domains $\mathcal{D}$. Surprisingly, [2] can even show that this high complexity cannot be reduced if paths are restricted to length one inside $\mathcal{ALCK}(\mathcal{D})$-concepts and key boxes. In analogy to Section 3.2, where this approach helped to overcome undecidability in the presence of general TBoxes, we call such concepts and key boxes *path-free*. The following theorem is proved by reduction of a NExpTime-complete variant of the PCP:

THEOREM 31 (Areces et al.) *For any arithmetic concrete domain $\mathcal{D}$, pure path-free $\mathcal{ALCK}(\mathcal{D})$-concept satisfiability and subsumption w.r.t. Boolean and path-free key boxes are NExpTime-hard.*

To devise a decision procedure for reasoning with key boxes, it does not suffice to assume admissibility of concrete domains: the concrete domain reasoner should not only tell us whether a given $\mathcal{D}$-conjunction is satisfiable, but also which variables in it must take the same value in solutions.

DEFINITION 32 (Key-admissible) A concrete domain $\mathcal{D}$ is called *key-admissible* iff there exists an algorithm that takes as input a $\mathcal{D}$-conjunction c, returns clash if c is unsatisfiable, and otherwise non-deterministically outputs an equivalence relation $\sim$ on the set of variables V used in c such that there exists a solution δ for c with the following property:

$$\delta(v) = \delta(v') \text{ iff } v \sim v' \text{ for all } v, v' \in V.$$

We say that *extended $\mathcal{D}$-satisfiability is in NP* if there exists an algorithm as above running in polynomial time.

This property is much less esoteric than it seems: as noted in [2], any concrete domain that is admissible and provides for an equality predicate is also key-admissible. This rather weak condition is satisfied by almost all (admissible) concrete domains proposed in the literature, c.f. Section 4. Using a tableau algorithm, Areces et al. [2] obtain a matching upper bound for Theorem 31:

THEOREM 33 (Areces et al.) *Let $\mathcal{D}$ be a concrete domain that is key-admissible. If extended $\mathcal{D}$-satisfiability is in* NP, *then pure $\mathcal{ALCOK}(\mathcal{D})$-concept satisfiability w.r.t. Boolean key boxes is in* NEXPTIME.

Note that, in contrast to Theorem 31, concepts do not have to be path-free. As usual, corresponding co-NEXPTIME results are obtained for concept subsumption.

Areces et al. also consider the extension of the description logic $\mathcal{SHOQ}(\mathcal{D}_n)$ (see Section 3.2) with key boxes. Since $\mathcal{SHOQ}(\mathcal{D}_n)$ provides only for the path-free variant of the concrete domain constructor, it is natural to require key boxes to also be path-free. Due to the fact that each path-free $\mathcal{ALCK}(\mathcal{D})$-concept is also a path-free $\mathcal{SHOQ}(\mathcal{D}_n)$-concept, Theorem 31 provides us with a lower NEXPTIME complexity bound. In [2], the corresponding upper bound is obtained by devising an appropriate tableau algorithm:

THEOREM 34 (Areces et al.) *Let $\mathcal{D}$ be a concrete domain that is key-admissible. If extended $\mathcal{D}$-satisfiability is in* NP, *then path-free $\mathcal{SHOQ}(\mathcal{D}_n)$-concept satisfiability w.r.t. path-free key boxes is in* NEXPTIME.

Note that the key boxes in Theorem 34 are not required to be Boolean! We obtain a corresponding co-NEXPTIME bound for concept subsumption.

4 Concrete domains

In this section, we discuss several concrete domains that have been proposed in the literature. We start with numerical concrete domains, which are useful in a wide range of application areas, and then consider more specific concrete domains such as temporal and spatial ones.

4.1 Numerical concrete domains

Let us start with reconsidering the concrete domain **Q** introduced in Section 2. It is based on the rational numbers $\mathbb{Q}$ and provides for the following predicates:

- (unary) predicates $<_q, \leq_q, =_q, \neq_q, \geq_q$, and $>_q$ for comparisons with rational numbers q;

- binary comparison predicates $<, \leq, =, \neq, \geq$, and $>$;

- a ternary addition predicate $+$ and its negation $\overline{+}$;

- unary predicates $\top_{\mathbf{Q}}$ and $\bot_{\mathbf{Q}}$ (for admissibility).

Note that we could drop some of the predicates since, e.g., $\exists u.<_7$ can be written as $\exists g.=_7 \sqcap \exists u, g.<$, and $\exists u_1, u_2.\geq$ can be written as $\exists u_1, u_2.= \sqcup$

	Complexity of $\mathcal{D}$-satisfiability	Complexity of (pure) $\mathcal{ALC}(\mathcal{D})$-concept satisfiability
Q + '∗' + 'int'	undecidable	undecidable
Q + '∗'	in EXPTIME	in NEXPTIME
Q + 'int'	NP-complete	PSPACE-complete
Q	in PTIME	PSPACE-complete

Figure 1. Numerical concrete domains and their complexity

$\exists u_2, u_1. <$.[5] It is not very hard to prove that Q-satisfiability is in PTIME using a reduction to linear programming (LP). More precisely, a *linear programming problem* has the form $Ax = b$, where A is an $m \times n$-matrix of rational numbers, x is an n-vector of variables, and b is an m-vector of rational numbers (see, e.g. [58]). A *solution* of $Ax = b$ is a mapping δ that assigns a rational number to each variable such that the equality $Ax = b$ holds. Deciding whether a given LP problem has a solution is well known to be in PTIME [58]. Details on the reduction of Q-satisfiability to linear programming can be found in [45].

There exist several interesting predicates that can be added to Q in order to extend its expressive power. From the viewpoint of many applications, the most useful ones are the following:

- ternary predicates ∗ and $\overline{\ast}$ with $(*)^Q = \{(q, q', q'') \in \mathbb{Q}^3 \mid q \cdot q' = q''\}$ and $(\overline{\ast})^Q = \mathbb{Q}^3 \setminus (*)^Q$;
- unary predicates int and $\overline{\text{int}}$ with $(\text{int})^Q = \mathbb{Z}$ (where $\mathbb{Z}$ denotes the integers) and $(\overline{\text{int}})^Q = \mathbb{Q} \setminus \mathbb{Z}$.

Adding different combinations of these predicates, we obtain three extensions of the concrete domain Q, which are listed in Figure 1, together with known complexity bounds on $\mathcal{D}$-satisfiability and pure $\mathcal{ALC}(\mathcal{D})$-concept satisfiability. Note that, since the obtained concrete domains should be admissible, we assume that the addition of the predicates '∗' and 'int' implies the addition of their negations. Let us discuss the given bounds in some more detail:

- It is easily proved that the concrete domain Q + '∗' + 'int' is undecidable using a reduction of Hilbert's 10-th problem. Clearly, the undecidability is inherited by $\mathcal{ALC}(\mathcal{D})$-concept satisfiability.

---

[5] It is sufficient to provide, for example, the predicates $\{=_q \mid q \in \mathbb{Q}\} \cup \{<, +\}$: all other predicates can be defined in terms of these. The corresponding concrete domain is, however, not closed under negation and thus not admissible.

- The EXPTIME upper bound for Q + '∗' stems from the fact that, for this concrete domains, finite predicate conjunctions can be translated into formulas of Tarski algebra (also known as the theory of real closed fields) without quantifier alternation. The satisfiability of such formulas has been proved to be decidable in EXPTIME [48; 26]. The NEXPTIME upper bound for $\mathcal{ALC}(\mathcal{D})$-concept satisfiability stems from a more general variant of Theorem 7 that is proved in [45].

  In [12], it has even been proposed to use all formulas of Tarski algebra (also those with quantifiers) as concrete domain predicates. For the concrete domain obtained in this way, $\mathcal{D}$-satisfiability is EXPSPACE-complete [48].

- Finally, NP-completeness of the concrete domain Q + 'int' can be shown via mutual reductions to and from mixed integer programming (MIP), i.e., linear programming with an additional type of variables that must take integer values in solutions. Deciding the existence of a solution for MIP problems is known to be NP-complete. More details on the reductions can be found in [45]. The complexity of $\mathcal{ALC}(\mathcal{D})$-concept satisfiability is then obtained from Theorem 7.

Note that some *fragments* of the concrete domain Q are also interesting, examples being the concrete domains C and C^+ discussed in Section 3.2: in contrast to Q itself, they can be combined with general TBoxes without losing decidability.

4.2 Other concrete domains

In this section, we present two examples for non-numerical concrete domains that have been proposed in the literature. The first example is concerned with representing time: since it is a natural idea to take into account temporal aspects when reasoning about conceptual knowledge, many temporal extensions of description logics have been proposed, see e.g. [56; 5; 62] and the survey [6]. As discussed in [45; 43], one possible approach for such an extension is to use an appropriate, temporal concrete domain which we introduce in the following.

In temporal reasoning, one of the most fundamental decisions to be made is whether to use time points or time intervals as the atomic temporal entity. Time points can obviously be represented using numerical concrete domains such as those from Section 4.1. If we choose time intervals as our atomic temporal entity, it seems appropriate to define an interval-based, temporal concrete domain. Usually, such concrete domains are based on the 13 *Allen relations*, which describe the possible relationships between any two intervals over some temporal structure. We refer to [1] for an exact definition

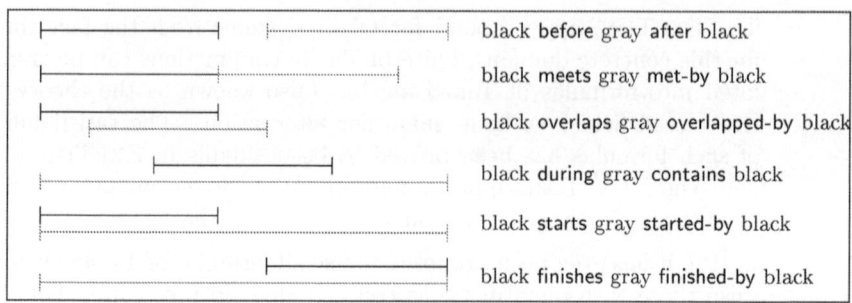

Figure 2. The Allen relations (without **equal**)

and confine ourselves with the graphical presentation of the relations given in Figure 2. The most important property of the Allen relations is that they are jointly exhaustive and pairwise disjoint, i.e., for each temporal structure $(T, \prec)$ and $t_1, t_2 \in T$, there exists exactly one relation r such that $t_1 \, \mathsf{r} \, t_2$. We now define a concrete domain I that is based on the temporal structure $(\mathbb{Q}, <)$:

- $\Delta_\mathrm{I} := \{(t_1, t_2) \mid t_1, t_2 \in \mathbb{Q} \text{ and } t_1 < t_2\}$;
- Φ_I contains unary predicates $\top_\mathrm{I}$ and $\bot_\mathrm{I}$ (with the obvious extension) and binary predicates rel and $\overline{\mathsf{rel}}$ for each Allen relation rel such that $(\mathsf{rel})^\mathrm{I} = \{(i_1, i_2) \in \Delta_\mathrm{I} \times \Delta_\mathrm{I} \mid i_1 \, \mathsf{rel} \, i_2\}$ and $(\overline{\mathsf{rel}})^\mathrm{I} = \Delta_\mathrm{I}^2 \setminus (\mathsf{rel})^\mathrm{I}$.

It is easily verified that I satisfies Property 1 of admissibility. Moreover, it follows from standard results in temporal reasoning that I-satisfiability is NP-complete—more details can be found in Section 2.4.3 of [43]. Thus, from Theorem 7 we obtain that $\mathcal{ALC}(\mathrm{I})$-concept satisfiability is PSPACE-complete.

Interestingly, there exists a polynomial reduction of $\mathcal{ALC}(\mathrm{I})$-concept satisfiability to $\mathcal{ALC}(\mathsf{C})$-concept satisfiability, where C is the concrete domain based on $\mathbb{Q}$ and the binary comparisons $<, \leq, =, \neq, \geq, >$ introduced in Section 3.2: intuitively, it is possible to represent intervals in terms of their endpoints and Allen relations in terms of comparisons between interval endpoints—details can again be found in Section 2.4.3 of [43]. Since this reduction also works in the presence of general TBoxes, Theorem 11 implies that $\mathcal{ALC}(\mathrm{I})$-concept satisfiability w.r.t. general TBoxes is decidable (and EXPTIME-complete). The usefulness of $\mathcal{ALC}(\mathrm{I})$ with general TBoxes for temporal reasoning is illustrated in [40] in a process engineering context. It should be noted that $\mathcal{ALC}(\mathrm{I})$ (without TBoxes) has been used to obtain

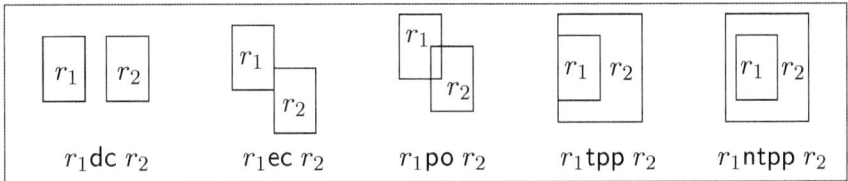

Figure 3. The RCC-8 relations in two-dimensional space

complexity results for an interval-based temporal description logic that is not based on concrete domains [7].

Although spatial aspects are as important for conceptual reasoning as are temporal aspects, until now only rather few spatial description logics have been proposed, see e.g. [27; 37; 38]. This is particularly surprising since, in the spatial case, the number of choices for atomic entities and relations/predicates is much larger than in the temporal case: as spatial primitives, we may use points in a metric, Euclidean, or topological space, sets of such points (to represent regions), or sets of such points with certain characteristics such as connectedness or definability by polytopes. For the predicates, we may for example choose distance relations, orientation relations, or the so-called RCC-8 relations. In the following, we take a closer look at the last possibility since this approach has been used in the only spatial description logic based on concrete domains that has been proposed in the literature [27].

The set of so-called RCC-8 relations is well-known from the area of qualitative spatial reasoning [52; 14; 53]. RCC-8 consists of eight jointly exhaustive and pairwise disjoint relations that describe the possible relationships between any two regular closed regions in a topological space.[6] For 2D space, the relations are illustrated in Figure 3, where the equality relation eq, the inverse tppi of tpp, and the inverse ntppi of ntpp have been omitted. We now define a spatial concrete domain S based on the standard topology of two-dimensional space:

- Δ_S is the set $\mathcal{RC}_{\mathbb{R}^2}$ of all regular closed subsets of $\mathbb{R}^2$;
- Φ_S contains unary predicates $\top_S$ and $\bot_S$ and binary predicates rel and $\overline{\text{rel}}$ for each topological relation rel such that $(\text{rel})^S = \{(r_1, r_2) \in \mathcal{RC}_{\mathbb{R}^2} \times \mathcal{RC}_{\mathbb{R}^2} \mid r_1 \text{ rel } r_2\}$ and $(\overline{\text{rel}})^S = \Delta_S^2 \setminus (\text{rel})^S$..

The concrete domain S obviously satisfies Condition 1 of admissibility. Using standard results from qualitative spatial reasoning, it is straightforward

---

[6]A region r is regular closed if it satisfies $ICr = r$, where C is the topological closure operator and I is the topological interior operator.

to show that S-satisfiability is in NP—details can be found in [45]. Thus, $\mathcal{ALC}(\mathsf{S})$-concept satisfiability is PSPACE-complete by Theorem 7. In [27], the concrete domain S has been used in the description logic $\mathcal{ALCRP}(\mathsf{S})$, i.e. $\mathcal{ALC}(\mathsf{S})$ extended with the concrete domain role constructor from Section 3.3, to reason about spatio-terminological knowledge. By Theorem 23, $\mathcal{ALCRP}(\mathsf{S})$-concept satisfiability is in NEXPTIME—the corresponding lower bound does not apply since S is not arithmetic. It is an interesting open question whether the description logic $\mathcal{ALC}(\mathsf{S})$ can be combined with general TBoxes without losing decidability.

Other sets of relations from the area of qualitative spatial reasoning (see e.g. [59]) may be used to define different spatial concrete domains. Interesting related work has been presented in [37; 38]: the authors propose to combine description logics with modal logics for metric spaces. The expressive power of the resulting spatial description logics seems to be orthogonal to the expressive power of spatial description logics based on concrete domains.

5 Final remarks

In this paper, we have given an overview over the research on description logics with concrete domains, focussing on decidability and complexity results. We have tried to cover most relevant results, but had to drop a few issues due to space limitation. For example, one omission concerns so-called ABoxes which are frequently used to describe states of the world. ABoxes are not an extension of the concept language but rather situated "outside of it", similar to TBoxes (nevertheless, ABoxes are closely related to nominals, though much weaker). It seems that there exists no natural description logic with concrete domains for which reasoning with ABoxes is of a different complexity than reasoning without ABoxes. Some results on the combination of ABoxes and concrete domains can be found in, e.g., [28; 45; 43].

We should like to note that the research on description logics with concrete domains has already led to first reasoning systems that are equipped with concrete domains: the **RACER** system offers a concrete domain based on linear equations and inequalities resembling the concrete domain Q discussed in Section 4.1 [29]. Moreover, there exist plans to extend the **FaCT** system [32] with concrete domains. Since both **RACER** and **FaCT** provide for general TBoxes, they only offer the path-free variant of the concrete domain constructor discussed in Section 3.2. Serious implementations of description logics that provide for the full constructor remain yet to be seen.

BIBLIOGRAPHY

[1] J. Allen. Maintaining knowledge about temporal intervals. *Communications of the ACM*, 26(11), 1983.

[2] C. Areces, I. Horrocks, C. Lutz, and U. Sattler. Keys, nominals, and concrete domains. LTCS-Report 02-04, Technical University Dresden, 2002. To appear.

[3] C. Areces and M. de Rijke. From description logics to hybrid logics, and back. In *Advances in Modal Logics Volume 3*. CSLI Publications, Stanford, CA, USA, 2001.

[4] C. Areces and C. Lutz. Concrete domains and nominals united. In *Proceedings of the fourth Workshop on Hybrid Logics (HyLo'02)*, 2002.

[5] A. Artale and E. Franconi. A temporal description logic for reasoning about actions and plans. *Journal of Artificial Intelligence Research (JAIR)*, 9:463–506, 1998.

[6] A. Artale and E. Franconi. Temporal description logics. In *Handbook of Time and Temporal Reasoning in Artificial Intelligence*. MIT Press, 2001. To appear.

[7] A. Artale and C. Lutz. A correspondence between temporal description logics. In *Proceedings of the International Workshop on Description Logics (DL'99)*, number 22 in CEUR-WS (http://ceur-ws.org/), pages 145–149, 1999.

[8] F. Baader, I. Horrocks, and U. Sattler. Description logics for the semantic web. *KI – Künstliche Intelligenz*, 3, 2002. To appear.

[9] F. Baader, C. Lutz, H. Sturm, and F. Wolter. Fusions of description logics and abstract description systems. *Journal of Artificial Intelligence Research (JAIR)*, 16:1–58, 2002.

[10] F. Baader. Logic-based knowledge representation. In *Artificial Intelligence Today, Recent Trends and Developments*, number 1600 in Lecture Notes in Computer Science, pages 13–41. Springer-Verlag, 1999.

[11] F. Baader and P. Hanschke. A scheme for integrating concrete domains into concept languages. In *Proceedings of the Twelfth International Joint Conference on Artificial Intelligence (IJCAI-91)*, pages 452–457, Sydney, Australia, 1991.

[12] F. Baader and P. Hanschke. Extensions of concept languages for a mechanical engineering application. In *Proceedings of the 16th German AI-Conference (GWAI-92)*, vol. 671 of *Lecture Notes in Computer Science*, pages 132–143. Springer-Verlag, 1992.

[13] F. Baader and U. Sattler. Description logics with aggregates and concrete domains. *Information Systems*, 2002. To appear.

[14] B. Bennett. Modal logics for qualitative spatial reasoning. *Journal of the Interest Group in Pure and Applied Logic*, 4(1), 1997.

[15] T. Berners-Lee, J. Hendler, and O. Lassila. The semantic web. *Scientific American*, 284(5):34–43, 2001.

[16] A. Borgida. On the relative expressiveness of description logics and predicate logics. *Artificial Intelligence*, 82(1 - 2):353–367, 1996.

[17] A. Borgida and G. E. Weddell. Adding uniqueness constraints to description logics (preliminary report). In *Proceedings of the 5th International Conference on Deductive and Object-Oriented Databases (DOOD97)*, vol. 1341 of *LNCS*, pages 85–102. Springer, 1997.

[18] R. J. Brachman, D. L. McGuinness, P. F. Patel-Schneider, L. A. Resnick, and A. Borgida. Living with classic: When and how to use a KL-ONE-like language. In J. F. Sowa, editor, *Principles of Semantic Networks – Explorations in the Representation of Knowledge*, chapter 14, pages 401–456. Morgan Kaufmann, 1991.

[19] D. Calvanese, M. Lenzerini, and D. Nardi. Description logics for conceptual data modeling. In *Logics for Databases and Information Systems*, pages 229–263. Kluwer Academic Publisher, 1998.

[20] D. Calvanese, G. De Giacomo, and M. Lenzerini. Keys for free in description logics. In *Proceedings of the 2000 International Workshop in Description Logics (DL2000)*, number 33 in CEUR-WS (http://ceur-ws.org/), pages 79–88, 2000.

[21] P. P.-S. Chen. The entity-relationship model-toward a unified view of data. *ACM Transactions on Database Systems*, 1(1):9–36, 1976.

[22] J. Edelmann and B. Owsnicki. Data models in knowledge representation systems: A case study. In *Proceedings of the Tenth German Workshop on Artificial Intelligence (GWAI'86) and the Second Austrian Symposium on Artificial Intelligence (ÖGAI'86)*, vol. 124 of *Informatik-Fachberichte*, pages 69–74. Springer Verlag, 1986.

[23] D. Fensel, I. Horrocks, F. van Harmelen, S. Decker, M. Erdmann, and M. Klein. OIL in a nutshell. In *Proceedings of the European Knowledge Acquisition Conference (EKAW 2000)*, vol. 1937 of *Lecture Notes In Artificial Intelligence*. Springer-Verlag, 2000.

[24] E. Franconi and G. Ng. The i.com tool for intelligent conceptual modeling. In *Proceedings of the Seventh International Workshop on Knowledge Representation Meets Databases (KRDB2000)*, number 29 in CEUR-WS (http://ceur-ws.org/), pages 45–53, 2000.

[25] G. D. Giacomo and M. Lenzerini. Boosting the correspondence between description logics and propositional dynamic logics. In *Proceedings of the Twelfth National Conference on Artificial Intelligence (AAAI'94). Volume 1*, pages 205–212. AAAI Press, 1994.

[26] D. Y. Grigorev. The complexity of deciding Tarski algebra. *Journal of Symbolic Computation*, 5(1,2):65–108, 1988.

[27] V. Haarslev, C. Lutz, and R. Möller. A description logic with concrete domains and role-forming predicates. *Journal of Logic and Computation*, 9(3):351–384, 1999.

[28] V. Haarslev, R. Möller, and M. Wessel. The description logic $\mathcal{ALCNH}_{R+}$ extended with concrete domains: A practically motivated approach. In *Proceedings of the First International Joint Conference on Automated Reasoning IJCAR'01*, number 2083 in Lecture Notes in Artifical Intelligence, pages 29–44. Springer-Verlag, 2001.

[29] V. Haarslev and R. Möller. Practical reasoning in racer with a concrete domain for linear inequations. In *Proceedings of the International Workshop in Description Logics 2002 (DL2002)*, number 53 in CEUR-WS (http://ceur-ws.org/), 2002.

[30] P. Hanschke. Specifying role interaction in concept languages. In W. Nebel, Bernhard; Rich, Charles; Swartout, editor, *Proceedings of the Third International Conference on Principles of Knowledge Representation and Reasoning (KR'92)*, pages 318–329. Morgan Kaufmann, 1992.

[31] I. Horrocks, U. Sattler, and S. Tobies. Practical reasoning for very expressive description logics. *Logic Journal of the IGPL*, 8(3):239–264, 2000.

[32] I. Horrocks. Using an expressive description logic: Fact or fiction? In *Proceedings of the Sixth International Conference on the Principles of Knowledge Representation and Reasoning (KR98)*, pages 636–647, 1998.

[33] I. Horrocks and P. Patel-Schneider. The generation of DAML+OIL. In *Proceedings of the International Workshop in Description Logics 2001 (DL2001)*, number 49 in CEUR-WS (http://ceur-ws.org/), pages 30–35, 2001.

[34] I. Horrocks and U. Sattler. A description logic with transitive and inverse roles and role hierarchies. *Journal of Logic and Computation*, 9(3), 1999.

[35] I. Horrocks and U. Sattler. Ontology reasoning in the $\mathcal{SHOQ}(D)$ description logic. In B. Nebel, editor, *Proceedings of the Seventeenth International Joint Conference on Artificial Intelligence (IJCAI'01)*, pages 199–204. Morgan-Kaufmann, 2001.

[36] V. L. Khizder, D. Toman, and G. E. Weddell. On decidability and complexity of description logics with uniqueness constraints. In *Proceedings of the 8th International Conference on Database Theory (ICDT2001)*, vol. 1973 of *LNCS*, pages 54–67. Springer, 2001.

[37] O. Kutz, F. Wolter, and M. Zakharyaschev. A note on concepts and distances. In *Proceedings of the 2001 International Workshop in Description Logics (DL'01)*, number 49 in CEUR-WS (http://ceur-ws.org/), pages 113–121, 2001.

[38] O. Kutz, F. Wolter, and M. Zakharyaschev. Connecting abstract description systems. In *Proceedings of the Eighth International Conference on Principles of Knowledge Representation and Reasoning (KR2002)*, pages 215–226. Morgan Kaufman, 2002.

[39] C. Lutz. Complexity of terminological reasoning revisited. In *Proceedings of the 6th International Conference on Logic for Programming and Automated Reasoning (LPAR'99)*, number 1705 in Lecture Notes in Artificial Intelligence, pages 181–200. Springer-Verlag, 1999.

[40] C. Lutz. Interval-based temporal reasoning with general TBoxes. In B. Nebel, editor, *Proceedings of the Seventeenth International Joint Conference on Artificial Intelligence (IJCAI'01)*, pages 89–94. Morgan-Kaufmann, 2001.

[41] C. Lutz. NExpTime-complete description logics with concrete domains. In *Proceedings of the First International Joint Conference on Automated Reasoning (IJCAR'01)*, number 2083 in Lecture Notes in Artifical Intelligence, pages 45–60. Springer-Verlag, 2001.

[42] C. Lutz. Adding numbers to the $\mathcal{SHIQ}$ description logic—First results. In *Proceedings of the Eighth International Conference on Principles of Knowledge Representation and Reasoning (KR2002)*, pages 191–202. Morgan Kaufman, 2002.

[43] C. Lutz. *The Complexity of Reasoning with Concrete Domains*. PhD thesis, LuFG Theoretical Computer Science, RWTH Aachen, Germany, 2002.

[44] C. Lutz. NExpTime-complete description logics with concrete domains. *ACM Transactions on Computational Logic*, 2002. to appear.

[45] C. Lutz. PSPACE reasoning with the description logic $\mathcal{ALCF}(\mathcal{D})$. *Logic Journal of the IGPL*, 10(5):535–568, 2002.

[46] C. Lutz. Reasoning about entity relationship diagrams with complex attribute dependencies. In *Proceedings of the International Workshop in Description Logics 2002 (DL2002)*, number 53 in CEUR-WS (http://ceur-ws.org/), pages 185–194, 2002.

[47] C. Lutz and R. Möller. Defined topological relations in description logics. In *Proceedings of the International Workshop on Description Logics (DL'97)*, pages 15–19, Gif sur Yvette (Paris), France, 1997. Université Paris-Sud, Centre d'Orsay.

[48] E. W. Mayr and A. R. Meyer. The complexity of the word problem for commutative semigroups and polynomial ideals. *Advanced Mathematics*, 46:305–329, 1982.

[49] R. Molitor. *Unterstützung der Modellierung verfahrenstechnischer Prozesse durch Nicht-Standardinferenzen in Beschreibungslogiken*. PhD thesis, LuFG Theoretical Computer Science, RWTH Aachen, Germany, 2000.

[50] B. Nebel. Terminological reasoning is inherently intractable. *Artificial Intelligence*, 43:235–249, 1990.

[51] J. Z. Pan and I. Horrocks. Reasoning in the $\mathcal{SHOQ}(\mathcal{D}_n)$ description logic. In *Proceedings of the International Workshop in Description Logics 2002 (DL2002)*, number 53 in CEUR-WS (http://ceur-ws.org/), pages 53–62, 2002.

[52] D. A. Randell, Z. Cui, and A. G. Cohn. A spatial logic based on regions and connection. In *Proceedings of the Third International Conference on Principles of Knowledge Representation and Reasoning (KR'92)*, pages 165–176. Morgan Kaufman, 1992.

[53] J. Renz and B. Nebel. On the complexity of qualitative spatial reasoning: A maximal tractable fragment of the region connection calculus. *Artificial Intelligence*, 108(1-2):69–123, 1999.

[54] U. Sattler. *Terminological knowledge representation systems in a process engineering application*. PhD thesis, LuFG Theoretical Computer Science, RWTH-Aachen, 1998.

[55] K. D. Schild. A correspondence theory for terminological logics: Preliminary report. In *Proceedings of the Twelfth International Joint Conference on Artificial Intelligence (IJCAI-91)*, pages 466–471. Morgan Kaufmann, 1991.

[56] K. D. Schild. Combining terminological logics with tense logic. In *Progress in Artificial Intelligence – 6th Portuguese Conference on Artificial Intelligence, EPIA'93*, vol. 727 of *Lecture Notes in Artificial Intelligence*, pages 105–120. Springer-Verlag, 1993.

[57] M. Schmidt-Schauß and G. Smolka. Attributive concept descriptions with complements. *Artificial Intelligence*, 48(1):1–26, 1991.

[58] A. Schrijver. *Theory of Linear and Integer Programming*. Wiley, Chichester, UK, 1986.

[59] O. Stock, editor. *Spatial and Temporal Reasoning*. Kluwer Academic Publishers, Dordrecht, Holland, 1997.

[60] T. J. Teorey. *Database Modeling and Design - the Entity-Relationship Approach*. Morgan Kaufmann, 1990.

[61] J. F. A. K. van Benthem. *Modal Logic and Classical Logic*. Bibliopolis, Naples, Italy, 1983.

[62] F. Wolter and M. Zakharyaschev. Temporalizing description logic. In *Frontiers of Combining Systems*, pages 379–402. Studies Press/Wiley, 1999.

Carsten Lutz
Dresden University of Technology
Department of Computer Science
Institute for Theoretical Computer Science
01062 Dresden, Germany
E-mail: lutz@tcs.inf.tu-dresden.de

14
Restricted Interpolation in Modal Logics

LARISA MAKSIMOVA

ABSTRACT. A new variant IPR of interpolation is introduced and its interrelations with the known interpolation properties and various variants of the Beth property are investigated. It is stated that on the class of modal logics IPR follows from the usual interpolation properties CIP and IPD; on the other hand, there are modal logics that have IPR and do not have IPD. It is proved that IPR follows from the projective Beth property PB2. An algebraic equivalent of IPR, namely, the restricted amalgamation property, is found.

Interpolation theorem proved by W. Craig [4] in 1957 for the first order logic was a source of a lot of investigations devoted to interpolation problem in various logical theories [1, 3]. Interpolation is considered as a desirable and "nice" property; also it has important practical applications in computer science [3].

The most known systems of modal logic, for instance, Lewis' systems S4 and S5, Grzegorczyk's logic Grz, the logic GL of provability, the logic K4 and the least normal modal logic K possess the interpolation property. On the other hand, it appeared that this important property is rather rare. For instance, there is a continuum of normal extensions of the propositional modal logic S4 but only finitely many of these logics possess the interpolation [7]. Moreover, there are two natural variants CIP and IPD of the interpolation property which are not equivalent on the class of modal logics, more exactly, IPD is weaker than CIP.

Due to great significance of the interpolation, its weaker versions may also be useful. In the present paper we introduce a new variant IPR of interpolation that seems to be very natural. One can show that there are modal logics with IPR and without IPD. For example, such is the logic which is characterized by one 3-element linearly ordered frame. On the other hand, IPR is not trivial, for instance, one can prove that the logic S4.3 does not possess this property.

Interpolation properties are closely connected with the Beth properties which have arisen due to the well-known theorem on implicit definability found by E. Beth [2]. Correlations of variants of the Beth properties and interpolation are widely presented [4, 1, 15, 8, 5].

In [9] we considered two versions PB1 and PB2 of the projective Beth property and proved that in normal modal logics the property PB1 is equivalent to the Craig interpolation property CIP, whereas the weaker versions IPD and PB2 are independent.

In investigations of the projective Beth property PB2 in superintuitionistic logics [11] some interesting connections of PB2 with interpolation and amalgamation properties were found. First of all, in these logics PB2 is implied by interpolation which is equivalent to the amalgamation property AP in associated varieties of Heyting algebras. On the other hand, PB2 implies some restricted variant of amalgamation that is proved to be weaker than AP.

We extend the notion of restricted amalgamation to arbitrary varieties. Also we prove that the restricted version IPR of interpolation property is equivalent to the restricted amalgamation in varieties of modal or Heyting algebras and show that IPR follows from PB2. In addition we find an algebraic criterion for PB2 with use of the restricted amalgamation property.

1 Beth properties and interpolation

If $\mathbf{p}$ is a list of variables, let $A(\mathbf{p})$ denote a formula whose all variables are in $\mathbf{p}$, and $\mathcal{F}(\mathbf{p})$ the set of all such formulas.

Let L be a logic, $\vdash_L$ deducibility relation in L. Suppose that $\mathbf{p}, \mathbf{q}, \mathbf{q}'$ are disjoint lists of variables that do not contain x and y, $\mathbf{q}$ and $\mathbf{q}'$ are of the same length, and $A(\mathbf{p}, \mathbf{q}, x)$ is a formula. We say that the logic L possesses *the projective Beth property PB1*, whenever $\vdash_L A(\mathbf{p}, \mathbf{q}, x) \& A(\mathbf{p}, \mathbf{q}', y) \to (x \leftrightarrow y)$ implies $\vdash_L A(\mathbf{p}, \mathbf{q}, x) \to (x \leftrightarrow B(\mathbf{p}))$ for some formula $B(\mathbf{p})$. Further, L has *the projective Beth property PB2*, if $A(\mathbf{p}, \mathbf{q}, x), A(\mathbf{p}, \mathbf{q}', y) \vdash_L x \leftrightarrow y$ implies $A(\mathbf{p}, \mathbf{q}, x) \vdash_L x \leftrightarrow B(\mathbf{p})$ for some $B(\mathbf{p})$; L possesses *the Beth property B1*, whenever $\vdash_L A(\mathbf{p}, x) \& A(\mathbf{p}, y) \to (x \leftrightarrow y)$ implies $\vdash_L A(\mathbf{p}, x) \to (x \leftrightarrow B(\mathbf{p}))$ for a suitable formula $B(\mathbf{p})$; L has *the Beth property B2* if $A(\mathbf{p}, x), A(\mathbf{p}, y) \vdash_L x \leftrightarrow y$ implies $A(\mathbf{p}, x) \vdash_L x \leftrightarrow B(\mathbf{p})$ for a formula $B(\mathbf{p})$.

The Beth properties are closely connected with *the interpolation properties CIP and IPD* defined as follows (where the lists $\mathbf{p}, \mathbf{q}, \mathbf{r}$ are disjoint):

CIP. If $\vdash_L A(\mathbf{p}, \mathbf{q}) \to B(\mathbf{p}, \mathbf{r})$, then there exists a formula $C(\mathbf{p})$ such that $\vdash_L A(\mathbf{p}, \mathbf{q}) \to C(\mathbf{p})$ and $\vdash_L ?\mathbf{p}) \to B(\mathbf{p}, \mathbf{r})$.

IPD. If $A(\mathbf{p}, \mathbf{q}) \vdash_L B(\mathbf{p}, \mathbf{r})$, then there exists a formula $C(\mathbf{p})$ such that $A(\mathbf{p}, \mathbf{q}) \vdash_L C(\mathbf{p})$ and $?\mathbf{p}) \vdash_L B(\mathbf{p}, \mathbf{r})$.

The most known modal logics such as Lewis' systems S4 and S5, Grzegorczyk's logic Grz, the logic GL of provability, the logic K4 and the least normal modal logic K have the properties CIP and PB1. We mean that the language contains at least one propositional constant $\top$ ("truth") or $\bot$ ("false").

A *normal modal logic* is any set of modal formulas containing all the tautologies of the two-valued propositional logic and the axioms $\Box(A \to B) \to (\Box A \to \Box B)$, and closed under the inference rules R1: $A, A \to B/B$ and R2: $A/\Box A$, and the substitution rule. The set of normal modal logics containing a normal modal logic L is denoted by $NE(L)$. If L is a normal modal logic, by $\vdash_L$ we denote deducibility in L by the rules R1 and R2. The well-known deduction theorem in normal modal logics says that

$$\Gamma, A \vdash_L B \iff \Gamma \vdash_L [n]A \to B \text{ for some } n \geq 0,$$

where $[n]A \rightleftharpoons A \& \Box A \& \ldots \& \Box^n A$.

Interrelations of interpolation with Beth properties are presented in the following

PROPOSITION 1 [9, 12] *In the family of normal modal logics*

  a) *PB1 is equivalent to each of the properties CIP and B1;*

  b) *PB1 implies the conjunction of B2 and IPD but the converse does not hold;*

  c) *the conjunction of B2 and IPD implies PB2;*

  d) *PB2 implies B2 but the converse is not true,*

  e) *PB2 and IPD are independent.*

We introduce a *restricted interpolation property*:

IPR. If $A(\mathbf{p}, \mathbf{q}), B(\mathbf{p}, \mathbf{r}) \vdash_L C(\mathbf{p})$, then there exists a formula $A'(\mathbf{p})$ such that $A(\mathbf{p}, \mathbf{q}) \vdash_L A'(\mathbf{p})$ and $A'(\mathbf{p}), B(\mathbf{p}, \mathbf{r}) \vdash_L C(\mathbf{p})$.

By the deduction theorem, one can easily show that IPD implies IPR in modal logics. On the other hand, in this paper we prove that PB2 implies IPR. Since PB2 does not imply IPD [12], we get that IPD does not follow from IPR.

2 Restricted interpolation and restricted amalgamation

In this section we find an algebraic equivalent of IPR. For the properties B1 and B2, and also for CIP and IPD it was done in [8].

It is well known that there exists a duality between normal modal logics and varieties of modal algebras. A *modal algebra* is an algebra $\mathbf{A} = (A, \to, \neg, \Box)$ that is a boolean algebra with respect to $\to$ and $\neg$ and, moreover, satisfies the conditions $\Box\top = \top$ and $\Box(x \to y) \le \Box x \to \Box y$. A modal algebra $\mathbf{A}$ is called *transitive* if it satisfies an inequality $\Box x \le \Box\Box x$; a *topoboolean algebra*, or *interior algebra* if it satisfies $\Box x \le x$; a *diagonalizable algebra* if it satisfies $\Box(\Box x \to x) = \Box x$. It is well known that the modal logic K4 can be characterized by the variety of all transitive algebras, S4 by the variety of topoboolean algebras, GL by diagonalizable algebras.

By a *valuation* in an algebra $\mathbf{A}$, we mean, as usually, a homomorphism from the algebra of all formulas into $\mathbf{A}$.

If A is a formula, $\mathbf{A}$ a modal algebra, then $\mathbf{A} \models A$ denotes that the identity $A = \top$ is satisfied in $\mathbf{A}$. We write $\mathbf{A} \models L$ instead of $(\forall A \in L)(\mathbf{A} \models A)$. We denote $V(L) = \{\mathbf{A} | \mathbf{A} \models L\}$. Each normal modal logic L is characterized by the variety $V(L)$.

A set $\Gamma \subseteq \mathcal{F}(\mathbf{p})$ is called an *open L-theory of the language* $\mathcal{F}(\mathbf{p})$ whenever it is closed with respect to $\vdash_L$, i.e. for every formula $F(\mathbf{p})$, $\Gamma \vdash_L F(\mathbf{p})$ implies $F(\mathbf{p}) \in \Gamma$. With each open L-theory T of the language $\mathcal{F}(\mathbf{p})$, one can associate an equivalence relation on the set $\mathcal{F}(\mathbf{p})$:

$$A \sim_T B \rightleftharpoons (A \leftrightarrow B) \in T.$$

Due to the replacement lemma, $\sim_T$ is a congruence on $\mathcal{F}(\mathbf{p})$. It makes possible to define *the Lindenbaum-Tarski algebra*

$$\mathbf{A}(\mathbf{p}, T) = \mathcal{F}(\mathbf{p})/\sim_T$$

as a quotient-algebra of $\mathcal{F}(\mathbf{p})$. Let us denote $\|A\| = \{B | A \sim_T B\}$ for each formula $A = A(\mathbf{p})$. The following statement holds (see, for instance,[14]):

LEMMA 2 *Let $L \in NE(K)$. For each open L-theory T of the language $\mathcal{F}(\mathbf{p})$, the algebra $\mathbf{A}(\mathbf{p}, T)$ is in $V(L)$; the canonical mapping $\varkappa : \mathcal{F} \to \mathbf{A}(\mathbf{p}, T)$, where for every formula $A = A(\mathbf{p})$*

$$\varkappa(A) = \|A\| = A/\sim_T,$$

is a homomorphism and, moreover, $A \in T \Leftrightarrow \|A\| = \top$ in $\mathbf{A}(\mathbf{p}, T)$.

In particular, $\mathbf{A}(\mathbf{p}, L)$ is a free algebra of $V(L)$, with free generators $\|p\|$, where $p \in \mathbf{p}$.

All varieties of modal algebras possess such important properties as congruence-distributivity and congruence extension property:

CEP. If $\mathbf{A}$ is a subalgebra of $\mathbf{B}$ then every congruence Φ on $\mathbf{A}$ can be extended to a congruence Ψ on $\mathbf{B}$ such that $\Psi \cap \mathbf{A}^2 = \Phi$.

A logic L has CIP if and only if $V(L)$ is super-amalgamable, hence B1 and PB1 are also equivalent to the super-amalgamation property of the corresponding variety; IPD is equivalent to the amalgamation property AP [8]. We introduce a notion of the restricted amalgamation property RAP and prove its equivalence to IPR. Moreover, in Section 3 we find an algebraic criterion for PB2 with use of RAP.

Remind that a modal algebra is *subdirectly irreducible (finitely indecomposable)* if it can not be represented as a subdirect product (finite subdirect product) of its proper quotient algebras. For any class V of algebras, by $FI(V)$ and $SI(V)$ we denote the classes of finitely indecomposable and subdirectly irreducible algebras in V respectively, $FG(V)$ stands for finitely generated algebras in V.

A class V has *Amalgamation Property* if it satisfies the condition

AP. For each $\mathbf{A}, \mathbf{B}, \mathbf{C} \in V$ such that $\mathbf{A}$ is a common subalgebra of $\mathbf{B}$ and $\mathbf{C}$, there exist an algebra $\mathbf{D}$ in V and monomorphisms $\delta : \mathbf{B} \to \mathbf{D}$ and $\varepsilon : \mathbf{C} \to \mathbf{D}$ such that $\delta(x) = \varepsilon(x)$ for all $x \in \mathbf{A}$.

A class V has *Strong Amalgamation Property (StrAP)* if it satisfies AP and, moreover, $\delta(\mathbf{B}) \cap \varepsilon(\mathbf{C}) = \delta(\mathbf{A})$.

We say that a class V has *Restricted Amalgamation Property* if it satisfies the condition:

RAP. For each $\mathbf{A}, \mathbf{B}, \mathbf{C} \in V$ such that $\mathbf{A}$ is a common subalgebra of $\mathbf{B}$ and $\mathbf{C}$ there exist an algebra $\mathbf{D}$ in V and homomorphisms $\delta : \mathbf{B} \to \mathbf{D}$, $\varepsilon : \mathbf{C} \to \mathbf{D}$ such that $\delta(x) = \varepsilon(x)$ for all $x \in \mathbf{A}$ and the restriction δ' of δ onto $\mathbf{A}$ is a monomorphism.

In the same method as in [8] we prove

THEOREM 3 *A normal modal logic L has IPR iff $V(L)$ has RAP iff $FG(V(L))$ has RAP.*

Proof. Assume that L has IPR. We prove that $V(L)$ has RAP. Let $\mathbf{A}, \mathbf{B}, \mathbf{C} \in V(L)$ and $\mathbf{A}$ be a subalgebra of both $\mathbf{B}$ and $\mathbf{C}$. We take the sets of propositional variables $\mathbf{p} = \{p_a | a \in \mathbf{A}\}$, $\mathbf{q} = \{q_b | b \in \mathbf{B} - \mathbf{A}\}$, $\mathbf{r} = \{r_c | c \in \mathbf{C} - \mathbf{A}\}$ and define the valuations $v' : \mathbf{p} \cup \mathbf{q} \to \mathbf{B}$ and $v'' : \mathbf{p} \cup \mathbf{r} \to \mathbf{C}$ by

$$v'(p_a) = v''(p_a) = a, \ v'(q_b) = b, \ v''(r_c) = c.$$

We denote

$$T_1 = \{B | B \in \mathcal{F}(\mathbf{p}, \mathbf{q}) \text{ and } v'(B) = \top_\mathbf{B}\},$$
$$T_2 = \{C | C \in \mathcal{F}(\mathbf{p}, \mathbf{r}) \text{ and } v''(C) = \top_\mathbf{C}\},$$
$$T = \{D | D \in \mathcal{F}(\mathbf{p}, \mathbf{q}, \mathbf{r}) \text{ and } T_1 \cup T_2 \vdash_L D\}.$$

Then T_1, T_2 and T are open L-theories of the languages $\mathcal{F}(\mathbf{p}, \mathbf{q}), \mathcal{F}(\mathbf{p}, \mathbf{r})$ and $\mathcal{F}(\mathbf{p}, \mathbf{q}, \mathbf{r})$ respectively.

Let $\mathcal{F} = \mathcal{F}(\mathbf{p}, \mathbf{q}, \mathbf{r})$ and $\mathbf{D} = \mathbf{A}(\mathbf{p} \cup \mathbf{q} \cup \mathbf{r}, T) = \mathcal{F}/\sim_T$ be a Lindenbaum-Tarski algebra, where $D \sim D' \rightleftharpoons T_1, T_2 \vdash D \leftrightarrow D'$.

Due to $T_1, T_2 \subseteq T$, the mappings $\delta: \mathbf{B} \to \mathbf{D}$ and $\varepsilon: \mathbf{C} \to \mathbf{D}$ defined by $\delta(a) = \varepsilon(a) = \|p_a\|$ for $a \in \mathbf{A}$, $\delta(b) = \|q_b\|$ for $b \in \mathbf{B} - \mathbf{A}$, and $\varepsilon(c) = \|r_c\|$ for $c \in \mathbf{C} - \mathbf{A}$, are homomorphisms that coincide on $\mathbf{A}$. Assume that $\delta(a) = \delta(a')$ for some $a, a' \in \mathbf{A}$. Then we have

$$T_1, T_2 \vdash_L p_a \leftrightarrow p_{a'}.$$

Since all derivations are finite, it follows that there are some formulas $B(\mathbf{p}, \mathbf{q}) \in T_1$ and $C(\mathbf{p}, \mathbf{r}) \in T_2$ such that

$$B(\mathbf{p}, \mathbf{q}), C(\mathbf{p}, \mathbf{r}) \vdash_L p_a \leftrightarrow p_{a'}.$$

By IPR there exists $B'(\mathbf{p})$ such that

$$B(\mathbf{p}, \mathbf{q}) \vdash B'(\mathbf{p}) \text{ and } B'(\mathbf{p}), C(\mathbf{p}, \mathbf{r}) \vdash_L p_a \leftrightarrow p_{a'}.$$

We get that $v'(B(\mathbf{p}, \mathbf{q})) = v'(B'(\mathbf{p})) = \top_\mathbf{B}$, so $v''(B'(\mathbf{p})) = v'(B'(\mathbf{p})) = \top_\mathbf{A} = \top_\mathbf{C}$ and $B'(\mathbf{p}) \in T_2$. It follows that $(p_a \leftrightarrow p_{a'}) \in T_2$, hence $v''(p_a \leftrightarrow p_{a'}) = \top_\mathbf{C}$ and $a = v''(p_a) = v''(p_{a'}) = a'$. Thus $V(L)$ has RAP.

It is clear that $FG(V(L))$ has RAP whenever $V(L)$ has RAP.

Assume that $FG(V(L))$ has RAP. Suppose that $A(\mathbf{p}, \mathbf{q}), B(\mathbf{p}, \mathbf{r}) \vdash_L C(\mathbf{p})$. We consider

$$T = \{A'(\mathbf{p}) \mid A(\mathbf{p}, \mathbf{q}) \vdash_L A'(\mathbf{p})\}$$

and prove that

(1) $T, B(\mathbf{p}, \mathbf{q}) \vdash_L C(\mathbf{p})$.

Let us define

$$T_2 = \{F(\mathbf{p}, \mathbf{r}) \mid T, B(\mathbf{p}, \mathbf{r}) \vdash_L F(\mathbf{p}, \mathbf{r})\}, \quad T_0 = T_2 \cap \mathcal{F}(\mathbf{p}),$$

$$T_1 = \{F(\mathbf{p}, \mathbf{q}) \mid A(\mathbf{p}, \mathbf{q}), T_0 \vdash_L F(\mathbf{p}, \mathbf{q})\}.$$

We show that $T_1 \cap \mathcal{F}(\mathbf{p}) = T_0$.

Assume $F(\mathbf{p}) \in T_1$. Then $A(\mathbf{p}, \mathbf{q}), T_0 \vdash_L F(\mathbf{p})$, so by the deduction theorem we have $A(\mathbf{p}, \mathbf{q}) \vdash_L [n]A_1(\mathbf{p}) \to F(\mathbf{p})$ for some $n \geq 0$ and some $A_1(\mathbf{p}) \in T_0$. It follows that $([n]A_1(\mathbf{p}) \to F(\mathbf{p})) \in T \subseteq T_0$, $[n]A_1(\mathbf{p}) \in T_0$, and so $F(\mathbf{p}) \in T_0$. Thus we have

$$F(\mathbf{p}) \in T_2 \iff F(\mathbf{p}) \in T_0 \iff F(\mathbf{p}) \in T_1$$

for any formula $F(\mathbf{p})$.

We set $\mathbf{B} = \mathcal{F}(\mathbf{p},\mathbf{q})/\sim_{T_1}$, $\mathbf{C} = \mathcal{F}(\mathbf{p},\mathbf{r})/\sim_{T_2}$; let $\mathbf{A}$ be a subalgebra of $\mathbf{B}$ generated by $\mathbf{p}/T_1$. Then $\mathbf{A}$ is embeddable into $\mathbf{C}$ by a monomorphism $\varphi(p/\sim_{T_1}) = p/\sim_{T_2}$ for $p \in \mathbf{p}$. It is clear from Lemma 2 that $\mathbf{A}, \mathbf{B}, \mathbf{C} \in V(L)$. By RAP there exist $\mathbf{D} \in V(L)$ and homomorphisms $\delta : \mathbf{B} \to \mathbf{D}$, $\varepsilon : \mathbf{C} \to \mathbf{D}$ such that for every $p \in \mathbf{p}$ we have $\delta(p/\sim_{T_1}) = \varepsilon\varphi(p/\sim_{T_1}) = \varepsilon(p/\sim_{T_2})$ and the restriction δ' of δ onto $\mathbf{A}$ is a monomorphism.

Let us consider the valuation

$$v(p) = \delta(p/\sim_{T_1}) \text{ for } p \in \mathbf{p},$$
$$v(q) = \delta(q/\sim_{T_1}) \text{ for } q \in \mathbf{q},$$
$$v(r) = \varepsilon(r/\sim_{T_2}) \text{ for } r \in \mathbf{r}.$$

We note that $A(\mathbf{p},\mathbf{q}) \in T_1$, so $v(A(\mathbf{p},\mathbf{q})) = \delta(A(\mathbf{p},\mathbf{q})/\sim_{T_1}) = \top_{\mathbf{D}}$. In addition, $B(\mathbf{p},\mathbf{r}) \in T_2$, so $v(B(\mathbf{p},\mathbf{r})) = \varepsilon(B(\mathbf{p},\mathbf{r})/\sim_{T_2}) = \top_{\mathbf{D}}$. Due to $A(\mathbf{p},\mathbf{q}), B(\mathbf{p},\mathbf{r}) \vdash_L C(\mathbf{p})$, we obtain $v(C(\mathbf{p})) = \delta(C(\mathbf{p})/\sim_{T_1}) = \top_{\mathbf{D}}$. Since the restriction of δ onto $\mathbf{A}$ is a monomorphism, we get $C(\mathbf{p})/\sim_{T_1} = \top_{\mathbf{B}}$ and $C(\mathbf{p}) \in T_1$, so $C(\mathbf{p}) \in T_2$. Thus $T, B(\mathbf{p},\mathbf{q}) \vdash_L C(\mathbf{p})$, and we proved (1).

Since all derivations are finite, there exist a finite subset Γ of T such that $\Gamma, B(\mathbf{p},\mathbf{q}) \vdash_L C(\mathbf{p})$. Taking the conjunction of all formulas in Γ as $A'(\mathbf{p})$, we obtain $A(\mathbf{p},\mathbf{q}) \vdash_L A'(\mathbf{p})$ and $A'(\mathbf{p}), B(\mathbf{p},\mathbf{q}) \vdash_L C(\mathbf{p})$. ∎

We bring out various equivalents of RAP in

PROPOSITION 4 *For every variety V of modal algebras the following are equivalent:*

(1) V has RAP;

(2) for any finitely indecomposable $\mathbf{A}, \mathbf{B}, \mathbf{C}$ in V such that $\mathbf{A}$ is a subalgebra of both $\mathbf{B}$ and $\mathbf{C}$ there exist $\mathbf{D}$ in V and homomorphisms $g : \mathbf{B} \to \mathbf{D}$, $h : \mathbf{C} \to \mathbf{D}$ such that $g(z) = h(z)$ for all z in $\mathbf{A}$ and the restriction of g onto $\mathbf{A}$ is a monomorphism;

(3) for any subdirectly irreducible $\mathbf{A}, \mathbf{B}, \mathbf{C}$ in V such that $\mathbf{A}$ is a subalgebra of both $\mathbf{B}$ and $\mathbf{C}$ there exist $\mathbf{D}$ in V and homomorphisms $g : \mathbf{B} \to \mathbf{D}$, $h : \mathbf{C} \to \mathbf{D}$ such that $g(z) = h(z)$ for all z in $\mathbf{A}$ and the restriction of g onto $\mathbf{A}$ is a monomorphism;

(4) $SI(V)$ has RAP;

(5) for any finitely generated and finitely indecomposable $\mathbf{A}, \mathbf{B}, \mathbf{C}$ in V such that $\mathbf{A}$ is a subalgebra of both $\mathbf{B}$ and $\mathbf{C}$ and for any $b, c \in \mathbf{A}$, $b \neq c$, there exist a subdirectly irreducible algebra $\mathbf{D}$ in V and homomorphisms $g : \mathbf{B} \to \mathbf{D}$, $h : \mathbf{C} \to \mathbf{D}$ such that $g(z) = h(z)$ for all z in $\mathbf{A}$ and $g(b) \neq g(c)$.

Proof. The proposition can be proved by analogy with the proof of the similar statement concerning the amalgamation property AP ([8], Theorem 2). It is clear that $(1) \Rightarrow (2) \Rightarrow (3)$. We prove

(3) ⇒ (4). Let $\mathbf{A}, \mathbf{B}, \mathbf{C} \in SI(V)$ such that $\mathbf{A}$ is a subalgebra of both $\mathbf{B}$ and $\mathbf{C}$. Then there exist an algebra $\mathbf{D}$ in V and homomorphisms $g : \mathbf{B} \to \mathbf{D}$, $h : \mathbf{C} \to \mathbf{D}$ such that $g(z) = h(z)$ for all z in $\mathbf{A}$ and the restriction g' of g onto $\mathbf{A}$ is a monomorphism. Since $g(\mathbf{A})$ is subdirectly irreducible, there exists a pair (a, a') of elements in $g(\mathbf{A})$ such that $a \neq a'$ and every non-zero congruence on $g(\mathbf{A})$ contains (a, a'). Take any congruence Φ on $\mathbf{D}$ which is maximal among the congruences on $\mathbf{D}$ not containing (a, a'). Then the algebra $\mathbf{D}' = \mathbf{D}/\Phi$ is subdirectly irreducible and the homomorphisms $g_1 : \mathbf{B} \to \mathbf{D}'$ and $h_1 : \mathbf{C} \to \mathbf{D}$, where $g_1(b) = g(b)/\Phi$ and $h_1(c) = h(c)/\Phi$, coincide on $\mathbf{A}$. Moreover, the restriction of g_1 onto $\mathbf{A}$ is a monomorphism since $(a, a') \notin \Phi$.

(4) ⇒ (5). Let $\mathbf{A}, \mathbf{B}, \mathbf{C} \in V$, $\mathbf{A}$ be a subalgebra of both $\mathbf{B}$ and $\mathbf{C}$, and $b, c \in \mathbf{A}$, $b \neq c$. We take any congruence Φ on $\mathbf{B}$ which is maximal among the congruences on $\mathbf{B}$ not containing (b, c). Then $\Phi_0 = \Phi \cap \mathbf{A}^2$ is a congruence on $\mathbf{A}$. By CEP it may be extended to a congruence Ψ on $\mathbf{C}$ such that $\Psi \cap \mathbf{A}^2 = \Phi_0$. By Zorn's lemma we can take Ψ to be maximal among the congruences with this property. We note that $(b, c) \notin \Psi$ and any proper extension of Ψ must contain the pair (b, c). It means that the algebras $\mathbf{A}' = \mathbf{A}/\Phi_0$, $\mathbf{B}' = \mathbf{B}/\Phi$ and $\mathbf{C}' = \mathbf{C}/\Psi$ are subdirectly irreducible. Moreover, $\mathbf{A}'$ is embeddable into $\mathbf{B}'$ by a monomorphism $\varphi(a/\Phi_0) = a/\Phi$ and into $\mathbf{C}'$ by a monomorphism $\psi(a/\Phi_0) = a/\Psi$. Since $SI(V)$ has RAP, there exist $\mathbf{D} \in SI(V)$ and homomorphisms $g : \mathbf{B}' \to \mathbf{D}$, $h : \mathbf{C}' \to \mathbf{D}$ such that $g\varphi(z) = h\psi(z)$ for all z in $\mathbf{A}'$ and the restriction of $g\varphi$ onto $\mathbf{A}'$ is a monomorphism.

We define $g'(y) = g(y/\Phi)$ for $y \in \mathbf{B}$ and $h'(z) = h(z/\Psi)$ for $z \in \mathbf{C}$. Then g' and h' are the required homomorphisms.

(5) ⇒ (1). Due to Theorem 3 it is sufficient to prove that $FG(V)$ has RAP. Let $\mathbf{A}, \mathbf{B}, \mathbf{C} \in FG(V)$ and $\mathbf{A}$ be a subalgebra of both $\mathbf{B}$ and $\mathbf{C}$. For any $b, c \in \mathbf{A}$, $b \neq c$, we take a congruence $\Phi(b, c)$ on $\mathbf{B}$ that is maximal among the congruences on $\mathbf{B}$ not containing the pair (b, c). By analogy with the proof of (4)⇒(5) we construct a congruence $\Psi(b, c)$ on $\mathbf{C}$ that is maximal among the congruences Ψ on $\mathbf{C}$ such that $\Psi \cap \mathbf{A}^2 = \Phi(b, c) \cap \mathbf{A}^2 = \Phi_0$. Then the algebra $\mathbf{A}' = \mathbf{A}/\Phi_0$ is embeddable into both $\mathbf{B}' = \mathbf{B}/\Phi(b, c)$ and $\mathbf{C}' = \mathbf{C}/\Psi(b, c)$, and all the three algebras are in $FGFI(V)$. Also we have $b/\Phi_0 \neq c/\Phi_0$. By (5) there exist an algebra $\mathbf{D}_{bc}$ and homomorphisms $g_{bc} : \mathbf{B}' \to \mathbf{D}_{bc}$ and $h_{bc} : \mathbf{C}' \to \mathbf{D}_{bc}$ such that

$g_{bc}(x/\Phi(b, c)) = h_{bc}(x/\Psi(b, c))$ for all $x \in \mathbf{A}$ and
$g_{bc}(b/\Phi(b, c)) \neq g_{bc}(c/\Phi(b, c))$.

We take $\mathbf{D} = \prod_{b,c \in \mathbf{A}; b \neq c} \mathbf{D}_{bc}$ and define $g : \mathbf{B} \to \mathbf{D}$ and $h : \mathbf{C} \to \mathbf{D}$ as follows:

$$g(y)(b, c) = g_{bc}(y/\Phi(b, c)) \text{ for } y \in \mathbf{B},$$

$$h(z)(b,c) = h_{bc}(z/\Psi(b,c)) \text{ for } z \in \mathbf{C}.$$

Then $g(x) = h(x)$ for all $x \in \mathbf{A}$. Moreover, $g(x) \neq g(x')$ for all $x, x' \in \mathbf{A}$ such that $x \neq x'$, since $g_{xx'}(x/\Phi(x,x')) \neq g_{xx'}(x'/\Phi(x,x'))$, and so $g(x)(x,x') \neq g(x'(x,x'))$. Thus g and h are the required homomorphisms. ∎

For transitive modal algebras we bring out some additional equivalents of RAP in Section 4.

3 A criterion for PB2 in modal logics

In [10] we found an algebraic equivalent of PB2 that is so-called strong epimorphisms surjectivity SES. It was proved in [11] that SES implies RAP in varieties of Heyting algebras. The analogous statement holds for varieties of transitive modal algebras [12]. In this section we extend this result to all modal algebras and find a useful criterion for PB2 with use of RAP.

We say that a class V has *Strong Epimorphisms Surjectivity* if it satisfies
(SES) For each $\mathbf{A}, \mathbf{B}$ in V, for every monomorphism $\alpha : \mathbf{A} \to \mathbf{B}$ and for every $x \in \mathbf{B} - \alpha(\mathbf{A})$ there exist $\mathbf{C} \in V$ and homomorphisms $\beta : \mathbf{B} \to \mathbf{C}$, $\gamma : \mathbf{B} \to \mathbf{C}$ such that $\beta\alpha = \gamma\alpha$ and $\beta(x) \neq \gamma(x)$.

We introduced this property in [9] and proved

THEOREM 5 [10] *Let L be a normal modal logic. Then L possesses the projective Beth property PB2 iff $V(L)$ has the property SES.*

It was proved in [5] that this equivalence holds for a large class of algebraizable logics. Also we need

THEOREM 6 ([10], Theorem 3.8) *For each normal modal logic L the following are equivalent:*
(1) L has the property PB2;
(2) for any finitely generated and finitely indecomposable $\mathbf{A}, \mathbf{B}, \mathbf{C}$ in $V(L)$, for every monomorphisms $\beta : \mathbf{A} \to \mathbf{B}$, $\gamma : \mathbf{A} \to \mathbf{C}$ and for any $b \in \mathbf{B}, c \in \mathbf{C}$, if $\neg(\exists z \in \mathbf{A})(b = \beta(z) \& \gamma(z) = c)$ then there exist a subdirectly irreducible algebra $\mathbf{D}$ in $V(L)$ and homomorphisms $\delta : \mathbf{B} \times \mathbf{C} \to \mathbf{D}$, $\varepsilon : \mathbf{B} \times \mathbf{C} \to \mathbf{D}$ such that $\delta(b,c) \neq \varepsilon(b,c)$ and $\delta(\beta(z),\gamma(z)) = \varepsilon(\beta(z),\gamma(z))$ for all z in $\mathbf{A}$.

The following characterization of finitely indecomposable modal algebras was found in [8], Lemma 2.3 (where $[n]x = x \& \Box x \& \ldots \& \Box^n x$).

LEMMA 7 *A modal algebra $\mathbf{A}$ is finitely indecomposable if and only if for any elements $a \neq \top$ and $b \neq \top$ in $\mathbf{A}$ there exists some n such that $[n]a \vee [n]b \neq \top$.*

We prove the following

LEMMA 8 *Let* $\mathbf{B}, \mathbf{C}, \mathbf{D}$ *be modal algebras,* $\mathbf{D}$ *subdirectly irreducible and* $h : \mathbf{B} \times \mathbf{C} \to \mathbf{D}$ *a homomorphism. Then just one of the following conditions is fulfilled:*

1) $h(\top, \bot) = \top$, $h(x, y) = h(x, \top)$ *for all* $x \in \mathbf{B}, y \in \mathbf{C}$, *and* $\alpha(x) = h(x, \top)$ *is a homomorphism from* $\mathbf{B}$ *into* $\mathbf{D}$,

2) $h(\bot, \top) = \top$, $h(x, y) = h(\top, y)$ *for all* $x \in \mathbf{B}, y \in \mathbf{C}$, *and* $\beta(y) = h(\top, y)$ *is a homomorphism from* $\mathbf{C}$ *into* $\mathbf{D}$.

Proof. Suppose $h(\top, \bot) \neq \top$ and $h(\bot, \top) \neq \top$. Since $\mathbf{D}$ is finitely indecomposable, by Lemma 7 there exists n such that

$$[n]h(\top, \bot) \vee [n]h(\bot, \top) < \top,$$

where $[n]x = x \& \Box x \& \ldots \& \Box^n x$. On the other hand,
$[n]h(\top, \bot) \vee [n]h(\bot, \top) = h([n](\top, \bot) \vee [n](\bot, \top)) = h((\top, \bot) \vee (\bot, \top)) = h(\top, \top) = \top$,
because of $[n](\top, \bot) = ([n]\top, [n]\bot) = (\top, \bot)$, and so we get a contradiction.

Assume $h(\top, \bot) = \top$. Then for every $y \in \mathbf{C}$ we have $h(\top, y) \geq h(\top, \bot) = \top$, and for every $x \in \mathbf{B}$:

$$h(x, y) = h(x, \top) \& h(\top, y) = h(x, \top).$$

It is easy to check that in this case $\alpha(x) = h(x, \top)$ is a homomorphism. Another case $h(\bot, \top) = \top$ is considered by analogy. ∎

The lemma 8 is essentially used in the proof of the following

THEOREM 9 *For any variety* V *of modal algebras, SES is equivalent to conjunction of two conditions:*

(a) *$FI(V)$ has SES;*

(b) *for any finitely generated and finitely indecomposable* $\mathbf{A}, \mathbf{B}, \mathbf{C}$ *in* V *such that* $\mathbf{A}$ *is a subalgebra of both* $\mathbf{B}$ *and* $\mathbf{C}$ *and for any* $b, c \in \mathbf{A}$, $b \neq c$, *there exist a subdirectly irreducible algebra* $\mathbf{D}$ *in* V *and homomorphisms* $g : \mathbf{B} \to \mathbf{D}$, $h : \mathbf{C} \to \mathbf{D}$ *such that* $g(z) = h(z)$ *for all* z *in* $\mathbf{A}$ *and* $g(b) \neq g(c)$.

Proof. In Theorem 6(2) take β and γ to be identical embeddings so that $\mathbf{A}$ is a subalgebra of both $\mathbf{B}$ and $\mathbf{C}$. Then this condition is equivalent to:

(2') for any finitely generated and finitely indecomposable $\mathbf{A}, \mathbf{B}, \mathbf{C}$ in V such that $\mathbf{A}$ is a subalgebra of both $\mathbf{B}$ and $\mathbf{C}$ and for any $b \in \mathbf{B}, c \in \mathbf{C}$, if $\neg(\exists z \in \mathbf{A})(b = z \& z = c)$, then there exist a subdirectly irreducible algebra

D in V and homomorphisms $\delta : \mathbf{B} \times \mathbf{C} \to \mathbf{D}$, $\varepsilon : \mathbf{B} \times \mathbf{C} \to \mathbf{D}$ such that $\delta(b,c) \neq \varepsilon(b,c)$ and $\delta(z,z) = \varepsilon(z,z)$ for all z in $\mathbf{A}$.

First note that conjunction of (a) and (b) implies (2'). Assume the premise of (2') is satisfied. If $b \in \mathbf{B} - \mathbf{A}$ or $c \in \mathbf{C} - \mathbf{A}$, then one easily constructs homomorphisms δ and ε by (a). One should take two homomorphisms from $\mathbf{B}$ or, respectively, from $\mathbf{C}$ into a suitable algebra in $SI(V)$ and then take their compositions with projections from $\mathbf{B} \times \mathbf{C}$ onto $\mathbf{B}$ or $\mathbf{C}$.

Now let $b, c \in \mathbf{A}$. Then we have $b \neq c$, so by (b) there exist a subdirectly irreducible algebra $\mathbf{D}$ in V and homomorphisms $g : \mathbf{B} \to \mathbf{D}$, $h : \mathbf{C} \to \mathbf{D}$ such that $g(b) \neq g(c)$, and $g(z) = h(z)$ for all z in $\mathbf{A}$. We set

$$\delta(x,y) = g(x) \text{ and } \varepsilon(x,y) = h(x)$$

for all $x \in \mathbf{B}$, $y \in \mathbf{C}$. Then we have

$$\delta(b,c) = g(b) \neq g(c) = h(c) = \varepsilon(b,c)$$

and for all $z \in \mathbf{A}$:

$$\delta(z,z) = g(z) = h(z) = \varepsilon(z,z),$$

and thus (2') is satisfied.

Conversely, assume V has SES. It is clear that SES in V implies SES in $FI(V)$, so (1) is satisfied. Prove that (2') implies (b). We take finitely generated and finitely indecomposable $\mathbf{A}, \mathbf{B}, \mathbf{C}$ in V such that $\mathbf{A}$ is a subalgebra of both $\mathbf{B}$ and $\mathbf{C}$. Let $b \neq c$ for some $b, c \in \mathbf{A}$. By (2') there exist a subdirectly irreducible algebra $\mathbf{D}$ in V and homomorphisms $\delta : \mathbf{B} \times \mathbf{C} \to \mathbf{D}$, $\varepsilon : \mathbf{B} \times \mathbf{C} \to \mathbf{D}$ such that $\delta(b,c) \neq \varepsilon(b,c)$ and $\delta(z,z) = \varepsilon(z,z)$ for all z in $\mathbf{A}$. Then by Lemma 8 we have two possibilities:

Case 1. $\delta(\top, \bot) = \top$, $\delta(x,y) = \delta(x,\top)$ for all $x \in \mathbf{B}, y \in \mathbf{C}$, and $g(x) = \delta(x,\top)$ is a homomorphism of $\mathbf{B}$ into $\mathbf{D}$.

Case 2. $\delta(\bot, \top) = \top$, $\delta(x,y) = \delta(\top,y)$ for all $x \in \mathbf{B}, y \in \mathbf{C}$, and $h(y) = \delta(\top,y)$ is a homomorphism from $\mathbf{C}$ into $\mathbf{D}$.

The analogous alternative holds for ε.

In Case 1 we have $\delta(b,c) = \delta(b,b) = \varepsilon(b,b) \neq \varepsilon(b,c)$, so there exists x such that $\varepsilon(x,b) \neq \varepsilon(x,c)$. By Lemma 8 it follows that $h(y) = \varepsilon(x,y)$ is a homomorphism of $\mathbf{C}$ into $\mathbf{D}$, and $h(z) = \varepsilon(z,z) = \delta(z,z) = g(z)$ for all $z \in \mathbf{A}$. Moreover, $g(b) = \delta(b,c) \neq \varepsilon(b,c) = h(c) = g(c)$, i.e., $g(b) \neq g(c)$.

In Case 2 we have $\delta(b,c) = \delta(c,c) = \varepsilon(c,c) \neq \varepsilon(b,c)$. So $g(x) = \varepsilon(x,\top)$ is a homomorphism of $\mathbf{B}$ into $\mathbf{D}$ and, moreover, $h(c) = \delta(b,c) \neq \varepsilon(b,c) = g(b)$. In addition, $h(z) = \delta(z,z) = \varepsilon(z,z) = g(z)$ for all $z \in \mathbf{A}$, so $g(c) = h(c) \neq g(b)$. It completes the proof. ∎

From Proposition 4 and Theorem 9 we immediately get

COROLLARY 10 *SES implies RAP in any variety of modal algebras.*

By Proposition 4 and Theorems 5 and 9 we obtain the following criterion

THEOREM 11 *Let L be a normal modal logic. Then L has PB2 if and only if $SI(V)$ has RAP and $FI(V)$ has SES.*

4 Two definitions of restricted amalgamation

In [11] we defined RAP for Heyting algebras in another way and proved that this property follows from PB2. An analogous notion was introduced in [12] for transitive modal algebras which characterize the modal system K4 = K+($\Box A \to \Box\Box A$). We state that for varieties of transitive algebras, the definition of restricted amalgamation introduced in Section 2 is equivalent to the definition in [12].

Recall that a modal algebra is *transitive* if it satisfied the condition $\Box x \leq \Box\Box x$. Let $[*]x = x \& \Box x$. We say that an element a of a transitive algebra $\mathbf{A}$ is an *opremum* of $\mathbf{A}$ if $a \neq \top$, $[*]a = a$ and $(\forall x \in \mathbf{A})(x \neq \top \Rightarrow [*]x \leq a)$. Remind that a transitive algebra is subdirectly irreducible iff it has an opremum; it is easy to see that an opremum is unique.

The following two lemmas were proved in [12]:

LEMMA 12 *Let $\mathbf{A}$ be a transitive algebra, $a, b \in \mathbf{A}$ and $a = [*]a$, $b = [*]b$. If $a \not\leq b$ then there exist a subdirectly irreducible $\mathbf{C}$ with an opremum ω and a homomorphism $f : \mathbf{A} \to \mathbf{C}$ such that $f(a) = \top$ and $f(b) = \omega$.*

LEMMA 13 *Let $\mathbf{A}$ be a subdirectly irreducible transitive algebra with an opremum ω and h a homomorphism from $\mathbf{A}$ into $\mathbf{B}$ such that $h(\omega) < \top$. Then h is a monomorphism.*

Say that a monomorphism $\alpha : \mathbf{A} \to \mathbf{B}$ of two subdirectly irreducible transitive algebras is *o-preserving* if it maps the opremum of $\mathbf{A}$ into opremum of $\mathbf{B}$. The *Restricted Amalgamation Property* of a class V was defined in [12] as follows:

(RAP^*) for any o-preserving monomorphisms $\beta : \mathbf{A} \to \mathbf{B}$ and $\gamma : \mathbf{A} \to \mathbf{C}$ of subdirectly irreducible algebras in V there exist an algebra $\mathbf{D}$ in V and monomorphisms $\delta : \mathbf{B} \to \mathbf{D}$, $\varepsilon : \mathbf{C} \to \mathbf{D}$ such that $\delta\beta = \varepsilon\gamma$.

It is evident that AP implies implies RAP^*, and for any variety V of transitive algebras, RAP^* is equivalent to the property:

for each subdirectly irreducible $\mathbf{A}, \mathbf{B}, \mathbf{C} \in V$ such that $\mathbf{A}$ is a common subalgebra of $\mathbf{B}$ and $\mathbf{C}$ and all three algebras have the same opremum, there exist an algebra $\mathbf{D}$ in V and monomorphisms $\delta : \mathbf{B} \to \mathbf{D}$, $\varepsilon : \mathbf{C} \to \mathbf{D}$ such that $\delta(x) = \varepsilon(x)$ for all $x \in \mathbf{A}$.

Also we proved in [12] that the following lemma holds:

LEMMA 14 *For all varieties of transitive algebras, RAP^* is equivalent to the property:*
for any o-preserving monomorphisms $\beta : \mathbf{A} \to \mathbf{B}$ and $\gamma : \mathbf{A} \to \mathbf{C}$ of subdirectly irreducible algebras in V there exist a subdirectly irreducible algebra $\mathbf{D}$ in V and o-preserving monomorphisms $\delta : \mathbf{B} \to \mathbf{D}$, $\varepsilon : \mathbf{C} \to \mathbf{D}$ such that $\delta\beta = \varepsilon\gamma$.

We use this lemma and Proposition 4 in order to prove

THEOREM 15 *For any variety of Heyting algebras or transitive modal algebras, RAP and RAP^* are equivalent.*

Proof. Let V have RAP. Assume that $\mathbf{A}, \mathbf{B}$ and $\mathbf{C}$ are subdirectly irreducible algebras in V such that $\mathbf{A}$ is a common subalgebra of $\mathbf{B}$ and $\mathbf{C}$ and all three algebras have the same opremum ω. Then there exist $\mathbf{D} \in V$ and two homomorphisms $\delta : \mathbf{B} \to \mathbf{D}$, $\varepsilon : \mathbf{C} \to \mathbf{D}$ such that $\delta(x) = \varepsilon(x)$ for all $x \in \mathbf{A}$ and the restriction of δ to $\mathbf{A}$ is a monomorphism, so $\delta(\omega) = \varepsilon(\omega) < \top$. Then δ and ε are monomorphisms by Lemma 13, and thus V satisfies RAP^*.

Conversely, assume V has RAP^*. We prove that $SI(V)$ has RAP, then V will possess RAP by Proposition 4(3). Let us take $\mathbf{A}, \mathbf{B}$ and $\mathbf{C}$ in $SI(V)$ such that $\mathbf{A}$ is a common subalgebra of $\mathbf{B}$ and $\mathbf{C}$ and let ω be an opremum of $\mathbf{A}$. By Lemma 12 there exists a homomorphism h from $\mathbf{B}$ onto a subdirectly irreducible algebra $\mathbf{B}'$ such that $h(\omega)$ is an opremum in $\mathbf{B}'$. Similarly, there exists a homomorphism g from $\mathbf{C}$ onto some $\mathbf{C}'$ in $SI(V)$ such that $g(\omega)$ is an opremum in $\mathbf{C}'$. By Lemma 13, the restrictions h' of h to $\mathbf{A}$ and g' of g to $\mathbf{A}$ are monomorphisms of $\mathbf{A}$ into $\mathbf{B}'$ and $\mathbf{C}'$ respectively.

By Lemma 14 there exist $\mathbf{D} \in SI(V)$ and o-preserving monomorphisms $\delta : \mathbf{B}' \to \mathbf{D}$, $\varepsilon : \mathbf{C}' \to \mathbf{D}$ such that $\delta h' = \varepsilon g'$. Then δh and εg are the homomorphisms from $\mathbf{B}$ and $\mathbf{C}$ into $\mathbf{D}$ as required. ∎

5 Concluding remarks

Summarizing the results of this paper together with those of [8] and [10], we see that for all normal modal logics and varieties of modal algebras the following hold:
  (1) IPD & B2 $\iff$ StrAP,
  (2) IPD & B2 $\Rightarrow$ PB2 $\iff$ SES,
  (3) PB2 $\not\Rightarrow$ IPD $\iff$ AP, IPD $\not\Rightarrow$ PB2, PB2 $\Rightarrow$ B2,
  (4) PB2 $\Rightarrow$ IPR $\iff$ RAP,

(5) IPD $\Rightarrow$ IPR, IPR & B2 $\not\Rightarrow$ IPD,
(6) IPR $\not\Rightarrow$ B2, B2 $\not\Rightarrow$ IPR.

In fact, most of these implications are valid for arbitrary varieties which are congruence-distributive, have congruence extension property and satisfy an additional condition: any subalgebra of a subdirectly irreducible algebra is finitely indecomposable. The proofs will be given in a separate paper. In particular, all of (1)-(6) hold for logics without contraction characterized by varieties of residuated lattices which are investigated in [6].

It is known that all varieties of Heyting algebras, of relatively pseudo-complemented lattices and of transitive modal algebras have the Beth property B2. For those varieties and corresponding logics we obtain:

IPD $\iff$ AP $\iff$ StrAP,
IPD $\Rightarrow$ PB2 $\iff$ SES,
PB2 $\not\Rightarrow$ IPD,
PB2 $\Rightarrow$ IPR $\iff$ RAP.

We note that the implication from PB2 to IPR in modal logics follows from purely algebraic Corollary 3.6. It would be interesting to find a direct proof in logical terms.

Acknowledgements

The research was supported by Russian Foundation for Basic Research, grant no. 03-06-80178, and by Ministry for Education of Russian Federation, grant no. PD02-1.1-460.

BIBLIOGRAPHY

[1] *J.Barwise, S.Feferman, eds*. Model-Theoretic Logics. New York: Springer-Verlag, 1985.

[2] *E.W.Beth*. On Padoa's method in the theory of definitions. Indagationes Math. 15, no. 4 (1953), 330-339.

[3] *J.Bicarregui, T.Dimitrakos, D.Gabbay, T.Maibaum*. Interpolation in Practical Formal Development. Logic Journal of the IGPL, 9, no. 2 (2001), 247-259.

[4] *W.Craig*. Three uses of Herbrand-Gentzen theorem in relating model theory and proof theory. J. Symbolic Logic, 22 (1957), 269-285.

[5] *E.Hoogland*. Definability and Interpolation. Model-theoretic investigations. ILLC Dissertation Series DS-2001-05, Amsterdam, 2001.

[6] *T.Kowalski, H.Ono*. Splittings in the variety of residuated lattices. Algebra Universalis, 44 (2000), 283-298.

[7] *L.L.Maksimova*. Interpolation theorems in modal logics and amalgamable varieties of topoboolean algebras. Algebra i Logika, 18, no. 5 (1979), 556–586.

[8] *L.L.Maksimova*. Modal Logics and Varieties of Modal Algebras: the Beth Property, Interpolation and Amalgamation. Algebra and Logic, 31, no.2 (1992), 145-166.

[9] *L.Maksimova*. Explicit and Implicit Definability in Modal and Related Logics. Bulletin of Section of Logic, 27, no. 1/2 (1998), 36–39.

[10] *L.L.Maksimova*. Projective Beth properties in modal and superintuitionistic logics. Algebra and Logic, 38 (1999), 316-333.

[11] *L.L.Maksimova*. Intuitionistic Logic and Implicit Definability. Annals of Pure and Applied Logic, 105 (2000), 83–102.
[12] *L. Maksimova*. Projective Beth's Properties in Infinite Slice Extensions of K4. In: F.Wolter, H.Wansing, M. de Rijke, M.Zakharyaschev, eds. Advances in Modal Logics, Volume 3, World Scientific, Singapore, 2002, 349-363.
[13] *A.I.Maltsev*. Algebraic systems. Moscow: Nauka, 1970.
[14] *H.Rasiowa, R.Sikorski*. The Mathematics of Metamathematics. Warszawa: PWN, 1963.
[15] *I.Sain*. Beth's and Craig's Properties via Epimorphisms and Amalgamation in Algebraic Logic. In: *C.H.Bergman, R.D.Maddux, D.I.Pigozzi (ed)*, Algebraic Logic and Universal Algebra in Computer Science. Lecture Notes in Computer Science, 425 (1990), Springer Verlag, 209–226.

Larisa Maksimova
Institute of Mathematics,
Siberian Branch of
Russian Academy of Sciences
630090, Novosibirsk, Russia
E-mail: lmaksi@math.nsc.ru

15
Binary Logics, Orthologics and their Relations to Normal Modal Logics
Yutaka Miyazaki

ABSTRACT. We study the relation between the class EXT(**O**) of orthologics and the class NEXT(**KTB**) of normal modal logics over **KTB** by means of embeddings of logics. First we introduce *binary logics*, which are generalizations of orthologics, as logics that are embeddable into some normal modal logics. We investigate the embeddability relation between the class of binary logics and the class of all normal modal logics, and establish some preservation results. Then we analyze the relation between EXT(**O**) and NEXT(**KTB**) and show that there exists a continuum of normal modal logics over **KTB**, and that any tabular orthologic is embeddable into infinitely many normal modal logics over **KTB**.

1 Binary logics as generalizations of orthologics

In 1974, R.I.Goldblatt showed that the orthologic **O** can be embedded into the normal modal logic **KTB** by a translation of formulas [2]. Our aim is to investigate the structure of the lattice of orthologics and that of normal modal logics by means of embeddability relations between these classes of logics.

The "basic" orthologic **O** can be thought of as the logic of *ortholattices*. The term "ortholattice" is an abbreviation of *ortho-complemented lattice* and a prime example of an ortholattice is the lattice of all subspaces of a finite dimensional vector space with orthogonal complement.

The most characteristic feature of ortholattices is that they do not in general have the distributive law. Because of this feature, we face some serious difficulties in studying ortholattices and orthologics. One of such difficulties in syntactical analysis of orthologics is the lack of a good implication connective [6].

Although the success of Kripke semantics brought about remarkable developments in the research of modal logics, we do not know as much about normal modal logics containing the axiom **B** ($q \supset \Box \Diamond q$) as about intermediate logics and modal logics containing **S4**. This is mainly because *symmetric* Kripke frames are less tractable than *transitive* ones. A completely new idea seems to be needed to investigate the modal logics containing **B**.

In [2], Goldblatt formulated orthologics as sets of pairs of formulas, in his words, as *binary logics*, introduced Kripke-type semantics of orthologics, and proved Kripke completeness of the smallest orthologic **O**. He showed moreover that **O** has the finite model property and hence it is decidable, and that **O** is embeddable into the normal modal logic **KTB** by a translation of formulas which we call here η_1.

In view of this embeddability, we may ask the following questions.

(1) What logic is embeddable into **K** by a translation similar to η_1? Further, can we establish an embeddability relation between the class of extensions of that logic and the class NEXT(**K**)?

(2) Can we extend the embeddability of **O** into **KTB** to the embeddability relation between the class EXT(**O**) of extensions of **O** and the class NEXT(**KTB**) of normal extensions of **KTB**?

To answer question (1) above, we will define a notion of *binary logics* in the same fashion as orthologics in [2], where the orthocomplement is not only a kind of negation-like connective, but it behaves like an antitone modal operator. By introducing a new translation η_0, we can show that the smallest binary logic **L₀** is embeddable into **K**. Then we will discuss the relation between EXT(**L₀**) and NEXT(**K**) by Kripke-type semantics to some extent, and establish some preservation theorems by this embeddability relation.

To attack question (2), we use the translation η_1 to discuss the relation between EXT(**O**) and NEXT(**KTB**) in general. In particular, we can show that there exists a continuum of normal modal logics over **KTB**, and that any tabular orthologic is embeddable into at least countably infinitely many normal modal logics over **KTB** .

In the remaining part of this section, we will introduce the two classes of propositional logics that we are concerned with. First of all, we define the notion of *binary logics* in our sense.

The language $\mathcal{L}$ for binary logics consists of:(1) a set of propositional variables $\{p_1, p_2, \ldots\}$, (2) a propositional constant $\bot$, (3) conjunction $\wedge$, (4) a unary connective $\boxminus$, and (5) a pair of parentheses (,). The set Φ of formulas of $\mathcal{L}$ is defined as usual. Binary logics are defined as sets of pairs of formulas on Φ as below:

DEFINITION 1 (Binary logics). A *binary logic* **L** on the set Φ of formulas is a subset of the product $\Phi \times \Phi$ (we write $\alpha \vdash_\mathbf{L} \beta$, instead of $\langle \alpha, \beta \rangle \in \mathbf{L}$) which includes the following axiom schemes and is closed under the following inference rules

Axiom schemes:

(Ax1) $\alpha \vdash_{\mathbf{L}} \alpha$

(Ax2) $\alpha \wedge \beta \vdash_{\mathbf{L}} \alpha$

(Ax3) $\alpha \wedge \beta \vdash_{\mathbf{L}} \beta$

(Ax4) $\bot \vdash_{\mathbf{L}} \alpha$

(Ax5) $\alpha \vdash_{\mathbf{L}} \Box \bot$

Inference Rules:

(R1) $\dfrac{\alpha \vdash_{\mathbf{L}} \beta \quad \beta \vdash_{\mathbf{L}} \gamma}{\alpha \vdash_{\mathbf{L}} \gamma}$

(R2) $\dfrac{\alpha \vdash_{\mathbf{L}} \beta \quad \alpha \vdash_{\mathbf{L}} \gamma}{\alpha \vdash_{\mathbf{L}} \beta \wedge \gamma}$

(R3) $\dfrac{\alpha \vdash_{\mathbf{L}} \beta}{\Box \beta \vdash_{\mathbf{L}} \Box \alpha}$

(R4) $\dfrac{\alpha \vdash_{\mathbf{L}} \beta}{\alpha[\sigma/p] \vdash_{\mathbf{L}} \beta[\sigma/p]}$

In (R4), $[\sigma/p]$ is a uniform substitution operator. We sometimes write $\alpha \vdash \beta$, instead of $\alpha \vdash_{\mathbf{L}} \beta$ if no confusion occurs. For a subset Γ of Φ, we write $\Gamma \vdash_{\mathbf{L}} \alpha$ to mean that there exist finitely many formulas $\beta_1, \beta_2, \ldots, \beta_n \in \Gamma$ such that $\langle \beta_1 \wedge \beta_2 \wedge \cdots \wedge \beta_n, \alpha \rangle \in \mathbf{L}$. The *smallest* binary logic is denoted by $\mathbf{L_0}$.

Note that our language $\mathcal{L}$ does not have the disjunction connective, and that our system for binary logics does not include any axiom scheme or any inference rule for it. If we introduce disjunction to our logic and interpret it classically (as *or*), we obtain essentially the same logic as the classical logic with a modal operator. If we introduce disjunction as an abbreviation of $\Box(\Box \alpha \wedge \Box \beta)$, we obtain the logics over the orthologic $\mathbf{O}$. Our aim however is to find a class of logics which are weaker than orthologics and which correspond to members of the class of all normal modal logics by a translation similar to Goldblatt's η_1. Thus, there can be no room for disjunction in our language.

For a binary logic $\mathbf{L}$, $\langle \alpha, \beta \rangle \in \mathbf{L}$ means, roughly speaking, that "β is *deducible from* α in $\mathbf{L}$", that is, we use pairs of formulas to compensate for the lack of implication symbol. $\mathbf{L}$ is said to be *consistent* if there exists a pair of formulas $\langle \alpha, \beta \rangle$ such that $\langle \alpha, \beta \rangle \notin \mathbf{L}$. For a set of pairs of formulas $\Sigma = \{\langle \alpha_\lambda, \beta_\lambda \rangle \in \Phi^2 \mid \lambda \in \Lambda\}$, the smallest binary logic which includes both $\mathbf{L_0}$ and Σ as axiom schemes, is denoted by $\mathbf{L_0} + \Sigma$. For a binary logic $\mathbf{L}$, the class of binary logics containing $\mathbf{L}$ is denoted by $\text{EXT}(\mathbf{L})$. In our terminology, the smallest orthologic $\mathbf{O}$ is a binary logic which is defined as: $\mathbf{O} := \mathbf{L_0} + \{\langle p, \Box \Box p \rangle, \langle \Box \Box p, p \rangle, \langle p \wedge \Box p, \bot \rangle\}$.

For normal modal logics, we will follow basic notions and nomenclature from [1]. We recall here only a few bits of notation we use in this paper.

The modal language $\mathcal{ML}$ consists of: (1) a set of propositional variables $\{q_1, q_2, \ldots\}$, (2) a propositional constant $\bot$, (3) conjunction $\wedge$, (4) negation $\neg$, (5) necessity $\Box$, and (6) a pair of parentheses $(,)$. The set of formulas of this language is denoted by Ψ. The smallest normal modal logic is denoted by $\mathbf{K}$. For a set Γ of formulas the smallest normal modal logic that contains Γ is denoted by $\mathbf{K} \oplus \Gamma$. For a modal logic $\mathbf{M}$, the class of all normal modal logics above $\mathbf{M}$ is denoted by $\text{NEXT}(\mathbf{M})$.

2 Embedding of binary logics into normal modal logics

A map η from Φ to Ψ is called a *translation* of formulas. For a binary logic $\mathbf{L}$ and a normal modal logic $\mathbf{M}$, we say that $\mathbf{L}$ is *embeddable* into $\mathbf{M}$ by a translation η when for any $\alpha, \beta \in \Phi$, $\langle \alpha, \beta \rangle \in \mathbf{L}$ if and only if $\eta(\alpha) \supset \eta(\beta) \in \mathbf{M}$. In this case, $\mathbf{M}$ is called a *modal companion* of $\mathbf{L}$.

In this section, we consider the embeddability relation between $\text{EXT}(\mathbf{L_0})$ and $\text{NEXT}(\mathbf{K})$ by a translation of formulas $\eta_0 : \Phi \to \Psi$ defined inductively as follows: $\eta_0(\bot) := \bot$, $\eta_0(p_i) := q_i$, $\eta_0(\alpha \wedge \beta) := \eta_0(\alpha) \wedge \eta_0(\beta)$, and $\eta_0(\boxminus \alpha) := \Box \neg \eta_0(\alpha)$. We use algebraic semantics first, to establish a general embeddability result.

DEFINITION 2 (Algebra for binary logic). $\mathfrak{A} = \langle A, \cap, J, 0, 1 \rangle$ is an *algebra for binary logic* (ABL for short) if and only if $\langle A, \cap, 0, 1 \rangle$ is a bounded meet semilattice and J is a unary operation on A satisfying

1. $J(0) = 1$

2. for any $x, y \in A$, $x \leq y$ implies $J(y) \leq J(x)$

In an ABL $\mathfrak{A}$, a *partial order* $\leq$ can be defined as: $x \leq y$ if and only if $x \cap y = x$. A *valuation* on $\mathfrak{A}$ is an assignment $v : \Phi \to \mathfrak{A}$ such that $v(\bot) = 0$, $v(\alpha \wedge \beta) = v(\alpha) \cap v(\beta)$, and $v(\boxminus \alpha) = J(v(\alpha))$. A pair of formulas $\langle \alpha, \beta \rangle$ is *valid* in an ABL $\mathfrak{A}$ ($\mathfrak{A} \models \langle \alpha, \beta \rangle$ in symbols) if and only if $v(\alpha) \leq v(\beta)$ holds for any valuation $v : \Phi \to \mathfrak{A}$. Then, the algebraic completeness for the smallest binary logic $\mathbf{L_0}$ can be shown in the usual way. That is, for any formulas α, β, $\langle \alpha, \beta \rangle \in \mathbf{L_0}$ if and only if $\mathfrak{A} \models \langle \alpha, \beta \rangle$ for any ABL $\mathfrak{A}$. In a similar fashion, any class of ABLs can determine a binary logic. More precisely, for any ABL $\mathfrak{A}$, the binary logic characterized by this $\mathfrak{A}$ is denoted by $\mathbf{L}(\mathfrak{A})$, that is, $\mathbf{L}(\mathfrak{A}) := \{\langle \alpha, \beta \rangle \mid \mathfrak{A} \models \langle \alpha, \beta \rangle\}$. For a class $\mathcal{C}$ of ABLs, the binary logic $\mathbf{L}(\mathcal{C})$, that is characterized by the class $\mathcal{C}$, is defined as: $\mathbf{L}(\mathcal{C}) := \bigcap_{\mathfrak{A} \in \mathcal{C}} \mathbf{L}(\mathfrak{A})$.

DEFINITION 3 (Modal algebra). $\mathfrak{B} = \langle \mathsf{B}, \cap, \cup, -, I, 0, 1\rangle$ is a *modal algebra* if and only if it satisfies the following:

(1) $\langle \mathsf{B}, \cap, \cup, -, 0, 1\rangle$ is a Boolean algebra.

(2) I is a unary operation on B satisfying: for all $x, y \in \mathsf{B}$,

   (a) $I(1) = 1$
   (b) $I(x \cap y) = I(x) \cap I(y)$

Let $\mathfrak{B}$ be a modal algebra. A *modal valuation* $u : \Psi \to \mathfrak{B}$ is a function which satisfies the conditions: $u(\bot) = 0$, $u(A \wedge B) = u(A) \cap u(B)$, $u(\Box A) = I(u(A))$. A modal formula A is *valid* in a modal algebra $\mathfrak{B}$ ($\mathfrak{B} \models A$ in symbol) if and only if $u(A) = 1$ for any modal valuation $u : \Psi \to \mathfrak{B}$. Then it is easy to see that the logic **K** is characterized by the class of all modal algebras. That is, for any modal formula A, $A \in \mathbf{K}$ if and only if $\mathfrak{B} \models A$ for every modal algebra $\mathfrak{B}$. Any class of modal algebras determines a normal modal logic. The modal logic that is determined by a modal algebra $\mathfrak{B}$ is denoted by $\mathbf{M}(\mathfrak{B})$. For a class $\mathcal{D}$ of modal algebras, we put
$$\mathbf{M}(\mathcal{D}) := \bigcap_{\mathfrak{B} \in \mathcal{D}} \mathbf{M}(\mathfrak{B}).$$

Note that for any binary logic **L**, there exists a class $\mathcal{C}$ of ABLs such that $\mathbf{L} = \mathbf{L}(\mathcal{C})$. This is due to the fact that the Lindenbaum algebra for **L** characterizes the logic **L**. Of course, such a class of ABLs is not uniquely determined. The situation is the same for any normal modal logic.

For a modal algebra $\mathfrak{B}$, we define an algebra $\widehat{\mathfrak{A}_0}(\mathfrak{B}) := \langle \mathsf{B}, \cap, J_\mathfrak{B}, 0, 1\rangle$, where $J_\mathfrak{B}(x) := I(-x)$. It is easily seen that $\widehat{\mathfrak{A}_0}(\mathfrak{B})$ is an ABL for any $\mathfrak{B}$. Then the following holds.

LEMMA 4. *For $\alpha, \beta \in \Phi$, $\widehat{\mathfrak{A}_0}(\mathfrak{B}) \models \langle \alpha, \beta \rangle$ if and only if $\mathfrak{B} \models \eta_o(\alpha) \supset \eta_o(\beta)$.*

Proof. Suppose $\mathfrak{B} \not\models \eta_o(\alpha) \supset \eta_o(\beta)$. Then there exists a modal valuation $u : \Psi \to \mathsf{B}$ such that $u(\eta_o(\alpha)) \not\leq u(\eta_o(\beta))$. Define a valuation $v : \Phi \to \mathsf{B}$ as: $v(p_i) := u(q_i)$. Then we can show by induction that $v(\varphi) = u(\eta_o(\varphi))$ for any $\varphi \in \Phi$. Therefore we have $v(\alpha) \not\leq v(\beta)$ in $\widehat{\mathfrak{A}_0}(\mathfrak{B})$. Similarly suppose $\widehat{\mathfrak{A}_0}(\mathfrak{B}) \not\models \langle \alpha, \beta \rangle$. Then there exists a valuation $v : \Phi \to \mathsf{B}$ such that $v(\alpha) \not\leq v(\beta)$. Define a modal valuation $u : \Psi \to \mathsf{B}$ as $u(q_i) := v(p_i)$. Then by the fact that $v(\varphi) = u(\eta_o(\varphi))$ for any $\varphi \in \Phi$, we have $u(\eta_o(\alpha)) \not\leq u(\eta_o(\beta))$. ∎

Note that in the proof above, the fact that $v(\varphi) = u(\eta_o(\varphi))$ for any $\varphi \in \Phi$ shows that the construction $\widehat{\mathfrak{A}_0}(\cdot)$ is closely related to the translation η_o. By this lemma, the following theorems are proved.

THEOREM 5. *Let* **L** *be a binary logic and* **M** *a modal logic. Then the following two conditions are equivalent.*

(1) **L** *is embeddable into* **M** *by* η_0.

(2) *There exists a class* $\mathcal{D}$ *of modal algebras such that:*
 (a) $\mathbf{M} = \mathbf{M}(\mathcal{D})$ *and (b)* $\mathbf{L} = \mathbf{L}(\{\widehat{\mathfrak{A}_0}(\mathfrak{B}) \mid \mathfrak{B} \in \mathcal{D}\})$.

Proof. From (2) to (1), it is immediate. For the converse direction, we take as $\mathcal{D}$ the class of modal algebras that determines the logic **M**. This means that $\mathbf{M} = \mathbf{M}(\mathcal{D})$. Then we have to show (b). Take any $\langle \alpha, \beta \rangle \in \mathbf{L}$ and any $\mathfrak{B} \in \mathcal{D}$. Since **L** is embeddable into **M** by η_0, we have $\eta_0(\alpha) \supset \eta_0(\beta) \in \mathbf{M}$; that means $\mathfrak{B} \models \eta_0(\alpha) \supset \eta_0(\beta)$, and by Lemma 4, this is equivalent to $\widehat{\mathfrak{A}_0}(\mathfrak{B}) \models \langle \alpha, \beta \rangle$. For $\langle \alpha, \beta \rangle \notin \mathbf{L}$, by a similar argument, we can show that there exists a modal algebra $\mathfrak{B}$ such that $\widehat{\mathfrak{A}_0}(\mathfrak{B}) \not\models \langle \alpha, \beta \rangle$. ∎

Given any ABL $\mathfrak{A}$, if we could construct a modal algebra $\mathfrak{B}$ such that $\widehat{\mathfrak{A}_0}(\mathfrak{B})$ is isomorphic to $\mathfrak{A}$, it would follow that every binary logic has a modal companion. Unfortunately, we have not been able to construct such a $\mathfrak{B}$ and thus it is still unknown whether modal companions always exist. In the other direction, however, every normal modal logic **M** has a corresponding binary logic which is embeddable into **M** by η_0, and this binary logic is determined uniquely by **M**.

THEOREM 6. *For any modal logic* **M**, *there exists a unique binary logic* **L** *such that* **L** *is embeddable into* **M** *by* η_0.

Proof. Put $\mathcal{D}_\mathbf{M} := \{\mathfrak{B} \mid \mathfrak{B} \models A \text{ for any } A \in \mathbf{M}\}$. Then it is easily seen by the above theorem that $\mathbf{L} = \mathbf{L}(\{\widehat{\mathfrak{A}_0}(\mathfrak{B}) \mid \mathfrak{B} \in \mathcal{D}_\mathbf{M}\})$ is embeddable into **M** by η_0. If a binary logic $\mathbf{L}'$ is also embeddable into **M** by η_0, then for any $\alpha, \beta \in \Phi$, $\langle \alpha, \beta \rangle \in \mathbf{L}'$ if and only if $\eta_0(\alpha) \supset \eta_0(\beta) \in \mathbf{M}$ if and only if $\langle \alpha, \beta \rangle \in \mathbf{L}$. This means that $\mathbf{L}' = \mathbf{L}$. ∎

It is clear that if a binary logic **L** has a modal companion, then the set of all modal companions of **L** has the smallest element. But the largest element in this set does not always exist. As for a characterization of logics for which it does exist, we only know the following.

PROPOSITION 7. *Let* **L** *be a binary logic and* $MC_{\eta_0}(\mathbf{L})$ *the set of all modal companions of* **L**. *Suppose* $MC_{\eta_0}(\mathbf{L}) \neq \emptyset$. *Then the greatest normal modal logic exists in* $MC_{\eta_0}(\mathbf{L})$ *if and only if* $\langle MC_{\eta_0}(\mathbf{L}), \subseteq \rangle$ *is directed.*

Proof. If the greatest element exists in $MC_{\eta_0}(\mathbf{L})$, then it is immediate that $\langle MC_{\eta_0}(\mathbf{L}), \subseteq \rangle$ is directed. Conversely, suppose $\langle MC_{\eta_0}(\mathbf{L}), \subseteq \rangle$ is directed. Put $\overline{\mathbf{M}} := \bigcup MC_{\eta_0}(\mathbf{L})$. Then it is trivial that $\mathbf{L}$ is embeddable into $\overline{\mathbf{M}}$ by η_0 and that $\overline{\mathbf{M}}$ is the greatest. And we can also show that $\overline{\mathbf{M}}$ is a normal modal logic because of directedness. ∎

3 Kripke complete binary logics and their embedding into normal modal logics

3.1 Kripke-type semantics for binary logics

Apart from algebraic semantics, we can also employ a Kripke-type semantics for binary logics, and indeed we can prove that the binary logic $\mathbf{L_0}$ is complete with respect to it. Our semantics is a modification of orthoframes and orthomodels used by Goldblatt in [2]. Using this type of semantics we analyze the embeddability of *Kripke complete* binary logics into normal modal logics, and we prove the preservation of Kripke completeness and finite model property between them.

DEFINITION 8 (B-frame and B-model). A *B-frame* is a structure $\mathcal{G} = \langle W, S, Q \rangle$, where W is a non-empty set, S is a binary relation on W, and Q is a subset of $\mathcal{P}(W)$ that is closed under $\cap$ (intersection) and a unary operation J_S on $\mathcal{P}(W)$ defined as: $J_S(X) := \{a \in W \mid \forall b \in W, (b \in X \text{ implies } aSb) \}$ for $X \in \mathcal{P}(W)$. A *B-model* on a frame $\mathcal{G}$ is $\mathfrak{N} = \langle \mathcal{G}, U \rangle$, where U is a map that assigns to each propositional variable an element in Q. In a B-model $\mathfrak{N}$, U generates a relation $\models$ on $W \times \Phi$ as follows. Below, $\mathfrak{N} \models_a \alpha$ means that a formula α is true at a in a model $\mathfrak{N}$.

(1) $\mathfrak{N} \models_a p_i$ iff $a \in U(p_i)$.

(2) $\mathfrak{N} \not\models_a \bot$ for any $a \in W$.

(3) $\mathfrak{N} \models_a \alpha \wedge \beta$ iff $\mathfrak{N} \models_a \alpha$ and $\mathfrak{N} \models_a \beta$.

(4) $\mathfrak{N} \models_a \boxminus \alpha$ iff for any $b \in W$, $\mathfrak{N} \models_b \alpha$ implies aSb

In particular, any B-frame of the form $\langle W, S, \mathcal{P}(W) \rangle$ is called a *Kripke B-frame*, and denoted by $\langle W, S \rangle$. Let Γ be a set of formulas and α a formula. Γ *implies* α at $a \in W$ in $\mathfrak{N}$ ($\mathfrak{N} : \Gamma \models_a \alpha$) if and only if there exist $\beta_1, \beta_2, \ldots, \beta_n \in \Gamma$ such that $\mathfrak{N} \not\models_a \beta_1 \wedge \beta_2 \wedge \cdots \wedge \beta_n$, or $\mathfrak{N} \models_a \alpha$. Γ *implies* α in $\mathfrak{N}$ ($\mathfrak{N} : \Gamma \models \alpha$) if $\mathfrak{N} : \Gamma \models_a \alpha$ for all $a \in W$. Γ *implies* α in a B-frame $\mathcal{G}$ ($\mathcal{G} : \Gamma \models \alpha$) if and only if $\mathfrak{N} : \Gamma \models \alpha$ for any B-model $\mathfrak{N}$ on $\mathcal{G}$. Let $\mathcal{C}$ be a class of B-frames. Γ *implies* α in $\mathcal{C}$ ($\mathcal{C} : \Gamma \models \alpha$) if and only if $\mathcal{G} : \Gamma \models \alpha$ for any B-frame $\mathcal{G}$ in $\mathcal{C}$. We sometimes write $\mathfrak{N} \models_a \langle \alpha, \beta \rangle$ instead of $\mathfrak{N} : \{\alpha\} \models_a \beta$. $\mathcal{G} \models \langle\alpha,\beta\rangle$ and $\mathcal{C} \models \langle\alpha,\beta\rangle$ are used similarly.

We can prove the following completeness theorem of the binary logic $\mathbf{L_0}$ with respect to the class of all Kripke B-frames.

THEOREM 9 (Completeness for $\mathbf{L_0}$). *Let Θ be the set of all Kripke B-frames. For any set of formulas Γ and any formula α, $\Theta : \Gamma \models \alpha$ if and only if $\Gamma \vdash_{\mathbf{L_0}} \alpha$.*

The proof is omitted except for the construction of the canonical B-model. For a consistent binary logic $\mathbf{L}$, a non-empty set Γ of formulas is $\mathbf{L}$-*full* if Γ satisfies the following:

(a) there exists a formula α such that $\Gamma \nvdash_{\mathbf{L}} \alpha$,

(b) for $\alpha, \beta \in \Phi$, if $\alpha \in \Gamma$ and $\alpha \vdash_{\mathbf{L}} \beta$, then $\beta \in \Gamma$,

(c) for $\alpha, \beta \in \Phi$, if $\alpha, \beta \in \Gamma$, then $\alpha \wedge \beta \in \Gamma$.

Then the canonical B-model for a consistent binary logic is defined as follows.

DEFINITION 10 (Canonical model for binary logics). Let $\mathbf{L}$ be a consistent binary logic. $\mathfrak{N}_{\mathbf{L}} = \langle W_{\mathbf{L}}, S_{\mathbf{L}}, Q_{\mathbf{L}}, U_{\mathbf{L}}\rangle$ is the *canonical B-model for* $\mathbf{L}$ if it satisfies the following:

(1) $W_{\mathbf{L}} := \{a \subseteq \Phi(V) \mid a \text{ is } \mathbf{L}\text{-full}\}$.

(2) $S_{\mathbf{L}}$ is a binary relation on $W_{\mathbf{L}}$ defined as: $xS_{\mathbf{L}}y$ if and only if there exists $\varphi \in \Phi$, such that $\boxminus\varphi \in x$ and $\varphi \in y$.

(3) $Q_{\mathbf{L}} := \{\{x \in W_{\mathbf{L}} \mid \alpha \in x\} \mid \alpha \in \Phi\}$.

(4) $U_{\mathbf{L}}(p_i) := \{x \in W_{\mathbf{L}} \mid p_i \in x\}$

For any consistent binary logic, we can prove the completeness theorem with respect to a suitable class of B-frames by the standard technique of general frames for normal modal logics. That is, the following holds.

THEOREM 11. *Let $\mathbf{L} := \mathbf{L_0} + \Sigma$ be a consistent binary logic. Then there exists a class $\mathcal{C}_{\mathbf{L}}$ of B-frames for $\mathbf{L}$ such that for $\Gamma \subseteq \Phi$ and $\alpha \in \Phi$, $\mathcal{C}_{\mathbf{L}} : \Gamma \models \alpha$ if and only if $\Gamma \vdash_{\mathbf{L}} \alpha$.*

To choose suitable B-frames for $\mathbf{L}$, we have to adjust Q in a B-frame depending on Σ.

We can also introduce the notion of Kripke completeness for binary logics as follows.

DEFINITION 12 (Kripke completeness for binary logics). A binary logic **L** is called *Kripke complete* if and only if the following are equivalent. For any pair of formulas $\langle \alpha, \beta \rangle$,

(1) $\langle \alpha, \beta \rangle \in \mathbf{L}$.

(2) For any Kripke B-frame $\mathcal{G}$, if $\mathcal{G} : \varphi \models \psi$ for any $\langle \varphi, \psi \rangle \in \mathbf{L}$, then $\mathcal{G} : \alpha \models \beta$

In other words, **L** is Kripke complete if and only if there exists a class $\mathcal{C}_\mathbf{L}$ of Kripke B-frames for **L** such that for any pair $\langle \alpha, \beta \rangle$ of formulas, $\langle \alpha, \beta \rangle \in \mathbf{L}$ is equivalent to $\mathcal{C}_\mathbf{L} \models \langle \alpha, \beta \rangle$.

Now we discuss the embeddability via Kripke-type semantics. Frames and models for normal modal logics are defined as usual. We also use standard notation, for example, $\mathcal{F} = \langle W, R, P \rangle$ for a (general) frame, $\mathfrak{M} = \langle \mathcal{F}, V \rangle$ for a model, and $\mathcal{F} = \langle W, R, \mathcal{P}(W) \rangle = \langle W, R \rangle$ for a Kripke frame. Further, we denote $\mathfrak{M} \models_a A$ to mean that a formula A is true at a in a model $\mathfrak{M}$. A formula A is *valid* in a frame $\mathcal{F}$ ($\mathcal{F} \models A$) if and only if for any model $\mathfrak{M}$ on $\mathcal{F}$, $\mathfrak{M} \models_a A$ for any $a \in W$. A normal modal logic **M** is called *Kripke complete* when for each formula A, $A \in \mathbf{M}$ if and only if $\mathcal{F} \models A$ in any Kripke frame $\mathcal{F}$ such that $\mathcal{F} \models B$ for any $B \in \mathbf{M}$.

To understand how η_0 works, it is helpful to use the following pair of *transformations* of Kripke frames and Kripke B-frames. Let $\mathcal{K}$ be the class of all Kripke frames. For a binary relation R on a set W, $\check{R}$ denotes its complement as a subset $W \times W$. Transformations $\xi_0 : \Theta \to \mathcal{K}$ and $\xi_0^{-1} : \mathcal{K} \to \Theta$ are defined as follows: for Kripke B-model $\mathfrak{N} = \langle W, S, U \rangle$, $\xi_0(\mathfrak{N}) := \langle W, \check{S}, V \rangle$, where $V(q_i) := U(p_i)$. Conversely, for Kripke model $\mathfrak{M} = \langle W, R, V \rangle$, $\xi_0^{-1}(\mathfrak{M}) := \langle W, \check{R}, U \rangle$, where $U(p_i) := V(q_i)$. Clearly, ξ_0 and ξ_0^{-1} are mutually inverse, namely, $\xi_0^{-1} \circ \xi_0(\mathfrak{N}) = \mathfrak{N}$ for any $\mathfrak{N}$ and $\xi_0 \circ \xi_0^{-1}(\mathfrak{M}) = \mathfrak{M}$ for any $\mathfrak{M}$. Then η_0, ξ_0 and ξ_0^{-1} together have the following nice property.

LEMMA 13. *Let $\mathfrak{N} = \langle W, S, U \rangle$ be a Kripke B-model, and $\mathfrak{M} = \langle W, R, V \rangle$ a Kripke model.*

(1) *For $\varphi \in \Phi$ and for any $x \in W$, $\mathfrak{N} \models_x \varphi$ if and only if $\xi_0(\mathfrak{N}) \models_x \eta_0(\varphi)$.*

(2) *For $\varphi \in \Phi$ and for any $x \in W$, $\mathfrak{M} \models_x \eta_0(\varphi)$ if and only if $\xi_0^{-1}(\mathfrak{M}) \models_x \varphi$.*

Proof. Both can be proved by induction on the formula φ. We show only the case $\varphi \equiv \boxdot \psi$ for (1). Similar argument goes through for (2).

(1): By definition, $\mathfrak{N} \models_x \boxdot \psi$ if and only if for any $y \in W$, $\mathfrak{N} \models_y \psi$

implies xSy. Equivalently, for any $y \in W$, $x\check{S}y$ implies $\mathfrak{N} \not\models_y \psi$. By induction hypothesis, this is equivalent to the fact that for any $y \in W$, $x\check{S}y$ implies $\xi_0(\mathfrak{M}) \not\models_y \eta_0(\psi)$. In a normal modal logic, this means that $\xi_0(\mathfrak{M}) \models_x \Box\neg\eta_0(\psi)$, that is, $\xi_0(\mathfrak{M}) \models_x \eta_0(\boxminus\psi)$. ∎

Preservation results for Kripke complete logics follow from the above lemma. For a Kripke B-frame $\mathcal{G} = \langle W, S\rangle$, we put $\xi_0(\mathcal{G}) := \langle W, \check{S}\rangle$ and for a class $\mathcal{C}$ of Kripke B-frames, and $\xi_0(\mathcal{C}) := \{\xi_0(\mathcal{G}) \mid \mathcal{G} \in \mathcal{C}\}$. For a Kripke complete binary logic **L**, there exists a class $\mathcal{C}_\mathbf{L}$ of Kripke B-frames such that **L** is complete with respect to this $\mathcal{C}_\mathbf{L}$. For this **L**, define a modal logic $\delta_{\xi_0}(\mathbf{L}) := \{A \in \Psi \mid \xi_0(\mathcal{C}_\mathbf{L}) \models A\}$. Then we have the following.

THEOREM 14. *Let **L** be a Kripke complete binary logic. Then,*

(1) **L** *is embeddable into* $\delta_{\xi_0}(\mathbf{L})$ *by* η_0.

(2) $\delta_{\xi_0}(\mathbf{L})$ *is Kripke complete with respect to the class* $\xi_0(\mathcal{C}_\mathbf{L})$.

*(3) If **L** has the finite model property, then so does* $\delta_{\xi_0}(\mathbf{L})$.

This theorem says that δ_{ξ_0} maps every Kripke complete binary logics to a Kripke complete modal logic as its modal companion, and it preserves finite model property.

On the other hand, for a Kripke frame $\mathcal{F} = \langle W, R\rangle$, we put $\xi_0^{-1}(\mathcal{F}) := \langle W, \check{R}\rangle$ and for a class $\mathcal{D}$ of Kripke frames, $\xi_0^{-1}(\mathcal{D}) := \{\xi_0^{-1}(\mathcal{F}) \mid \mathcal{F} \in \mathcal{D}\}$. For a Kripke complete normal modal logic **M**, there exists a class $\mathcal{D}_\mathbf{M}$ of Kripke frames such that **M** is complete with respect to $\mathcal{D}_\mathbf{M}$. For this **M**, define a binary logic $\varepsilon_{\xi_0^{-1}}(\mathbf{M}) := \{\langle\alpha,\beta\rangle \in \Phi^2 \mid \xi_0^{-1}(\mathcal{D}_\mathbf{M}) \models \langle\alpha,\beta\rangle\}$. Then similarly to Theorem 14, we have the following.

THEOREM 15. *Let **M** be a Kripke complete normal modal logic. Then,*

(1) $\varepsilon_{\xi_0^{-1}}(\mathbf{M})$ *is embeddable into* **M** *by* η_0.

(2) $\varepsilon_{\xi_0^{-1}}(\mathbf{M})$ *is Kripke complete with respect to the class* $\xi_0^{-1}(\mathcal{D}_\mathbf{M})$.

*(3) If **M** has the finite model property, then so does* $\varepsilon_{\xi_0^{-1}}(\mathbf{M})$.

Both theorems follow from Lemma 13. As we have seen above, Kripke complete logics possess very nice properties with respect to the embeddability relation. The reason is that transformations ξ_0 and ξ_0^{-1} are mutually inverse.

3.2 Correspondence theory for binary logics

Like *correspondence theory* for modal logic, we establish some correspondences between axiom schemes of binary logics and first order conditions on the relation S of B-frames. Let $\mathcal{G}$ be a B-frame and Form a first order formula in the language of frames. We write $\mathcal{G} \models$ Form to mean that $\mathcal{G}$ satisfies the condition Form. First we give some examples of correspondence between axioms and first order conditions on Kripke B-frames.

PROPOSITION 16. *Let* $\mathcal{G} = \langle W, S \rangle$ *a Kripke B-frame.*

(1) $\mathcal{G} \models \langle p, \boxminus\boxminus p \rangle$ *if and only if* $\mathcal{G} \models \forall x, y (xSy \to ySx)$.

(2) $\mathcal{G} \models \langle p \wedge \boxminus p, \bot \rangle$ *if and only if* $\mathcal{G} \models \forall x \sim(xSx)$.

(3) $\mathcal{G} \models \langle p, \boxminus p \rangle$ *if and only if* $\mathcal{G} \models \forall x, y (xSy)$.

(4) $\mathcal{G} \models \langle \boxminus p, \boxminus\boxminus p \rangle$ *if and only if* $\mathcal{G} \models \forall x, y, z (xSy \,\&\, zSy \to xSz)$.

(5) $\mathcal{G} \models \langle p \wedge \boxminus p, \boxminus\boxminus p \rangle$ *if and only if* $\mathcal{G} \models \forall x (xSx \to \forall y (ySx \to xSy))$.

(6) $\mathcal{G} \models \langle \boxminus p \wedge \boxminus\boxminus p, \boxminus\boxminus\boxminus p \rangle$ *if and only if*
$\mathcal{G} \models \forall x [\forall y \{\forall z (xSz \to ySz) \to xSy\} \to$
$\forall u \{\forall v (\forall w (xSw \to vSx) \to uSv) \to xSu\}]$

Proof. We show only (5). Others can be proved similarly. Suppose that $\mathcal{G} \models \forall x (xSx \to \forall y (ySx \to xSy))$. Consider an arbitrary valuation U on $\mathcal{G}$ and a Kripke B-model $\mathfrak{N} = \langle \mathcal{G}, U \rangle$. Assume $\mathfrak{N} \models_a p \wedge \boxminus p$ for an element $a \in W$. Then $\mathfrak{N} \models_a p$ and $\mathfrak{N} \models_a \boxminus p$, and so, we have aSa. Take any $b \in W$, such that $\mathfrak{N} \models_b \boxminus p$. Then bSa holds, and by our supposition aSb also holds. Therefore we have $\mathfrak{N} \models_a \boxminus\boxminus p$. Thus we have $\mathcal{G} : p \models \boxminus\boxminus p$. Conversely, suppose $\mathcal{G} \not\models \forall x (xSx \to \forall y (ySx \to xSy))$. Then there exist $a, b \in W$ such that aSa, bSa, but $\sim(aSb)$. Define a valuation U such that $U(p) := \{a\}$. Then in the Kripke B-model $\mathfrak{N} = \langle \mathcal{G}, U \rangle$, $\mathfrak{N} \models_a p$ and $\mathfrak{N} \models_a \boxminus p$, and $\mathfrak{N} \models_b \boxminus p$. But since $\sim(aSb)$, $\mathfrak{N} \not\models_a \boxminus\boxminus p$. Thus we have $\mathcal{G} : p \wedge \boxminus p \not\models \boxminus\boxminus p$. ∎

To say more for the case (5), let $\mathbf{L} = \mathbf{L_0} + \langle p \wedge \boxminus p, \boxminus\boxminus p \rangle$. Then it can be shown that the canonical frame for $\mathbf{L}$ satisfies the first order condition on the right hand side in (5). For, consider the canonical Kripke B-frame $\mathcal{G}_\mathbf{L} = \langle W_\mathbf{L}, S_\mathbf{L} \rangle$. Suppose that $xS_\mathbf{L}x$ for $x \in W_\mathbf{L}$, then there exists $\alpha \in \Phi$ such that $\boxminus \alpha \in x$ and $\alpha \in x$. Further assume that $yS_\mathbf{L}x$ for $y \in W_\mathbf{L}$, then there exists $\beta \in \Phi$ such that $\boxminus \beta \in y$ and $\beta \in x$. Here we have $\alpha \wedge \beta \in x$ and $\boxminus(\alpha \wedge \beta) \in x$, and so, $\boxminus\boxminus(\alpha \wedge \beta) \in x$ by the extra axiom for $\mathbf{L}$. On the other hand, $\boxminus(\alpha \wedge \beta) \in y$ also holds. Therefore $xS_\mathbf{L}y$. Thus, we can conclude that the binary logic **L** is Kripke complete with respect to the class of Kripke B-frames that satisfy $\forall x(xSx \to \forall y(ySx \to xSy))$.

Similarly, for all other cases we can show that each binary logic which is defined by adding to $\mathbf{L_0}$ the pair of formulas on the left hand side is Kripke complete with respect to the class of Kripke B-frames that satisfy the corresponding first order condition on the right hand side.

Note that, in (2) the pair $\langle p \wedge \boxminus p, \bot \rangle$ defines the *irreflexivity* of Kripke B-frames. This fact is in a sharp contrast to first order definability in modal logics, where the irreflexivity of a Kripke frame can not be defined by any modal formula.

Below, we rewrite every statement in Proposition 16 for normal modal logics by η_0 and ξ_0. Then we obtain the following.

(1^∂) $\mathcal{F} \models q \supset \Box\Diamond q$ if and only if $\mathcal{F} \models \forall x, y(yRx \to xRy)$.

(2^∂) $\mathcal{F} \models q \supset \Diamond q$ if and only if $\mathcal{F} \models \forall x(xRx)$.

(3^∂) $\mathcal{F} \models \Diamond q \supset \neg q$ if and only if $\mathcal{F} \models \forall x, y(\sim(xRy))$.

(4^∂) $\mathcal{F} \models \Box\neg q \supset \Box\Diamond q$ if and only if $\mathcal{F} \models \forall x, y, z(xRz \to ((xRy) or (zRy)))$.

(5^∂) $\mathcal{F} \models (q \wedge \Box\neg q) \supset \Box\Diamond q$ if and only if
$\mathcal{F} \models \forall x\{\exists y(xRy \ \& \sim(yRx)) \to xRx\}$.

(6^∂) $\mathcal{F} \models (\Box\neg q \wedge \Box\Diamond q) \supset \Box\neg\Box\Diamond q$ if and only if
$\mathcal{F} \models \forall x[\forall y\{xRy \to \exists z(yRz \ \& \sim(xRz))\} \to \forall u\{xRu \to \exists v(\forall w(vRx \to xRw) \& uRv)\}]$.

For example, again for the case (5) and (5^∂), let $\mathbf{L} := \mathbf{L_0} + \{\langle p \wedge \boxminus p, \boxminus\boxminus p\rangle\}$. Then, since **L** is Kripke complete, by Theorem 14 and correspondences (5) in Proposition 16 and (5^∂), we obtain that **L** can be embedded into the normal modal logic $\mathbf{K} \oplus \{(q \wedge \Box\neg q) \supset \Box\Diamond q\}$ by η_0. (1^∂) and (2^∂) are well known as **B** and **T** respectively, whereas others are not familiar.

These facts are only a first step of correspondence theory for binary logics. To determine the class of formulas which characterizes some first order properties of Kripke B-frames will be an interesting problem. Also, to investigate the expressive power of formulas of binary logics by model theoretic arguments will be also another challenging project.

4 Goldblatt's translation of EXT(O) into NEXT(KTB)

In [2], Goldblatt used another translation of formulas to show that **O** is embeddable into **KTB**, namely, η_1 defined as: $\eta_1(\bot) := \bot$, $\eta_1(p_i) := \Box\Diamond q_i$, $\eta_1(\alpha \wedge \beta) := \eta_1(\alpha) \wedge \eta_1(\beta)$, and $\eta_1(\boxminus\alpha) := \Box\neg\eta_1(\alpha)$.

According to Proposition 16 and the discussion following it, the binary logic $\mathbf{L_0} + \{\langle p, \boxminus \boxminus p\rangle, \langle p \wedge \boxminus p, \bot\rangle\}$ can also be embedded into the modal logic **KTB** by η_0. The difference between $\eta_0(p_i)$ and $\eta_1(p_i)$ arises from the axiom scheme $\langle \boxminus \boxminus p, p\rangle$ of orthologics. There seems to be no first order condition on Kripke B-frames to characterize this axiom scheme. In order to overcome this disadvantage, Goldblatt uses a little trick (see [2]).

In this section, we discuss the relation between $\mathrm{EXT}(\mathbf{O})$ and $\mathrm{NEXT}(\mathbf{KTB})$ by the translation η_1 with the help of algebraic semantics and Kripke-type semantics. Since we have the *double negation law* $(\langle p, \boxminus \boxminus p\rangle, \langle \boxminus \boxminus p, p\rangle)$ in this case, we can go further to obtain finer results on embeddability of orthologics into normal modal logics over **KTB**, that are presented in the next section.

We begin with the relation between the algebraic semantics first. An *ortholattice* is an ABL $\mathfrak{A} = \langle \mathsf{A}, \cap, \cup, J, 0, 1\rangle$ in which the operation J satisfies the following conditions. For any $x \in \mathsf{A}$, (1): $J(J(x)) = x$, and (2): $x \cap J(x) = 0$. Note that in an ortholattice, a operation $\cup$ (join) can be defined as $x \cup y := J(J(x) \cap J(y))$ and so, an ortholattice is a lattice. An *orthologic* is a binary logic which contains **O**. Any orthologic is characterized by a class of ortholattices.

A **KTB**-algebra is a modal algebra $\mathfrak{B} = \langle \mathsf{B}, \cap, \cup, -, I, 0, 1\rangle$ in which the operation I satisfies (1): $I(x) \leq x$, and (2): $x \leq I(-I(-x))$. The modal logic **KTB** is characterized by the class of all **KTB**-algebras. The conditions (1) and (2) correspond to **T** and **B** respectively. We call a normal modal logic over **KTB** a **KTB**-*logic*.

For a **KTB**-algebra $\mathfrak{B}$, we can also define an algebra $\widehat{\mathfrak{A}_1}(\mathfrak{B}) := \langle \mathcal{R}(\mathsf{B}), \cap, \cup, J_\mathfrak{B}, 0, 1\rangle$, where $\mathcal{R}(\mathsf{B}) := \{a \in \mathsf{B} \mid I(-I(-a)) = a\}$ and $J_\mathfrak{B}(x) = I(-x)$. It is easily seen that this $\widehat{\mathfrak{A}_1}(\mathfrak{B})$ is an ortholattice for any **KTB**-algebra $\mathfrak{B}$. An element $a \in \mathcal{R}(\mathsf{B})$ is called *regular*. Now we have the following.

LEMMA 17. *For $\alpha, \beta \in \Phi$, $\widehat{\mathfrak{A}_1}(\mathfrak{B}) \models \langle \alpha, \beta\rangle$ if and only if $\mathfrak{B} \models \eta_1(\alpha) \supset \eta_1(\beta)$.*

Proof. This proof is basically the same as that for Lemma 4. Note that for a modal valuation $u : \Psi \to \mathfrak{B}$, if we define a valuation $v : \Phi \to \mathcal{R}(\mathsf{B})$ by $v(p_i) := u(\eta_1(p_i)) = u(\square \diamond q_i)$, then we can show by induction that $v(\alpha) = u(\eta_1(\alpha)) \in \mathcal{R}(\mathsf{B})$ for all $\alpha \in \Phi$. Therefore in our proof, when a modal valuation $u : \Psi \to \mathfrak{B}$ is given, v may be defined as $v(p_i) := u(\eta_1(p_i))$. On the other hand, when a valuation $v : \Phi \to \mathcal{R}(\mathsf{B})$ is given, it is enough to define u as: $u(q_i) := v(p_i)$, because $v(p_i) = I - (I - v(p_i)) = I - (I - u(q_i)) = u(\square \diamond q_i) = u(\eta_1(p_i))$. ∎

From the above lemma, similar results as in the general case of binary logics follow, that is, (1) a necessary and sufficient condition for an orthologic to be embeddable into a **KTB**-logic, (2) the uniqueness of the orthologic that is embeddable into a given **KTB**-logic, and (3) a necessary and sufficient condition for an orthologic to have the greatest **KTB**-logic in its modal companions. Their proofs are quite the same as those of Theorem 5, 6, and Proposition 7.

However, we can get more in the case of orthologics. First, for an ortholattice $\mathfrak{A} = \langle \mathsf{A}, \cap, \cup, J, 0, 1 \rangle$, a modal algebra $\widehat{\mathfrak{B}}(\mathfrak{A})$ is defined as a structure $\langle \mathcal{P}(F(\mathsf{A})), \cap, \cup, -, I_R, \emptyset, F(\mathsf{A}) \rangle$, where

(1) $F(\mathsf{A}) := \{F \subseteq \mathsf{A} \mid F \text{ is a proper filter on } \mathfrak{A}\}$,

(2) $\cap, \cup, -$ are the set-theoretic intersection, union, and complement respectively,

(3) R is a binary relation on $F(\mathsf{A})$ defined as: $_F R_G$ if and only if $J(a) \in F$ implies $a \notin G$ for any $a \in \mathsf{A}$, and

(4) $I_R(X) := \{F \in F(\mathsf{A}) \mid \forall G (_F R_G \text{ implies } G \in X)\}$ for $X \in \mathcal{P}(F(\mathsf{A}))$.

It is easy to see that $\widehat{\mathfrak{B}}(\mathfrak{A})$ is a **KTB**-algebra. Define a map $\varphi : \mathsf{A} \to \mathcal{P}(F(\mathsf{A}))$ as: $\varphi(a) := \{F \in F(\mathsf{A}) \mid a \in F\}$. This map φ has the following property.

LEMMA 18. *For $a, b \in \mathsf{A}$,*

(1) $\varphi(J(a)) = I_R(-\varphi(a))$.   *(3) $\varphi(a \cap b) = \varphi(a) \cap \varphi(b)$.*

(2) $\varphi(a) = I_R(-I_R(-\varphi(a)))$.   *(4) $a \leq b$ if and only if $\varphi(a) \subseteq \varphi(b)$.*

This means that there is an isomorphic copy $\varphi(\mathfrak{A})$ in $\widehat{\mathfrak{B}}(\mathfrak{A})$. This construction is a modification of Jónsson-Tarski representation of modal algebras [4]. It can also be applied for general ABLs and modal algebras, but in that case, we cannot continue the following discussion.

For an ortholattice $\mathfrak{A}$ whose universe is A, the regular elements of $\widehat{\mathfrak{B}}(\mathfrak{A})$ is $\mathcal{R}(\widehat{\mathfrak{B}}(\mathfrak{A})) := \{X \in \mathcal{P}(F(\mathsf{A})) \mid I_R(-I_R(-X)) = X\}$. We put $\varphi(\mathsf{A}) := \{\varphi(a) \mid a \in \mathsf{A}\}$. If A is finite, then we have the following.

LEMMA 19. *Let $\mathfrak{A}$ be an ortholattice whose universe A is finite. Then $\varphi(\mathsf{A}) = \mathcal{R}(\widehat{\mathfrak{B}}(\mathfrak{A}))$.*

Proof. It is straightforward to show that $\varphi(\mathsf{A}) \subseteq \mathcal{R}(\widehat{\mathfrak{B}}(\mathfrak{A}))$ by Lemma 18 (2). To show the converse, suppose $X \in \mathcal{R}(\widehat{\mathfrak{B}}(\mathfrak{A}))$. If $X = F(\mathsf{A})$, then of course, $X = \varphi(1) \in \varphi(\mathsf{A})$. Therefore consider the case where $X \neq F(\mathsf{A})$.

Take an arbitrary $x \in F(\mathsf{A}) - X$, that is, $x \notin X = I_R(-I_R(-X))$. Then there exists $y \in F(\mathsf{A})$ such that xRy and $y \notin -I_R(-X)$. The latter is equivalent to that for any $z \in X$, there exists an element $a(z) \in \mathsf{A}$ such that $J(a(z)) \in y$ and $z \in \varphi(a(z))$. Since z is an arbitrary element of X, we obtain $X \subseteq \bigcup_{z \in X} \varphi(a(z))$. Because A is finite, X is also finite and so, X can be expressed as $\{z_i \mid 1 \leq i \leq n\}$ for some n. Therefore $X \subseteq \bigcup_{i=1}^{n} \varphi(a(z_i))$. Putting $a(x) := a(z_1) \cup \cdots \cup a(z_n)$, we have

$$X \subseteq \varphi(a(x)). \quad (\star)$$

On the other hand, since $J(a(z_i)) \in y$ for $i = 1, 2, \ldots, n$, $\bigcap_{i=1}^{n} J(a(z_i)) \in y$. Because $J(\bigcap_{i=1}^{n} J(a(z_i))) = \bigcup_{i=1}^{n} a(z_i) = a(x)$, if $a(x) \in x$, then $J(J(a(x))) \in x$ and $J(a(x)) \in y$. This is a contradiction since xRy. Thus $a(x) \notin x$, that is, $x \in -\varphi(a(x))$. Since x is an arbitrary element of $-X$, we obtain

$$-X \subseteq \bigcup_{x \in -X} -\varphi(a(x)) \quad (\sharp)$$

By $(\star)$, we have also $-X \supseteq -\varphi(a(x))$, and so, $-X \supseteq \bigcup_{x \in -X} -\varphi(a(x))$. This, together with $(\sharp)$ implies that $-X = \bigcup_{x \in -X} -\varphi(a(x))$. Since $-X$ is finite, and so $-X = \{x_j \mid 1 \leq j \leq m\}$ for some m, we obtain $-X = \bigcup_{j=1}^{m} -\varphi(a(x_j))$. Therefore $X = \bigcap_{j=1}^{m} \varphi(a(x_j)) = \varphi(\bigcap_{j=1}^{m} a(x_j))$. Thus $X \in \varphi(\mathsf{A})$. ∎

This proof is based on representation of ortholattices by Goldblatt [3]. In the case of ABLs, we do not have a nice representation theorem for them, and therefore no similar result seems to be in sight. Now we can prove that for a finite ortholattice $\mathfrak{A}$, $\mathbf{L}(\mathfrak{A})$ has a modal companion by η_1.

LEMMA 20. *Let $\mathfrak{A}$ be any finite ortholattice. Then for any $\alpha, \beta \in \Phi$, $\mathfrak{A} \models \langle \alpha, \beta \rangle$ if and only if $\widehat{\mathfrak{B}}(\mathfrak{A}) \models \eta_1(\alpha) \supset \eta_1(\beta)$.*

Proof. The direction from right to left is trivial by Lemma 18. Suppose $\widehat{\mathfrak{B}}(\mathfrak{A}) \not\models \eta_1(\alpha) \supset \eta_1(\beta)$. Then there exists a modal valuation $u : \Psi \to \widehat{\mathfrak{B}}(\mathfrak{A})$ such that $u(\eta_1(\alpha)) \not\subseteq u(\eta_1(\beta))$. Since for every p_i, $u(\eta_1(p_i))$ is regular in $\widehat{\mathfrak{B}}(\mathfrak{A})$, there exists the unique element $a_i \in \mathsf{A}$ such that $u(\eta_1(p_i)) = \varphi(a_i)$

by Lemma 19. Now define a valuation $v : \Phi \to \mathfrak{A}$ by $v(p_i) := a_i$ for every p_i. Then, of course, $\varphi(v(p_i)) = u(\eta_1(p_i))$. It can be shown easily by induction that for any $\gamma \in \Phi$, $\varphi(v(\gamma)) = u(\eta(\gamma))$. Then by our supposition, we have $\varphi(v(\alpha)) \not\sqsubseteq \varphi(v(\beta))$, which means that $v(\alpha) \not\leq v(\beta)$. Thus we have $\mathfrak{A} \not\models \langle \alpha, \beta \rangle$. ∎

THEOREM 21.

(1) Let $\mathfrak{A}$ be a finite ortholattice. Then, for any $\alpha, \beta \in \Phi$, $\langle \alpha, \beta \rangle \in \mathbf{L}(\mathfrak{A})$ if and only if $\eta(\alpha) \supset \eta(\beta) \in \mathbf{M}(\widehat{\mathfrak{B}}(\mathfrak{A}))$, that is, every tabular ortholongic is embeddable into a tabular modal logic in $\text{NEXT}(\mathbf{KTB})$.

(2) Every orthologic that has the finite model property is embeddable by η_1 into a normal modal logic in $\text{NEXT}(\mathbf{KTB})$, which also has the finite model property.

Next we deal with Kripke-type semantics of orthologics and $\mathbf{KTB}$-logics, and the embeddability relation between these logics. An *orthoframe* is a Kripke B-frame $\mathcal{G} = \langle W, S \rangle$, where S is irreflexive and symmetric, and an *orthomodel* on $\mathcal{G}$ is a Kripke B-model $\mathfrak{N} = \langle \mathcal{G}, U \rangle$, where the valuation U fulfills the following restriction. For any variable p_i,

$$J_S(J_S(U(p_i))) = U(p_i) \qquad (*)$$

We use $\models^*$ for an orthomodel to express that its valuation satisfies $(*)$. Then the following holds.

THEOREM 22. Let Ω be the class of all orthoframes. For any set of formulas Γ and any formula α, $\Omega : \Gamma \models^* \alpha$ if and only if $\Gamma \vdash_O \alpha$. □

Kripke frames which characterize the modal logic $\mathbf{KTB}$, that is, *reflexive* and *symmetric* Kripke frames, are called $\mathbf{KTB}$-Kripke frames. Kripke models on them are called $\mathbf{KTB}$-Kripke models.

A pair of transformations ξ_1 and ξ_1^{-1} between orthomodels and $\mathbf{KTB}$-Kripke models is defined as follows: For an orthomodel $\mathfrak{N} = \langle W, S, U \rangle$, a $\mathbf{KTB}$-Kripke model $\xi_1(\mathfrak{N}) := \langle W, \check{S}, V \rangle$, where $V(q_i) := U(p_i)$. For a $\mathbf{KTB}$-Kripke model $\mathfrak{M} = \langle W, R, V \rangle$, an orthomodel $\xi_1^{-1}(\mathfrak{M}) := \langle W, \check{R}, U \rangle$, where $U(p_i) := \{x \in W \mid \mathfrak{M} \models_x \Box \Diamond q_i\}$. ξ_1 and ξ_1^{-1} are mutually inverse in frame part, and we also have the following, quite similar to Lemma 13.

LEMMA 23.

(1) Let $\varphi \in \Phi$ and $\mathfrak{N} = \langle W, S, U \rangle$ be an orthomodel. For any $x \in W$, $\mathfrak{N} \models^*_x \varphi$ if and only if $\xi_1(\mathfrak{N}) \models_x \eta_1(\varphi)$.

(2) Let $\varphi \in \Phi$ and $\mathfrak{M} = \langle W, R, V \rangle$ be a **KTB**-*Kripke model*. For any $x \in W$, $\mathfrak{M} \models_x \eta_1(\varphi)$ if and only if $\xi_1^{-1}(\mathfrak{M}) \models_x^* \varphi$.

By this lemma, we can prove analogue of Theorems 14 and 15. But for correspondence theory on orthoframes, things are more delicate because of the restriction (∗).

5 Two results on the embeddability of orthologics into normal logics over KTB

We finish off with two more results which are closely related to the embeddability relation between EXT(**O**) and NEXT(**KTB**) by η_1.

5.1 A continuum of normal modal logics over KTB

Recently the author proved [8] that there exists a continuum of orthologics by employing a class of ortholattices of special type. From this fact together with our embeddability of tabular orthologics, the existence of a continuum of **KTB**-logics follows. Our ortholattices are shown in the figure below. In this figure, the left and the right a_0's should be identified. So should the left and the right b_0's. We can see that the greatest element 1 is surrounded by $2n - 1$ a_k's, and that the least element 0 is surrounded by $2n - 1$ b_k's. The operation J is defined in the following way, so as to meet the requirement of definition for ortholattices.

$$J(a_0) := b_{n-1}, J(a_1) := b_n, \ldots\ldots, J(a_n) := b_0, \ldots\ldots, J(a_{2n-2}) := b_{n-2}$$

Of course, $J(1) = 0$ and $J(J(x)) = x$ for every element x. We name this type of ortholattices *chandelier type*, and denote the ortholattice of chandelier type with $2n-1$ atoms and $2n-1$ coatoms by $\mathfrak{A}_{2n-1}$. Its atoms and coatoms are denoted by b_k's and a_k's ($k = 0, 1, \ldots 2n - 2$) respectively. The class of all ortholattices of chandelier type is denoted by $\mathcal{C}h$, more precisely, $\mathcal{C}h := \{\mathfrak{A}_{2k-1} \mid k \geq 3\}$. It is easily seen that no member of $\mathcal{C}h$ is orthomodular. To see that, notice that each member of $\mathcal{C}h$ has a subortholattice of benzene-type, and thus it is not orthomodular, by using the well-known criterion for orthomodular lattices [5]. This type of lattice was used by McKenzie [7] to prove that there exists a continuum of equational theories of lattices. The author modified his lattices to be ortholattices.

Let ℓ be any odd number greater than or equal to 5. For propositional variables $p_0, p_1, \ldots, p_{\ell-1}$, define two formulas α_ℓ, β_ℓ as follows:

$$\alpha_\ell := \bigwedge_{i \neq j, i \neq k, j \equiv k+1} p_i \vee (p_j \wedge p_k), \qquad \beta_\ell := \bigvee_{s \neq t \pm 1} (p_s \wedge p_t),$$

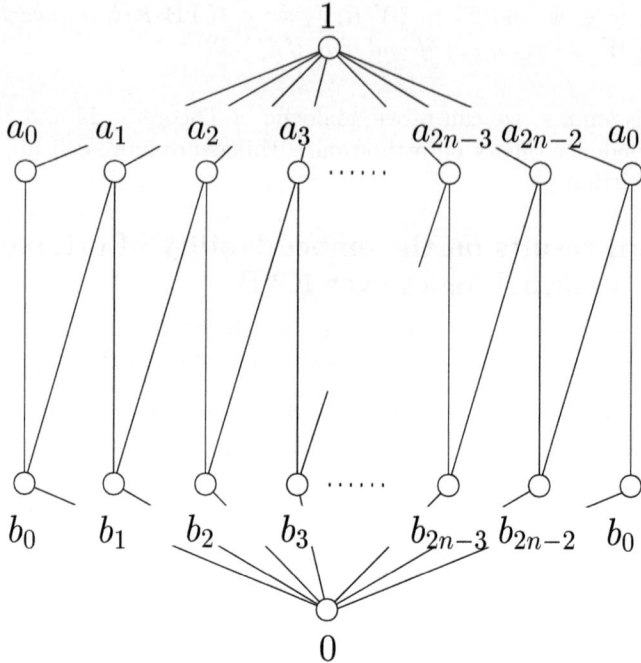

Figure 1. The ortholattice of chandelier type $\mathfrak{A}_{2n-1}$

where $\equiv$ expresses the congruence relation modulo ℓ. On these ortholattices and formulas we have the following [8].

LEMMA 24. *Let ℓ, m be odd numbers greater than or equal to 5.*
(1) $\mathfrak{A}_\ell \not\models \langle \alpha_\ell, \beta_\ell \rangle$.
(2) *For $m \neq \ell$, $\mathfrak{A}_m \models \langle \alpha_\ell, \beta_\ell \rangle$.*

By the above lemma, the following holds.

THEOREM 25.
(1) *For different subclasses $\mathcal{C}_1, \mathcal{C}_2 \subseteq Ch$, $\mathbf{L}(\mathcal{C}_1) \neq \mathbf{L}(\mathcal{C}_2)$.*
(2) *There exists a continuum of orthologics.*

Proof. (1): We may assume that $C_1 \not\subseteq C_2$. Then there exists a number ℓ such that $\mathfrak{A}_\ell \in C_1$ but $\mathfrak{A}_\ell \notin C_2$. By Lemma 24, we have that $C_1 \not\models \langle \alpha_\ell, \beta_\ell \rangle$ and that $C_2 \models \langle \alpha_\ell, \beta_\ell \rangle$. This means that $\mathbf{L}(C_2) \not\subseteq \mathbf{L}(C_1)$.
(2): Since the cardinality of Ch is $\aleph_0$, the cardinality of all subclasses of Ch is $2^{\aleph_0}$, which is, by (1) above, the same cardinality of orthologics which are determined by some subclass of Ch. ∎

Because every member of Ch is a finite ortholattice, we can show the following by Lemma 20.

LEMMA 26. *Let ℓ, m be odd numbers greater or equal to 5. Then,*

(1) $\mathfrak{B}(\mathfrak{A}_\ell) \not\models \eta_1(\alpha_\ell) \supset \eta_1(\beta_\ell)$.

(2) *For $m \neq \ell$, $\mathfrak{B}(\mathfrak{A}_m) \models \eta_1(\alpha_\ell) \supset \eta_1(\beta_\ell)$.*

Denote $\mathfrak{B}(Ch) := \{\mathfrak{B}(\mathfrak{A}_{2k-1}) \mid k \geq 3\}$. Then now we can show our result.

THEOREM 27. *For all $\mathcal{D}_1, \mathcal{D}_2 \subseteq \mathfrak{B}(Ch)$, if $\mathcal{D}_1 \neq \mathcal{D}_2$, then $\mathbf{M}(\mathcal{D}_1) \neq \mathbf{M}(\mathcal{D}_2)$. Therefore, there exists a continuum of $\mathbf{KTB}$-logics.*

5.2 Infinitely many modal companions of every tabular orthologic

We consider orthologics which are characterized by single finite orthoframes here. We show that every such orthologic has infinitely many modal companions. For an orthoframe $\mathcal{G}$, we denote $\mathbf{L}(\mathcal{G}) := \{\langle \alpha, \beta \rangle \mid \mathcal{G} \models \langle \alpha, \beta \rangle\}$. Similarly, for a $\mathbf{KTB}$-Kripke frame $\mathcal{F}$, we denote $\mathbf{M}(\mathcal{F}) := \{A \mid \mathcal{F} \models A\}$. Then by Lemma 23, we have the following.

LEMMA 28. *Let $\mathcal{G}$ be an orthoframe and $\mathcal{F}$ a $\mathbf{KTB}$-Kripke frame. For any $\alpha, \beta \in \Phi$,*

(1) $\langle \alpha, \beta \rangle \in \mathbf{L}(\mathcal{G})$ *if and only if* $\eta_1(\alpha) \supset \eta_1(\beta) \in \mathbf{M}(\xi_1(\mathcal{G}))$.

(2) $\langle \alpha, \beta \rangle \in \mathbf{L}(\xi_1^{-1}(\mathcal{F}))$ *if and only if* $\eta_1(\alpha) \supset \eta_1(\beta) \in \mathbf{M}(\mathcal{F})$.

Let $\mathcal{G} = \langle W, S \rangle$ be an orthoframe in which $|W| = n$ $(n \in \omega)$. For this $\mathcal{G}$, define $\mathcal{G}^{(+)} := \langle W \cup \{a\}, S \rangle$, where a is a new irreflexive point and nothing is changed except this. Then the following holds.

PROPOSITION 29.

(1) $\mathbf{L}(\mathcal{G}) = \mathbf{L}(\mathcal{G}^{(+)})$.

(2) There exists a formula $A \in \Psi$ such that $A \in \mathbf{M}(\xi_1^{-1}(\mathcal{G}))$ and that $A \notin \mathbf{M}(\xi_1^{-1}(\mathcal{G}^{(+)}))$

Proof. (1): This can be proved by using the fact that all regular subsets in W coincide with all regular subsets in $W \cup \{a\}$ since a is not related to any points in W by S.
(2): In $\xi_1^{-1}(\mathcal{G}^{(+)})$, a is the point that is related to every point (including a itself) by $\check{S}$, therefore a has $n+1$ *arms*, whereas $\xi_1^{-1}(\mathcal{G})$ has no such point because it has only n points. Consider formulas $F_0, F_1, \ldots, F_n \in \Psi$ such that each of them is satisfiable but that every couple of them cannot be satisfied at the same one point. In fact, for any $k \in \omega$, such k formulas can be effectively picked up. (For example, for $n = 3$, we can pick up $F_0 = q_1 \wedge q_2$, $F_1 = \neg q_1 \wedge q_2$, $F_2 = q_1 \wedge \neg q_2$, and $F_3 = \neg q_1 \wedge \neg q_2$.) Define $B := F_0 \wedge \Diamond F_1 \wedge \cdots \wedge \Diamond F_n$. For an appropriate valuation V, $\langle \xi_1^{-1}(\mathcal{G}^{(+)}), V \rangle \models_a B$, but there is no such point in $\xi_1^{-1}(\mathcal{G})$ that satisfies B. Therefore we have $\xi_1^{-1}(\mathcal{G}) \models \neg B$ and $\xi_1^{-1}(\mathcal{G}^{(+)}) \not\models \neg B$, which means that $\neg B \in \mathbf{M}(\xi_1^{-1}(\mathcal{G}))$ and that $\neg B \notin \mathbf{M}(\xi_1^{-1}(\mathcal{G}^{(+)}))$. ∎

Now we have our result.

THEOREM 30. *Let $\mathcal{G}$ be a finite orthoframe. Then there exists an infinite descending chain of modal logics in* NEXT(**KTB**) *into each of which* $\mathbf{L}(\mathcal{G})$ *can be embedded by η_1.*

Proof. Suppose $\mathcal{G}$ has n points and is denoted by $\mathcal{G}_{(n)}$. For each k, $\mathcal{G}_{(n+k)}$ denotes the orthoframe by only adding k irreflexive points to $\mathcal{G}_{(n)}$ with everything else untouched. Then by the previous Proposition, $\mathbf{L}(\mathcal{G}_{(n)}) = \mathbf{L}(\mathcal{G}_{(n+k)})$ for any k. Put $\mathbf{M}_\ell := \bigcap_{k=0}^{\ell} \mathbf{M}(\xi_1^{-1}(\mathcal{G}_{(n+k)}))$, then trivially, $\mathbf{M}_0 \supseteq \mathbf{M}_1 \supseteq \cdots$. Moreover, by a similar argument as in the proof of the proposition above, for every ℓ, there exists a formula $A_{\ell+1}$ such that $A_{\ell+1} \in \mathbf{M}_\ell$ and that $A_{\ell+1} \notin \mathbf{M}_{\ell+1}$. Thus $\mathbf{M}_\ell \supsetneq \mathbf{M}_{\ell+1}$. Finally, by Lemma 28, $\langle \alpha, \beta \rangle \in \mathbf{L}(\mathcal{G}_{(n)})$ if and only if $\eta_1(\alpha) \supset \eta_1(\beta) \in \mathbf{M}_\ell$ for any ℓ. ∎

This work is only a first step of studying the relation between EXT($\mathbf{L}_0$) and NEXT(**K**), and that between EXT(**O**) and NEXT(**KTB**). We have not yet reached complete answers to problems (1) and (2) in the first section. But we hope this study will give a framework in which the structure of the lattices of these classes of logics are clarified.

BIBLIOGRAPHY

[1] A. Chagrov and M. Zakharyaschev. *Modal logic*, Oxford University Press, 1997.
[2] R. I. Goldblatt. Semantic analysis of orthologic, *Journal of Philosophical Logic*, **3**, 19–35, 1974.
[3] R. I. Goldblatt. The Stone space of an ortholattice, *Bull. of London Math. Soc.*, **7**, 45–48, 1975.
[4] B. Jónsson and A. Tarski. Boolean algebras with operators, *Amer. Jour. of Math.*, **73**, 891–939, 1951.
[5] G. Kalmbach. *Orthomodular Lattices*, Academic Press, 1983.
[6] J. Malinowski. The deduction theorem for quantum logic – some negative results, *Jour. of Symbolic Logic*, **55**, 615–625, 1990.
[7] R. McKenzie. Equational bases for lattice theories, *Math. Scand.*, **27**, 24–38, 1970.
[8] Y. Miyazaki. Some properties on orthologics, To appear in *Studia Logica*.

Yutaka Miyazaki
Meme Media Laboratory
Hokkaido University
Kita 13, Nishi 8, Kita-ku
Sapporo, 060-8628, Japan
E-mail: y-miya@meme.hokudai.ac.jp

16
Completions of Algebras and Completeness of Modal and Substructural Logics

HIROAKIRA ONO

ABSTRACT. We will give here a brief overview of completions of algebras, and of completeness results on both modal logics and substructural logics which are obtained by using completions. Our aim is to discuss completions of algebras and completeness theorems of logics of various kinds in a uniform way, and to try to find common features among them. As will be shown in the last two sections, this is attained successfully in giving an algebraic proof of cut elimination theorems for modal and substructural logics, where a generalization of the MacNeille completions, called *quasi-completions* of algebras, plays a key role.[1]

1 Complete extensions and MacNeille completions

In the present paper, we will discuss completions of algebras for modal logics and substructural logics, and completeness theorems for them, which are obtained as logical consequences of completions. Though both canonical extensions and the MacNeille completions are considered, we will be concerned mainly about the latter. Completions of algebras are discussed in the first three sections. Many of completeness theorems of various logics, not only propositional but predicate ones, with respect to both Kripke-type semantics and algebraic one, are obtained from results on completions. The last two sections are devoted to an algebraic proof of cut elimination theorems for modal and substructural logics. We take the sequent system **LJ** of intuitionistic logic as an example. Algebraic structures for the sequent system **LJ** without cut rule are introduced in § 5, and then the completeness of this cut-free system with respect to algebraic semantics determined by these structures is shown in § 6 by using *quasi-completions* of algebras. Our final remark says that quasi-completions are generalizations

---
[1]This work is partly supported by Grant-in-Aid for Scientific Research (B)(2) No.13430001 from Japan Society for Promotion of Science.

of the MacNeille completions. Completions of residuated lattices and complete embeddings are discussed already in Ono [26], and algebraic proofs of cut elimination theorems will be studied comprehensively in the forthcoming paper Belardinelli-Jipsen-Ono [1].

Throughout the paper, we consider algebras with lattice structures. The examples are Boolean algebras with operators and residuated lattices. The formers are algebras for modal logics and the latters are for substructural logics. By a *complete* algebra, we mean an algebra whose lattice reduct is complete, i.e. an algebra in which both infinite joins and infinite meets exist always. An algebra **Q** is a *completion* of an algebra **P** if **Q** is a complete extension of **P**, i.e. **Q** is a complete algebra into which **P** is embeddable.

As is described in Givant–Venema [7], there are two different ways of getting completions of a given algebra **P**. The first is the *canonical extension* of **P**: the Boolean algebra consisting of all subsets of the set of all ultrafilters of **P** when **P** is a Boolean algebra, or the distributive lattice consisting of all upward closed subsets of the set of all prime filters of **P** when **P** is a distributive lattice. They arise from Stone representation theorem. The notion of canonical extensions is extended also to Boolean algebras with operators, and to bounded distributive lattices with operators etc. (see e.g. [13, 12, 8, 6]). The characteristics of canonical extensions are that they are atomic, but they do not always preserve infinite joins.

The second way is the Dedekind-MacNeille completion of a given algebra **P**. This completion is sometimes called simply *the completion* of **P**, since it is the minimal complete extension detemined uniquely by **P** up to isomorphism (see Sikorski [33] § 35 Theorem 35.2 for Boolean algebras, and also Ono [22] Lemma 4.5 for residuated lattices). To avoid confusions, we will call this, the *MacNeille completion* in the present paper. The MacNeille completion are discussed in e.g. [33, 18, 7] for Boolean algebras (with operators), and in e.g. [22] for residuated lattices (though a different word is used for "residuated lattices" in it). The special feature of the MacNeille completion is that it preserves all existing infinite joins and infinite meets while it is not always atomic. Also, it should be noticed that the MacNeille completion of a distributive lattice is not always distributive.

We will mainly concern with the MacNeille completion of Boolean algebras with oparators, in particular of modal algebras, and of residuated lattices in this paper. So, we start our paper by giving a brief account of the MacNeille completion of partially ordered sets and of Boolean algebras, which is defined here by using *cuts* (see § 35 Example F in [33]).

Let $\mathbf{P} = \langle P, \leq \rangle$ be a partially ordered set with the greatest element $\top$ and the least $\bot$. For each $X \subseteq P$, define $X^{\rightarrow}$ and $X^{\leftarrow}$ to be the set of upper bounds of X and of lower bounds of X, respectively. Define a map C on $\wp(\mathbf{P})$ by $CX = ((X^{\rightarrow})^{\leftarrow})$, for each $X \subseteq P$. The map C is in fact a *closure operator* on $\wp(\mathbf{P})$, i.e. it satisfies that for all X and Y in $\wp(\mathbf{P})$,

1. $X \subseteq CX$,
2. $X \subseteq Y$ implies $CX \subseteq CY$,
3. $CCX \subseteq CX$.

We say that a subset X of P is *C-closed* if $X = CX$. It is easy to see that for C-closed subsets X and Y, the intersection $X \cap Y$ of X and Y is C-closed. Now, define $X \cup_C Y$ by $C(X \cup Y)$ where $\cup$ means the union. Then, the set $C_\mathbf{P}$ of all C-closed subsets of P forms a lattice with respect to $\cap$ and $\cup_C$. The lattice $C_\mathbf{P}$ is complete since it has always the meet $\bigcap_i X_i$ and the join $C(\bigcup_i X_i)$ for any set $\{X_i\}_i$ of C-closed sets. We define a map h from $\mathbf{P}$ to $C_\mathbf{P}$ by $h(a) = (a]$ for each element $a \in P$, where $(a] = \{x \in P : x \leq a\}$ which is in fact C-closed. Then, h is an order isomorphism. We can show that if $\mathbf{P}$ has moreover a lattice structure, then h is a lattice isomorphism which preserves all existing infinite joins and infinite meets in P. When $\mathbf{P}$ is a Boolean algebra, then $C_\mathbf{P}$ becomes a complete Boolean algebra with the minimum element $\{\bot\} (= C\emptyset)$ and h is a Boolean isomorphism, if we define the complement $-X$ of a C-closed subset X by

$$-X = \{x \in P : x \leq -b \text{ for any } b \in X\},$$

where $-b$ denotes the complement of b in $\mathbf{P}$. We note that $-(-X) = CX$ holds for any subset X of P. The complete algebra $C_\mathbf{P}$ defined in this way is called the MacNeille completion of an algebra $\mathbf{P}$.

Let $\mathbf{P}^+$ be the MacNeille completion of a lattice $\mathbf{P}$. Then, each element X of $\mathbf{P}^+$ is a subset of P such that

1. X is downward closed and is closed under the finite join,
2. if the join $\vee X$ exists in P, then $X = \{a \in P : a \leq \vee X\}$ (and hence $\vee X \in X$).

Since each element X of $\mathbf{P}^+$ can be expressed as $X = \bigcup\{(a] : a \in X\}$, by identifying the C-closed set $(a]$ with an element $a \in P$ we can show the following:

> Each element c of the MacNeille completion of a lattice $\mathbf{P}$ is represented as $c = \vee\{a \in P : a \leq c\}$.

2 Completions of residuated lattices

To make a comparison, we will next consider the MacNeille completions of residuated lattices. Residuated lattices are algebraic structures for substructural logics. For general information on substructural logics, see e.g. [28, 30, 22, 14]. To simplify the discussion, here we consider only commutative residuated lattices, i.e. algebras for substructural logics with exchange rule, or equivalently, algebras for extensions of intuitionistic linear logic (without exponentials). The contents of this section are discussed in details in Ono [22, 26].

An algebra $\mathbf{P} = \langle P, \wedge, \vee, \cdot, \rightarrow, 0, 1, \bot, \top \rangle$ is a *bounded commutative residuated lattice* if it satisfies the following:

1. $\langle P, \wedge, \vee, \bot, \top \rangle$ is a bounded lattice,
2. $\langle P, \cdot, 1 \rangle$ is a commutative monoid with the unit 1,
3. $x \cdot y \leq z$ iff $x \leq (y \rightarrow z)$.

For simplicity's sake, we omit the word "bounded" in the rest of paper. In the above definition, 0 is an arbitrary element of P by which the "negation" $-x$ of x is defined as $x \rightarrow 0$. The third condition is called the law of residuation between the monoid operation $\cdot$, sometimes called *fusion*, and the *residual* $\rightarrow$. It implies the following distributivity of the monoid operation $\cdot$ over join :

(1) $\quad (\vee x_i) \cdot y = y \cdot (\vee x_i) = \vee (x_i \cdot y)$ for any x_i, y.

Equalities in (1) mean that if $\vee x_i$ exists, then $\vee(x_i \cdot y)$ exists also and the equalities hold. A commutative residuated lattice is *integral* if the unit 1 is equal to the greatest element $\top$ and 0 is equal to $\bot$, and it is *weakly idempotent* if $x \leq x \cdot x$ for each x. It is easy to see that a commutative residuated lattice $\mathbf{P}$ is both integral and weakly idempotent if and only if $x \cdot y = x \wedge y$ for all x, y, i.e. $\mathbf{P}$ is a Heyting algebra. Note that syntactically, integrality and weak idempotency correspond to weakening rules and contraction rules, respectively, in sequent calculi.

When we consider completions of residuated lattices, we need to manage also the monoid operation and the residual. Since residuated lattices are not always distributive, Stone representation theorem does not work. Because of this, we don't have any suitable notion of canoncal extensions of these algebras at present.

First, we will show a standard way of constructing complete commutative residuated lattices. See e.g. [22] for the details. Suppose that $\mathbf{M} = \langle M, \cdot, 1 \rangle$ is a commutative monoid. We define two operations $*$ and $\Rightarrow$ on the set of all subsets of M by

Completions of Algebras and Completeness of Modal and Substructural Logics   339

i. $X * Y = \{a \cdot b \in M : a \in X \text{ and } b \in Y\}$,
ii. $X \Rightarrow Y = \{a \in M : c \cdot a \in Y \text{ for each } c \in X\}$.

Then, we have the following proposition.

PROPOSITION 1 *For any given commutative monoid* **M** *and any subset* O *of* M, $\wp(\mathbf{M}) = \langle \wp(M), \cap, \cup, *, \Rightarrow, O, \{1\}, \emptyset, M \rangle$ *is a complete commutative residuated lattice, where* $\cap$ *and* $\cup$ *denote set intersection and union, respectively.*

A map C on a commutative residuated lattice **P** is a *closure operator* if for all $x, y \in P$

1. $x \leq Cx$,
2. $x \leq y$ implies $Cx \leq Cy$,
3. $CCx \leq Cx$,
4. $Cx \cdot Cy \leq C(x \cdot y)$.

Suppose that C is a closure operator on a commutative residuated lattice **P**. An element x of P is C-*closed* when $x = Cx$ holds. Let $C(P)$ be the set of all C-closed elements of P. Though $C(P)$ is closed under both $\wedge$ and $\rightarrow$, it is not always closed under $\vee$ nor $\cdot$. So, define $\vee_C$ and $\cdot_C$ by

$$x \vee_C y = C(x \vee y) \text{ and } x \cdot_C y = C(x \cdot y).$$

Then, we have the following.

PROPOSITION 2 *Suppose that* **P** *is a commutative residuated lattice and* C *is a closure operator on* P. *Then, the algebra* $C(\mathbf{P}) = \langle C(P), \wedge, \vee_C, \cdot_C, \rightarrow, d, C1, C\bot, \top \rangle$ *forms a residuated lattice where* d *is an arbitrary* C-*closed element. If* **P** *is integral then so is* $C(\mathbf{P})$ *if we take* $C\bot$ *for* d, *if* **P** *is weakly idempotent then so is* $C(\mathbf{P})$, *and also if* **P** *is complete then so is* $C(\mathbf{P})$.

By Propositions 1 and 2, we have the following.

COROLLARY 3 *If* **M** *is a commutative monoid and* C *is a closure operator on* $\wp(\mathbf{M})$, *then* $\mathbf{C_M} = \langle C(\wp(M)), \cap, \cup_C, *_C, \Rightarrow, D, C(\{1\}), C(\emptyset), M \rangle$ *is a complete commutative residuated lattice where* D *is any* C-*closed subset of* M.

Now that we have a general way of constructing complete commutative residuated lattice from commutative monoids, we will show how to get a completion when a residuated lattice is given. For this purpose, we assume that a given commutative monoid is partially ordered. More precisely, a structure $\mathbf{M} = \langle M, \cdot, 1, \leq \rangle$ is called a *partially ordered commutative monoid* (or simply, a *po-monoid*), if $\langle M, \cdot, 1 \rangle$ is a commutative monoid with the unit element 1 and $\langle M, \leq, 1 \rangle$ a partially ordered set such that for any $x, y, z \in M$, $x \leq y$ implies $x \cdot z \leq y \cdot z$. We say that a po-monoid $\mathbf{M}$ is integral when the unit element 1 is equal to the greatest element, and it is weakly idempotent when $x \leq x \cdot x$ holds for every x.

Suppose that C is a closure operator on $\wp(\mathbf{M})$ for a po-monoid $\mathbf{M}$, in which every C-closed subset X is *downward closed*, i.e. $x \in X$ and $y \leq x$ imply $y \in X$ for any $x, y \in M$. In this case, both the integrality and the weak idempotency are preserved under the construction of $\mathbf{C_M}$ from $\mathbf{M}$. More precisely, when $\mathbf{M}$ is integral, $C(\{1\}) = M$ holds, and hence the residuated lattice $\mathbf{C_M}$ with $D = C(\emptyset)$ becomes also integral.

For a given commutative residuated lattice $\mathbf{P}$, consider its po-monoid reduct $\mathbf{P}^*$. Then this determines a complete commutative residuated lattice $\mathbf{C_{P^*}}$. The next theorem tells us when a complete embedding from $\mathbf{P}$ to $\mathbf{C_{P^*}}$ exists. We show the theorem in a more general setting.

For a given po-monoid $\mathbf{M}$ and a closure operator C on $\wp(\mathbf{M})$, define a subset U of M by $U = \{b \in M : (b] \text{ is } C\text{-closed}\}$, where $(b] = \{x \in M : x \leq b\}$. We assume now that the following *closure condition* holds for C: For any C-closed subset X,

1. X is downward closed,
2. if $a_i \in U$ for each $i \in I$ and moreover if $\vee_i a_i$ exists and belongs to U, then $\vee_i a_i \in X$ whenever $a_i \in X$ for each $i \in I$.

THEOREM 4 *Suppose that $\mathbf{M}$ is a po-monoid and C is a closure operator on $\wp(\mathbf{M})$ for which closure conditions hold. Then the map $h : U \to C(\wp(M))$ defined by $h(b) = (b]$ is a complete embedding, i.e. an order isomorphism which preserves all existing products and residuals in U, and also all existing (infinite) joins and meets in U.*

For the proof, see [26]. Let us consider the MacNeille completion of a given commutative residuated lattice $\mathbf{P}$. That is, take its po-monoid reduct $\mathbf{P}^*$ and define a map C^0 on $\wp(\mathbf{P}^*)$ by $C^0 X = ((X^\to)^\leftarrow)$ for each $X \subseteq P$. Then, we can show that C^0 becomes a closure operator and that the set U is equal to P. In this case, the second item of the closure condition becomes equal to the condition that if $a_i \in X$ for each $i \in I$ and $\vee_i a_i$ exists then $\vee_i a_i \in X$. Thus, we can show that C^0 satisfies the closure condition. Therefore, we have the following by Theorem 4.

THEOREM 5 *Each (integral, weakly idempotent) commutative residuated lattice can be embedded into its MacNeille completion by a residuated lattice isomorphism which preserves all existing infinite joins and meets.*

We note that the converse of Corollary 3 holds. That is, any complete commutative residuated lattice **P** is isomorphic to $\mathbf{C_M}$ for a commutative monoid **M** with a closure operator C on $\wp(\mathbf{M})$. In fact, it suffices to take $\mathbf{C^0_{P^*}}$ for $\mathbf{C_M}$ since the MacNeille completion of a complete commutative residuated lattice **P** is isomorphic to **P**.

The above result on the MacNeille completion of commutative residuated lattices can be extended to that of commutative residuated lattices with *exponentials*, where exponentials are particular modal operations (see [22, 25] for the details). The result is extended also to commutative residuated lattices with weaker modalities (see Bucalo [4]).

THEOREM 6 *Each commutative residuated lattice with exponentials has the MacNeille completion.*

We remark that results mentioned in this section hold for Heyting algebras, and in particular for Boolean algebras, since Heyting algebras are just commutative, integral and weakly idempotent residuated algebras. Note that $x \cdot y = x \wedge y$ holds for x, y in any Heyting algebra and therefore $X * Y = X \cap Y$ holds for all downward closed sets X and Y.

3 Complete additivity, conjugates and residuated pairs

By an *operator* on a Boolean algebra we mean an operation which preserves finite joins. An operator is *completely additive* if it preserves all existing joins. Boolean algebras with completely additive operators play key roles in the study of Boolean algebras with operators. For each Boolean algebra with operators **P** there exists a complete and atomic Boolean algebra with completely additive operators **Q** which is a canonical extension of **P**. Any complete and atomic Boolean algebra with completely additive operators can be represented as a complex algebra determined by its atom structure (see Jónsson-Tarski [13], Jónsson [12]).

Also, any Boolean algebra with complete additive operators **P** has the MacNeille completion $\mathbf{P}^+$, since each complete additive operator h of **P** can be extended to a complete additive operator h^+ of $\mathbf{P}^+$ (see Monk [18], Givant-Venema [7]). Here, for a given n-ary operation h of **P**, the operation h^+ of **P** is defined by

$$h^+(\mathbf{c}) = \vee \{h(\mathbf{a}) : \mathbf{a} \leq \mathbf{c} \text{ with } \mathbf{a} \in P^n\}$$

and is called the *complete extension* of h. Preservation theorems under canonical extensions and the MacNeille completions are discussed in [13, 12, 18, 7]. See also [9].

In this section we will give some remarks on the complete additive operations. First recall that the monoid operation · in the previous section satisfies the equality (1) in the previous section, which means the complete additivity of ·. The complete extension of this monoid operation · becomes the monoid operation in the MacNeille completion. But, to show the complete additivity of this complete extension, the standard argument used in Boolean algebras with complete additive operators does not work, since the proof of complete additivity in Boolean algebras needs the *complete distributivity* of (infinite) joins over meet (see the proof of Lemmas 12 and 13 in [7]), which does not hold always in residuated lattices. The complete additivity of the monoid operation comes in fact from the law of residuation.

We can explain this in a general setting, based on a result from *residuation theory* (see Blyth-Janowitz [3]). Suppose that **M** and **N** are partially ordered sets and that $f : M \to N$ and $g : N \to M$ are maps. If

$$f(x) \leq y \Leftrightarrow x \leq g(y) \text{ for all } x \in M \text{ and } y \in N,$$

then f and g are said to form a *reduated pair*, and the left member f is said to be *residuated*. Then, we can show the following.

PROPOSITION 7 *Suppose that $f : M \to N$ and $g : N \to M$ form a residuated pair. Then,*

1. $f(\vee_i x_i) = \vee_i f(x_i)$ *for $x_i \in M$,*
2. $g(\wedge_j y_j) = \wedge_j g(y_j)$ *for $y_j \in N$.*

Here, the equlities mean that if $\vee_i x_i$ ($\wedge_j y_j$) exists, then $\vee_i f(x_i)$ ($\wedge_j g(y_j)$, respectively) exists and these equlities hold.

Recall that the law of residuation says that $x \cdot y \leq z$ iff $x \leq (y \to z)$ for all x, y, z. For a fixed y, define maps f and g by $f(x) = x \cdot y$ and $g(z) = y \to z$. Then f and g form a residuated pair, and therefore f preserves all existing joins, i.e. it is completely additive, and g preserves all existing meets.

Let us compare the notion of residuated pairs with that of *conjugates* introduced in Jónsson-Tarski [13]. Two unary operations h and k on a Boolean algebra are *conjugates* if

$$x \wedge h(y) = 0 \Leftrightarrow k(x) \wedge y = 0 \text{ for all } x, y.$$

In Boolean algebras, this is equivalent to say that h and k^d form a residuated pair, where k^d is the *dual* of k, which is defined as $k^d(x) = -k(-x)$ for all x. Thus, by Proposition 7 we can derive the following well-known result (see e.g. Lemma 19 in [7]).

COROLLARY 8 *If h and k are conjugates then they are completely additive.*

Lemma 20 in [7] says that if h and k are conjugates then so are their complete extensions h^+ and k^+. On the other hand, while the monoid operation $*$ of the MacNeille completion of a given commutative residuated lattice **P** is the complete extension of the monoid operation $\cdot$ in **P**, the residual of $*$ is not always equal to the complete extension of the residual of $\cdot$.

4 Completeness theorems

As consequences of results on completions of algebras in the previous sections, we can derive various types of completeness theorems for modal and substructural logics. A clue to the completeness is that desired properties of the original algebra, e.g. the validity of some axioms, can be transferred from it to its completions. In the following, $\supset, \wedge, \vee$ and $\neg$ denote logical connectives implication, conjunction, disjunction and negation, respectively.

Needless to say, preservation theorems under canonical extensions of modal algebras offer Kripke completeness for many of modal propositional logics. (See e.g. [5] for general information.) The argument goes as follows: Let $\mathcal{V}$ be the class of all modal algebras for a given modal logic **L**. If φ is not provable in **L** then it is not valid in an algebra **P** in $\mathcal{V}$ (for instance in the Lindenbaum algebra of **L**) by the algebraic completeness of **L**. Here the validity of formulas is defined by using *valuations* on **P**, in which each logical connective is interpreted as its corresponding algebraic operation. Let $\tilde{\mathbf{P}}$ be the canonical extension of **P**. Since **P** is embedded into $\tilde{\mathbf{P}}$, φ is neither valid in $\tilde{\mathbf{P}}$. Since the validity of formulas in $\tilde{\mathbf{P}}$ is the same as the validity of those in the *canonical Kripke structure* of **P**, φ is not valid in the canonical Kripke structure of **P**. Moreover, when $\tilde{\mathbf{P}}$ belongs to $\mathcal{V}$, the canonical Kripke structure of **P** becomes a Kripke structure for **L**. Thus, we have the Kripke completeness of **L**.

The similar argument does not work for Kripke completeness of modal predicate logics, since quantifiers are usually interpreted as infinite joins and meets in algebras but canonical extensions do not preserve infinite joins. In fact, we have a lot of examples of modal predicate logics which are Kripke incomplete (see [21, 31])

However, in a restricted case, canonical extensions in a modified form still work for modal and intermediate predicate logics. The basic idea used here is usually stated as *Rasiowa-Sikorski lemma*. Originally the lemma is proved for Boolean algebras (see e.g. § 24 of [33]). It is extended in [10, 29] to Heyting algebras satisfying the following distributivity: $\bigwedge_i(a_i \vee b) = (\bigwedge_i a_i) \vee b$ for any b and any a_i such that $\bigwedge_i a_i$ exists. Rasiowa-Sikorski lemma says that if Q is a fixed *countable* set of infinite joins and meets in a Boolean algebra **P**, there exists a complete Boolean algebra **P**′ and an embedding f from **P** into **P**′ which preserves all infinite joins and meets in Q. To make a given first-order formula φ invalid in **P**′, when φ is invalid in **P**, preserving countably many infinite joins and meets is enough since φ contains finitely many quantifiers. In Tanaka-Ono [34], we succeeded to extend this embedding result to modal algebras with a completely additive modal operator $\Diamond$.

For a normal modal propositional logic **M** over **K**, let **Q-M** be the minimum predicate extension of **M**. Let BF be the Barcan formula $\forall x \Box \psi \supset \Box \forall x \psi$, which expresses the complete additivity of $\Diamond$ in its dual form. Let **Q-M**+BF be the modal predicate logic obtained from **Q-M** by adding the axiom scheme BF. We say that **M** is *universal* if the class of Kripke structures for **M** is representable by a set of universal first-order sentences. Then we have the following [34].

THEOREM 9 *If a normal modal propositional logic **M** over **K** is Kripke complete and universal then **Q-M**+BF is Kripke complete.*

By using Rasiowa-Sikorski Lemma for Heyting algebras with the distributivity mentioned above, we can also give an alternative proof of the Kripke completeness result by Shimura [32] on minimal predicate extensions of some intermediate logics with the axiom of constant domain (see also [34]). Here, the axiom of constant domain means formulas of the form $\forall x(\varphi(x) \vee \psi) \supset (\forall x \varphi(x) \vee \psi)$, where ψ does not contain any free occurrence of x. Shimura showed the following.

THEOREM 10 *If an intermediate propositional logic **J** is either tabular or a subframe logic, the minimal predicate extension with the axiom of constant domain **Q-J**+CD is Kripke complete.*

In contrast with propositional logics, predicate logics are not always complete with respect to algebraic semantics (see [21, 31, 25]). But, using preservation theorems in [7], we can derive also a similar result on the algebraic completeness of modal predicate logics of the form **Q-M**+BF where

M is axiomatized by a set of Sahlqvist formulas obtained from Sahlqvist equations discussed in [7] (see [34]).

Next we will show how completeness results for substructural logics can be derived from completions of residuated lattices. First, we discuss Kripke-type semantics for substructural logics introduced in Ono-Komori [27].

A structure $\mathbf{M} = \langle M, \cdot, 0, 1, \vee, \leq \rangle$ is a *semilattice-ordered, commutative integral monoid* (or, simply a *so-monoid*) if it satisfies the following:

1. $\langle M, \cdot, 1, \leq \rangle$ is a po-monoid with the smallest element 0,
2. the join $x \vee y$ exists for every $x, y \in M$,
3. the monoid operation $\cdot$ is finitely additive, i.e. $x \cdot (y \vee z) = (x \cdot y) \vee (x \cdot z)$ holds.

Note here that so-monoids in [27] are defined by using reverse orders. A nonempty subset X of a so-monoid **M** is an *ideal*, if (1) X is downward closed, and (2) if both a and b are in X, then $a \vee b \in X$. We define a operation C on $\wp(M)$ by the condition that $C(X)$ is the smallest ideal containing X. Then, we can show that C is a closure operator on the complete commutative residuated lattice $\wp(\mathbf{M})$.

Let **FL**$_\mathbf{ew}$ be the substructural logic obtained from intuitionistic propositional logic by deleting contraction rule. In algebraic terms, **FL**$_\mathbf{ew}$ is the logic complete with respect to the class of all *integral* commutative residuated lattices. *Kripke frames* for **FL**$_\mathbf{ew}$ introduced in [27] are nothing but integral so-monoids. A *valuation* on a Kripke frame **M** is any map v from the set of all propositional variables to $C(\wp(M))$. Then, v can be naturally extended to a map from the set of all formulas to $C(\wp(M))$, since $\mathbf{C_M}$ forms an integral commutative residuated lattice. Note that in [27], the notation $a \models \varphi$ is used instead of writing $a \in v(\varphi)$, where $\models$ corresponds to a given valuation v, and also that the words *strong valuations* and *total strong frames* are used instead of the words valuations and Kripke frames, respectively. A formula φ is *valid* in a Kripke frame **M** (**M** $\models \varphi$, in symbol) if $a \models \varphi$ for every $a \in M$.

From the definition, it is obvious that for any Kripke frame **M** and any formula φ, **M** $\models \varphi$ if and only if φ is valid in the integral commutative residuated lattice $\mathbf{C_M}$. Now, we take the so-monoid reduct $\mathbf{P}^\dagger$ of a given integral commutative residuated lattice **P** for **M**, and apply Theorem 4 in § 2. Note that U is equal to P in the present case and each C-closed set satisfies the closure condition (for finite joins).

COROLLARY 11 *Every integral commutative residuated lattice* **P** *can be embedded into a complete, commutative integral residuated lattice* $\mathbf{C_{P^\dagger}}$.

This is the essence of the proof of the completeness of $\mathbf{FL_{ew}}$ with respect to Kripke frames given in [27]. Now, we have the following.

COROLLARY 12 *The substructural propositional logic $\mathbf{FL_{ew}}$ is Kripke-complete.*

Another type of completeness results on substructural predicate logics follows from Theorems 5 and 6 in § 2, on the MacNeille completions of commutative residuated lattices (with exponentials). In fact, we can show the following (see Ono [22] for the details).

COROLLARY 13 *Intuitionistic linear predicate logic $\mathbf{FL_e}$ (with exponentials) is complete with respect to the class of complete, commutative residuated lattices (with exponentials). Similarly, the substructural predicate logic $\mathbf{FL_{ew}}(\mathbf{FL_{ec}})$ is complete with respect to the class of complete, integral (weakly idempotent) commutative residuated lattices, where $\mathbf{FL_{ew}}(\mathbf{FL_{ec}})$ is obtained from $\mathbf{FL_e}$ by adding weakening rule (contraction rule, respectively). This holds also for the case with exponentials.*

5 Algebras for cut-free systems

In this section, we will show how techniques of completions of algebras can be applied to algebraic proof of cut elimination theorems for modal and substructural logics. Our idea is based on results by Maehara [16] and Okada [19] (see also Jipsen-Tsinakis [11]), and the topics in this section will be discussed comprehensively in our forthcoming paper [1] with Belardinelli and Jipsen.

Our method works well for a wide variety of sequent systems of nonclassical logics, including Gentzen's systems **LK** and **LJ** for classical logic and intuitionistic one, not only for propositional logics but for predicate logics. We will give here a proof of cut elimination theorem for **LJ** as a typical example. By a slight modification of it, we can get a similar algebraic proof of cut elimination theorems for modal logics.

To lighten a burden to readers, we will try to minimize explanations of sequent systems and their syntactic properties. When necessary, consult a survey article [24], written from a syntactic point of view, for more information on cut elimination theorems for nonclassical logics.

Each sequent of **LJ** is of the form $\Gamma \Rightarrow \alpha$, where Γ is a (possibly empty) sequence of formulas separated by "commas", and α a formula. The negation $\neg \alpha$ is defined to be the abbreviation of $\alpha \supset 0$, where 0 denotes the constant symbol expressing the falsehood. The system **LJ** consists of initial sequents and rules of inference. Each initial sequent is one of the following:

1) $\alpha \Rightarrow \alpha$, 2) $0, \Gamma \Rightarrow \delta$. There are two kinds of rules of inference. The first is rules for logical connectives. For instance, rules for $\supset$ and $\wedge$ consist of the following:

$$\frac{\Gamma \Rightarrow \alpha \quad \beta, \Sigma \Rightarrow \delta}{\alpha \supset \beta, \Gamma, \Sigma \Rightarrow \delta} \ (\supset\Rightarrow) \qquad \frac{\Gamma, \alpha \Rightarrow \beta}{\Gamma \Rightarrow \alpha \supset \beta} \ (\Rightarrow\supset)$$

$$\frac{\alpha, \Gamma \Rightarrow \delta}{\alpha \wedge \beta, \Gamma \Rightarrow \delta} \ (\wedge 1 \Rightarrow) \qquad \frac{\beta, \Gamma \Rightarrow \delta}{\alpha \wedge \beta, \Gamma \Rightarrow \delta} \ (\wedge 2 \Rightarrow)$$

$$\frac{\Gamma \Rightarrow \alpha \quad \Gamma \Rightarrow \beta}{\Gamma \Rightarrow \alpha \wedge \beta} \ (\Rightarrow \wedge)$$

The second is structural rules, consisting of weakening rule, contraction rule, exchange rule, and also cut rule. Here, we show weakening rule and cut rule.

$$\frac{\Gamma \Rightarrow \delta}{\alpha, \Gamma \Rightarrow \delta} \ (w \Rightarrow) \qquad \frac{\Gamma \Rightarrow \alpha \quad \alpha, \Sigma \Rightarrow \delta}{\Gamma, \Sigma \Rightarrow \delta} \ (cut)$$

Cut elimination theorem for **LJ** says that if a sequent $\Gamma \Rightarrow \alpha$ is provable in **LJ** then it is provable in **LJ** without using cuts. Many important syntactic properties, including the decidability for the case of propositional logics, follows from cut elimination theorem (see e.g. [24]).

It is well-known that intuitionistic propositonal logic is complete with respect to the class of Heyting algebras. Corresponding to the sequent system **LJ** *without cut*, we introduce algebraic structures, called *semi-Heyting structures*. For a given nonempty set Q, let Q^* be the set of all finite (possibly empty) subsets of Q. The empty set is denoted here by ε. For elements x and y of Q^*, xy denotes the set-theoretic union of x and y. When y is in particular a singleton set $\{c\}$ with $c \in Q$, xy is also written as xc. Obviously, Q^* forms a commutative idempotent monoid with the unit ε. In the following, letters x, y, z, u, v are used for expressing elements in Q^*, and letters a, b, c, d are used for expressing elements in Q.

A structure $\mathbf{Q} = \langle Q, \preceq, \wedge, \vee, \rightarrow, 0 \rangle$ is a *semi-Heyting structure* if it satisfies the following conditions: For all $x, y \in Q^*$ and all $a, b, c \in Q$,

- $a \preceq a$,
- $0x \preceq c$,
- $x \preceq c$ implies $yx \preceq c$,
- $x \preceq a$ and $by \preceq c$ imply $(a \rightarrow b)xy \preceq c$,

- $ax \preceq b$ implies $x \preceq a \to b$,
- $ax \preceq c$ and $bx \preceq c$ imply $(a \vee b)x \preceq c$,
- $x \preceq a$ implies $x \preceq a \vee b$,
- $x \preceq b$ implies $x \preceq a \vee b$,
- $ax \preceq c$ implies $(a \wedge b)x \preceq c$,
- $bx \preceq c$ implies $(a \wedge b)x \preceq c$,
- $x \preceq a$ and $x \preceq b$ imply $x \preceq a \wedge b$.

Each of these conditions corresponds to either one of initial sequents or one of rules of inference except cut rule, and vice versa. Exchange rule and contraction rule are incorporated into the definition of Q^*, while weakening rule is expressed as the third condition. Thus, we have the following by considering the *free* semi-Heyting structure for the cut-free **LJ**. Here, t_α denotes the algebraic term corresponding to a given formula α. An *inequality* is an expression of the form $\langle s_1, \ldots, s_m \rangle \preceq t$ for terms $s_1, \ldots, s_m, t$. An inequality $\langle s_1, \ldots, s_m \rangle \preceq t$ is said to hold in a semi-Heyting structure **Q**, if for every valuation h on **Q** $\{h(s_1), \ldots, h(s_m)\} \preceq h(t)$ holds in **Q**.

LEMMA 14 *For any sequent* $\alpha_1, \ldots, \alpha_m \Rightarrow \beta$, $\alpha_1, \ldots, \alpha_m \Rightarrow \beta$ *is provable in* **LJ** *without using cut rule if and only if* $\langle t_{\alpha_1}, \ldots, t_{\alpha_m} \rangle \preceq t_\beta$ *holds in every semi-Heyting structure.*

Suppose that $\mathbf{P} = \langle P, \wedge, \vee, \to, 0 \rangle$ is a Heyting algebra with the lattice order $\leq$. We define the relation $\preceq$ by the condition that $\{a_1, \ldots, a_m\} \preceq c$ if and only if $(a_1 \wedge \ldots \wedge a_m) \leq c$ holds in **P**. Then, $\mathbf{P}' = \langle P, \preceq, \wedge, \vee, \to, 0 \rangle$ becomes a semi-Heyting structure. Thus, each Heyting algebra can be regarded naturally as a semi-Heyting structure by identifying **P** with **P**'.

Conversely, suppose that the relation $\preceq$ of a given semi-Heyting structure **Q** satisfies moreover the following *transitivity condition*:

$x \preceq a$ and $ay \preceq c$ imply $xy \preceq c$,

which is an algebraic expression of cut rule. Then, we have the following.

1. The relation $\preceq_0$ which is the restriction of $\preceq$ between two elements of Q is transitive and hence is a quasi order,
2. $\{a_1, \ldots, a_m\} \preceq c$ if and only if $(a_1 \wedge \ldots \wedge a_m) \preceq c$,
3. $ax \preceq b$ if and only if $x \preceq a \to b$.

To show 3, it suffices to show the if-part. From $a \preceq a$ and $b \preceq b$, $(a \to b) \wedge a \preceq b$ follows. So, if $x \preceq a \to b$ then we can show that $ax = xa \preceq (a \to b) \wedge a$. Then, by the transitivity condition we have $ax \preceq b$. Now, if the relation $\preceq$ of a semi-Heyting structure **Q** satisfies the transitivity condition and moreover if $\preceq_0$ is a partial order, $\mathbf{Q}_0 = \langle Q, \wedge, \vee, \to, 0 \rangle$ becomes a Heyting algebra with the lattice order $\preceq_0$.

Since any Heyting algebra can be regarded as a semi-Heyting structure in this way, we can see that for all terms $s_1, \ldots, s_m$ and t, if an inequality $\langle s_1, \ldots, s_m \rangle \preceq t$ holds for every semi-Heyting structure then $(s_1 \wedge \ldots \wedge s_m) \leq t$ holds for every Heyting algebra. Our Theorem 16, proved in the following, says that the converse holds also. This is the algebraic content of cut elimination theorem for **LJ**.

6 Quasi-completions and cut elimination theorems

Now suppose that an inequality $\langle s_1, \ldots, s_m \rangle \preceq t$ fails in a semi-Heyting structure **Q** under a valuation f, i.e. $\{f(s_1), \ldots, f(s_m)\} \preceq f(t)$ does not hold in **Q**. Since Q^* forms a commutative monoid, we have that $\mathbf{D}_{\mathbf{Q}^*}$ is a complete commutative residuated lattice for any closure operator D on $\wp(Q^*)$ by using Corollary 3 in § 2.

Now we define a special closure operator C on $\wp(Q^*)$ as follows. For $x \in Q^*$ and $a \in Q$, define

$$[x; a] = \{w \in Q^* : xw \preceq a\}.$$

Define an operation C on $\wp(Q^*)$ as follows: For each subset X of Q^*,

$$C(X) = \bigcap \{[x; a] : X \subseteq [x; a] \text{ for } x \in Q^* \text{ and } a \in Q \}.$$

We can show that C is actually a closure operator in our sense, which satisfies $C(\{\varepsilon\}) = Q^*$. Moreover, we can show that $X \cap Y = X * Y$ whenever both X and Y are C-closed. In fact, it is obvious that $X \cap Y \subseteq X * Y$. Conversely, suppose that $xy \in X * Y$ where $x \in X$ and $y \in Y$. Suppose that $X = \bigcap_i \{[z_i; a_i] : i \in I\}$. Then, for each $i \in I$, $xz_i \preceq a_i$ and hence $xyz_i \preceq a_i$. This implies that $xy \in X$. Similarly, $xy \in Y$. Hence, $xy \in X \cap Y$. Thus, it turns out that $\mathbf{C}_{\mathbf{Q}^*}$ is a complete Heyting algebra. The Heyting algebra $\mathbf{C}_{\mathbf{Q}^*}$ is uniquely determined by a given semi-Heyting structure **Q**. So we call it the *quasi-completion* of **Q**. Corollary 19 below will explain the reason why we call it in this way.

Let us define a map $k : Q \to C(\wp(Q^*))$ by $k(a) = [a]$, where $[a] = [\varepsilon; a]$, i.e. $[a] = \{w \in Q^* : w \preceq a\}$. The following theorem can be shown essentially in the same way as the proof of Lemma 7.3 in [11] (see also Maehara [16] and Okada [19]).

THEOREM 15 *Suppose that $a, b \in Q$ and that U and V are arbitrary C-closed subsets of Q^* such that $a \in U \subseteq k(a)$ and $b \in V \subseteq k(b)$. Then for each $\star \in \{\wedge, \vee, \rightarrow\}$, $a \star b \in U \star_C V \subseteq k(a \star b)$, where $\star_C$ denotes $\cap, \cup_C$ and $\Rightarrow$, respectively. Thus, in particular $a \star b \in k(a) \star_C k(b) \subseteq k(a \star b)$.*

Now we will show the following theorem.

THEOREM 16 *For all terms $s_1, \ldots, s_m$ and t, if $(s_1 \wedge \ldots \wedge s_m) \leq t$ holds in every Heyting algebra, then $\langle s_1, \ldots, s_m \rangle \preceq t$ holds in every semi-Heyting structure.*

Suppose that $\langle s_1, \ldots, s_m \rangle \preceq t$ doesn't hold in a semi-Heyting structure **Q**. Then, there exists a valuation f on **Q** such that $\{f(s_1), \ldots, f(s_m)\} \preceq f(t)$ is not true in **Q**. We define a valuation g on the quasi-completion $C_{\mathbf{Q}^*}$ of **Q** as follows. For each variable q appearing in one of $s_1, \ldots, s_m$ and t, $g(q) = k(f(q))$. (For the constant 0 we suppose that $f(0) = 0$ for f, and define that $g(0) = C\emptyset$ ($\subseteq k(0)$).) Then, using Theorem 15, we can prove with the unit ε the following by induction on the length of the term r. with the unit ε

LEMMA 17 *For each subterm r of one of $s_1, \ldots, s_m$ and t, $\{f(r)\} \in g(r) \subseteq k(f(r))$.*

Suppose that $(s_1 \wedge \ldots \wedge s_m) \leq t$ holds in $\mathbf{C}_{\mathbf{Q}^*}$. Then $(g(s_1) \cap \ldots \cap g(s_m)) \subseteq g(t)$, in particular. Using Lemma 17, $\{f(s_i)\} \in g(s_i)$ for each i and hence $\{f(s_1), \ldots, f(s_m)\} \in g(s_i)$. Thus,

$$\{f(s_1), \ldots, f(s_m)\} \in (g(s_1) \cap \ldots \cap g(s_m)) \subseteq g(t) \subseteq k(f(t)),$$

by Lemma 17. But this implies $\{f(s_1), \ldots, f(s_m)\} \preceq f(t)$, which is a contradiction. Therefore, the inequality $(s_1 \wedge \ldots \wedge s_m) \leq t$ doesn't hold in the Heyting algebra $\mathbf{C}_{\mathbf{Q}^*}$. This completes the proof.

COROLLARY 18 *For all terms $s_1, \ldots, s_m$ and t, $(s_1 \wedge \ldots \wedge s_m) \leq t$ holds in every Heyting algebra if and only if $\langle s_1, \ldots, s_m \rangle \preceq t$ follows from the system of basic "inequalities" by using rules for "inequalities" described in the definition of semi-Heyting structures. Thus, the sequent system **LJ** without cut rule is complete with respect to the class of all Heyting algebras.*

As shown essentially in [16], in order to discuss sequent systems like **LK** where each sequent is of the form $\Gamma \Rightarrow \Delta$ with sequences of formulas Γ and Δ, we need to make an obvious modification of the definition of $\preceq$. That is, $\preceq$ must be defined as a binary relation on Q^*. Also, instead of taking $[x; a]$ we need to define $[x; y]$ for $x, y \in Q^*$ by

$$[x; y] = \{\langle u, w \rangle \in (Q^*)^2 : xu \preceq yw\},$$

and then define a closure operator C on $\wp((Q^*)^2)$ by using them. For **LK**, we define $\sim X$ for each subset X of $(Q^*)^2$ by

$$\sim X = \bigcap\{[x; y] : \langle x, y \rangle \in X\},$$

which plays a role of the complement of X. In fact, we can show that $\sim\sim X = C(X)$ for any X, and therefore the law of double negation holds when each C-closed X. For modal logics, we introduce $\Diamond X$ for a subset X of $(Q^*)^2$ by

$$\Diamond X = \bigcap\{[\Box x; \Diamond y] : X \subseteq [x; y]\},$$

where $\Box x$ ($\Diamond x$) is the element $\{\Box a_1, \ldots, \Box a_m\}$ ($\{\Diamond a_1, \ldots, \Diamond a_m\}$, respectively) of Q^* when x is $\{a_1, \ldots, a_m\}$. By using this, we can get an algebraic proof of cut elimination theorem for sequent systems of some of basic modal logics, including **K**, **KT** and **S4**.

By modifying our algebraic proof of cut elimination theorems and using ideas developed by Lafont and Okada-Terui [15, 20], we can show moreover the finite model property of a logic which has a cut-free sequent system, if every unprovable formula of it has a finite *proof search tree* in the sequent system. The details will be discussed in [1]. Usually, the finite model property is shown in order to derive the decidability. So, the above statement may sound strange, as we use the existence of decision procedures to show the finite model property. But, this has happened in the study of substructural logics, where decidability results on most of basic substructural logics were shown much earlier than those of the finite model property (see [23, 17, 15, 20]). Another promising way of obtaining the finite model property of substructural logics is given in the paper Blok-van Alten [2], where the *finite embeddability property* of the class of algebras for a given substructural logic is used.

In the last, we will remark that the quasi-completion of any Heyting algebra **P** is nothing but the MacNeille completion. As mentioned above, each Heyting algebra $\mathbf{P} = \langle P, \wedge, \vee, \to, 0, 1 \rangle$ with the lattice order $\leq$ determines naturally a a semi-Heyting structure $\mathbf{P}'$, by identifying each element $z = \{a_1, \ldots, a_m\}$ in P^* with an element $\tilde{z} = a_1 \wedge \ldots \wedge a_m$ in P, and taking $\leq$ for $\preceq$.

Under this identification, each set of the form $[x; a]$ will be regarded as a subset $\{b \in P : b \wedge \tilde{x} \leq a\}$, or equivalently $\{b \in P : b \leq \tilde{x} \to a\}$, which we write also as $[\tilde{x} \to a]$ by abuse of symbols. Of course, each set of the form $[c]$ for $c \in P$ is downward closed since $\leq$ is transitive. Thus, we can define an operation C on $\wp(P)$, instead of $\wp(P^*)$ by

$C(X) = \bigcap \{[c] : X \subseteq [c]\}$ for an element $c \in P\}$ for each subset X of P.

This is equivalent to the condition that $C(X) = ((X)^{\rightarrow})^{\leftarrow}$. Thus, the quasi-completion of a Heyting algebra **P** is identified with the Heyting algebra $\mathbf{C_P}$, which is the MacNeille completion of **P**. Moreover, in the present case, by using $a \star b \in k(a) \star_C k(b)$ in Theorem 15, we have $k(a \star b) \subseteq k(a) \star_C k(b)$ and hence $k(a \star b) = k(a) \star_C k(b)$. Since k is injective, we have the following when identifying a Heyting algebra **P** with the corresponding semi-Heyting structure **P'**.

COROLLARY 19 *The quasi-completion of a Heyting algebra is equal to the MacNeille completion, and moreover the map k defined by $k(a) = [a]$ for each $a \in P$ is an embedding of* **P** *into* $\mathbf{C_P}$.

BIBLIOGRAPHY

[1] F. Belardinelli, P. Jipsen and H. Ono, *An algebraic proof of cut elimination theorem*, in preparation.
[2] W.J. Blok and C.J. van Alten, *The finite embeddability property for residuated lattices, pocrims and BCK-algebras*, draft, 2000.
[3] T.S. Blyth and M.F. Janowitz, *Residuation Theory*, International Series of Monographs in Pure and Applied Mathematics 102, Pergamon Press, 1972.
[4] A. Bucalo, *Modalities in linear logic weaker than the exponential "of course": algebraic and relational semantics*, Journal of Logic, Language and Information 3, (1994), 211-232.
[5] A.V. Chagrov and M.V. Zakharyaschev, *Modal Logic*, Oxford University Press, 1997.
[6] M. Gehrke and B. Jónsson, *Bounded distributive lattices with operators*, Mathematica Japonica 40, (1994), 207-215.
[7] S. Givant and Y. Venema, *The preservation of Sahlqvist equations in completions of Boolean algebras with operators*, Algebra Universalis 41, (1999), 47-84.
[8] R. Goldblatt, *Varieties of complex algebras*, Annals of Pure and Applied Logic 44, (1989), 173-242.
[9] R. Goldblatt, *Persistence and atomic generation for varieties of Boolean algebras with operators*, Studia Logica 68, (2001), 155-171.
[10] S. Görnemann, *A logic stronger than intuitionism*, Journal of Symbolic Logic 36, (1971), 249-261.
[11] P. Jipsen and C. Tsinakis, *A survey of residuated lattices*, in: Ordered Algebraic Structures, ed. by J. Martinez, Kluwer Academic Publishers, (2002), 19-56.
[12] B. Jónsson, *On the canonicity of Sahlqvist identities*, Studia Logica 53, (1994), 473-491.
[13] B. Jónsson and A. Tarski, *Boolean algebras with operators. Part I*, American Journal of Mathematics 73, (1951), 891-939.
[14] T. Kowalski and H. Ono, *Residuated Lattices: An algebraic glimpse at logics without contraction*, monograph, March, 2001.
[15] Y. Lafont, *The finite model property for various fragments of linear logic*, Journal of Symbolic Logic 62, (1997), 1202-1208.
[16] S. Maehara, *Lattice-valued representation of the cut-elimination theorem*, Tsukuba Journal of Mathematics 15, (1991), 509-521.

[17] R.K. Meyer and H. Ono, *The finite model property for BCK and BCIW*, Studia Logica 53, (1994), 107-118.
[18] J.D. Monk, *Completions of Boolean algebras with operators*, Mathematische Nachrichten 46, (1970), 47-55.
[19] M. Okada, *Phase semantics for higher order completeness, cut-elimination and normalization proofs (extended abstract)*, Electronic Notes in Theoretical Computer Science 3, (1996).
[20] M. Okada and K. Terui, *The finite model property for various fragments of intuitionistic linear logic*, Journal of Symbolic Logic 64, (1999), 790-802.
[21] H. Ono, *A study of intermediate predicate logics*, Publications of RIMS, Kyoto University 8, (1973), 619-649.
[22] H. Ono, *Semantics for substructural logics*, in: Substructural Logics, eds. by K. Došen and P. Schroeder-Heister, Oxford University Press, (1993), 259-291.
[23] H. Ono, *Decidability and the finite model property of substructural logics*, in: Tbilisi Symposium on Logic, Language and Computation: Selected Papers (Studies in Logic, Language and Information), eds. by J. Ginzburg et.al, CSLI, (1998), 263-274.
[24] H. Ono, *Proof-theoretic methods for nonclassical logic — an introduction*, in: Theories of Types and Proofs (MSJ Memoirs 2), eds. by M. Takahashi, M. Okada and M. Dezani-Ciancaglini, Mathematical Society of Japan, (1998), 207-254.
[25] H. Ono, *Algebraic semantics for predicate logics and their completeness*, in: Logic at Work; Essays dedicated to the memory of Helena Rasiowa (Studies in Fuzziness and Soft Computing 24), ed. by E. Orlowska, Physica-Verlag, (1999), 637-650.
[26] H. Ono, *Closure operators and complete embeddings of residuated lattices*, to appear in Studia Logica 74, (2003).
[27] H. Ono and Y. Komori, *Logics without the contraction rule*, Journal of Symbolic Logic 50, (1985), 169-201.
[28] F. Paoli, *Substructural Logics: A Primer*, Trends in Logic 13 – Studia Logica Library, Kluwer Academic Publishers, 2002.
[29] C. Rauszer and B. Sabalski, *Remarks on distributive pseudo-Boolean algebra*, Bulletin de l'Académie Polonaise des Sciences, Série des Sciences Mathematiques, Astronomiques et Physiques 23, (1975), 123-129.
[30] G. Restall, *An Introduction to Substructural Logics*, Routledge, 2000.
[31] V.B. Shehtman and D.P. Skvortsov, *Semantics of non-classical first order predicate logics*, in: Mathematical Logic, ed. by P.P. Petkov, Plenum Press, (1990), 105-116.
[32] T. Shimura, *Kripke completeness of some intermediate predicate logics with the axiom of constant domain and a variant of canonical formulas*, Studia Logica 52, (1993), 23-40.
[33] R. Sikorski, *Boolean Algebras*, third edition, Ergebnisse der Mathematik ind ihrer Grenzgebiete 25, Springer Verlag, 1969.
[34] Y. Tanaka and H. Ono, *Rasiowa-Sikorski lemma and Kripke completeness of predicate and infinitary modal logics*, in: Advances in Modal Logic 2, eds. by M. Zakharyaschev et al, CSLI Publications, (2001), 401-419.

Hiroakira Ono
Graduate School of Information Science,
Japan Advanced Institute of Science and Technology (JAIST),
Tatsunokuchi, Ishikawa, 923-1292, Japan
Email: ono@jaist.ac.jp

An Axiomatization of Prior's Ockhamist Logic of Historical Necessity

MARK REYNOLDS

ABSTRACT. We present a complete axiomatization of Prior's Ockhamist branching time logic of historical necessity. This solves a long standing open problem.

1 Introduction

Logics of *historical necessity*, the most natural combination of tense with the modality of necessity and possibility, have a long and venerable history of development, especially in connection with arguments of human freedom versus determinism. Their variations and close relatives (such as the discrete branching time logics CTL[5], CTL*[8] and PCTL*[14, 18]), also have important recent applications including many in AI, agent reasoning and robotics [25, 1, 17], software engineering, reactive systems, and data communication protocols, [7, 21], and natural language understanding and philosophy[15].

These logics allow for reasoning about processes over time when the development is not necessarily determined. To achieve this, indeterminist time is usually modelled by a tree with each point having one linear past but branching towards the future. These future branches represent several alternative possibilities. The unique past represents the necessity of history: it can not be changed and there are no choices in that direction. Thus such logics are often termed logics of historical necessity.

Arthur Prior [16] recognized two main alternative approaches to historical necessity. The Peircean approach only allows statements of the form "p will occur" to be true at a point of indeterminist time if on all possible futures, p occurs at some time. In this paper we will concentrate on the other approach. This is named the Ockhamist after William of Ockham who is supposed to have held that it is meaningful to make statements whose truth is not yet known to us. The idea is that the truth of statements is evaluated with respect to a point on a particular whole linear history through time. The Ockhamist logic is the more expressive [20].

Prior started to formalize these ideas in logics with the standard temporal G and H connectives in combination with a modality which quantifies over all histories through the point of evaluation. We might call this the *alethic* modality and we symbolize it by $\Box$. The semantics of Prior's Ockhamist (and Peircean) logic were provided in [3] where we also find the definition of a related "bundled" version of the logic. The bundled Ockhamist logic simply restricts evaluation and modal quantification to within a certain specified and fixed set or "bundle" of histories. Since then there has been much work in the area. Complete axiom systems have been presented for many branching time logics: CTL [6], Prior's Peircean logic [4] (or in [27] without a Gabbay-style IRR-rule), bundled CTL* [21], the bundled version of Prior's Ockhamist logic [26] (although an unpublished IRR-rule based axiomatization by Gabbay has been cited), bundled Ockhamist logic with "Until" and "Since" [28], CTL* [19], and PCTL* [18].

The decidability of many of these logics follows from the decidability of the second order monadic logic of trees [12]. Complexity results have been established for some of the variations.

This is an active area of research. Recent related work includes development of hybrid branching time logics [2], calculi [24] for first-order versions, decidability of fragments of the first-order versions [13], and decision procedures and axiomatizations for the versions allowing quantification over propositional variables[9].

Despite this effort and interest, technical difficulties have left the presentation of a sound and complete axiom system for Prior's most basic original Ockhamist logic of historical necessity as an open problem. In section 6 below we present a complete axiom system for the logic.

This conference paper also gives a brief sketchy overview of the long, complex and quite interesting completeness proof. The proof starts off in a traditional way but brings in new "banning" ideas from [19] and an unusual trans-countable construction. A full version of the proof (of over 100 pages) is in preparation.

Future work will include the extension of the technique to the language with "until" and "since" connectives, to combinations with epistemic logics, to the first-order version and to that including quantification of propositions. There is also need for the development of applications in reasoning about agents and about complex reactive systems. We will also try determining the complexity of decision procedures, and implementing efficient decision procedures.

2 The logic

A *tree* is a pair $(T, <)$ where T is a non-empty set, $<$ is a binary relation on T such that:
1. $<$ is irreflexive and transitive;
2. $<$ is past-linear, i.e. ,
   if $y < x$ and $z < x$ then either $y < z$, $z < y$ or $y = z$; and
3. $(T, <)$ is connected, i.e. ,
   for all $x, y \in T$, there is $z \in T$ with $z \leq x$ and $z \leq y$.

A *history* of $(T, <)$ is a maximal linearly $<$-ordered subset of $(T, <)$. Let $\mathbf{B}(T, <)$ be the set of all histories of $(T, <)$. A set of histories $B \subseteq \mathbf{B}(T, <)$ is called a *bundle* if for every $t \in T$ there is some $b \in B$ with $t \in b$. If B is a bundle on $(T, <)$ then $(T, <, B)$ is called a *bundled frame*. $(T, <, B)$ is *complete* iff $B = \mathbf{B}(T, <)$.

Fix a countable set $\mathcal{L}$ of atoms. *Bundled tree structures* $\mathcal{T} = (T, <, B, h)$ will have a bundled tree frame $(T, <, B)$ and a valuation h for the atoms i.e. for each atom $p \in \mathcal{L}$, $h(p) \subseteq T$. We have an *Ockhamist structure* iff $(T, <, B)$ is complete.

The set of formulas contains the atoms and recursively includes $\neg \alpha$, $\alpha \wedge \beta$, $G\alpha$, $H\alpha$ and $\Box \alpha$.

Formulas are evaluated at points on bundle histories in bundled tree structures:

$\mathcal{T}, \sigma, x \models p$ iff $x \in h(p)$, for p atomic;
$\mathcal{T}, \sigma, x \models \neg \alpha$ iff $\mathcal{T}, \sigma, x \not\models \alpha$;
$\mathcal{T}, \sigma, x \models \alpha \wedge \beta$ iff   both $\mathcal{T}, \sigma, x \models \alpha$ and $\mathcal{T}, \sigma, x \models \beta$;
$\mathcal{T}, \sigma, x \models G\alpha$ iff   for all $y > x$ in σ we have $\mathcal{T}, \sigma, y \models \alpha$;
$\mathcal{T}, \sigma, x \models H\alpha$ iff   for all $y < x$ in σ we have $\mathcal{T}, \sigma, y \models \alpha$;
$\mathcal{T}, \sigma, x \models \Box \alpha$ iff   for every history $\pi \in B$ containing x we have $\mathcal{T}, \pi, x \models \alpha$.

As well as the usual classical abbreviations, we have linear temporal ones $F\alpha \equiv \neg G \neg \alpha$, $P\alpha \equiv \neg H \neg \alpha$, $F_{\leq}\alpha \equiv \alpha \vee F\alpha$, $G_{\leq}\alpha \equiv \alpha \wedge G\alpha$, $P_{\leq}\alpha \equiv \alpha \vee P\alpha$ and $H_{\leq}\alpha \equiv \alpha \wedge H\alpha$. There is also the branching modal diamond, $\Diamond \alpha \equiv \neg \Box \neg \alpha$.

We say that α is *valid* in the bundled logic (and we write $\models_B \alpha$) iff for all bundled tree structures $\mathcal{T} = (T, <, B, h)$, for all histories $\sigma \in B$, for all points $x \in \sigma$, we have $\mathcal{T}, \sigma, x \models \alpha$.

Prior's Ockhamist logic is the logic of complete bundled structures. We write $(T, <, h), \sigma, x \models \alpha$ for $(T, <, \mathbf{B}(T, <), h), \sigma, x \models \alpha$. We say that α is *valid* in Prior's Ockhamist logic (and we write $\models_O \alpha$) iff for all complete bundled structures $\mathcal{T}$, for all histories σ in $\mathcal{T}$, for all points $x \in \sigma$, we have $\mathcal{T}, \sigma, x \models \alpha$.

It is worth noting that for us formulas are evaluated at points on histories but the valuation or assignment of atoms is determined only by the point. Thus the truth of an atom agrees at two histories through the same point. We can justify this prescription by assuming that atomic propositions do not contain any trace of futurity: an atom is either true now or false now regardless of how the future is to evolve. This assumption is arguable. Prior discusses the issue in [16] on page 124, and describes two different logics: one using two sorts of atomic propositions and the other (on page 126) being the one we look at, having only one sort of proposition with a single truth value at each point. Our approach is also taken in [22]. The alternative approach, with assignment depending on point and history is taken in [3] and [29] which paper also contains further interesting discussion on the question. In the completeness proof the non-futurity asssumption is relied on at various important places (see discussion later).

3 A system for bundled validity

An axiom system $\vdash_B$ for bundled validity will form part of our final axiom system. The inference rules are modus ponens, temporal and branching generalization, a Gabbay-style IRR-rule:

$$\frac{\alpha, \alpha \to \beta}{\beta} \qquad \frac{\alpha}{G\alpha} \qquad \frac{\alpha}{H\alpha} \qquad \frac{\alpha}{\Box\alpha} \qquad \frac{(p \wedge H\neg p) \to \alpha}{\alpha} \text{ if } p \text{ does not appear in } \alpha,$$

and a special rule (atomic non-futurity) saying that the truth of propositional atoms only depends on the present:

ANF $\quad p \to \Box p$, for each atomic proposition p.

The axiom schemes include all substitution instances of propositional tautologies along with all instances of the following axioms.

Those for general linear temporal logic from the complete system given by [3]:
L1 $\quad G(\alpha \to \beta) \to (G\alpha \to G\beta)$ and mirror image;
L2 $\quad G\alpha \to GG\alpha$
L3 $\quad \alpha \to GP\alpha$ and mirror image;
L4 $\quad (F\alpha \wedge F\beta) \to (F(\alpha \wedge \beta) \vee F(\alpha \wedge F\beta) \vee F(\beta \wedge F\alpha))$
   and mirror image.

We add axioms ensuring that the history modality $\Box$ behaves as in the modal logic S5:
BK $\quad \Box(\alpha \to \beta) \to (\Box\alpha \to \Box\beta)$,
BT $\quad \Box\alpha \to \alpha$,
BE $\quad \Diamond\alpha \to \Box\Diamond\alpha$.

We need just one axiom to specify the interaction between modalities:

HN $P\alpha \to \Box P \Diamond \alpha$.

Finally, we require maximality of histories:
MB $G\bot \to \Box G\bot$.

Note that we do not use a substitution rule as it is not valid for $\models_B$. For example, $\Diamond p \to p$ is valid for each atomic proposition p (and can be derived easily) but $\Diamond \alpha \to \alpha$ is not generally valid: it is easy to construct a model of $\Diamond Fp \wedge \neg Fp$.

4 Bundled completeness and Kamp frames

By following the general idea of proofs in [11], [26], [10] and [29], we could show that

LEMMA 1 $\vdash_B$ *is sound and complete for* $\models_B$.

The proof is based on the traditional step by step construction of a perfect chronicle attributed to Kripke and Lemmon. A chronicle is a frame with points labelled by maximally consistent sets of formulas. The chronicle is perfect if the labels respect the accessibility relations between points in terms of both box and diamond formulas.

In the Kripke-Lemmon style axiomatic completeness proof for the linear temporal logic (see [3]), a chronicle is built on a linear frame. Each point is labelled with a maximal consistent set of formulas and we eventually ensure that the final chronicle is perfect. We can define a valuation on this frame by checking for the existence of an atom (rather than its negation) inside the label at each point. The perfection of the chronicle allows us to prove a "truth lemma" showing the equivalence of the truth of a formula at a point in the model to the appearance of that formula in the label at that point.

In the branching case there is an immediate problem with this basic strategy. Truth of formulas is not evaluated at points in a tree frame but instead at history-point pairs. Thus our chronicle will need to allow several different labels at each point of the tree. Fortunately, this problem has been neatly solved by the use of what is now called Kamp frames which were invented by Hans Kamp in a 1979 unpublished manuscript.

There are variations on this idea ([22], [23], [29]) but the general idea is that we make enough copies of each tree point and separate the various histories in the tree. We end up with a vaguely grid-like structure with the temporal <-relation increasing vertically and an equivalence relation $\equiv$ relating various nodes on a horizontal level. A whole $\equiv$-class of Kamp frame points correspond to just one point of a tree. Details can be found in [20]; see figures 1 and 2 for an example.

In our proof we use our own version of the idea:

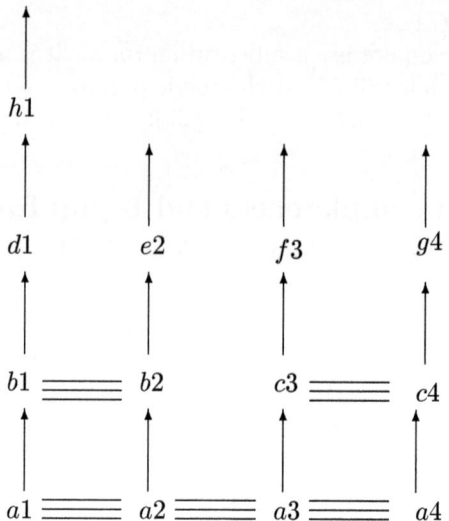

Figure 1. A Kamp frame

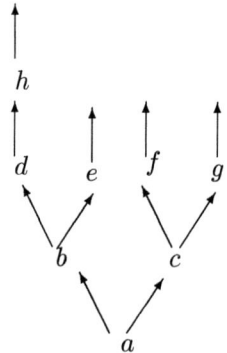

Figure 2. The corresponding tree

DEFINITION 2 A *Kamp* frame is $(T, <, \equiv)$ where:
- KF1) $<$ is a transitive, irreflexive linear relation on T
- KF2) $\equiv$ is an equivalence relation such that:
  - KF2.1) if $x \equiv y$ then we do not have $x < y$; and
  - KF2.2) if $x \equiv y$ and $u < x$ then there is $v < y$ such that $u \equiv v$;
- KF3)   the structure is connected, i.e.
  for all $x, y \in T$ there is $u \leq x$ and $v \leq y$ such that $u \equiv v$; and
- KF4)   the columns are separated, i.e.
  for each $x \equiv y$ with $x \neq y$ there exists $v > y$ such that there is no $u > x$ with $u \equiv v$.

A chronicle for the bundled logic is a Kamp frame with points labelled by maximally consistent sets of formulas. As we build the chronicle, starting from a one point one, we ensure that the labels of points respect the three box modalities, G, H and $\Box$. Thus, if $x < y$ and $G\alpha$ is in the label of x then α is in the label of y. Similarly, if $x \equiv y$ and $\Box\alpha$ is in the label of x then α is in the label of y.

Building a perfect chronicle requires us to reach, in the limit of the step by step construction, a structure in which the three modal diamonds F, P and $\Diamond$ are respected as well. Thus, for example, if $F\alpha$ is in the label of x then we require that there is a point $y > x$ with α in the label of y. If, at some finite stage, $F\alpha$ is in the label of x but there is no point $y > x$ with α in the label of y then we say that $(x, F\alpha)$ is a *defect*. One step in the construction consists in curing a defect by extending the chronicle slightly to add the needed y.

By enumerating the defects and being fair in curing them we can ensure that the limit is perfect. The Kamp frame corresponds directly with a bundled tree structure via a process of collapsing $\equiv$-classes (see [20] for details) and the truth lemma follows.

Note that the IRR rule is used in the construction to provide "names" for labels and hence to make it easy to match up labels around a square $x > y \equiv y' < x' \equiv x$. There is a complicated axiom in [26] which provides an alternative to using the IRR rule to axiomatize this logic.

5   Limit closure

The example $\gamma = \Box G(p \to \Diamond Fp) \to \Diamond G(p \to Fp)$ from [20] shows that $\models_B$ and $\models_O$ are distinct. A simple transfinite induction shows that $\models_O \gamma$. But now consider $T = \{(n,m) \in \mathbb{N} \times \mathbb{N} | 0 < n \leq m \text{ or } n = m = 0\}$. Put $(a,b) < (c,d)$ iff ($a = c$ and $b < d$) or $a = b < c$. Let $D = \{(n,n) | n \in \mathbb{N}\} \in B(T, <)$ and define the bundle to be $B = B(T, <) \setminus \{D\}$. Finally, define g so that $(n,m) \in g(p)$ iff $n = m$. We have $(T, <, B, g), (0,0) \models \neg\gamma$. Thus $\not\models_B \gamma$.

There is a discussion of this difference in [20] in which it is related to a similar observation concerning CTL*. The difference in validity between bundled and complete versions of branching logics is connected with what is sometimes called the *limit closure* property of the complete structures: the complete bundle is closed under taking the (in our case transfinite) limit of a process of keeping on extending initial segments of histories towards the future.

It follows that the system for $\vdash_B$ is incomplete for $\models_O$. It is interesting to see where the completeness proof for $\vdash_B$ as above goes wrong for $\models_O$. In the limit of the step by step construction of a perfect Kamp frame chronicle we only construct a countable number of vertical lines (or columns) which correspond to histories when we factor out by $\equiv$. However, the resulting tree generally also has an uncountable number of other histories which do not correspond to columns. These may be called "emergent" histories and it is they which generally prevent the truth lemma from holding: they do not even have any labels constructed for them. More details of this problem can be found in [20].

Despite being incomplete the system $\vdash_B$ is, of course, sound for $\models_O$. There have been many attempts to "find the missing axiom". Here we are able to introduce a new axiom schema to allow us to derive the extra validities.

We solve the limit closure problem by the addition of what we call a limit closure schema. It is an infinite sequence of axioms: one for each $n > 0$. Suppose $\alpha_0, ..., \alpha_{n-1}$ are any formulas. Put $\alpha_n = \alpha_0$. The schema allows

LC $\quad \Box G_\leq \bigwedge_{i=0}^{n-1} (\Diamond \alpha_i \to \Diamond F \Diamond \alpha_{i+1}) \quad \to \quad \Diamond G_\leq \bigwedge_{i=0}^{n-1} (\Diamond \alpha_i \to F \Diamond \alpha_{i+1}).$

LEMMA 3 *LC is sound.*

Proof. The soundness proof is a straightforward transfinite induction. Basically, we cycle through the indicies gradually building an increasing sequence of points in the tree where the $\Diamond \alpha_i$ hold.

To proceed more formally suppose t is a point in a history b of a structure $(T, <, h)$ such that $(T, <, h), b, t \models \beta$ where $\beta = \Box G_\leq \bigwedge_{i=0}^{n-1} (\Diamond \alpha_i \to \Diamond F \Diamond \alpha_{i+1})$.

We observe the following. If there is some $s \geq t$, some history $b' \ni s$ and some $i = 0, 1, ..., n-1$, such that $(T, <, h), b', s \models \Diamond \alpha_i$ then there is a point $s' > s \geq t$ and a history $b'' \ni s'$ such that $(T, <, h), b'', s' \models \Diamond \alpha_{i+1}$.

Recalling that $\alpha_n = \alpha_0$, a short induction thus tells us that if there is some $s \geq t$, some history $b' \ni s$ and some $i = 0, 1, ..., n-1$, such that $(T, <, h), b', s \models \Diamond \alpha_i$ then there is a point $s' > s \geq t$ and a history $b'' \ni s'$ such that $(T, <, h), b'', s' \models \Diamond \alpha_0$.

If there is no $s \geq t$ with some history $b' \ni s$ and some $i = 0, 1, ..., n-1$, such that $(T, <, h), b', s \models \Diamond \alpha_i$ then clearly $(T, <, h), b, t \models G_\leq \bigwedge_{i=0}^{n-1}(\Diamond \alpha_i \to F \Diamond \alpha_{i+1})$ and we are done.

So now suppose that there is some $s \geq t$, some history $b' \ni s$ and some $i = 0, 1, ..., n-1$, such that $(T, <, h), b', s \models \Diamond \alpha_i$.

We find by transfinite induction (using the observation above) an ordinal Λ and a possibly transfinite sequence $\{t_{\lambda,i} | \lambda < \Lambda, i = 0, 1, ..., n-1\}$ of points and some histories $b_{\lambda,j}$ such that $t \leq t_{0,0}$ and for each $\lambda < \Lambda$, for each $i = 0, 1, 2, ..., n-1$, we have $t_{\lambda,i} \in b_{\lambda,i}$ and $(T, <, h), b, t_{\lambda,i} \models \Diamond \alpha_i$. We also ensure that if $\lambda < \mu < \Lambda$ then $t_{\lambda,i} < t_{\mu,j}$ and if $i < j < n$ then $t_{\lambda,i} < t_{\lambda,j}$.

The case of a successor ordinal is straightforward. Consider a limit ordinal λ such that we have defined $t_{\mu,i}$ for each $\mu < \lambda$ and each $i = 0, 1, ..., n-1$. If there is no history b', point $s \in b'$ and $i = 0, 1, ..., n-1$ such that each $t_{\mu,i} < s$ and $(T, <, h), b', s \models \Diamond \alpha_i$ then we let $\Lambda = \lambda$ and we are finished the construction. If there are such b', s and i then we know that there is also $t_{\lambda,0} > s$ and $b_{\lambda,0} \ni s$ such that $(T, <, h), b_{\lambda,0}, t_{\lambda,0} \models \Diamond \alpha_0$. This continues our construction.

Now suppose we have finished our construction at Λ: the construction can obviously not continue forever. Let

$$\pi = \{x \in T | \text{ there is some } \lambda, i \text{ such that } x \leq t_{\lambda,i}\}.$$

This is clearly linearly ordered by $<$. Thus there is a history $\sigma \supseteq \pi$.

We have $(T, <, h), \sigma, t \models G_\leq \bigwedge_{i=0}^{n-1}(\Diamond \alpha_i \to F \Diamond \alpha_{i+1})$.

As $t \leq s = t_{0,0}$ we have $t \in \pi \subseteq \sigma$ and so

$$(T, <, h), b, t \models \Diamond G_\leq \bigwedge_{i=0}^{n-1}(\Diamond \alpha_i \to F \Diamond \alpha_{i+1})$$

as required. ∎

6 The axiom system for Ockamist validity

Let $\vdash_H$ be derivability in the axiom system with the schema LC added to $\vdash_B$.

THEOREM 4 $\vdash_H$ *is sound and complete for the logic* **HN** *of historical necessity.*

Soundness is the usual induction. To show completeness we have fixed a formula ϕ which is supposed to be $\vdash_H$-consistent, i.e. $\not\vdash_H \neg \phi$. We are to show that it has a model. Most of the rest of the paper sketches an intricate construction of a model of ϕ: the full version is in preparation.

7 Overview of the completeness proof

The completeness proof proceeds in two stages. The first stage is based on the traditional step by step construction of a perfect Kamp frame chronicle described above. Stage two, which we return to briefly below, is a transfinite process of what we call grafting, copying the latter part of branches after "matching ends".

The solution by matching ends

The aim of stage one is to make a perfect Kamp frame chronicle in which the ends of emergent branches match initial segments of constructed branches in terms of the "colours" of labels (see below) which appear cofinally. In order to achieve the matching in stage one we use a banning technique along the lines of that used in [19]. To do this requires finite sets as labels, the IRR rule and the new LC rule.

Finite labels: hues and colours

During our construction we are really only interested in making a chronicle which is relatively perfect in respecting the finite set of subformulas of ϕ (and a few other formulas): this is enough to establish the truth lemma. This allows us to join parts of the chronicle together at points which have labels which agree on those formulas but maybe not on others. We say that the labels are of the same *hue* (defined below). Our banning procedure needs to do this and it also needs to control which hues recur along branches during the construction.

Recall that in an Ockhamist frame the equivalence relation $\equiv$ relates points which collapse to a state under the usual branching time semantics. Thus we need to ensure that equivalent points' labels agree on *state* formulas, i.e. boolean combinations of those of the form $\Box\beta$.

The hue of a point will also indicate which other hues will hold at other points which are $\equiv$-related to that point. A set of hues sported by the members of a whole $\equiv$-class of points will be called a colour. All equivalent points will share a colour and so it is not surprising that corresponding to a colour will be a state formula γ_c: the colour is a property of the whole equivalence class. This fact will ensure that we will be able to talk about the sequence of colours (but not hues) along the emergent histories in the limit of our construction.

The first of several closure sets we use is $\mathbf{cl}\phi = \{\psi, \neg\psi | \psi \leq \phi\}$ where $\psi \leq \phi$ denotes that ψ is a subformula of ϕ. Let

$$\mathbf{fcl}\phi = \{\psi, \neg\psi, \Diamond\psi, \neg\Diamond\psi, \Diamond\neg\psi, \neg\Diamond\neg\psi | \psi \leq \phi\}.$$

For $a \in \wp(\mathbf{fcl}\phi)$, define

$$\delta_a = \bigwedge_{\delta \in a} \delta \wedge \bigwedge_{\delta \in (\mathbf{fcl}(\phi) \setminus a)} \neg \delta.$$

Let $C = \wp(\wp(\mathbf{fcl}\phi))$: the set of ϕ-*colours*. For $c \in C$ define

$$\gamma_c = \bigwedge_{a \in c} \Diamond \delta_a \wedge \bigwedge_{a \in (\wp(\mathbf{fcl}\phi) \setminus c)} \neg \Diamond \delta_a.$$

Given a colour $c \in C$ the various $a \in c$ determine what we will call the *hues* of c. So c has at most $|c|$ different hues. The hue $h(a,c)$ of c corresponding to $a \in c$ is given by

$$h(a,c) = \{\delta | \delta \in a\} \cup \{\neg\delta | \delta \in \mathbf{fcl}\phi \setminus a\} \cup \{\Diamond \delta_b | b \in c\} \cup \{\neg \Diamond \delta_b | b \in \wp(\mathbf{fcl}\phi) \setminus c\}.$$

The set $H(\phi)$ of all hues of ϕ is $H(\phi) = \{h(a,c) | a \in c \in C\}$. Each hue h is a hue of exactly one colour h^*.

Maximal consistent sets (MCSs) extend (consistent) hues: Let $\mathbf{hcl}\phi = \{\delta, \neg\delta | \delta \in \mathbf{fcl}\phi\} \cup \{\Diamond \delta_b, \neg \Diamond \delta_b | b \in \wp(\mathbf{fcl}\phi)\}$. Then for each MCS Δ, the set $h = \Delta \cap \mathbf{hcl}\phi$ is a hue. Furthermore, for each Δ this is the only hue which satisfies $h \subseteq \Delta$.

We define two useful relations on $H(\phi)$ as well as the traditional ones between MCSs. $\Gamma R_G \Delta$ iff for all α, if $G\alpha \in \Gamma$ then $\alpha \in \Delta$; and $\Gamma R_N \Delta$ iff for all α, if $\Box \alpha \in \Gamma$ then $\alpha \in \Delta$. Say that $hR_G h'$ iff there are MCSs Γ and Γ' such that $h \subseteq \Gamma$, $h' \subseteq \Gamma'$ and $\Gamma R_G \Gamma'$. Say that $hR_N h'$ iff there are MCSs Γ and Γ' such that $h \subseteq \Gamma$, $h' \subseteq \Gamma'$ and $\Gamma R_N \Gamma'$.

In our chronicle, points $x \equiv y$ have labels which extend hues of the same colour and points $x < y$ have respective labels Γ_x, Γ_y with $\Gamma_x R_G \Gamma_y$.

IRMCSs

Here we use the IRR rule for two reasons: matching squares $x > y \equiv y' < x' \equiv x$ as in the bundled case and as part of the banning procedure which we describe below. We just use a slight extension of the idea of IRR-theories as given in definition 6.2.3 of [10]: we allow $\Diamond$ as well as F and P. Points will be labelled with these what we call IRMCS sets which each have a unique name in that they contain some formula $p \wedge H \neg p$. Many useful results follow: eg, if $F\alpha$ is in IRMCS Γ then there is an IRMCS Δ with $\Gamma R_G \Delta \ni \alpha$.

The step by step

There is no space for a full definition of a chronicle but it is essentially a Kamp frame with each point labelled by two IRMCS and a ban and a banning query as defined shortly.

One of the IRMCS labels at each point is thought of as facing downwards, the other upwards. The two IRMCS at each point agree on hues, i.e. on formulas in **hcl**(ϕ). In a traditional chronicle we would only have one label at a point but having two allows us to join together pieces of different (traditional) chronicles provided they agree enough at the joins on the finite set of formulas of interest to us. This is important to the banning process. At most points the two labels agree exactly.

At all stages all three box modalities are respected in labels. Each step in the construction is a cure of one of the three types defects corresponding to the three modal diamonds. Because of the joins in the chronicles where labels may only match huewise, we are mostly only concerned with the cure of defects involving formulas in **cl**ϕ but there are a few complications which can not be discussed here.

Banning

The only time that banning comes in to play is when we cure an alethic defect, i.e. when $\Diamond \alpha$ is in the label of some point and we create a new vertical column of points containing a cure for the defect. The new point will be a joint and its downward facing label just any IRMCS cure for the defect will not necessarily agree with its upward facing label.

The upward facing label at this new point will have to take account of any new ban.

In order to make sure that the matching ends property results from this construction we use banning to prevent certain colours from occuring above a joint unless certain conditions are met. Ironically for historical necessity, banning is a way of gradually eliminating indeterminancy during the construction process. We want to only allow a colour to turn up as a label if it has to. This will mean, thanks to the LC rule, that the colours which turn up along an emergent history also turn up along a constructed history.

To do this we use three types of bans. Each of the three types has a corresponding formula of the language.

Suppose that we have a $\Diamond$ defect of $\Diamond \alpha$ in the label at x. We add a new column and a new point y such that $x \equiv y$ and α is in the label at y. Say that we choose the hue h of colour a for y. Now we consider making one of the three types of bans. We may decide whether to ban colour a from appearing again at any point up ($<$) and across ($\equiv$) from y. Suppose τ represents the current ban at x. We will ban a unless $\vdash (a \wedge \tau) \rightarrow \Diamond F a$. If this query succeeds, i.e. if $\vdash (a \wedge \tau) \rightarrow \Diamond F a$ then we do nothing: a may recur. Otherwise we add $\Box G \neg a$ to our bans and labels from y upwards. In this case colour a will not recur.

The second type of ban is to disallow a colour b from coming up again above y unless we then ban a from occurring again above that.

The third type of ban concerns a sequence $[a = e_0, e_1, e_2, ..., e_n]$ of colours and a disjoint set D of colours. We consider whether there must be a sequence of points $y \equiv y_0 < y_1 < ... < y_n$ such that y_i has colour e_i and there are no points between y_0 and y_n coloured from D. To capture the query and ban in this case needs to make use of the unique "names" that points have when we construct a model using the IRR rule. The query and ban formulas are lengthy.

At each joint, each time we cure an alethic defect, we schedule one of these ban queries in a fair way, and add the appropriate ban depending on the outcome of the query.

The limit chronicle as a tree

In the ω-limit of our construction we have a chronicle which corresponds to a bundled Ockhamist tree model $(S_0, <_0, B_0, g_0)$ of ϕ. It is relatively perfect in respecting formulas from $\mathbf{cl}(\phi)$.

This will not generally be able to be completed to an Ockhamist model of ϕ because of the problem of emergent branches mentioned above.

However, the banning procedure will have ensured that we have the important property of "matching ends" as follows. Suppose that η_0 is the labelling of $\{(b, s) | s \in b \in B_0\}$ by hues in the limit chronicle. We say that $\pi \subset S_0$ is an initial segment (of history σ) iff $\pi \subseteq \sigma$ and if $x < y \in \pi$ then $x \in \pi$. For each initial segment π of a history in $(S_0, <_0)$ we define infcol(π) to be the set of colours which appear cofinally as the colours of labels of points in π. We also let ccreg(π) be the set of points $x \in \pi$ such that if $x \leq y \in \pi$ then the colour of the label of y is in infcol(π). Then we have:

LEMMA 5 *Let σ be any history of $(S_0, <_0, B_0, \eta_0)$ which is not in B_0.*
 Then for all $x \in \sigma$ there is an initial segment π of a history $\rho \in B_0$ such that:
  M1) $x \in \pi \cap \sigma$;
  M2) ccreg(π) $\cap$ ccreg(σ) $\neq \emptyset$;
  M3) infcol(π) = infcol(σ); and
  M4) $\pi \not\subseteq \sigma$; and
  M5) either $\pi = \rho$ or colours not in infcol(π) appear arbitrarily soon after π in ρ.

Grafting trees

A grafting tree is $(S, <, B, \eta)$ where $(S, <, B)$ is a bundled tree, $\eta : \{(b, s) \in B \times S | s \in b\} \to H(\phi)$ and η perfectly labels $(S, <, B)$ in the sense that both box and diamond formulas in hues are respected.

Starting with the grafting tree created by the limit chronicle, we proceed via a transfinite induction until the $\aleph_1$ stage. At each stage we extend the previous grafting trees to a new one.

The basic idea is that the grafting tree $\mathcal{S}_\kappa$ at stage κ only has non-bundled histories which each end just like some specific identified non-bundled history of $\mathcal{S}_0$. Consider a non-bundled history σ of $\mathcal{S}_\kappa$. Suppose that the end of it is just a copy of non-bundled history ρ of $\mathcal{S}_0$. By the matching ends property of $\mathcal{S}_0$ we can find am initial segment π of a bundled history of $\mathcal{S}_0$ which matches in terms of cofinal colours. To create $\mathcal{S}_{\kappa+1}$ we extend each such non-bundled σ as follows. Take each (bundled or not) history τ of $\mathcal{S}_0$ which extends π and copy all the points (with their labels) in $\tau \setminus \pi$ over to lie after σ. We create a copy of the branching sub-tree after π and put it after σ: this is what we call grafting. It can be shown that again we have a perfect grafting tree.

At limit ordinal stages we just take unions of the grafting trees at all earlier stages.

The truth lemma

During our transfinite grafting process we also record which points in $\mathcal{S}_0$ are copied to make the points in each $\mathcal{S}_\kappa$.

Certain technical arguments to do with the sets of cofinal colours on the matching ends of extended histories allow us to find, for each unbundled history σ in $\mathcal{S}_{\aleph_1}$ a gradually strictly increasing sequence of initial segments in $\mathcal{S}_0$ which are copied across to contribute to making σ. In fact there will be a set of ordinals, cofinal in $\aleph_1$, which enumerate a strictly increasing sequence of segments and hence points in $\mathcal{S}_0$.

As no countable set of ordinals can be cofinal in $\aleph_1$ this would contradict the countability of $\mathcal{S}_0$. The only conclusion is that there are no unbundled histories in $\mathcal{S}_{\aleph_1}$.

Thus we finally have a complete and perfect bundled grafting tree. Define a valuation g for the atoms using the hue labels $\eta_{\aleph_1}$ and the truth lemma for the resulting Ockhamist tree structure $\mathcal{S} = (S_{\aleph_1}, <_{\aleph_1}, g)$ follows directly.

8 Extensions

Follow on work could include investigating the version of the logic in which the truth values of atoms may depend on the history of evaluation, ie in which we do not make the assumption that atoms contain no trace of futurity. It is not clear whether the techniques in this paper will go through easily to this other logic. There is a particular problem with the copying and grafting technique in which we use the fact that points of the same colour sufficiently far into the cofinal closure region of branches agree on all formulas up to a given complexity. This would not hold if the colour of a point did not determine the truth of atoms.

Other useful related axiomatizations which are left as future work include the Ockhamist logic with until and since connectives, and Ockhamist logics over trees in which all histories have particular properties such as denseness or being the real numbers.

BIBLIOGRAPHY

[1] N. Belnap. Agents in branching time. In *Logic and Reality: Essays in Pure and applied Logic, in Memory of Arthur Prior*. OUP, 1996.
[2] P. Blackburn and V. Goranko. Hybrid ockhamist temporal logic. In C. Bettini and A. Montanari, editors, *Proc. of the Eighth International Symposium on Temporal Representation and Reasoning (TIME-01, Cividale del Friuli, Italy, June 14-16, 2001)*,, pages 183–188. IEEE Computer Society Press, 2001.
[3] J. P. Burgess. Logic and time. *J. Symbolic Logic*, 44:566–582, 1979.
[4] J. P. Burgess. Decidability for branching time. *Studia Logica*, 39:203–218, 1980.
[5] E. Clarke and E. Emerson. Synthesis of synchronization skeletons for branching time temporal logic. In *Proc. IBM Workshop on Logic of Programs, Yorktown Heights, NY*, pages 52–71. Springer, Berlin, 1981.
[6] E. Emerson and J. Halpern. Decision procedures and expressiveness in the temporal logic of branching time. *J. Comp and Sys. Sci*, 30(1):1–24, 1985.
[7] E. Emerson and J. Halpern. 'Sometimes' and 'not never' revisited: on branching versus linear time. *J. ACM*, 33, 1986.
[8] E. Emerson and A. Sistla. Deciding full branching time logic. *Information and Control*, 61:175 – 201, 1984.
[9] T. French. Decidability of quantified propositional branching time logics. In M. Stumptner, D. Corbett, and M. Brooks, editors, *AI 2001: Advances in Artificial Intelligence*. Springer, 2001.
[10] D. Gabbay, I. Hodkinson, and M. Reynolds. *Temporal Logic: Mathematical Foundations and Computational Aspects, Volume 1*. Oxford University Press, 1994.
[11] D. M. Gabbay. An irreflexivity lemma with applications to axiomatizations of conditions on tense frames. In U. Monnich, editor, *Aspects of Philosophical Logic*, pages 67–89. Reidel, Dordrecht, 1981.
[12] Y. Gurevich and S. Shelah. The decision problem for branching time logic. *J. of Symbolic Logic*, 50:668–681, 1985.
[13] I. Hodkinson, F. Wolter, and M. Zakharyashev. Decidable and undecidable fragments of first-order branching temporal logics. Proceedings of the 17th Annual IEEE Symposium on Logic in Computer Science, LICS 2002, pages 393-402.
[14] F. Laroussinie and Ph. Schnoebelen. Specification in CTL+past for verification in CTL. *Information and Computation*, 156:236–263, 2000.
[15] P. Øhrstrøm and P. F. V. Hasle. *Temporal Logic: From Ancient Ideas to Artificial Intelligence*, volume 57 of *Studies in Linguistic and Philosophy*. Dordrecht: Kluver Academic Publishers, 1996.
[16] A. Prior. *Past, Present and Future*. Oxford University Press, 1967.
[17] A. Rao and M. Georgeff. Decision procedures for BDI logics. *Journal of Logic and Computation*, 8(3):293–342, 1998.
[18] M. Reynolds. More past glories. In *Proceedings of Fifteenth Annual IEEE Symposium on Logic in Computer Science (LICS'2000), Santa Barbara, California, USA, June 26-28, 2000*, pages 229–240. IEEE, Los Alamitos, California, USA, 2000.
[19] M. Reynolds. An axiomatization of full computation tree logic. *J. of Symbolic Logic*, 66(3):1011–1057, 2001.
[20] M. Reynolds. Axioms for branching time. *J. Logic and Computation*, 2002. to appear.
[21] C. Stirling. Modal and temporal logics. In S. Abramsky, D. Gabbay, and T. Maibaum, editors, *Handbook of Logic in Computer Science, Volume 2*, pages 477–563. OUP, 1992.

[22] R. Thomason. Combinations of tense and modality. In D. Gabbay and F. Guenthner, editors, *Handbook of Philosophical Logic, Vol II: Extensions of Classical Logic*, pages 135–165. Reidel, Dordrecht, 1984.

[23] F. von Kutschera. $T \times W$ completeness. *Journal of Philosophical Logic*, 26(3):241–250, 1997.

[24] S. Woelfl. Combinations of Tense and Modality for Predicate Logic. *Journal of Philosophical Logic*, 28(4):371–398, 1999.

[25] M. Xu. Decidability of deliberative *stit* theories with multiple agents. In D. Gabbay and H. Ohlbach, editors, *Temporal Logic, Proceedings of ICTL '94*, number 827 in LNAI, pages 332–348. Springer-Verlag, 1994.

[26] A. Zanardo. A finite axiomatization of the set of strongly valid Ockamist formulas. *J. of Philosophical Logic*, 14:447–468, 1985.

[27] A. Zanardo. Axiomatization of 'Peircean' branching-time logic. *Studia Logica*, 49:183–195, 1990.

[28] A. Zanardo. A complete deductive system for since-until branching time logic. *J. Philosophical Logic*, 1991.

[29] A. Zanardo. Branching-time logic with quantification over branches: the point of view of modal logic. *J. Symbolic Logic*, 61:1–39, 1996.

Mark Reynolds
School of Information Technology
Murdoch University
Murdoch, Perth, Western Australia 6150
Australia
E-mail: m.reynolds@murdoch.edu.au

Combining Dynamic Logic with Doxastic Modal Logics

RENATE A. SCHMIDT AND DMITRY TISHKOVSKY

ABSTRACT. We prove completeness and decidability results for a family of combinations of propositional dynamic logic and unimodal doxastic logics in which the modalities may interact. The kind of interactions we consider include three forms of commuting axioms, namely, axioms similar to the axiom of perfect recall and the axiom of no learning from temporal logic and a Church-Rosser axiom. We investigate the influence of the substitution rule on the properties of these logics and propose a new semantics for the test operator to avoid unwanted side effects caused by the interaction of the classic test operator with the extra axioms.

1 Introduction

Propositional dynamic logic (*PDL*) [6] is a powerful and convenient logical tool for reasoning about programs or actions. It has found applications in various branches of computer science, ranging from program verification to such areas as agent-based systems. Agent-based systems are formalised and analysed with a variety of logical methods. Often combinations of *PDL* and modal logics form the basis of agent theories which are intended to formalise the behaviour of autonomous agents [2, 7, 10, 12]. While *PDL* provides the apparatus for modelling and reasoning about the dynamic activities of the agents, informational aspects of the agents, in particular, the properties of the agents' belief and knowledge spaces, are typically modelled in the doxastic modal logics *K(D)45* and *S5* [2]. Combinations of *PDL* with *K(D)45* and *S5* are thus meaningful in agent applications.

Generally, there are different ways of combining of modal logics. Certain forms of combinations of modal logics are rather well-behaved, while other forms of combinations are known to pose problems in proving any results about them. The simplest form of combination of two (or more) logics is their fusion, or independent join. It is well-known [8, 9, 16] that fusions of modal logics inherit properties such as soundness, completeness, the finite model property and decidability, from the individual logics.

In modelling intelligent agents, actions and belief and also knowledge are normally linked in some form. Well-known connections between informational attitudes and actions are the properties of no learning and prefect recall (see e.g. [14]). No learning is given by the axiom schema

$$[a]\Box p \to \Box[a]p$$

(where $[a]$ is a *PDL* modality and $\Box$ is the epistemic modality). The schema says that the agent knows the result of her action in advance; in other words, there is no learning. Perfect recall is commonly formulated by the axiom schema

$$\Box[a]p \to [a]\Box p.$$

It expresses the persistence of the agent's knowledge after the execution of an action. In this paper we also allow the interaction of modalities via the Church-Rosser axiom

$$\Diamond[a]p \to [a]\Diamond p.$$

The axiom says that if an agent has a possibility to perform an action a in such way that she will believe p, then in the current state she believes that it is possible to obtain the result p by doing a.

Whereas fusions of well-behaved modal logics inherit the good properties from the individual logics, the situation of fusions of modal logics extended with interaction axioms like the above is not as straightforward. For such combinations there is no preservation theorem of the generality as for fusions. The particular type of interaction between the modal operators and the *PDL* operators makes it more difficult (but not impossible) to obtain positive results regarding completeness, the finite model property and decidability as we will see in this paper.

A particular problem we encounter in the study of combinations of *PDL* and doxastic modal logics with the above interaction axioms is the collapse of certain combinations of logics to one of their constituents. Our analysis shows that there are at least two explanations for this collapse. One explanation is the strength of the standard substitution rule. Normal modal logics are closed under substitution, which means that arbitrary propositional formulae may be substituted for propositional variables in axioms. *PDL* is also closed under substitution. Because *PDL* has a two-sorted language over actions and propositions, this means arbitrary propositional formulae may be substituted for propositional variables in axioms, and likewise, arbitrary action formulae may be substituted for action variables. In the combinations of *PDL* and informational logics such as $K(D)45$ and $S5$ with any of the axioms of no learning, perfect recall and the Church-Rosser axiom, closure under substitution, particularly, closure under substitution for actions, can have unwanted effects as is shown in this paper. We consider therefore the question as to how substitutivity should be defined in such logics based on *PDL*. The question is the following: In axiom schemata, do we allow substitution of all action formulae into action variables, or do we allow substitution of atomic action formulae into action variables only? If the answer is 'yes' we speak of *full substitutivity*, whereas if the answer is 'no' we speak of *weak substitutivity*. *PDL* is closed under full substitution. It means propositions and actions are treated uniformly in *PDL*. However in extensions of *PDL* and *K(D)45* by any of the axioms no learning, perfect recall and Church-Rosser axiom this (good) property implies unnatural properties of an agent. More dramatically, we show that under full substitutivity the *S5* operator in the analogous combination of *PDL* and *S5* is vacuous.

This paper focuses on the fusion of *PDL* and *K45*, *KD45* or *S5* extended with various choices of the axioms of no learning, perfect recall and the Church-Rosser axiom. We explore all the possible definitions of such extensions with respect to full and weak substitutivity. For each definition we consider the problems of completeness, the small model property, decidability and the admissibility of the full substitution rule (Sections 3 and 4). Our proofs combine the Gabbay-Shehtman filtration method for products of *S5* in [5] and the Fischer-Ladner filtration technique developed for *PDL* [3]. We prove that in combinations of *PDL* with *K45*, *KD45* or *S5* full substitutivity does not follow from weak substitutivity. We prove further that full substitutivity in the extensions of the fusion of *PDL* and *S5* causes this logic to collapse to *PDL*.

This collapse is not witnessed for test-free versions of the logics (Section 5). This is a pointer to the negative influence of the test operator of *PDL* in the interactions. Closer analysis shows indeed a second explanation for the collapse to *PDL* is: the interaction between the test operator of *PDL* with reflexive modalities. This negative result is perhaps not surprising because the test operator includes propositions into the semantics at the level of frames. Since the omission of the test operator from *PDL* would imply the agents have no means to program, or reason about programs, we use a modification of the definition and semantics of the test operator introduced in [13]. The new test operator has informational character rather than universal character; it confirms the belief (or knowledge) of an agent as opposed to confirming the properties of the actual state as the classical test operator does in *PDL*. Completeness and the small model property (therefore also decidability) are shown for the logics with the informational test operator (Section 6). These logics are shown to coincide under both notions of substitutivity, and we show that even though the definition and semantics of the test operator has been changed, classic *PDL* can still be simulated in the fusion of the introduced informational variation of *PDL* and modal logic *S5*.

The next section (Section 2) introduces the logics under consideration and provides the definitions of important concepts needed for the main part of the paper (Sections 3–6). The paper concludes with a mention of some open problems and observations relating to products of the considered logics (Section 7).

2 Main definitions and preliminary observations

The language $\mathcal{L}$ we consider is an extension of the language of *PDL* [6] with a new modal operator $\Box$. Formally, the language $\mathcal{L}$ is defined over the following primitive types: a countable set $\text{Var} = \{p, q, r, \ldots\}$ of propositional variables and a countable set $\text{AtAc} = \{a, b, c, \ldots\}$ of atomic action variables. The connectives in $\mathcal{L}$ are the Boolean connectives, $\rightarrow$, $\bot$, the dynamic logic connectives, $\cup$, ;, $*$, ?, and the modal operators $[\_]$ and $\Box$. The set For of formulae and the set Ac of action formulae in $\mathcal{L}$ are the smallest sets that satisfy the following conditions. (i) $\text{AtAc} \subseteq \text{Ac}$, $\text{Var} \cup \{\bot\} \subseteq \text{For}$. (ii) If ϕ and ψ are formulae in For and α and β are action formulae in Ac then $\phi?$, α^*, $\alpha \cup \beta$, $\alpha;\beta$ are action formulae in Ac, and $\phi \rightarrow \psi$, $\Box\phi$, $[\alpha]\phi$, are formulae in For. The connectives $\neg$, $\vee$, $\wedge$, $\leftrightarrow$, $\langle\_\rangle$ are defined as usual.

By definition, an *atomic action* is an action variable, and a *semi-atomic action* is an atomic action or a test action $\phi?$.

By a *theory* in $\mathcal{L}$ we understand any subset of For closed under the following standard rules:

$$\phi, \phi \rightarrow \psi \vdash \psi \qquad \phi \vdash [\alpha]\phi \qquad \phi \vdash \Box\phi$$

Generally axioms and theorems of a logic are assumed true for all instantiations for the atomic symbols. However, in this paper we distinguish between two variants of the substitution rule. The *weak substitution rule* allows the substitution of arbitrary formulae for the atomic propositional symbols but does not allow substitution for atomic action symbols. By contrast, the *full substitution rule* allows both kinds of substitutions, i.e. both propositional substitutions and action substitutions.

A *logic* in $\mathcal{L}$ is a theory which is closed under the full substitution rule and a *weak logic* in $\mathcal{L}$ is a theory closed under the weak substitution rule. Weak logics are notationally discerned by a subscript w.

For an axiomatisation of *PDL* we refer to [6], and axiomatisations of *K45*, *KD45* and *S5* can be found in [1, 16].

Let Γ and Δ be any subsets of For. By $\Gamma \oplus \Delta$ (resp. $\Gamma \oplus \Delta_w$) we denote the least logic (resp. the least weak logic) which contains both Γ and Δ. The *fusion*, or *independent join*, of *PDL* and a unimodal logic L is therefore denoted by $PDL \oplus L$. By an *extension* of a logic L (resp. a weak logic L_w)

Axiom		Correspondence property
(NL) $[a]\Box p \to \Box[a]p$	(com^r)	$R \circ Q(a) \subseteq Q(a) \circ R$
(PR) $\Box[a]p \to [a]\Box p$	(com^l)	$Q(a) \circ R \subseteq R \circ Q(a)$
(CR) $\Diamond[a]p \to [a]\Diamond p$	(cr)	$Q(a)^{\smile} \circ R \subseteq R \circ Q(a)^{\smile}$

Figure 1. The interaction axioms and their correspondence properties.

by a set of axioms Δ we understand the logic $L \oplus \Delta$ (resp. $L \oplus \Delta_w$). We say the full substitution rule is *admissible* in a weak logic L_w iff $L_w = L$.

PROPOSITION 1 *Let L be any normal unimodal logic. Then $PDL \oplus L_w = PDL \oplus L$, i.e. the full substitution rule is admissible in $PDL \oplus L_w$.*

For any unimodal logic L, standard $PDL \oplus L$ models are combinations of the familiar standard models for PDL and Kripke models for L. That is, a *standard $PDL \oplus L$ model* is a tuple $\langle S, Q, R, \models \rangle$ such that $\langle S, Q, \models_0 \rangle$ is a standard PDL model [6] and $\langle S, R, \models_1 \rangle$ is an L model [1] where $\models_0$ and $\models_1$ are obvious restrictions of $\models$ on PDL-formulae and unimodal formulae respectively. Here, S is a non-empty set of states, R is a binary relation on S, $\models$ is the usual truth relation on the model and Q is a mapping from the set of actions to the set of binary relations on S which satisfies the following conditions.

$$Q(\alpha \cup \beta) = Q(\alpha) \cup Q(\beta) \qquad Q(\alpha;\beta) = Q(\alpha) \circ Q(\beta)$$
$$Q(\alpha^*) = Q(\alpha)^* \qquad Q(\phi?) = \{(s,s) \in S^2 \mid s \models \phi\}$$

($\circ$ denotes relational composition, and $*$ is the reflexive and transitive closure operator on relations.)

This paper investigates the family of all possible extensions of $PDL \oplus L$ and test-free $PDL \oplus L$, where L is any of the logics in $\{K45, KD45, S5\}$, by the axioms NL, PR and CR. NL is short for no learning, PR for perfect recall, and CR for Church-Rosser. These axioms are listed in Figure 1 together with their Sahlqvist correspondence properties. The symbol $\smile$ in the figure denotes converse. In the case of logics with full substitution we assume that a ranges over all the actions, whereas for weak logics we assume that a ranges over the set of atomic actions only. In the latter case we refer to the correspondence properties as com^r_w, com^l_w and cr_w.

Our aim is to prove soundness and completeness results, so, the properties com^l, com^r, cr and the weak versions of them determine the intended standard models for the logics. For instance, by definition, a *standard $PDL \oplus L \oplus \{NL, PR, CR\}$ model* (resp. a *standard test-free $PDL \oplus L \oplus \{NL, PR, CR\}$*

model) is a standard $PDL \oplus L$ model (resp. a standard test-free $PDL \oplus L$ model) which satisfies the properties com^r, com^l and cr. The definitions of the models for the other logics and their weak versions are similar and can be obtained by substitution of the appropriate model properties which correspond to the extra axioms of a given logic. Soundness of all these logic with respect to the corresponding classes of standard models can be proved in the usual manner.

For $\Delta \subseteq \{NL, PR, CR\}$, if a structure $\langle S, Q, R, \models \rangle$ satisfies all the properties of a standard $PDL \oplus L \oplus \Delta_w$ model except possibly the property $Q(\alpha^*) = (Q(\alpha))^*$, but still validates all the formulae of $PDL \oplus L \oplus \Delta_w$ then we call this structure a *non-standard model* for $PDL \oplus L \oplus \Delta_w$. This notion is similarly defined for full logics. It is obvious that, for any of the considered logics, the class of all standard model for this logic is included in the class of all non-standard models for the logic. It can be proved that the axioms NL, PR and CR are canonical in the presence of the full substitution rule for the properties com^r, com^l and cr, respectively. Similarly, these axioms are canonical for com^r_w, com^l_w and cr_w in the presence of the weak substitution rule. It is also a well-known fact that 4, 5, D and T are canonical, respectively, for transitivity, Euclideaness, seriality and reflexivity of the doxastic accessibility relation. In the canonical model of the logic $PDL \oplus K45 \oplus \{NL, PR\}$, for instance, any action relation commutes with the doxastic relation and the doxastic relation is transitive and Euclidean. Using a standard argument for PDL, it can be shown that canonical models for all the considered logics are not standard and belong to the class of non-standard models. In fact, the following (more general) completeness theorem holds.

THEOREM 2 *Let L be any canonical unimodal logic and suppose* $\Delta \subseteq \{NL, PR, CR\}$. *Then* $PDL \oplus L \oplus \Delta_w$ *and* $PDL \oplus L \oplus \Delta$ *are strongly complete with respect to the corresponding classes of non-standard models.*

It is easy to prove the following proposition which states that the axioms NL and CR are inter-derivable in the presence of the axiom $p \to \Box \Diamond p$.

PROPOSITION 3 *Let L_0 be either PDL or test-free PDL, and let L_1 be any normal unimodal logic extending* $KB = K \oplus \{p \to \Box \Diamond p\}$. *Then* $L_0 \oplus L_1 \oplus \{NL\}_w = L_0 \oplus L_1 \oplus \{CR\}_w$ *and* $L_0 \oplus L_1 \oplus \{NL\} = L_0 \oplus L_1 \oplus \{CR\}$.

Moreover, a consequence of soundness is the following.

PROPOSITION 4 $PDL \oplus S5 \oplus \{PR\}_w \not\supseteq PDL \oplus S5 \oplus \{CR\}_w$ *and test-free* $PDL \oplus S5 \oplus \{PR\}_w \not\supseteq$ *test-free* $PDL \oplus S5 \oplus \{CR\}_w$.

The result is also true for the combinations with $K45$ and $KD45$ in place of $S5$. Furthermore, it is easy to prove the next two propositions.

PROPOSITION 5 $PDL \oplus KD45 \oplus \{NL, PR\}_w \not\supseteq PDL \oplus KD45 \oplus \{CR\}_w$ and test-free $PDL \oplus KD45 \oplus \{NL, PR\}_w \not\supseteq$ test-free $PDL \oplus KD45 \oplus \{CR\}_w$.

PROPOSITION 6 $PDL \oplus KD45 \oplus \{PR, CR\}_w \not\supseteq PDL \oplus KD45 \oplus \{NL\}_w$ and test-free $PDL \oplus KD45 \oplus \{PR, CR\}_w \not\supseteq$ test-free $PDL \oplus KD45 \oplus \{NL\}_w$.

By these four propositions, all the axioms PR, NL and CR are independent of each other already with respect to the fusion of PDL and $KD45$ closed under the full or weak substitution rule. Thus, we need to consider all combinations of interaction axioms producing extensions of $PDL \oplus K45$, $PDL \oplus KD45$ and their weak versions. In the case of $PDL \oplus S5$ and $PDL \oplus S5_w$, by Propositions 3 and 4, in order to get all the possible extensions by all of the interaction axioms it is enough to consider only two of the axioms, one of which is PR, i.e., for instance, PR and CR.

In the remainder of the paper we will consider all possible extensions of $PDL \oplus L$ and test-free $PDL \oplus L$ (where $L \in \{K45, KD45, S5\}$) by the axioms NL, PR and CR for completeness (w.r.t. classes of the intended standard models), the admissibility of full substitution rule and the *small model property* which will be stated in the following form:

If a formula ϕ is satisfiable in a *non-standard* model for a logic then it is satisfiable in a finite *standard* model for this logic and the upper bound for the size of the latter is effectively computable from the length of ϕ.

It is well-known and quite easy to see that the small model property implies the existence of a decision algorithm for a logic. Thus, any logic with the small model property is decidable. Immediate consequences of the small model theorems proved in this paper are therefore decidability.

Because the canonical model is not standard, completeness with respect to the classes of standard models satisfying the frame correspondence properties of the additional axioms is not a straightforward consequence of the canonicity of the logic L and the extra axioms. To get completeness we modify the canonical models constructed by the method of Fischer and Ladner [3, 6], while taking care that left and right commutativity and the Church-Rosser property are preserved.

Subsequently when we talk about a model without explicit use of the words 'standard' or 'non-standard', we mean a standard model.

3 Weak extensions of $PDL \oplus L$

This section proves results about extensions $PDL \oplus L_w$ by axioms NL, PR and CR closed under weak substitutivity. The first results show that in combinations of PDL with extensions of $K45$, it is not the case that weak substitutivity implies full substitutivity.

THEOREM 7 *Let L be any normal unimodal logic which is contained in the logic of two-element clusters. Then any weak extension of the logic $PDL \oplus L$ by any combination of the axioms NL, PR and CR do not admit the rule of full substitution.*

Proof. Let M_0 be a model $\langle S, Q, R, \models \rangle$ with $S = \{0, 1\}$, $Q(a) = \emptyset$ for any atomic action a, R is the universal relation on S, $0 \models p$ and $1 \models \neg p$. It is easy to see that NL and PR are true in M_0 for all atomic actions. But all the formulae $[\neg p?]\Box p \to \Box[\neg p?]p$, $\Box[p?]p \to [p?]\Box p$ and $\Diamond[p?]\bot \to [p?]\Diamond\bot$ are false in the state 0. ∎

Standard examples of logics contained in the logic of two-element clusters are KB, KT, KD, $K4$, $K5$, $S5$, $KD45$, $K45$, etc. Thus, in particular:

COROLLARY 8 *None of the following logics admit the rule of full substitution: $PDL \oplus K45 \oplus \{NL, PR, CR\}_w$, $PDL \oplus KD45 \oplus \{NL, PR, CR\}_w$ and $PDL \oplus S5 \oplus \{NL, PR\}_w$.*

By the way we note that the example in the proof of Theorem 7 is also suitable for proving that the full substitution rule is not admissible in the product of PDL and L (as defined in [15]).

As said in the previous section, the canonical model for any of the considered logics is a non-standard model. Following the common line of argument in proving completeness results our goal is to modify non-standard models to obtain standard models while preserving the satisfiability of the given *finite* set of formulae. Our method essentially repeats the filtration method of Fischer and Ladner [3], but because of the modal interaction axioms, special considerations are needed which depart from the standard route to ensure the preservation of the corresponding properties through the filtration. What is required, as we show, is a combination of the filtration method of Fischer and Ladner with the filtration method of Gabbay and Shehtman [5] for products of modal logics with $S5$.

The following definition is standard.

DEFINITION 11 Let $\prec$ be the smallest transitive relation over $\{0,1\} \times \mathrm{For}$ satisfying the following.

$$(0,\phi),(0,\psi) \prec (0,\phi \to \psi) \qquad (1,[\alpha]\phi),(1,[\beta]\phi) \prec (1,[\alpha \cup \beta]\phi)$$
$$(0,\phi) \prec (0,\Box\phi) \qquad (1,[\alpha][\beta]\phi),(1,[\beta]\phi) \prec (1,[\alpha;\beta]\phi)$$
$$(0,\phi),(1,[\alpha]\phi) \prec (0,[\alpha]\phi) \qquad (1,[\alpha][\alpha^*]\phi) \prec (1,[\alpha^*]\phi)$$
$$(0,\phi) \prec (1,[\phi?]\psi)$$

Let $\preceq$ be the reflexive closure of $\prec$. For a set Σ of formulae, the *Fischer-Ladner closure* of Σ is defined by

$$FL(\Sigma) =^{\mathrm{def}} \{\psi \mid \exists j \in \{0,1\} \, \exists \phi \in \Sigma \ (j,\psi) \preceq (0,\phi)\}.$$

Subsequently, we write $FL(\phi)$ instead of $FL(\{\phi\})$.

It is easy to see that $\prec$ is well-founded. Consequently, $FL(\Sigma)$ is finite whenever Σ is finite. The next lemma allows us to find an upper bound for the cardinality of $FL(\Sigma)$ in terms of the lengths of formulae in Σ.

We denote by $|\phi|$ and $|\alpha|$ the lengths of ϕ and α (as number of the symbols excluding parentheses), respectively.

LEMMA 12 $\mathsf{Card}(FL(\phi)) \leq |\phi|$.

The next definition is inspired by [5, Proposition 12.5].

DEFINITION 13 Let $M = \langle S, Q, R, \models \rangle$ be a (non-standard) model for $PDL \oplus L$, let Σ be a finite set of formulae. For any s in M we denote by $\Sigma(s)$ the set $\{\phi \in FL(\Sigma) \mid s \models \phi\}$.

Define a relation $\sim_\Sigma$ on S by:

$$s \sim_\Sigma t \iff \Sigma(s) = \Sigma(t) \text{ and}$$
$$\{\Sigma(u) \mid (s,u) \in R\} = \{\Sigma(u) \mid (t,u) \in R\}.$$

Let $\|s\|_\Sigma$ (or, simply, $\|s\|$) denote the equivalence class for s, i.e. $\{t \mid s \sim_\Sigma t\}$. Let S^Σ be the set of equivalence classes over S, i.e. $S^\Sigma = \{\|s\| \mid s \in S\}$.

Define a filtrated model $M^\Sigma = \langle S^\Sigma, Q^\Sigma, R^\Sigma, \models \rangle$ by the *least filtration* of M, that is, the relations on the M^Σ are defined by the following. For any atomic action a,

$$Q^\Sigma(a) =^{\mathrm{def}} \{(\|s\|, \|t\|) \mid \exists s' \in \|s\| \, \exists t' \in \|t\| \ (s',t') \in Q(a)\},$$
$$R^\Sigma =^{\mathrm{def}} \{(\|s\|, \|t\|) \mid \exists s' \in \|s\| \, \exists t' \in \|t\| \ (s',t') \in R\}$$

Furthermore, for any propositional variable p,

$$\|s\| \models p \iff s \models p \text{ and } p \in FL(\Sigma).$$

The definition of the truth relation is extended to all formulae and the definition of Q^Σ is extended to all actions so that all the conditions on actions for standard $PDL \oplus L$ model are satisfied. This completes the definition of the filtrated model M^Σ.

The following lemma follows immediately from the definition of R^Σ.

LEMMA 14 *If R is reflexive then R^Σ is reflexive and if R is serial then R^Σ is serial.*

The proofs of the next two lemmata follow the proof of Proposition 12.5 in [5].

LEMMA 15 *If R is transitive and Euclidean then $\sim_\Sigma \circ R \subseteq R \circ \sim_\Sigma$.*

Proof. Suppose that $s \sim_\Sigma t$ and $(t, u) \in R$. Then, by the definition of $\sim_\Sigma$, there is some t' with $(s, t') \in R$ and $\Sigma(t') = \Sigma(u)$. We need to show that $\{\Sigma(v) \mid (t', v) \in R\} = \{\Sigma(v) \mid (u, v) \in R\}$. Let $(t', v) \in R$. Hence, $(s, v) \in R$ by the transitivity of R. It follows from $s \sim_\Sigma t$ that there exists v' such that $(t, v') \in R$ and $\Sigma(v) = \Sigma(v')$. And from $(t, u), (t, v') \in R$ we obtain $(u, v') \in R$ because R is a Euclidean relation. Thus, $\{\Sigma(v) \mid (t', v) \in R\} \subseteq \{\Sigma(v) \mid (u, v) \in R\}$. For the backward inclusion fix some v and assume that $(u, v) \in R$. Hence, $(t, v) \in R$ by the transitivity of R and, consequently, there exists v' such that $(s, v') \in R$ and $\Sigma(v') = \Sigma(v)$. And again, because R is Euclidean we get (t', v'). Thus, $t' \sim_\Sigma u$ and because s, t, u were picked arbitrarily, $\sim_\Sigma \circ R \subseteq R \circ \sim_\Sigma$. ∎

LEMMA 16 *If R is transitive and Euclidean then R^Σ is transitive and Euclidean.*

Proof. To prove R^Σ is transitive suppose that $(\|s\|, \|t\|), (\|t\|, \|u\|) \in R^\Sigma$. Without loss of the generality we can assume that $(s, t) \in R$, $t \sim_\Sigma t'$ and $(t', u) \in R$. By Lemma 15 there exists a u' such that $(t, u') \in R$ and $\|u'\| = \|u\|$. From the transitivity of R we have $(s, u') \in R$ and, hence, $(\|s\|, \|u\|) \in R^\Sigma$.

To prove that R^Σ is Euclidean suppose that $(\|s\|, \|t\|), (\|s\|, \|u\|) \in R^\Sigma$. Again, without loss of the generality we can assume that $(s, t) \in R$, $s \sim_\Sigma s'$ and $(s', u) \in R$. By Lemma 15 there exists t' such that $(s', t') \in R$ and $\|t'\| = \|t\|$. Because R is Euclidean, we obtain $(t', u') \in R$ and, hence, $(\|t\|, \|u\|) \in R^\Sigma$. ∎

LEMMA 17 *If R is transitive and Euclidean then the properties com_w^l and cr_w persist under the filtration of Definition 13.*

Proof. Let $(||s||, ||t||) \in Q(a)^\Sigma$ and $(||t||, ||u||) \in R^\Sigma$, where a is an atomic action. By definition of R^Σ and $Q^\Sigma(a)$, there exist s', t', t'', u' such that $s' \sim_\Sigma s$, $t' \sim_\Sigma t'' \sim_\Sigma t$, $u' \sim_\Sigma u$, $(s', t') \in Q(a)$ and $(t'', u') \in R$. By Lemma 15 there is a u'' such that $u' \sim_\Sigma u''$ and $(t', u'') \in R$. Thus, because com_w^l is true in M, we obtain that (s', u'') belongs to $R \circ Q(a)$ and consequently, $(||s||, ||u||) \in R^\Sigma \circ Q^\Sigma(a)$.

To prove the statement of the lemma for cr_w assume that $(||s||, ||t||) \in Q(a)^\Sigma$ and $(||s||, ||u||) \in R^\Sigma$. That is, without loss of the generality, $(s, t) \in Q(a)$, $(s', u) \in R$ and $s' \sim_\Sigma s$. By Lemma 15 there exists a u' such that $(s, u') \in R$ and $u' \sim_\Sigma u$. Because cr_w holds in M, there exists a v such that $(u', v) \in Q(a)$ and $(t, v) \in R$. Thus, $(||u||, ||v||) \in Q^\Sigma(a)$ and $(||t||, ||u||) \in R^\Sigma$. ∎

PROPOSITION 18 *Let L be any of the logics K45, KD45, or S5, and assume $\Delta \subseteq \{PR, CR\}$. Suppose that M is a non-standard model for $PDL \oplus L \oplus \Delta_w$. Then M^Σ is a standard model for $PDL \oplus L \oplus \Delta_w$.*

Proof. By construction M^Σ is a standard model for $PDL \oplus K_w$. Hence, we only need to prove that M^Σ satisfies all the semantical properties which correspond to the extra axioms of the logic $PDL \oplus L \oplus \Delta_w$. But M satisfies these properties because it is assumed to be a non-standard model for the logic, and all the properties which correspond to 4, 5, D, T, PR and CR are persistent under the filtration by Lemmata 14, 16 and 17. ∎

It is important in the filtration method of Fischer and Ladner that for a finite set Σ any equivalence class $||s||$ is formula definable in the original model M. This is where the argument in this paper crucially departs from the argument usually used in proving the completeness of *PDL*. Usually, what is required in proving the completeness of *PDL* and *PDL* related logics is that any union of the equivalence classes in the original model is formula definable. This allows to apply the standard syntactical argument [6] in Lemma 20 below. So, we need to prove the following lemma to make the method work.

LEMMA 19 *If Σ is finite then each equivalence class $||s||$ is associated with a formula $\psi(||s||)$ which uniquely determines $||s||$ in the original model M.*

Proof. Let $\Gamma_0 = \Sigma(s)$ and $\{\Gamma_1, \ldots, \Gamma_k\} = \{\Sigma(u) \mid (s, u) \in R\}$. The set $\mathcal{P}(FL(\Sigma))$ of all subsets of $FL(\Sigma)$ is finite, because Σ is finite. Let

$\{\Delta_1, \ldots, \Delta_l\} = \mathcal{P}(FL(\Sigma)) \setminus \{\Gamma_1, \ldots, \Gamma_k\}$. Let γ_0 be the conjunction of formulae from Γ_0 together with the negations of the formulae in $FL(\Sigma) \setminus \Gamma_0$. Similarly, formulae γ_j and δ_j are introduced for Γ_j and Δ_j, respectively. Define $\psi(\|s\|)$ as follows.

$$\psi(\|s\|) =^{\text{def}} \gamma_0 \wedge \bigwedge_{1 \le j \le k} \Diamond \gamma_j \wedge \bigwedge_{1 \le j \le l} \Box \neg \delta_j$$

Now it is routine to check that $\psi(\|s\|)$ is defined correctly and the only equivalence class in which it is true is $\|s\|$, i.e. $\|s\|$ coincides with $\{t \mid t \models \psi(\|s\|)\}$. ■

LEMMA 20 *Let Σ be a finite set of formulae.*

1. *For all $(0, \phi)$ such that $(0, \phi) \preceq (0, \sigma)$ for some $\sigma \in \Sigma$,*

    $\|s\| \models \phi \iff s \models \phi$.

2. *For all $(0, \Box\phi)$ such that $(0, \Box\phi) \preceq (0, \sigma)$ for some $\sigma \in \Sigma$,*

    if $(\|s\|, \|t\|) \in R^\Sigma$ and $s \models \Box\phi$ then $t \models \phi$.

3. *For all $(1, [\alpha]\eta)$ such that $(1, [\alpha]\eta) \prec (0, \sigma)$ for some $\sigma \in \Sigma$,*

    (a) *if $(s, t) \in Q(\alpha)$ then $(\|s\|, \|t\|) \in Q^\Sigma(\alpha)$;*

    (b) *if $(\|s\|, \|t\|) \in Q^\Sigma(\alpha)$ and $s \models [\alpha]\eta$ then $t \models \eta$.*

This lemma can be proved by induction on the well-founded relation $\prec$ (see [6]), using Lemma 19 in the case 3a for $\alpha = \beta^*$ which is actually the most difficult case in the induction.

Now the Filtration Lemma follows easily.

COROLLARY 21 (Filtration lemma) *If the set Σ of formulae is finite then for all $\phi \in FL(\Sigma)$, $\|s\| \models \phi \iff s \models \phi$.*

As a consequence, with the help of Lemmata 14, 16 and 17, we obtain the following theorem.

THEOREM 22 (Small model theorem) *Let L be any of the logics K45, KD45, or S5, and assume $\Delta \subseteq \{PR, CR\}$. Let ϕ be a formula and $|\phi| = n$. If ϕ is satisfiable in a possibly non-standard model for the logic $PDL \oplus L \oplus \Delta_w$ then it is satisfiable in a standard model for this logic with no more than $2^n \cdot 2^{2^n}$ states.*

Proof. Suppose that M is a (non-standard) model for the given logic. By the Filtration Lemma, ϕ is satisfiable in a (non-standard) model M if and only if it is satisfiable in the filtrated model M^Σ whenever $\phi \in FL(\Sigma)$. By Proposition 18, M^Σ is a standard model for the logic. Further, it is easy to see that every state of the filtrated model, i.e. any equivalence class $\|s\|$ is uniquely determined by the pair of sets $\langle \Sigma(s), \{\Sigma(u) \mid (s,u) \in R\}\rangle$. Then there are 2^k possibilities for $\Sigma(s)$ and 2^{2^k} possibilities for $\{\Sigma(u) \mid (s,u) \in R\}$, where $k = \mathsf{Card}(FL(\Sigma))$. Therefore, $\mathsf{Card}(S^\Sigma) \leq 2^k \cdot 2^{2^k}$. Thus, if $\Sigma = \{\phi\}$, then $\mathsf{Card}(S^\Sigma) \leq 2^n \cdot 2^{2^n}$, by Lemma 12. ∎

THEOREM 23 (Completeness) *Let L be any of the logics K45, KD45, or S5, and assume $\Delta \subseteq \{PR, CR\}$. Then the logic $PDL \oplus L \oplus \Delta_w$ is complete with respect to the corresponding class of standard models.*

Proof. If a formula ϕ is not derivable in any of the logics then the set $\{\neg\phi\}$ is consistent with respect to this logic and, consequently, $\neg\phi$ is satisfiable in the canonical model M for the logic. The canonical model is a non-standard model for the logic and, hence, by the Small Model Theorem $\neg\phi$ is satisfiable in a standard model for the logic. Thus, ϕ is not valid in any standard model of the logic. ∎

Observe that the logics $PDL \oplus K45 \oplus \{PR, NL, CR\}_w$ and $PDL \oplus KD45 \oplus \{PR, NL, CR\}_w$ are not covered by the above theorems. It is open whether these extensions with NL have the small model property and are complete with respect to a class of standard models.

4 Extensions of $PDL \oplus L$ with full substitution

This section considers the logics closed under the full substitution rule.

LEMMA 24 *Let L be a normal unimodal logic. If one of the formulae $\Box(p \to q) \to (p \to \Box q)$, $(p \to \Box q) \to \Box(p \to q)$, or $\Diamond(p \to q) \to (p \to \Diamond q)$ belongs to L, then $p \to \Box p$ belongs to L.*

Proof. (i) Assume the first formula belongs to L. As L is closed under the substitution rule, the formula $\Box(p \to p) \to (p \to \Box p)$ belongs to L. The left side of the implication is always true. Thus, this formula is classically equivalent to $p \to \Box p$. (ii) If the second formula holds in L then substitute $\neg q$ for p. So, $(\neg q \to \Box q) \to \Box(\neg q \to q)$ is in L. This formula is equivalent to $(\neg q \to \Box q) \to \Box q$. The following equivalences complete the proof. $((\neg q \to \Box q) \to \Box q) \leftrightarrow (\neg(q \lor \Box q) \lor \Box q) \leftrightarrow ((\neg q \land \neg \Box q) \lor \Box q) \leftrightarrow ((\neg q \lor \Box q) \land (\neg \Box q \lor \Box q)) \leftrightarrow (q \to \Box q)$. (iii) Let the third formula be in L. Then $\Diamond(p \to \bot) \to (p \to \Diamond \bot)$ too belongs to L. Hence, we have $\Diamond \neg p \to \neg p$, because $\Diamond\bot \leftrightarrow \bot$ and $(p \rightarrow \bot) \leftrightarrow \neg p$. Now observe that $p \rightarrow \Box p$ is the contrapositive of $\Diamond \neg p \rightarrow \neg p$. ∎

COROLLARY 25 *Let L be any normal unimodal logic. Then $p \rightarrow \Box p$ belongs to $PDL \oplus L \oplus \{NL\}$, $PDL \oplus L \oplus \{PR\}$, and $PDL \oplus L \oplus \{CR\}$.*

Proof. Take the axiom NL, $[a]\Box p \rightarrow \Box[a]p$, and substitute a test expression $q?$ for a. This gives $[q?]\Box p \rightarrow \Box[q?]p$. Since $[q?]p \leftrightarrow (q \rightarrow p)$ belongs to PDL, $(q \rightarrow \Box p) \rightarrow \Box(q \rightarrow p)$ is in the logic. In the cases of PR or CR being valid $\Box(q \rightarrow p) \rightarrow (q \rightarrow \Box p)$, resp. $\Diamond(p \rightarrow q) \rightarrow (p \rightarrow \Diamond q)$, is valid in the logic. Now apply Lemma 24. ∎

From a philosophical perspective the axiom $p \rightarrow \Box p$ is not appropriate as an axiom for belief or knowledge. Informally, this axiom says that an agent believes that all properties are true in the current state. But, in general, agents are not aware of everything that is happening in the world and, hence, cannot build a complete picture about the current state. Thus, this axiom expresses some divine property which is uncharacteristic of a typical agent. In the combinations with $S5$ the situation is even more dramatic.

We say that a modal logic L *admits the elimination of* $\Box$, if the formula $p \leftrightarrow \Box p$ belongs to L. Let KT be the weakest reflexive normal unimodal logic.

THEOREM 26 *If an extension L of $PDL \oplus KT$ contains any of the axioms NL, PR or CR and the full substitution rule is admissible in L, then L admits the elimination of the $\Box$ modality of L.*

Proof. For the proof it is enough to apply Corollary 25 and note that $\Box p \rightarrow p$ belongs to KT. ∎

Consequently:

THEOREM 27 *All possible extensions of $PDL \oplus S5$ by the axioms NL, PR, and CR with the full substitution rule are equal to $PDL \oplus K \oplus \{\Box p \leftrightarrow p\}$ and, hence, deductively equivalent to PDL.*

Proof. All the axioms of $S5$ are derivable from $\Box p \leftrightarrow p$. Thus, $S5 \oplus \{\Box p \leftrightarrow p\} = K \oplus \{\Box p \leftrightarrow p\}$. For the rest it is enough to refer to Theorem 26. ∎

It follows from this theorem that an analogue of Proposition 4 for the logics under full substitutivity does not hold.

It is not hard to see that in any model for the logic $PDL \oplus K \oplus \{\Box p \leftrightarrow p\}$ the relation R corresponding to the $\Box$-operator is an identity relation on the set of states. Denote by $\mathcal{C}_{triv}$ the class of $PDL \oplus K$ models in which the relation R is an identity relation on the set of states. Then the following theorems can be easily proved.

THEOREM 28 (Small model theorem) *Let ϕ be a formula of $\mathcal{L}$ and let n be the number of symbols in ϕ. If ϕ is satisfiable in a $\mathcal{C}_{triv}$ model then it is satisfiable in a $\mathcal{C}_{triv}$ model with no more than 2^n states.*

THEOREM 29 (Completeness) $PDL \oplus K \oplus \{\Box p \leftrightarrow p\}$ *and, consequently, any extension of $PDL \oplus S5$ by the axioms NL, PR, and/or CR is complete with respect to the class $\mathcal{C}_{triv}$.*

That is, all the considered extensions of the fusion of *PDL* and *S5* closed under the full substitution rule collapse to *PDL* and, hence, inherit all the good properties of *PDL* such as completeness, small model property and decidability.

5 Test-free extensions

Omitting the test operator from the considered logics we can prove the following results.

LEMMA 31 *Let L be any normal unimodal logic. Let P denote one of the properties com^r, com^l, cr. Let M be a standard $PDL \oplus L$ model which satisfies P_w. Then, P is true in M for all test-free actions.*

By Lemma 31 the two classes of the intended standard models for the test-free weak logics and the corresponding test-free full logics coincide. Thus, repeating the arguments of Section 3 we obtain the following results.

THEOREM 35 (Small model theorem) *Let L be any of the logics K45, KD45, or S5, and let $\Delta \subseteq \{PR, CR\}$. Let ϕ be a formula and $|\phi| = n$. If ϕ is satisfiable in a (non-standard) model for test-free $PDL \oplus L \oplus \Delta$ then it is satisfiable in a standard model for this logic with no more than $2^n \cdot 2^{2^n}$ states.*

THEOREM 36 (Completeness) *Let L be any of the logics K45, KD45, or S5, and let $\Delta \subseteq \{PR, CR\}$. Then test-free $PDL \oplus L \oplus \Delta$ is complete with respect to the corresponding class of standard models.*

THEOREM 37 *Let L be any of the logics K45, KD45, or S5, and let $\Delta \subseteq \{PR, CR\}$. Then test-free $PDL \oplus L \oplus \Delta_w$ admits the rule of full substitution.*

6 Logics with informational test

The results of Section 4 show that the formula $p \to \Box p$ is valid in full-substitutional extensions of the fusion of *PDL* and any normal unimodal logic by any of the axioms *NL*, *PR* and *CR*. As pointed out above, for agent applications this property is inappropriate for modelling and reasoning about beliefs of agents. Also, from a logical perspective the elimination of the *S5* operator in the combination of *PDL* and *S5* is unsatisfactory. The reason for the elimination of the *S5* operator is the implicit connection between the test operator and $\Box$ in the commutativity axioms under full substitutivity. This raises the question, whether, and how *PDL* and *S5* can be integrated in a formalism without any of the connectives becoming *trivially* superfluous. One possibility is to rely on the weaker form of substitutivity. Inspection of the proof of Lemma 24 suggests assuming full substitutivity is inappropriate for the commutativity axioms. However, full substitution gives us the possibility to reason uniformly about all actions, in the same way as we reason about all propositions in any logic. *PDL* is closed under the full substitution rule and, thus, fits this paradigm. So, perhaps, the problem is with the definition of the semantics of the test operator when interacting with $\Box$. A solution we propose here is to define an alternative semantics for test such that in the resulting logic, $\Box$ and test interact in a way so that weak substitutivity implies full substitutivity.

Therefore, as replacement for the standard test operator we use a new operator, denoted by $?$, first defined in [13]. This operator is intended to remedy the problem of the standard test in the presence of an informational modal operator. The new operator is called the *informational test* operator. The intuition of $p?$ is an action which can be successfully accomplished only if p is *believed* (or *known*) in the current state. Thus, $p?$ is the action of confirming the agent's own beliefs (or knowledge). In contrast, with the usual test operator the agent has the capability to confirm truths rather than beliefs (knowledge), which is a strong property of an agent. Thus in agent based applications the new interpretation of the test operator appears more suitable than the classic interpretation.

The logical apparatus is the same as previously with the obvious changes. The symbol $?$ is used in the superscript to indicate the replacement of the operator $?$ by $?$. Let $(PDL \oplus L)^?$ be the smallest logic in the language $\mathcal{L}^?$ containing L and the axioms of *PDL*, but the usual test axiom $[p?]q \leftrightarrow (p \to q)$ is replaced by the axiom

(*) $\qquad\qquad\qquad [p?]q \leftrightarrow \Box(\Box p \to q).$

In accordance with this axiom, the formula $[p?]q$ can be read as 'q is believed with respect to p being believed' (or with 'known' replacing 'believed' as appropriate). Consequently, we may think of the operator [_?] as the modal operator of *relative information*.

A standard (resp. non-standard) $(PDL \oplus L)^?$ model is a tuple $\langle S, Q, R, \models \rangle$ satisfying all the properties of a standard (resp. non-standard) $PDL \oplus L$ model, except the meaning of ? is specified by:

(**) $$Q(\phi?) =^{\text{def}} \{(s,t) \in R \mid t \models \Box\phi\}.$$

(This property can be easily found from (*) using the SCAN tool [11].) The definitions of standard and non-standard models for the extensions of $(PDL \oplus L)^?$ by the axioms *NL*, *PR*, and *CR* (with the full or weak substitution rule) are the expected generalisations of the definitions of standard and non-standard $PDL \oplus L$ models, respectively.

The definition of ? still allows the elimination of $\Box$ but this time the elimination is not trivial.

PROPOSITION 38 $\Box p \leftrightarrow [\top?]p \in (PDL \oplus L)^?$ *for any normal unimodal logic L.*

PROPOSITION 39 *The following formulae are derivable from the axioms of* $(PDL \oplus K45)^?$ *with the full or weak substitution rule.*

$$[p?]\Box q \to \Box[p?]q \qquad \Box[p?]q \to [p?]\Box q \qquad \Diamond[p?]q \to [p?]\Diamond q$$

To prove the completeness theorem we need to show that the canonical model for the given logic is a non-standard model for this logic. We did not prove this previously because the proof is standard. Because the semantics of the test operator has been modified, it is necessary to show that the canonical models for the considered logics with the informational test operator satisfy the property (**).

Let $\mathcal{L}$ be a (weak) logic in the language $\mathcal{L}^?$. The canonical model $M_\mathcal{L} = \langle S_\mathcal{L}, Q_\mathcal{L}, R_\mathcal{L}, \models_\mathcal{L}\rangle$ for the logic $\mathcal{L}$ is defined as follows. $S_\mathcal{L}$ is a set of all maximal theories Γ in the language $\mathcal{L}^?$ such that $\Gamma \supseteq \mathcal{L}$ and $\bot \notin \Gamma$. The accessibility relations $R_\mathcal{L}$ and $Q_\mathcal{L}(\alpha)$ for any action α are defined by

$$(\Gamma, \Delta) \in R_\mathcal{L} \iff \forall \phi\, (\Box\phi \in \Gamma \Rightarrow \phi \in \Delta)$$
$$(\Gamma, \Delta) \in Q_\mathcal{L}(\alpha) \iff \forall \phi\, ([\alpha]\phi \in \Gamma \Rightarrow \phi \in \Delta)$$

The truth relation $\models_\mathcal{L}$ is a membership relation: $\Gamma \models_\mathcal{L} \phi \iff \phi \in \Gamma$.

LEMMA 40 *Let $\mathcal{L}$ be arbitrary extension of* $(PDL \oplus K)^?_w$. *Then the canonical model $M_\mathcal{L}$ satisfies* (**), *i.e.* $Q_\mathcal{L}(\phi?) = \{(\Gamma, \Delta) \in R_\mathcal{L} \mid \Delta \models_\mathcal{L} \Box\phi\}$ *for any formula ϕ.*

Proof. Suppose $(\Gamma, \Delta) \in Q_{\mathcal{L}}(\phi?)$, i.e. $[\phi?]\psi \in \Gamma$ implies $\psi \in \Delta$ for any formula ψ. Take an arbitrary ψ such that $\Box\psi \in \Gamma$. The formula $\Box\psi \to \Box(\Box\phi \to \psi)$ is derivable in modal logic K and, therefore, in $\mathcal{L}$. Hence, from the axiom (∗) we obtain that $\Box\psi \to [\phi?]\psi$ is in $\mathcal{L}$. Therefore, Γ contains $[\phi?]\psi$ and, consequently, $\Delta \ni \psi$. Thus, $(\Gamma, \Delta) \in R_{\mathcal{L}}$. Further, $\Box(\Box\phi \to \Box\phi)$ is valid in K, and, hence, $\Box(\Box\phi \to \Box\phi) \in \mathcal{L} \subseteq \Gamma$. Hence, by the axiom (∗) $[\phi?]\Box\phi \in \Gamma$ and, therefore, $\Delta \ni \Box\phi$. Thus, $Q_{\mathcal{L}}(\phi?) \subseteq \{(\Gamma, \Delta) \in R_{\mathcal{L}} \mid \Delta \models_{\mathcal{L}} \Box\phi\}$.

For the backward inclusion, suppose that $(\Gamma, \Delta) \in R_{\mathcal{L}}$ and $\Box\phi \in \Delta$. Assume that $[\phi?]\psi \in \Gamma$. By use of (∗) we obtain $\Box(\Box\phi \to \psi) \in \Gamma$. Therefore, $\Box\phi \to \psi$ is in Δ because $(\Gamma, \Delta) \in R_{\mathcal{L}}$. Hence, $\psi \in \Delta$ because $\Box\phi \in \Delta$. Consequently, $(\Gamma, \Delta) \in Q_{\mathcal{L}}(\phi?)$. ∎

It turns out, we do not need to modify the definition of the Fischer-Ladner closure to apply the filtration method described previously. So, all the definitions of Section 3 are preserved. The proof of the Filtration Lemma is the same as earlier with a slight modification of the induction step for the test actions in the proof of Lemma 20.3 which can be formulated as the following lemma.

LEMMA 41 *Let $[\xi?]\eta \in FL(\Sigma)$ and $s \models \xi \iff \|s\| \models \xi$ for all $s \in S$. Then*

1. *if $(s, t) \in Q(\xi?)$ then $(\|s\|, \|t\|) \in Q^{\Sigma}(\xi?)$;*

2. *if $(\|s\|, \|t\|) \in Q^{\Sigma}(\xi?)$ and $s \models [\xi?]\eta$ then $t \models \eta$.*

Proof. For 1, suppose $(s, t) \in Q(\xi?)$, then by the definition of the semantics of the new test operator, $(s, t) \in R$ and $t \models \Box\xi$. Hence, $(\|s\|, \|t\|) \in R^{\Sigma}$ by the definition of R^{Σ}. We need to show that $\|t\| \models \Box\xi$. Suppose that $(\|t\|, \|u\|) \in R^{\Sigma}$ for some $\|u\|$. This means that there exist $t' \in \|t\|$ and $u' \in \|u\|$ such that $(t', u') \in R$. By Lemma 15 there exists $u'' \in \|u'\| = \|u\|$ such that $(t, u'') \in R$. Hence, $u'' \models \xi$ and by the assumptions $\|u\| \models \xi$. Thus, $\|t\| \models \Box\xi$ and, consequently, $(\|s\|, \|t\|) \in Q^{\Sigma}(\xi?)$.

For 2, let $(\|s\|, \|t\|) \in Q^{\Sigma}(\xi?)$ and $s \models [\xi?]\eta$. Hence, $(\|s\|, \|t\|) \in R^{\Sigma}$ and $\|t\| \models \Box\xi$ by the definition of the semantics of the new test operator. Consequently, there are $s' \in \|s\|$ and $t' \in \|t\|$ such that $(s', t') \in R$. Suppose that $(t', u) \in R$ for some u. Then, $(\|t\|, \|u\|) \in R^{\Sigma}$ and, consequently, $\|u\| \models \xi$. By the lemma assumptions, $u \models \xi$, and, hence, $t' \models \Box\xi$. So, we have $(s', t') \in Q(\xi?)$. Because $[\xi?]\eta \in FL(\Sigma)$, η also belongs to $FL(\Sigma)$. The rest is simple. We have $s' \models [\xi?]\eta$ because $s \sim_{\Sigma} s'$ and, hence, $t' \models \eta$. And from $t \sim_{\Sigma} t'$ we conclude that $t \models \eta$. ∎

Applying the filtration technique and using Lemma 31 and Proposition 39 we can obtain completeness, the small model property and the admissibility of the full substitution rule.

THEOREM 42 (Small model theorem) *Let L be K45, KD45, or S5. Let ϕ be a formula and $|\phi| = n$. If ϕ is satisfiable in a (non-standard) model for the logic $(PDL \oplus L)^?$, $(PDL \oplus L \oplus \{PR\})^?$, $(PDL \oplus L \oplus \{CR\})^?$ or $(PDL \oplus L \oplus \{PR, CR\})^?$ then it is satisfiable in a standard model for this logic with no more than $2^n \cdot 2^{2^n}$ states.*

THEOREM 43 (Completeness) *Let L be K45, KD45, or S5. Then $(PDL \oplus L)^?$ and any its extension by the axioms PR and/or CR is complete with respect to the corresponding class of standard models.*

THEOREM 44 *Let L be K45, KD45, or S5. Then any extension of $(PDL \oplus L)^?$ by the axioms PR and/or CR with the weak substitution rule admits the rule of full substitution.*

Finally let us consider how the standard test operator of *PDL* relates to the informational test operator. It turns out that there is a simulation of *PDL* in $(PDL \oplus S5)^?$. Define the translation mapping σ from formulae of *PDL* to *For$^?$* by the following:

$$\sigma p = \Box p \qquad \sigma \bot = \bot \qquad \sigma a = a$$
$$\sigma(\psi?) = (\sigma\psi)? \qquad \sigma(\alpha \cup \beta) = \sigma\alpha \cup \sigma\beta \qquad \sigma(\alpha;\beta) = \sigma\alpha;\top?;\sigma\beta$$
$$\sigma(\alpha^*) = (\sigma\alpha;\top?)^* \qquad \sigma(\phi \to \psi) = \Box(\sigma\phi \to \sigma\psi) \qquad \sigma([\alpha]\psi) = \Box[\sigma\alpha]\sigma\psi$$

THEOREM 45 *For any $\Box$-free formula ϕ in $\mathcal{L}$, $\phi \in PDL$ iff $\sigma\phi \in (PDL \oplus S5)^?$.*

Proof. Only the idea of the proof is given. It is easy to see that any model of *PDL* can be extended to a $(PDL \oplus S5)^?$ model such that R is the identity relation on the set of states. In any state s of such a model, any formula ϕ is valid if and only if $\Box\phi$ is valid in s. This proves the right to left direction of the theorem.

For the opposite direction, we observe that in any $(PDL \oplus S5)^?$ model $M = \langle S, Q, R, \models \rangle$, R is an equivalence relation. Now, define a *PDL* model $M' = \langle S', Q', \models' \rangle$ as follows. Let $\|s\| =^{\text{def}} \{t \in S \mid (s,t) \in R\}$, $S' =^{\text{def}} \{\|s\| \mid s \in S\}$, and

$$(\|s\|, \|t\|) \in Q'(\alpha) \iff \exists s_0 \in \|s\| \, \exists t_0 \in \|t\| \, (s_0, t_0) \in Q(\alpha),$$
$$M', \|s\| \models' p \iff M, s \models \Box p \quad \text{for any } p \in \text{Var}.$$

Now it is routine to check that $M', \|s\| \models' \phi$ iff $M, s \models \sigma\phi$, for any PDL-formula ϕ. ∎

7 Conclusion

The analysis in this paper of logics from agent-based systems with commuting dynamic and informational modalities provides some new results of completeness, the small model property and decidability for extensions of fusions of *PDL* with normal modal logics by the axioms *PR* and *CR*. The work also provides new insights into the interplay between the substitution rule, the test operator, the informational modalities and their interaction with action modalities. We conclude by mentioning some open problems arising from this work. One open problem already mentioned, is the completeness, the small model property and decidability of combinations of *PDL* and *K(D)45* with the *NL* axiom. A completeness result for the logics with all of the axioms *NL*, *PR* and *CR* has special significance in connection with the axiomatisation of products of *PDL* and modal logics. The threesome of the axioms *NL*, *PR* and *CR* is normally needed for axiomatising a product of two modal logics [4, 5]. For the axiomatisation of the product of *PDL* and *S5* it is however enough to consider any of the following pairs of the axioms $\{NL, PR\}$ or $\{CR, PR\}$. This is a consequence of Proposition 3. Now it follows from [4, Section 5.1] and from the completeness theorem for $PDL \oplus S5 \oplus \{PR, CR\}_w$ (Theorem 23) that the logic $PDL \oplus S5 \oplus \{PR, CR\}_w$ coincides with the corresponding product of the logics, in other words, the pair $(PDL, S5)$ is product matching [5].[1] Whether the pair $(PDL, K(D)45)$ is also product matching is still an open question. To prove this it would be enough to apply the same argument from [4, Section 5.1] as for $PDL \oplus S5 \oplus \{PR, CR\}_w$, provided it is possible to show the logic $PDL \oplus K(D)45 \oplus \{NL, PR, CR\}_w$ is complete with respect to the corresponding class of models satisfying com^r, com^l and cr.

The products of *PDL* (with converse) and normal modal logics have also been investigated in [15] where the quasi-model technique is applied to show the decidability of the logics. Later in [4], this method is used to show the decidability of the products of *PDL* with *S5* and *KD45*. This gives another proof of the decidability of $PDL \oplus S5 \oplus \{PR, CR\}_w$.

Acknowledgements

We are grateful to the anonymous referee for detailed comments and thank Ullrich Hustadt and Agnes Kurucz for valuable discussions and suggestions. The work is supported by research grant GR/M88761 from the UK Engineering and Physical Sciences Research Council.

---

[1] Agnes Kurucz pointed this out to us and showed us a proof.

BIBLIOGRAPHY

[1] P. Blackburn, M. de Rijke, and Y. Venema. *Modal Logic*. Cambridge Univ. Press, 2001.
[2] R. Fagin, J. Y. Halpern, Y. Moses, and M. Y. Vardi. *Reasoning about Knowledge*. MIT Press, 1995.
[3] M. J. Fischer and R. E. Ladner. Propositional dynamic logic of regular programs. *J. Comput. Syst. Sci.*, 18(2):194–211, 1979.
[4] D. Gabbay, A. Kurucz, F. Wolter, and M. Zakharyaschev. Many-dimensional modal logics: Theory and applications. Studies in Logic. Elsevier, 2003.
[5] D. Gabbay and V. Shehtman. Products of modal logics, part 1. *Logic J. IGPL*, 6(1):73–146, 1998.
[6] D. Harel, D. Kozen, and J. Tiuryn. *Dynamic Logic*. MIT Press, 2000.
[7] A. Herzig and D. Longin. Belief dynamics in cooperative dialogues. *J. Semantics*, 17(2), 2000.
[8] M. Kracht and F. Wolter. Properties of independently axiomatizable bimodal logics. *J. Symb. Logic*, 56:1469–1485, 1991.
[9] M. Kracht and F. Wolter. Simulation and transfer results in modal logic—a survey. *Studia Logica*, 59(2):149–177, 1997.
[10] J.-J. C. Meyer, W. van der Hoek, and B. van Linder. A logical approach to the dynamics of commitments. *Artificial Intelligence*, 113(1–2):1–40, 1999.
[11] H. J. Ohlbach. SCAN—Elimination of predicate quantifiers: System description. In *Automated Deduction: CADE-13*, vol. 1104 of *LNAI*, pp. 161–165. Springer, 1996.
[12] A. S. Rao and M. P. Georgeff. Modeling rational agents within a BDI-architecture. In *Proc. KR'91*, pp. 473–484. Morgan Kaufmann, 1991.
[13] R. A. Schmidt and D. Tishkovsky. On axiomatic products of PDL and S5: Substitution, tests and knowledge. *Bull. Section Logic*, 31(1):27–36, 2002.
[14] W. van der Hoek. Logical foundations of agent-based computing. In *Multi-Agent Systems and Applications*, vol. 2086 of *LNAI*, pp. 50–73. Springer, 2001.
[15] F. Wolter. The product of converse PDL and polymodal K. *J. Logic and Comp.*, 10(2):223–251, 2000.
[16] M. Zakharyaschev, F. Wolter, and A. Chagrov. *Advanced Modal Logic*. Kluwer, 1998.

Renate A. Schmidt
Department of Computer Science
University of Manchester
Manchester M13 9PL
United Kingdom
schmidt@cs.man.ac.uk

Dmitry Tishkovsky
Department of Computer Science
University of Manchester
Manchester M13 9PL
United Kingdom
dmitry@cs.man.ac.uk

19
The Complexity of Temporal Logic Model Checking
PH. SCHNOEBELEN

1 Introduction

Temporal logic. Logical formalisms for reasoning about time and the timing of events appear in several fields: physics, philosophy, linguistics, etc. Not surprisingly, they also appear in computer science, a field where logic is ubiquitous. Here temporal logics are used in automated reasoning, in planning, in semantics of programming languages, in artificial intelligence, etc.

There is one area of computer science where temporal logic has been unusually successful: the specification and verification of programs and systems, an area we shall just call *"programming"* for simplicity. In today's curricula, thousands of programmers first learn about temporal logic in a course on model checking!

Temporal logic and programming. Twenty five years ago, Pnueli identified temporal logic as a very convenient formal language in which to state, and reason about, the behavioral properties of parallel programs and more generally reactive systems [76, 77]. Indeed, correctness for these systems typically involves reasoning upon related events at *different moments* of a system execution [73]. Furthermore, when it comes to liveness properties, the expected behavior of reactive systems cannot be stated as a static property, or as an invariant one. Finally, temporal logic is well suited to expressing the whole variety of fairness properties that play such a prominent role in distributed systems [32].

For these applications, one usually restricts oneself to *propositional* temporal logic: on the one hand, this does not appear to be a severe limitation in practice, and on the other hand, this restriction allows decision procedures for validity and entailment, so that, at least in principle, the above-mentioned reasoning can be automated.

Model checking. Generally speaking, model checking is the algorithmic verification that a given logic formula holds in a given structure (the *model* that one *checks*). This concept is meaningful for most logics and classes of models but, historically, it was developed in the context of temporal logic formulae for finite Kripke structures, called *"temporal logic model checking"* in this survey. Temporal logic model checking has been a very active field of research for the last two decades because of its important applications in verification (see e.g. [20]).

A huge effort has been, and is being, devoted to the development of smarter and better model checking software tools, known as *model checkers*, that can verify ever larger models and deal with a wide variety of extended frameworks (real-time systems, stochastic systems, open systems, etc.). We refer to [42, 55, 16, 4] for more details on the practical aspects of model checking.

Model checking for modal logicians. Temporal logic can be seen as some brand of modal logic, but it seems fair to say that model checking is not a popular problem in the modal logic community: for example it is not mentioned in standard textbooks such as [6, 41, 10]. This is probably because model checking is too trivial a problem for the standard modal logics based on immediate successors (it is easy even for *PDL*, see [17]) and only becomes interesting when richer temporal logics are considered. However, standard texts on temporal logic aimed at logicians (e.g. [7, 89, 37]) just briefly mention that model checking is possible and do not really deal with the computational issues involved. A recent exception is [11] where a few pages are devoted to model checking for μ-calculi since, quoting [11, p. 315]:

> *Decidability and axiomatization are standard questions for logicians; but for the practitioner, the important question is model-checking.*

The complexity of model checking. Once decidability has been proved, the next basic problem in the theory of model checking is measuring its *complexity* [1].

The point is that, when the precise complexity of some computational problem has been established, it can be said that the *optimal algorithm* for the problem has been identified and proved optimal. Here "optimal" has a precise meaning: one only considers what computing resources are *asymptotically* necessary and sufficient for solving all instances, including the hardest ones (i.e., *in the worst case*). These simplifications and abstractions about the cost of algorithms lead to a surprisingly powerful theory that has been applied successfully in a huge number of fields.

---

[1] We assume the reader has some basic knowledge of the theoretical framework of computational complexity, and refer him to standard texts like [91, 46, 75] for more motivations and details.

In the field of temporal logic model checking, this research program led to a clearer understanding of why model checking works so well (or does not work). It also added a new dimension on which to compare different logical formalisms (alongside the more classical dimensions of expressive power and complexity of validity).

The goals of this survey. We present the main results on the complexity of model checking and the underlying algorithmic ideas. This covers the main temporal logics encountered in the programming literature: *LTL* (from [36]), *CTL* (from [13]), *CTL** (from [25]), their fragments, and their extensions with past-time modalities [2]. The presentation is mainly focused on complexity results, not on the usefulness, or elegance, or expressive power, of the temporal logics we consider [3]. However, when complexity of model checking is concerned, we (try to) explain the ideas behind the algorithms and hardness proofs, in the hope that these techniques can be useful in other subfields of modal logic.

Outline of the paper. We start with the basic concepts and definitions (temporal logics, their models, the model checking problem) in Section 2. Then Section 3 gives the main results on model checking for our three logics, before we consider fragments (in Section 4) and past-time extensions (in Section 5). Finally, we discuss more advanced questions (parameterized view of complexity in Section 6, and complexity of symbolic model checking in Section 7) that help bridge the gap between complexity theory and actual practice.

2 Basic notions

2.1 Temporal modalities

Temporal logic [78] is a brand of modal logic tailored to temporal reasoning, i.e. reasoning with modalities like "sometimes", "now", "often", "later", "while", "always", "inevitably", etc.

Temporal logics are usually interpreted in modal structures where the nodes (the modal *worlds*) are positions in time, often called *instants*. These need not be just points but can be, for example, time intervals (periods).

Usually, the modal relation relates two positions when the second lies in the future of the first. This provides a model for *qualitative* aspects of time, dealing with "before and after". More elaborate notions like, for example,

---

[2] We decided to omit mentioning μ-calculi because, even though they are popular in the programming community, we think they are less temporal logics than languages in which one can define temporal logics (much like monadic second-order logics). The interested reader may consult [3].

[3] We refer the reader to standard texts, like [27, 72], for motivation and examples.

metrics on time (durations) could be modeled as well but this survey limits itself to the simpler frameworks.

Thus a time frame is usually taken to be an ordered set $\langle T, \leq_T \rangle$, with $t \leq_T t'$ denoting that instant $t \in T$ precedes instant $t' \in T$ in time.

2.2 Linear-time and branching-time

A major distinction between temporal logics is whether they see time as *linear* or *branching*.[4] This is reflected in the classes of time frames they consider: linear orderings or trees.

In linear-time logics, all instants are linearly ordered from past to future and there is only one possible future: future is determined. In programming, this viewpoint is convenient for deterministic programs. For nondeterministic programs, the linear-time viewpoint applies to the runs of the system: any given run determines a single future. The nondeterminism of the system is taken into account at a different level, by considering all the runs.

In branching-time logics, the future is not determined and any given instant may have several distinct immediate successors,[5] hence the tree-like structure of the frame. This viewpoint is probably the more appropriate when it comes to nondeterministic systems [70, 35] but linear-time is often preferred for its simplicity, both notational (see below) and conceptual.

2.3 Kripke structures

A feature of model checking that explains its successes is that it mostly deals with finite structures displaying infinite behaviors: the finiteness of the structures entails the (efficient) decidability of model checking, while the non-finiteness of the behaviors allows one to model interesting situations.

Here it is crucial to distinguish between the *computational structure* and the *behavioral structure*.

The computational structure is a model of the program at hand, describing its possible configurations and the possible steps between them. Even if one only considers *finite* computational structures, it is still possible to model interesting programming problems,[6] as the last twenty years of model

---

[4] Not all structures for time fall into the linear or the branching category (see e.g. [105]) but these certainly are the two most often used in the programming literature.

[5] All through this survey we assume time is discrete.

[6] This is the case of *protocols*, where several small finite state machines interact in tricky combinatorial ways, of *reactive systems* where the system under study reacts to the stimuli of its environment and where temporal logic can state assumptions about the environment, of *hardware circuits* where finite-state gates are combined, and even of arbitrary *programs* after some abstraction (usually on their variables) has made them

checking have demonstrated.

The behavioral structure is a model of the behavior of the system modeled by a computational structure. It is usually infinite (systems are not supposed to terminate) but displays some regularity since it is obtained from the computational structure by some kind of traversal or unfolding.

Temporal logic is used to reason about the behavioral structure, where there is a clear notion of before and after along the runs of the system, in particular between different instants that correspond to a same, recurring, configuration. In summary, the temporal structure for our temporal logics is the behavioral structure, and the computational structure is just a finite-state model of the internal architecture of the system, from which the behavioral structure is derived by some kind of operational semantics.

With this in mind, it is unfortunate, and confusing for modal logicians, that computational structures are called *Kripke structures* in the model checking community![7] This usage is so widespread that we follow it in the rest of this survey, and hope that the previous paragraph is sufficient warning against possible misunderstandings.

Formally, given a set $AP = \{P_1, P_2, \ldots\}$ of *atomic propositions*, a *Kripke structure* over AP is a tuple $S = \langle Q, R, l, I \rangle$ where

- $Q = \{q, r, s, \ldots\}$ is a set of *states* (the configurations of the system).

- $R \subseteq Q \times Q$ is a *transition relation* (the possible steps). For simplification purposes, we require that R is *total*, i.e. for any $q \in Q$ there is at least one q' s.t. $q \, R \, q'$.

- $l : Q \to 2^{AP}$ is a *labeling* of states with propositions. $P \in l(q)$ encodes the fact that P holds in state q.

- $I \subseteq Q$ is a non-empty set of *initial configurations*. Often I is a singleton and we just write q_I for the initial state.

We say S is a *finite* Kripke structure when Q is finite.

2.4 Behaviors

A *path* through S is a sequence $q_0, q_1, q_2, \ldots$ of states s.t. $q_i \, R \, q_{i+1}$ for $i = 0, 1, \ldots$ A path may be finite or infinite. A *fullpath* is a maximal path and a *run* of S is a fullpath that starts from an initial state.

---

finite-state.

[7] Admittedly, it can be argued that Kripke structures are *bona fide* temporal structures for state-based branching-time pure-future logics like *CTL*. We do not favor this viewpoint since it leads to allow cycles in time.

We use π, π', etc. to denote fullpaths and write $\Pi_S(q)$, or just $\Pi(q)$, for the set of fullpaths that start from some $q \in Q$. Then $\Pi(S) \stackrel{\text{def}}{=} \bigcup_{q \in I} \Pi(q)$ is the set of runs of S. A run is a possible behavior of the Kripke structure, and $\Pi(S)$ is the "linear-time behavior" of S. Note that, since we only allowed Kripke structures with a total R, all fullpaths are infinite (which is a welcome simplification).

For a state $q \in Q$ of S, the *tree* rooted at $q \in Q$ is the infinite tree $T_S(q)$, often simply denoted $T(q)$, obtained by unfolding S from q (formally, the nodes of $T(q)$ are the finite paths starting from q ordered by the prefix relation). Then $T(S) \stackrel{\text{def}}{=} \{T(q) \mid q \in I\}$ is the set of *computation trees* of S. A tree $T(q)$ gives the full branching structure of the behaviors issued from q, and $T(S)$ is the "branching-time behavior" of S.

Figure 1 displays an example of a simple Kripke structure S_{exm} with its runs $\Pi(S_{\text{exm}})$ and its computation trees $T(S_{\text{exm}})$. In this figure, time flows from left to right along the runs and the computation trees.

The example uses a set $AP = \{\mathsf{a}, \mathsf{b}, \mathsf{c}, \mathsf{d}, \mathsf{e}\}$ of atomic propositions. S_{exm} has $Q = \{q_1, q_2, q_3, q_4\}$, $I = \{q_1, q_2\}$ (indicated by the incoming arrows) and l given by $l(q_1) = \{\mathsf{a}\}$, $l(q_2) = \emptyset$, $l(q_3) = \{\mathsf{b}, \mathsf{e}\}$, and $l(q_4) = \{\mathsf{c}\}$. The transitions in R are all the directed edges between states.

The runs and computation trees of a Kripke structure S are structures collecting states of S. This definition is convenient for algorithms (as we see later). However, from a semantical viewpoint, runs and computation trees are considered up-to isomorphism: what particular state of S appears at some position is irrelevant, only the labeling with propositions from AP and the ordering relation between positions are meaningful. This is the reason why Figure 1 does not carry state names in $T(S_{\text{exm}})$ and $\Pi(S_{\text{exm}})$.

REMARK 1 The dashed arrow in Figure 1 emphasizes the fact that the set $\Pi(S)$ of can be derived from $T(S)$. Since state names have been forgotten, it is not possible in general to reconstruct $T(S)$ from $\Pi(S)$: the branching-time semantics of a system provides more information about its behavior than the linear-time semantics does [35].

2.5 Three temporal logics: LTL, CTL^*, and CTL

2.5.1 LTL

LTL, for *Linear Temporal Logic*, is the temporal logic with Until and Next interpreted over runs, i.e. over any linearly ordered structure of type ω.

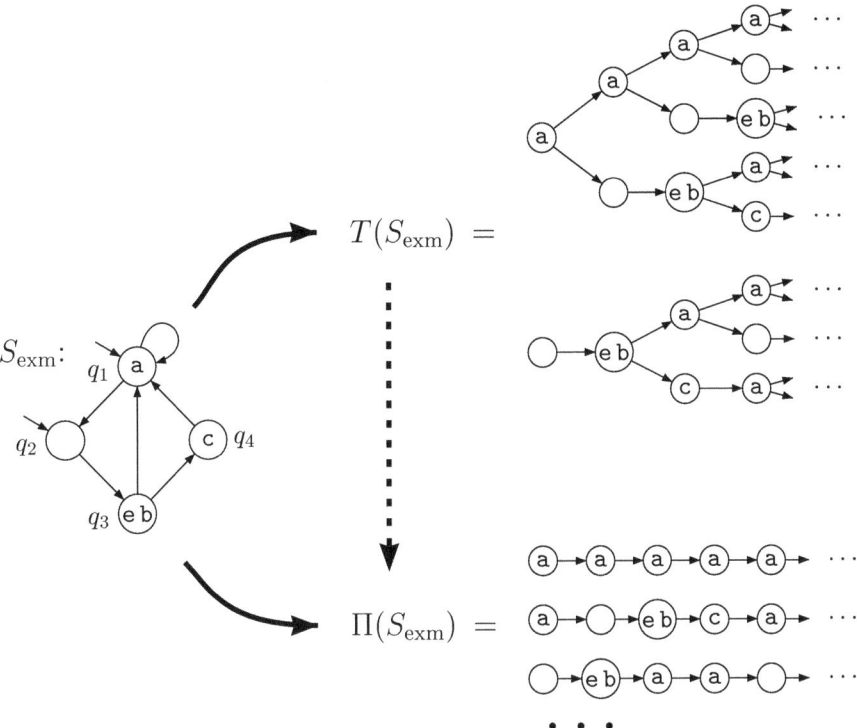

Figure 1. Branching-time and linear-time behaviors of S_{exm}

Syntax. Assuming a set $AP = \{P_1, P_2, \ldots\}$ of atomic propositions, the set of *LTL* formulae is given by the following abstract grammar:

$$\varphi, \psi ::= \varphi \, \mathsf{U} \, \psi \mid \mathsf{X} \varphi \mid \varphi \wedge \psi \mid \neg \varphi \mid P_1 \mid P_2 \mid \cdots \qquad (LTL \text{ syntax})$$

Other Boolean connectives $\top$, $\bot$, $\varphi \vee \psi$, $\varphi \Rightarrow \psi$ and $\varphi \Leftrightarrow \psi$ are defined via the usual abbreviations. A formula like $\mathsf{X}\,\mathrm{rain}$ reads *"it will rain (just next)"*, while $\mathrm{happy}\,\mathsf{U}\,\mathrm{rain}$ reads *"(I) will be happy until it (eventually) rains"*. The classical temporal modalities F (*"sometimes in the future"*) and G (*"always in the future"*) are obtained by

$$\mathsf{F}\varphi \equiv \top \, \mathsf{U} \, \varphi, \qquad \mathsf{G}\varphi \equiv \neg \mathsf{F} \neg \varphi. \qquad (1)$$

Semantics. Formally, a *run* π is an ω-sequence $\sigma_\pi = s_0, s_1, s_2, \ldots$ of *states* with a labeling $l_\pi : \{s_0, s_1, \ldots\} \to 2^{AP}$ with propositions from AP. One defines when an *LTL* formula φ holds at position i of π, written $\pi, i \models \varphi$, by induction over the structure of φ:

$$\pi, i \models \varphi \, \mathsf{U} \, \psi \stackrel{\text{def}}{\Leftrightarrow} \exists j \geq i \text{ s.t.} \begin{cases} \pi, j \models \psi \\ \text{and} \\ \pi, k \models \varphi \text{ for all } i \leq k < j, \end{cases} \quad \text{(S1)}$$

$$\pi, i \models \mathsf{X} \varphi \stackrel{\text{def}}{\Leftrightarrow} \pi, i+1 \models \varphi, \quad \text{(S2)}$$

$$\pi, i \models \varphi \wedge \psi \stackrel{\text{def}}{\Leftrightarrow} \pi, i \models \varphi \text{ and } \pi, i \models \psi, \quad \text{(S3)}$$

$$\pi, i \models \neg \varphi \stackrel{\text{def}}{\Leftrightarrow} \pi, i \not\models \varphi, \quad \text{(S4)}$$

$$\pi, i \models P \stackrel{\text{def}}{\Leftrightarrow} P \in l_\pi(s_i) \quad (\text{for } P \in AP). \quad \text{(S5)}$$

Observe that *LTL* is a *future-only* logic, i.e. a logic where whether $\pi, i \models \varphi$ only depends on the future $s_i, s_{i+1}, s_{i+2}, \ldots$ of the current situation.

REMARK 2 Our definition assumes a *reflexive* U (and F), where the present is part of the future. This is standard in programming, but some works in temporal logic consider a strict, irreflexive, U_{irr} (that would be equivalent to our X U) [48]. The irreflexive U_{irr} strives for minimality: it is a smart way of encoding both U and X in a single modality. The reflexive U is easier to use when writing real formulae, and allows considering the stutter-insensitive fragment of *LTL* formulae that do not use X [58].

REMARK 3 *LTL* being future-only, one often finds in the literature an equivalent definition for its semantics, where the pair π, i is replaced by the i-th suffix of π (seen as word).

There are two reasons why we did not follow that style of definitions and notations in this survey. Firstly, we preferred be as close as possible to the standard semantics of modal logics, where formulae are evaluated at a position in a structure, and where modalities refer to other positions in the same structure. Secondly, we want our definitions to be easily adapted when we consider logics with past-time modalities in Section 5.

REMARK 4 *LTL* being future-only, one often finds in the literature an equivalent definition for its semantics, where the pair π, i is replaced by the i-th suffix of π (seen as word).

There are two reasons why we did not follow that style of definitions and notations in this survey. Firstly, we preferred be as close as possible to the standard semantics of modal logics, where formulae are evaluated at a position in a structure, and where modalities refer to other positions in the same structure. Secondly, we want our definitions to be easily adapted when we consider logics with past-time modalities in Section 5.

The fundamental semantical definition leads to derived notions: we say that a *run π satisfies φ*, written $\pi \models \varphi$, when $\pi, 0 \models \varphi$. Furthermore, and since runs often come from a Kripke structure S, we say that *S satisfies φ*, written $S \models \varphi$, when $\pi \models \varphi$ for all $\pi \in \Pi(S)$, i.e. when φ holds in all runs issued from initial states of S. Finally, for a state $q \in Q$, we write $q \models \varphi$ when φ holds in the structure $S_{\{I \leftarrow q\}}$ obtained from S by assuming that q is the initial state.

REMARK 5 These notions of satisfaction "in a run" and "in a Kripke structure" consider that temporal specifications apply to the initial states. This viewpoint, sometimes called the "anchored viewpoint" [71], is the most natural for programming. A possible alternative, more natural for logicians, is to consider that φ holds in π when it holds at all positions along π (the "floating viewpoint"). It is easy to translate the floating viewpoint into the anchored viewpoint (with a G) and the anchored viewpoint into the floating viewpoint (with some extra labeling for the initial state, or with one of the past-time modalities we define in Section 5).

Other *LTL*-like logics. There is a convenient way of denoting linear-time temporal logics like *LTL*: we write $L(\mathsf{H}_1, \ldots)$ for the logic with $\mathsf{H}_1, \ldots$ as modalities. For example, *LTL* is $L(\mathsf{U}, \mathsf{X})$. This notation assumes that the H_is are equipped with a semantical definition like (S1) and (S2) above.

2.5.2 CTL^*

CTL^* extends *LTL* with quantification over runs. It is interpreted over computation trees, i.e. well-founded trees where the branches are ω-type runs.

Syntax. The set of CTL^* formulae is given by the following abstract grammar:

$$\varphi, \psi ::= \underline{\mathsf{E}\varphi} \mid \varphi \mathsf{U} \psi \mid \mathsf{X}\varphi \mid \varphi \wedge \psi \mid \neg \varphi \mid P_1 \mid P_2 \mid \cdots \qquad (CTL^* \text{ syntax})$$

where we underlined the extension of *LTL*.

E is the existential path quantifier (see semantical definition below) and the universal quantifier is defined as usual as an abbreviation: $\mathsf{A}\varphi \equiv \neg \mathsf{E} \neg \varphi$.

Semantics. We omit the formal definition of what is a computation tree T in general and assume that T is a tree $T(q) \in T(S)$. The branches of T are runs $\pi \in \Pi(q)$, called "runs in T". Below we let $\pi[0, \ldots, i]$ denote the sequence $s_0, \ldots, s_i$ of the first $i + 1$ states of π.

One defines when a CTL^* formula φ holds at position i of run π in computation tree T, written $T, \pi, i \models \varphi$ (or $\pi, i \models \varphi$ when T is clear from the context), by induction over the structure of φ: clauses (S1–S5) apply unchanged (thanks to our notational choices), and we add

$$\pi, i \models \mathsf{E}\varphi \stackrel{\text{def}}{\Leftrightarrow} \pi', i \models \varphi \text{ for some } \pi' \text{ in } T \text{ s.t. } \pi[0,\ldots,i] = \pi'[0,\ldots,i]. \quad (S6)$$

Then come the usual derived notions of satisfaction: a CTL^* formula φ holds in a tree T, written $T \models \varphi$, when $T, \pi, 0 \models \varphi$ for all runs π in T. We further say that φ holds in a Kripke structure S, written $S \models \varphi$, when $T \models \varphi$ for all $T \in T(S)$.

REMARK 6 Any LTL formula φ is also a CTL^* formula. The semantical definitions are coherent since

$$S \models_{LTL} \varphi \text{ iff } S \models^*_{CTL} \mathsf{A}\varphi \text{ iff } S \models^*_{CTL} \varphi$$

holds for any S.

Like LTL, CTL^* is a future-only logic. Therefore, it is possible to read $\mathsf{E}\varphi$ as "there exists a branch starting from the current situation, and where φ holds".

CTL^* is more expressive than LTL. The possibility of referring to the alternative runs that branch off from the current state is used e.g. in a formula like

$$\mathsf{A}\bigl[(\mathsf{G}\,\text{life}) \Rightarrow (\mathsf{G}\,\mathsf{E}\,\mathsf{X}\,\text{death})\bigr] \qquad (\varphi_{br})$$

stating that along all runs ("A") with eternal life ("G life") it is always ("G") possible ("E") to meet death at the next moment ("X death"). Obviously, this possible death assumes that we branch off and follow a different path!

Lamport [57] showed that a formula like φ_{br} has no LTL equivalent, that is, there is no LTL formula ψ such that $S \models \varphi_{br}$ iff $S \models \psi$ for all Kripke structures S. Here is why: consider the Kripke structures S_1 and S_2 from Figure 2. Compared to S_1, S_2 has additional runs with eternal life and no

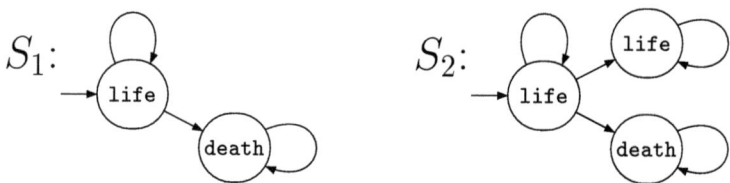

Figure 2. Two Kripke structures with $\Pi(S_1) = \Pi(S_2)$ and $T(S_1) \neq T(S_2)$

escape to death, thus $S_1 \models \varphi_{br}$ while $S_2 \not\models \varphi_{br}$. However, since the two structures have isomorphic sets of runs (see Remark 1), $S_1 \models \psi$ iff $S_2 \models \psi$ for any linear-time formula ψ.

Other CTL^*-like logics. Emerson and Halpern introduced a convenient notation, "$B(L)$", for branching-time logics like CTL^* that extend a linear-time logic L with quantification over branches. For example, CTL^* is $B(LTL)$, and we introduce other instances in the rest of this survey. The $B(L)$ notation assumes L was interpreted over runs, adds a "E" path quantifier, and interpret the resulting formulae over computation trees with the semantical clause (S6) above.

2.5.3 CTL

CTL is the fragment of CTL^* where every temporal modality (U or X) must be under the immediate scope of a path quantifier (E or A). The semantics is inherited from CTL^*.

An alternative view. While we have just completely defined CTL with the above paragraph, we should mention that this does not adopt the standard way of presenting CTL. For one thing, and as the names indicate, CTL was introduced before CTL^*, not as a fragment of it.

Requiring that temporal modalities be immediately under a path quantifier entails that they appear in pairs. Thus CTL can be defined with four modalities: EU, AU, EX and AX.

$$\varphi, \psi ::= \mathsf{E}(\varphi \mathsf{U} \psi) \mid \mathsf{A}(\varphi \mathsf{U} \psi) \mid \mathsf{EX}\, \varphi \mid \mathsf{AX}\, \varphi \qquad (CTL \text{ syntax})$$
$$\mid \varphi \wedge \psi \mid \neg \varphi \mid P_1 \mid P_2 \mid \cdots$$

Additional modalities like EF, AF, EG and AG are defined as abbreviations.

All CTL formulae are *state formulae*: whether φ holds in some T, π, i only depends on the current state, s_i, and not the future that is being considered, i.e. π. Consequently, we often simply write $T, s_i \models \varphi$, or $s_i \models \varphi$ when T is clear.

REMARK 7 In programming, temporal logic first considered linear runs. There F, the classical modality for "eventually" or "sometimes in the future", had the obvious definition. When tree-like structures were considered, it was not clear what "eventually" should mean, as pointed out by [57]. One possibility, perhaps the simplest, was to equate F with "in some future". The other possibility was to read F more as "eventually", i.e. as "inevitably", or "at some point in all futures". No single choice was expressive enough, and branching-time logics offer the two options, with EF for possibility, and AF for inevitability [9]. Applied to Until, this line of thought leads to CTL. (On the quirkiness of CTL combinators, see [84].)

EX and AX are dual modalities, and any one of them can be defined from the other. Observe that EF and AF are not dual modalities. The dual of EF is AG, a modality for "at all points in all futures", that could be rendered by "permanently" ("always" is possible but more ambiguous). The dual of AF is EG, a modality for "at all points along one future", which is hard to render in English (but perhaps "possibly always" is a decent attempt?). {EU, AF, EX} is a minimal complete set of modalities for CTL, while {AU, EF, EX} is not complete [59].

Other *CTL*-like logics. When $H_1, \ldots$ are linear-time modalities, we write $B(H_1, \ldots)$ to denote the branching-time logic where the modalities are EH_1, AH_1, ... Hence $B(H_1, \ldots)$ is the fragment of $B(L(H_1, \ldots))$ where every modality must appear under the immediate scope of a path quantifier. For example, CTL is $B(U, X)$ while CTL^* is $B(L(U, X))$, i.e. $B(LTL)$.

2.6 The model checking problem

The *standard model checking problem* for a temporal logic L is the decision problem associated with the language (set)

$$MC(L) \stackrel{\text{def}}{=} \{\langle S, \varphi \rangle \mid \varphi \in L \text{ and } S \models \varphi\}$$

where S ranges over all finite Kripke structures. Less formally, $MC(L)$ is the problem of deciding, for any inputs S and $\varphi \in L$, whether $S \models \varphi$ or not.

There exists many possible variants (checking that *some* state of a given S satisfies a given φ, ...) and/or restrictions (to *acyclic* structures, ...) motivated by practical or by theoretical considerations: all these problems can be called "model checking problems" and can often be addressed with the techniques we survey below.

For the huge majority of (propositional) temporal logics L found in the programming literature, $MC(L)$ is decidable.

It is not fair to say that "decidability is obvious since the Kripke structures are finite":[8] while S is finite, the models in which L is interpreted, $\Pi(S)$ or $T(S)$, are not! However, $\Pi(S)$ and $T(S)$ are ω-regular objects (in the automata-theoretic sense, see [93]) that admit general decidability results for monadic second-order logics in which our temporal logics are easily defined [80, 38].

Once the model checking problem for some temporal logic L is seen to be decidable, the next question is to find actual algorithms solving the problem,

---

[8]It is fair to say that model checking of first-order formulae over finite first-order structures is obviously decidable (and can be done in polynomial-space). Over finite structures, decidability for higher-order formulae is equally obvious.

to evaluate their cost (essentially, their running time), and to try to prove that no better algorithm exists.

As we explained in the introduction, computational complexity provides a powerful framework for proving optimality of algorithms, at the cost of some simplifying abstractions (asymptotic measures, comparing performance on the worst cases, up-to polynomial transformations).

The cost of algorithms deciding whether $S \models \varphi$ for some Kripke structure S and temporal formula φ is given in terms of the sizes $|S|$ and $|\varphi|$ of the inputs.

In practice we assume that $|\varphi|$ is the number of symbols in φ seen as a string, and $|S|$ is the size of the underlying graph, that is, the sum $|Q|+|R|$ of the number of nodes and the number of edges.[9]

2.7 Model checking vs. validity

Before we start explaining optimal algorithms for model checking, let us observe that model checking is usually easier than validity or satisfiability, and is almost never harder. The underlying reason is that, for many temporal logics, model checking reduces to validity since it is possible to describe a finite Kripke structure with a succinct temporal formula.

Formally, with a Kripke structure $S = \langle Q, R, l, I \rangle$ we associate a temporal formula φ_S that describe the runs of S. If the set of states is $Q = \{r, s, \ldots\}$, φ_S will use fresh propositions $P_r, P_s, \ldots$, one for each state of S, and is given by:

$$\bigvee_{q \in I} P_q \wedge \mathsf{G}\left(\bigvee_{q \in Q} P_q \wedge \bigwedge_{q \neq q' \in Q}(\neg P_q \vee \neg P_{q'})\right)$$
$$\wedge \bigwedge_{q \in Q} \mathsf{G}\left(P_q \Rightarrow \left(\mathsf{X} \bigvee_{q' \in R(q)} P_{q'}\right) \wedge \bigwedge_{P \in l(q)} P \wedge \bigwedge_{P \notin l(q)} \neg P\right) \tag{φ_S}$$

Now, for any *LTL* formula ψ

$$S \models \psi \text{ iff } \varphi_S \Rightarrow \psi \text{ is valid.}$$

This provides a logspace reduction from *LTL* model checking to *LTL* validity. Similar reductions exist for CTL^* (left as an exercise) and other logics.

---

[9]These definitions could be discussed (e.g. why not also consider the size of the node labeling when defining $|S|$?). Let us just claim authoritatively that they assume the right level of abstraction for the kind of complexity results we give in the rest of this survey.

For linear-time logics, validity can be reduced to model checking by considering a Kripke structure where all possible valuations are represented and connected. This construction helps one to understand why model checking and validity are so similar in linear-time logics. It does not *prove* they are equivalent since the Kripke structure has size exponential in the number of propositions we use. For branching-time logics, there is no such easy reduction, and validity is often much more complex than model checking.

3 Model checking for the main temporal logics

3.1 Upper bounds

We start with upper bounds, i.e. results stating that there exists some algorithm running inside the required complexity bounds.

3.1.1 CTL

The early model checkers from [79, 14] could only be implemented and successfully tackle non-trivial problems because they considered temporal logics with a polynomial-time model checking problem:

THEOREM 8 [14, 1] *The model checking problem for CTL can be solved in time* $O(|S|.|\varphi|)$.

An $O(|S|.|\varphi|)$ algorithm was published in [14] and is reproduced in [15, 16, 19]. It can be seen as a dynamic programming algorithm where one computes (and records in some array) whether $q \models \psi$ for all states q of S and all subformulae ψ of φ. The bilinear time is achieved since, using standard graph algorithms for reachability and strongly connected components, and treating the subformulae of ψ as mere additional propositions, one can decide whether $q \models \psi$ in linear-time.

Later, Arnold and Crubillé gave a more elegant algorithm for model checking *CTL* and more generally the alternation-free fragment of the branching-time μ-calculus [1, 17]: see [4, § 3.1] for an exposition geared towards *CTL*.

3.1.2 LTL

Model checking linear-time formulae is more difficult and this explains why *LTL* model checkers were not available immediately. It is interesting to note that decidability was proved as early as [76][10] since at that time model checking was certainly not yet widely recognized as a worthy problem.

THEOREM 9 [86] *The model checking problem for LTL is in* PSPACE.

---

[10]The result was given for the $L(\mathsf{F})$ fragment (the logic used in [76]) and relied on the now standard reduction to inclusion between ω-languages ("does $L(\varphi) \subseteq L(S)$?").

Sistla and Clarke show that satisfiability of *LTL* formulae is in PSPACE, and then obtain Theorem 9 via the reduction from model checking to satisfiability.

Their algorithm for satisfiability relies on a *small model theorem*: they show that a satisfiable *LTL* formula φ has an ultimately periodic model of size $2^{O(|\varphi|)}$. Their nondeterministic algorithm is simply to guess the model (a path) and check it step by step: one starts by guessing what subformulae of φ hold in the initial state. This set can be stored in polynomial-space. Then (the valuation of) the next state and the set of subformulae it satisfies are guessed. A local consistency check allows one to forget the subformulae of the previous set and go on to the next state. At some point, the algorithm guesses that the current state will be the one where we loop back to in the ultimately periodic path, and simply records it before going on with the next state. Eventually, the next state turns out to be what has been recorded: the loop is closed and φ has been proved satisfiable. The small-model theorem is important for unsatisfiable formulae: the algorithm needs some criterion to eventually terminate on negative instances without missing positive ones. A polynomial-space counter is enough for visiting at most $2^{O(|\varphi|)}$ states.

This gives an algorithm in NPSPACE, one concludes using NPSPACE = PSPACE.

The above algorithm is not practical: it is nondeterministic and (more importantly) it reduces model checking to satisfiability. This would lead to a deterministic algorithm running in time $2^{(|\varphi|+|S|)^{O(1)}}$.

A better method was needed:

THEOREM 10 [63, 100] *The model checking problem for LTL can be solved in time $2^{O(|\varphi|)} O(|S|)$.*

The first practical algorithm for *LTL* model checking was given in [63] and had the $2^{O(|\varphi|)} O(|S|)$ running time. Vardi and Wolper then described how their Büchi automata approach for modal logics could provide the same running time but with a clearer and conceptually simpler algorithm [100].

This approach is now well-known: one associates with any *LTL* formula φ a Büchi automaton $\mathcal{A}_\varphi$ that accepts exactly the models of φ (seen as infinite words of valuations). One can then use $\mathcal{A}_\varphi$ to check for satisfiability of φ or for the existence of a run satisfying φ in some Kripke structure S. The size of $\mathcal{A}_\varphi$ is $2^{O(|\varphi|)}$, hence Theorem 10. See [96, 106] for more details.

There now exists even more direct approaches based on alternating automata [95] but the algorithms implemented in popular *LTL* model checkers (such as Spin [43]) are based on standard nondeterministic automata.

3.2.3 CTL^*

In principle, model checking CTL^* is an easy adaptation of the model checking algorithms for LTL, as was observed by Emerson and Lei:

THEOREM 11 [26, 14] *For any linear-time future-only logic L, there is a polynomial-time Turing reduction from model checking $B(L)$ to model checking L.*

Hence if L has model checking in some complexity class $\mathcal{C}$, then $B(L)$ has model checking in $P^{\mathcal{C}}$.

This applies to CTL^* since CTL^* is $B(LTL)$: we only need to remember that LTL has PSPACE-complete model checking.

COROLLARY 12 [26, 14] *The model checking problem for CTL^* is in P^{PSPACE}, that is, in PSPACE.*

The algorithm underlying Theorem 11 is quite simple: one uses a dynamic programming approach à la CTL and first computes in which states the subformulae are satisfied before dealing with the superformula. For a formula of the form $\mathsf{E}\varphi_l$ where φ_l is a linear-time formula, it is enough to use an LTL model checking algorithm.

We illustrate this on a simple example: imagine φ is $\mathsf{AFG}\,\mathsf{E}[(P_1 \Rightarrow \mathsf{X}P_2)\mathsf{U}P_3]$. We replace the subformulae starting with a universal path quantifier by fresh propositions (a form of renaming). Here φ is rewritten as $\mathsf{AFG}\neg P$ where P stands for $\mathsf{A}\neg[(P_1 \Rightarrow \mathsf{X}P_2)\mathsf{U}P_3]$. Then an LTL model checking algorithm computes which states satisfies $\mathsf{A}\neg[(P_1 \Rightarrow \mathsf{X}P_2)\mathsf{U}P_3]$ and label them with the new proposition P. We then reuse the LTL model checker on the modified Kripke structure, checking where $\mathsf{AFG}\neg P$ holds. Finally, one can perform CTL^* model checking with $O(|S| \times |\varphi|)$ invocations of an LTL model checker on subformulae of φ:

COROLLARY 13 *The model checking problem for CTL^* can be solved in time $2^{O(|\varphi|)}O(|S|^2)$.*

The $2^{O(|\varphi|)}O(|S|^2)$ bound assume that we invoke the LTL model checker for all subformulae and all states of S. But the automata-theoretic method underlying Theorem 10 easily gives us *all the states* in S where a same subformula holds in just one invocation.

COROLLARY 14 [26, 56] *The model checking problem for CTL^* can be solved in time $2^{O(|\varphi|)}O(|S|)$.*

Hence model checking for CTL^* is really no harder than for LTL.

It is surprising that no real CTL^* model checker has yet been made available (but see [99]). The underlying reasons probably have to do with the hiatus between the pronouncements of theoretical complexity and what is observed in practice. Another factor is that, for non-specialists, branching-time logics are less natural than linear-time ones: witness the AX AF $P \neq$ AF AX P conundrum [98].

3.2 Lower bounds

Lower bounds on the complexity of a computational problem state that solving this problem requires at least a given amount of computing power. Such lower bounds are used to prove that a problem is inherently difficult, or that a known algorithm is "optimal" and cannot be improved (made more efficient) in an essential way.

3.2.1 CTL

Polynomial-time algorithms for model checking CTL are "optimal" since the problem is P-complete:

THEOREM 15 *The model checking problem for CTL is P-hard.*

While this result is not unknown in the model checking community, we failed to find any mention of it in the early literature.

In fact, model checking is already P-hard for the $B(X)$ and $B(F)$ fragments of CTL. A direct proof is by reduction from CIRCUIT-VALUE, well-known to be P-complete even when restricted to *monotone* (no negation) *synchronized* (connections between gates respect layers) circuits with *proper alternation* [34]. We illustrate the reduction on an example: Consider the circuit C from Fig. 3. C can be seen as a Kripke structure S_C where the initial state q_I is at the top and where transitions go downward. Then

$$C \text{ evaluates to } 1 \text{ iff } S_C \models \overbrace{\text{AX EX AX EX 1}}^{\varphi_C}.$$

Observe that, in this construction, the $B(X)$ formula φ_C depends on (the depth of) C.[11]

---

[11] This is inescapable in view of Theorem 43.2 below. A reduction using always the same formula is possible with the alternation-free fragment of the branching-time μ-calculus, an extension of CTL for which Theorem 43.2 does not apply.

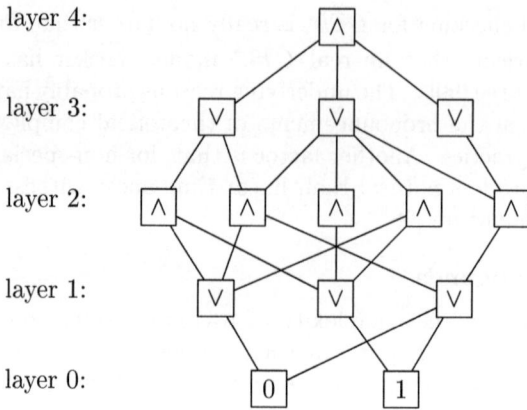

Figure 3. C, an instance of CIRCUIT-VALUE.

According to standard opinion in complexity theory, proving that model checking for CTL is P-hard means that, very probably,[12] it cannot be solved using only polylogarithmic space and does not admit efficient parallel algorithms: the problem is inherently sequential and requires storing a polynomial number of intermediary results.

3.2.2 LTL

Polynomial-space algorithms for model checking of LTL are "optimal" since the problem is PSPACE-complete:

THEOREM 16 [86] *The model checking problem for LTL is PSPACE-hard.*

We give a proof based on reduction from a tiling problem. This was inspired by Harel's proof (see [40]) that satisfiability of LTL formulae is PSPACE-hard (a result due to [44]): as is common with linear-time temporal logics, model checking and satisfiability are closely related problems, and proofs can often be transferred from one problem to the other.

Assume $C = \{c_1, c_2, \ldots, c_l\}$ is a set of *colors* and $D \subseteq C^4$ is a set of *tile types*: $d \in D$ has the form $d = \langle c_{\text{up}}, c_{\text{right}}, c_{\text{down}}, c_{\text{left}} \rangle$. A *tiling* of some region $\mathcal{R} \subseteq \mathbb{Z}^2$ is a mapping $t : \mathcal{R} \to D$ s.t. neighboring tiles have matching colors on shared edges. The PSPACE-complete problem we reduce from is: given some D of size n and two colors $c_0, c_1 \in C$, is there some $m \in \mathbb{N}$ and

---

[12] "Very probably" because it has not yet been proved that POLYLOG − SPACE or NC do not coincide with P (even though most researchers believe this is the case) [46, 75]. The situation here is like with the P ≠ NP? question.

a tiling of the $n \times m$ grid s.t. the bottom edge of the grid is colored with c_0 and the top edge with c_1.

With an instance $D = \{d_1, \ldots, d_n\}$ of the problem we associate the Kripke structure S_D depicted in Fig. 4. The states are labeled with tile

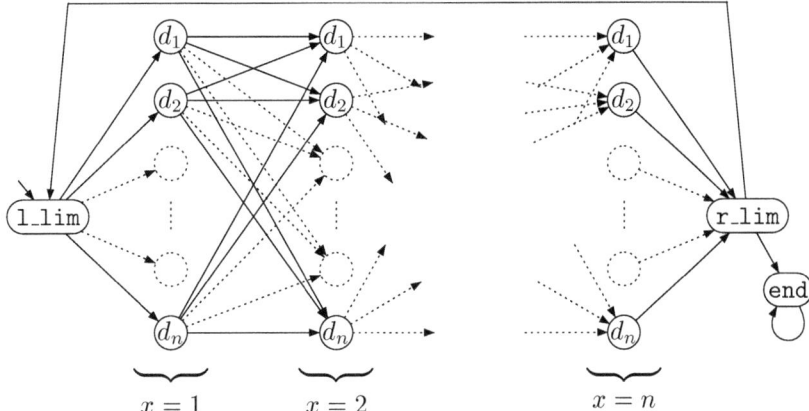

Figure 4. The Kripke structure S_D associated with a tiling problem D

types from D and with additional propositions like "$x = k$" (depending on position) or "up $= c$" (depending on tile type). A path π through S_D lists the tiles on the first row, then loops back to the leftmost state and lists the tiles on the second row, etc., until it perhaps decides to stop and loop forever in the end state. It remains to state that such a path is indeed a tiling:

1. The lower edge and the upper edge (exist and) have colors c_0 and c_1:

$$\bigwedge_{k=1}^{n} \mathsf{X}^k(\mathtt{down} = c_0) \wedge \mathsf{F}\left(\mathtt{l\_lim} \wedge \bigwedge_{k=1}^{n} \mathsf{X}^k(\mathtt{up} = c_1) \wedge \mathsf{X}^{n+2}\mathtt{end}\right) \quad (\varphi_1)$$

2. and neighboring tiles have matching edges:

$$\mathsf{G} \bigwedge_{c \in C} \left(\begin{array}{c} \mathtt{right} = c \Rightarrow \mathsf{X}\,(\mathtt{r\_lim} \vee \mathtt{left} = c) \\ \wedge \quad \mathtt{up} = c \Rightarrow \mathsf{X}^{n+2}\,(\mathtt{end} \vee \mathtt{down} = c) \end{array} \right) \quad (\varphi_2)$$

We now have the required reduction since obviously

$$D \text{ is not solvable iff } S_D \models \neg(\varphi_1 \wedge \varphi_2).$$

3.2.3 CTL^*

Since LTL is a fragment of CTL^*, Theorem 16 entails:

COROLLARY 17 *The model checking problem for CTL^* is* PSPACE-*hard.*

3.3 Completeness results

When lower bounds and upper bounds coincide we obtain completeness in a given complexity class. This is indeed the case with the three model checking problems we considered in Section 3, as summarized in Table 1.

CTL	P-complete
LTL	PSPACE-complete
CTL^*	PSPACE-complete

Table 1. Model checking the main temporal logics

The table shows a big contrast between CTL and LTL! In the early days of model checking, such results were used to argue that branching-time logics are preferable to linear-time logics.

In reality, the situation is not so clear-cut, as we see in Sections 6 and 7.

4 Model checking fragments of temporal logic

Once the cost of model checking the main temporal logics has been precisely measured, it is natural, for the theoretician and the practitioner alike, to look at *fragments*, i.e. sublanguages defined in some way or other. The goal is to better understand what makes model checking difficult or easy, and to ascertain the scope of the results we presented in the previous sections. For the practitioner, it is useful to identify fragments for which the complexity is reduced: if these fragments occur naturally in practice, the corresponding specialized algorithms may be worth implementing.

REMARK 18 *Clearly, similar motivations exist for considering* extensions. *For the practitioner, it is useful to know that an implemented algorithm can in fact be used for a richer and more expressive logic. Indeed there exists many proposals for logics that extend LTL (sometimes greatly) without being essentially more difficult for model checking. Four well-known examples are CTL^* (as seen in Section 3.2), LTL extended with (essentially) Büchi automata [104], LTL with existential quantification over propositions [92], and LTL + Past that we consider in Section 5 below.*

4.1 Fragments of LTL

4.1.1 Restricted set of modalities

Some fragments of LTL are obtained by restricting the allowed modalities. E.g. $L(U)$ is LTL where U is the only allowed modality (X is not allowed but F is since it can be expressed with U). Thus LTL is $L(U,X)$ and $L(F)$ is a fragment of $L(U)$ that coincides with $L(G)$ (since F and G are dual modalities).

THEOREM 19 [86] *The model checking problem for $L(F,X)$ is PSPACE-complete.*

Indeed our proof of Theorem 9 only used $L(F,X)$ formulae!

THEOREM 20 [86] *The model checking problem for $L(U)$ is PSPACE-complete.*

One just has to show that the lower bound still applies. For this we modify the reduction from Theorem 9 so that it uses $L(U)$ formulae:

$$(\text{l\_lim} \vee \text{down}=c_0) \: U \: \text{r\_lim} \qquad\qquad (\varphi_1')$$
$$\wedge \: F\Big(\text{l\_lim} \wedge (\text{l\_lim} \vee \text{up}=c_1 \vee \text{r\_lim}) \: U \: \text{end}\Big)$$

$$G \bigwedge_{k=1}^{n-1} x=k \Rightarrow \bigwedge_{c \in C} \left[\begin{array}{l} \text{up}=c \Rightarrow x=k \: U\Big[x \neq k \: U\Big[\begin{array}{l}\text{end} \vee \\ x=k \wedge \text{down}=c\end{array}\Big]\Big] \\ \wedge \: \text{right}=c \Rightarrow x=k \: U \: \text{left}=c \end{array} \right] \qquad (\varphi_2')$$

With Theorems 19 and 20 we have two instances of situations where considering strict fragments[13] of LTL does not simplify the model checking problem in an essential way. But further restricting the set of allowed modalities decreases the complexity of model checking, as the next two results show:

THEOREM 21 [86] *The model checking problem for $L(F)$ is coNP-complete.*

Membership in coNP is a consequence of a small model theorem: a satisfiable $L(F)$ formula φ has an ultimately periodic model of size $O(|\varphi|)$ [74, 86]. The reduction from model checking to satisfiability (section 2.6) translates this as: if $S \not\models \varphi$ then S has an ultimately periodic path of polynomial-size

---
[13]That $L(F,X)$ and $L(U)$ are less expressive than LTL is well-known. See [30, 54] for recent results on these issues.

that does not satisfy φ. It is enough to guess this counter-example path and check it.

Hardness for coNP can be explained by reduction from the validity problem for Boolean formulae in clausal form: assume θ is a 3SAT instance with variables among $X_n = \{x_1, \ldots, x_n\}$, e.g. θ is "$(x_1 \wedge \overline{x_2} \wedge \overline{x_4}) \vee (\overline{x_1} \wedge \cdots) \vee \cdots$". Consider the structure S_n from Fig. 5. There is a one-to-one correspondence between runs in S_n and valuations for X_n. Thus

$$\theta \text{ is valid iff } S_n \models (\mathsf{F}\,x_1 \wedge \mathsf{F}\,\overline{x_2} \wedge \mathsf{F}\,\overline{x_4}) \vee (\mathsf{F}\,\overline{x_1} \wedge \cdots) \vee \cdots,$$

providing the required reduction.

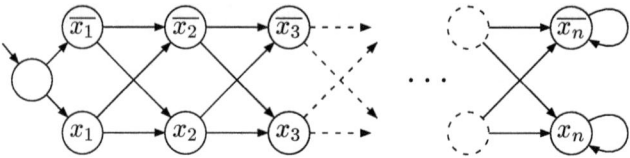

Figure 5. S_n, a structure for picking Boolean valuations of $\{x_1, \ldots, x_n\}$

THEOREM 22 [23] *The model checking problem for $L(\mathsf{X})$ is coNP-complete.*

Here the small model theorem is trivial: if φ has temporal depth k then only the k first states of a linear-time model for φ are relevant. For coNP-hardness, we proceed as we just did:

$$\theta \text{ is valid iff } S_n \models (\mathsf{X}^1 x_1 \wedge \mathsf{X}^2 \overline{x_2} \wedge \mathsf{X}^4 \overline{x_4}) \vee (\mathsf{X}^1 \overline{x_1} \wedge \cdots) \vee \cdots$$

It is possible to consider further natural "subsets" of temporal modalities among $\{\mathsf{X}, \mathsf{F}, \mathsf{U}\}$:

- $L(\overset{\infty}{\mathsf{F}})$ is the fragment of $L(F)$ where F can only be used in the form $\overset{\infty}{\mathsf{F}}$ (that is, GF). This fragment is useful for stating fairness properties and it does not have the full power of $L(\mathsf{F})$ (e.g., ordering constraints cannot be stated).

  We leave the reader to adapt the proof of Theorem 21 and show that the model checking problem for $L(\overset{\infty}{\mathsf{F}})$ is coNP-complete [26].

- $L(\mathsf{U}^-)$ is the fragment of $L(\mathsf{U})$ where only "flat Until" is allowed. Flat Until is Until where only propositional formulae (no temporal modality) is allowed in the left-hand side. E.g. $a\mathsf{U}(b\mathsf{U}c)$ uses flat Until but $(a\mathsf{U}b)\mathsf{U}c$

does not. In expressive power, flat Until lies strictly between F and U [21].

Since the proof of Theorem 20 only uses $L(\mathsf{U}^-)$ formulae, we conclude that model checking for $L(\mathsf{U}^-)$ is PSPACE-complete [23].

The results of this section are summarized in Table 2.

LTL	PSPACE-complete
$L(\mathsf{U})$ $L(\mathsf{U}^-)$ $L(\mathsf{F},\mathsf{X})$	PSPACE-complete
$L(\mathsf{F})$ $L(\overset{\infty}{\mathsf{F}})$ $L(\mathsf{X})$	coNP-complete

Table 2. Model checking fragments of *LTL* $(= L(\mathsf{U},\mathsf{X}))$

4.1.2 Restricted temporal depth

Let $\mathsf{H}_1, \ldots$ be some temporal modalities. For $k \in \mathbb{N}$, we write $L^k(\mathsf{H}_1, \ldots)$ for the fragment of $L(\mathsf{H}_1, \ldots)$ where only formulae of temporal depth at most k are allowed.

It is natural to ask whether enforcing such a depth restriction can make model checking easier. For example, for many modal logics considered in [39], satisfiability becomes polynomial-time when the modal depth is bounded.

The situation is different with *LTL*: for modalities like U and F, (satisfiability and) model checking is already at its hardest with a low temporal depth,[14] as stated by the next two results.

THEOREM 23 [23] *The model checking problem for* $L^2(\mathsf{U})$ *is* PSPACE-*complete.*

The proof of Theorem 20 uses $\varphi_1', \varphi_2' \in L^3(\mathsf{U})$, so we already know PSPACE-hardness for $L^3(\mathsf{U})$! We let the reader adapt this and prove PSPACE-hardness for $L^2(\mathsf{U})$ as well (or even for $L^2(\mathsf{U}^-)$).

THEOREM 24 [23] *The model checking problem for* $L^1(\mathsf{F})$ *is* coNP-*complete.*

---
[14]By contrast, expressive power always increases with temporal depth [30, 54].

Indeed the proof of Theorems 21 shows hardness for the $L^1(\mathsf{F})$ fragment. This also applies to $L^1(\overset{\infty}{\mathsf{F}})$.

Below these thresholds, complexity is decreased:

THEOREM 25 *The model checking problem for $L^1(\mathsf{U},\overset{\infty}{\mathsf{F}},\mathsf{X})$ is* coNP-*complete.*

This result is important since, as we see in Section 4.2 below, it has applications for branching-time logics like CTL and CTL^+, where path quantification only allows path formulae without nesting of temporal modalities, i.e. path formulae in some $L^1(\mathsf{H}_1,\ldots)$.

The ideas behind Theorem 25 are easy to summarize. Hardness was proved with Theorem 21. Membership in coNP comes from a small model theorem: a satisfiable $L^1(\mathsf{U},\overset{\infty}{\mathsf{F}},\mathsf{X})$ is satisfiable in a linear-sized model (which lets us proceed as in the proof of Theorem 21). For the $L^1(\mathsf{U},\mathsf{X})$ fragment, the small model theorem is proved in [23, Section 7]. As shown in [60, Section 6], this can further be extended to the $L^1(\mathsf{U},\mathsf{X},\overset{\infty}{\mathsf{F}})$ fragment, and in fact to any fragment that can be translated efficiently in $FO_2(<)$, the two-variables fragment of the first-order logic of linear orderings [85].

THEOREM 26 [23] *For any $k \in \mathbb{N}$, the model checking problem for $L^k(\mathsf{X})$ can be done in polynomial time.*

This result does not have much practical use. A possible polynomial-time (in fact, logspace) algorithm is simply to look at all tuples of $k+1$ consecutive states in the Kripke structure. This runs in time $O(|S|^{k+1} \times |\varphi|)$ which, for fixed k, is polynomial-time.

4.1.3 Other fragments of *LTL*

Other restrictions have been investigated but they do not seem as interesting as far as model checking is concerned. These may have to do with bounding the number of atomic propositions allowed in temporal formulae, or restricting the use of negations, or forbidding that a given modality occurs in the scope of another given modality (most notably, forbidding future-time modalities to appear under the scope of past-time modalities). Examples of such investigations are reported in [86, 23, 68].

4.2 Fragments of *CTL**

There does not seem to be much to say about model checking[15] fragments of *CTL*: they cannot be harder than *CTL* itself and the P-hardness proof of

---

[15] The situation is more interesting for the *satisfiability* problem. It is EXPTIME-complete for *CTL* [24] and this leaves much room for fragments that could be less intractable, see [29] for example.

Theorem 15 only needs EX and one proposition, or EF and two propositions!

Much more interesting are the fragments of CTL^*. In fact, a good deal of effort has been spent trying to extend CTL (because it lacks expressive power, most notably for fairness properties) without losing the efficient model checking algorithms that CTL permits.[16]

4.2.1 A selection of branching-time logics

Several natural fragments of CTL^* have been identified and named [28, 27]. Below we consider:

CTL^+: an extension of CTL, introduced in [25], and where Boolean combinations (not nesting) of temporal modalities are allowed under the scope of a path quantifier. E.g. CTL^+ allows formulae like $A(a U b \Rightarrow c U d)$. There exists a translation from CTL^+ to CTL [24] (hence the two logics have the same expressive power) but such a translation must produce exponential-size formulae [103, 2].

CTL^+ is $B(U, X, \wedge, \neg)$ in the notation of [25], but we prefer to see it as $B(L^1(U, X))$.

$ECTL$: an extension of CTL, introduced in [25], where the $\overset{\infty}{\mathsf{EF}}$ and $\overset{\infty}{\mathsf{AF}}$ modalities are allowed, providing some expressive power for fairness properties.[17] Thus $ECTL$ is $B(U, X, \overset{\infty}{\mathsf{F}})$ in the notation of [25, 27]. Emerson and Halpern showed that $ECTL$ is not sufficiently expressive for typical situations where fairness issues appear, as when one needs to write $A((\overset{\infty}{\mathsf{F}} a_1 \wedge \overset{\infty}{\mathsf{F}} a_2 \wedge \ldots \wedge \overset{\infty}{\mathsf{F}} a_n) \Rightarrow \mathsf{F}\ \mathit{end})$.

$ECTL^+$: was introduced in [25] and is $B(U, X, \overset{\infty}{\mathsf{F}}, \wedge, \neg)$ in Emerson's notation, or equivalently $B(L^1(U, X, \overset{\infty}{\mathsf{F}}))$. It allows one to relativize CTL formulae with any kind of fairness properties.

BT^*: is $B(L(\mathsf{F}))$, i.e. a fragment of CTL^* where F is the only allowed modality. Thus BT^* combines full branching-time power à la CTL^* but only based on Eventually, the classical temporal modality, instead of Until and Next. The name BT^* is from [14].

BX^*: is $B(L(\mathsf{X}))$, i.e. a fragment of CTL^* where X is the only allowed modality. It is even more basic than BT^* and, to the best of our knowledge, had not been identified in the literature.

---

[16]The choice of what are the CTL modalities is mostly a historical accident (see Remark 7) and it is possible to define more expressive logics for which essentially the same model checking algorithm still applies. Examples are $ECTL$ (see below), or the logic from [50].

[17]Only $\overset{\infty}{\mathsf{EF}}$ is a real addition to CTL since $\overset{\infty}{\mathsf{AF}}$ can be defined as AG AF.

4.2.2 Model checking fragments of CTL^*

As far as model checking is concerned, $ECTL$ behaves like CTL:

THEOREM 27 *The model checking problem for ECTL is P-complete, and can be solved in time $O(|S| \times |\varphi|)$.*

This is because the standard CTL model checking algorithm is easily extended to deal with the $\overset{\infty}{\mathsf{EF}} \varphi$ subformulae: one simply has to find the strongly connected components in the Kripke structure, select those where at least one node satisfies φ, and gather all states from which these connected components can be reached [14].

In fact, the $O(|S| \times |\varphi|)$ time can be achieved for any $B(\mathsf{H}_1, \mathsf{H}_2, \ldots, \mathsf{H}_k)$ logic based on a finite set of "existential path modalities" (see [84, 82]).

When it comes to the other branching-time logics in our selection, the situation is not so simple. They all have the form $B(L(\ldots))$ for a linear-time temporal logic $L(\ldots)$ with a coNP-complete model checking problem: (see Theorem 25 for CTL^+, and $ECTL^+$, Theorem 21 for BT^*, and Theorem 22 for BX^*). Thus Theorem 11 applies and gives a P^{coNP}, or equivalently P^{NP}, model checking procedure for these logics. Of course the coNP lower bound still applies. Hence

COROLLARY 28 *The model checking problem for CTL^+, $ECTL^+$, BT^* and BX^* is NP-hard, coNP-hard and can be solved in P^{NP}, i.e. in Δ_2^p.*

Δ_2^p is one of the levels in Stockmeyer's polynomial-time hierarchy [90]. Formally, a problem is in P^{NP} if it can be solved by a deterministic polynomial-time Turing machine making (adaptive) queries to an NP-complete set, e.g. to an oracle for satisfiability.[18] That model checking CTL^+ and BT^* is in Δ_2^p was already observed in [14, Theorem 6.2]. It is widely believed that Δ_2^p is strictly below PSPACE (inside which the whole polynomial-time hierarchy resides) so that temporal logics like $ECTL^+$ or BT^* are believed to have easier model checking problems than LTL or CTL^*.

The gap between the lower bound (NP-hard and coNP-hard) and the upper bound (in Δ_2^p) has only been recently closed:

THEOREM 29 [60] *The model checking problem for CTL^+, $ECTL^+$ and BT^* is Δ_2^p-complete.*

---

[18] Observe that the number of queries and their lengths are polynomially bounded since the algorithm has to produce them in polynomial-time. By "*adaptive* queries" we mean that the queries are produced and answered sequentially, so that the formulation of any query may depend on the answers that were received for the previous queries.

In fact the hardness proofs in [60] even apply to $B(L^1(\mathsf{F}))$ and $B(L^1(\overset{\infty}{\mathsf{F}}))$. Theorem 29 is mainly interesting for complexity theorists: there exist very few problems known to be complete for Δ_2^p (see [101, 53]), in particular none from temporal logic model checking,[19] and any addition from a new field helps understand the class. The techniques from [60] also have more general relevance: they have been used to show Δ_2^p-completeness of model checking for some temporal logics featuring quantitative informations about time durations [61].

In the case of BX^*, the Δ_2^p upper bound can be improved:

THEOREM 30 [88] *Model checking for BX^* is* $\mathrm{P}^{\mathrm{NP}[O(\log^2 n)]}$*-complete.*

Here $\mathrm{P}^{\mathrm{NP}[O(\log^2 n)]}$, from [18], is the class of problems that can be solved by a deterministic polynomial-time Turing machine that only makes $O(\log^2 n)$ queries to an NP-complete set (where n is the size of the input). It should be noted that, to the best of our knowledge, this is the first example of a problem complete for this class!

The results of Section 4.2 are summarized in Table 3. This does not

CTL^*	$= B(L(\mathsf{U},\mathsf{X}))$	PSPACE-complete
$ECTL^+$	$= B(L^1(\mathsf{U},\mathsf{X},\overset{\infty}{\mathsf{F}}))$	Δ_2^p-complete
CTL^+	$= B(L^1(\mathsf{U},\mathsf{X}))$	Δ_2^p-complete
BT^*	$= B(L(\mathsf{F}))$	Δ_2^p-complete
BX^*	$= B(L(\mathsf{X}))$	$\mathrm{P}^{\mathrm{NP}[O(\log^2 n)]}$-complete
$ECTL$	$= B(\mathsf{U},\mathsf{X},\overset{\infty}{\mathsf{F}})$	P-complete
CTL	$= B(\mathsf{U},\mathsf{X})$	P-complete

Table 3. Model checking fragments of CTL^*

provide the good compromise between expressive power and low model checking complexity we were looking for: $ECTL$ suffers from essentially the same expressivity limitations that plague CTL (the same is true of the alternation-free fragment of the branching-time μ-calculus).

5 Temporal logic with past

The temporal logics we have considered until now only had *future-time* modalities, dealing with future events. In fact, in most of programming, temporal logic, i.e. "logic of time", means "logic of the future".

---

[19] But recently some model checking problems for propositional default logics have been found Δ_2^p-complete [8].

At first sight, this situation seems a bit strange. For logicians and philosophers, temporal logic deals with both past and future, often treating them symmetrically. Furthermore, it is easy to observe that natural languages have a richer set of constructs (tenses, adverbs, ...) dealing with past than with future, and that our own sentences, in everyday speech or mathematical papers, use more past than future.

There are two explanations usually put forward when it comes to explaining this apparently slanted view from programming. Firstly, computer scientists deal with dynamical systems (e.g. Turing machines) that *move forward in time*, and the questions they are interested in concern what will happen in the future (e.g. will the Turing machines halt?). Secondly, a famous result by Gabbay states that past can be dispensed with, in a way that we make completely precise below, after the necessary definitions.

5.1 Past-time modalities

The past-time modalities most commonly used in programming are S ("Since"), F^{-1}, G^{-1} and X^{-1}, i.e. they are the past-time counterparts of U, F, G and X.[20]

Syntactically, one defines $PLTL$,[21] or $LTL + Past$, with the following grammar:

$$\varphi, \psi ::= \varphi\, U\, \psi \mid \underline{\varphi\, S\, \psi} \mid X\, \varphi \mid \underline{X^{-1}\, \varphi} \mid \varphi \wedge \psi \mid \neg \varphi \mid P_1 \mid \cdots$$
$$(PLTL \text{ syntax})$$

where we underlined what has been added to the LTL syntax from Section 2.5. One then defines F^{-1} and G^{-1} as abbreviations:

$$F^{-1}\varphi \equiv \top\, S\, \varphi \qquad\qquad G^{-1}\varphi \equiv \neg F^{-1} \neg \varphi \qquad (2)$$

Semantically, $PLTL$ formulae are still interpreted at positions along linear runs, and we complement (S1–5) from Section 2.5 with the following:

$$\pi, i \models \varphi\, S\, \psi \stackrel{\text{def}}{\Leftrightarrow} \exists j \in \{0, 1, \ldots, i\} \text{ s.t. } \begin{cases} \pi, j \models \psi \\ \text{and} \\ \pi, k \models \varphi \text{ for all } j < k \leq i, \end{cases}$$
(S7)

$$\pi, i \models X^{-1}\varphi \stackrel{\text{def}}{\Leftrightarrow} i > 0 \text{ and } \pi, i-1 \models \varphi. \qquad (S8)$$

---

[20] Sometimes the names P ("Past") and Y ("Yesterday") are used instead of F^{-1} and X^{-1} but we prefer the emphasis on symmetry. Admittedly, we would have been more consistent if we had used U^{-1} instead of S.

[21] Naming conventions are not yet 100% stable, and some works can be found where $PLTL$ stands for "**P**ropositional Linear temporal Logic", which is just plain LTL.

Thus S and X^{-1} behaves as mirror images of U and X, so that sad S rain reads *"(I) have been sad since it rained"*, X^{-1} rain reads *"it rained (just previously)"*, F^{-1} rain reads *"it rained (sometimes in the past)"*, and G^{-1} rain reads *"it has always rained"*.

A *pure-past* formula (resp. *pure-future*) is a formula that only uses the past-time modalities X^{-1} and S (resp. the future-time X and U). Hence *LTL* is the pure-future fragment if *LTL + Past*.

REMARK 31 *The only asymmetry between past and future is that, since the models are ω-words, there is a definite starting point (past is finite) and no end point (future is forever). This difference is imposed by the applications in programming, but it is unimportant for the main theoretical results.*

5.2 The expressive power of *LTL + Past*

Certainly, as was observed in [64], some natural language statements are easier to translate in *LTL + Past* than in *LTL*. For example, while a specification like *"any fault eventually raises the alarm"* is translated as the *LTL* formula

$$G\,(\texttt{fault} \Rightarrow F\,\texttt{alarm}), \qquad (\varphi_\rightarrow)$$

we would naturally translate *"any alarm is due to some (earlier) fault"* as

$$G\,(\texttt{alarm} \Rightarrow F^{-1}\,\texttt{fault}), \qquad (\varphi_\leftarrow)$$

an *LTL + Past* formula.

Observe that both formulae start with a G because our informal "any fault" and "any alarm" mean a fault at any time and an alarm at any time, and more precisely at any time along the runs (the future) of the system.

There is an equivalent way of stating the second property with only future-time constructs:

$$\neg\bigl[(\neg\texttt{fault})U(\texttt{alarm} \wedge \neg\texttt{fault})\bigr], \qquad (\varphi'_\leftarrow)$$

which could be read *"one never sees an alarm after no fault"*. This equivalent formula is only equivalent at the start of the system:

DEFINITION 32 (Global vs. initial equivalence)
1. Two PLTL formulae are (globally) equivalent, written $\varphi \equiv \psi$, when $\pi, i \models \varphi$ iff $\pi, i \models \psi$ for all linear-time models π, and positions $i \in \mathbb{N}$.
2. Two PLTL formulae are initially equivalent, written $\varphi \equiv_i \psi$, when $\pi, 0 \models \varphi$ iff $\pi, 0 \models \psi$ for all linear-time models π.

Clearly global equivalence entails initial equivalence but the converse is not true: e.g. $F^{-1} a \equiv_i a$ but $F^{-1} a \not\equiv a$. In our earlier example $\varphi_{\leftarrow} \equiv_i \varphi'_{\leftarrow}$ but the two formulae are not globally equivalent. Since $\varphi \equiv_i \psi$ entails ($S \models \varphi$ iff $S \models \psi$) for any S, initial equivalence is enough for comparing model checking specifications. Of course, for the future-only *LTL* logic, initial and global equivalence coincide.

Now Gabbay's Theorem compares *PLTL* and *LTL* with *initial equivalence*:

THEOREM 33 [33, 36] *Any PLTL formula is globally equivalent to a separated PLTL formula, i.e. a Boolean combination of pure-past and pure-future formulae.*

COROLLARY 34 (Gabbay's Theorem) *Any PLTL formula is initially equivalent to an LTL formula.*

As was alluded to in the introduction of Section 5, Gabbay's Theorem is one more explanation of why computer scientists often neglect past-time modalities in their temporal logics.

It turns out that there are good reasons to rehabilitate past-time modalities: not only does it make formulae easier to write (compare $\varphi_{\leftarrow}$ with the indirect $\varphi'_{\leftarrow}$), but it also gives added expressive power when one takes succinctness into account:

THEOREM 35 [62] *PLTL can be exponentially more succinct than LTL.*

More precisely, [62] considers the following sequence $(\psi_n)_{n=1,2,\ldots}$ of formulae:

$$\mathsf{G}\left[\left[\bigwedge_{i=1}^{n}(P_i \Leftrightarrow \mathsf{F}^{-1}(P_i \wedge \neg \mathsf{X}^{-1}\top))\right] \Rightarrow \left(P_0 \Leftrightarrow \mathsf{F}^{-1}(P_0 \wedge \neg \mathsf{X}^{-1}\top)\right)\right]. \quad (\psi_n)$$

ψ_n states that every future state that agrees with the initial state on propositions $P_1, \ldots, P_n$ also agree on P_0. The colloquialism $\mathsf{F}^{-1}(\ldots \wedge \neg \mathsf{X}^{-1}\top)$ is used to talk about the initial state, the only state where $\mathsf{X}^{-1}\top$ does not hold. Then, using results from [30], it can be shown that any *LTL* formulae initially equivalent to the ψ_ns have size in $\Omega(2^n)$.

At the moment, it is not known if *PLTL* formulae can be translated into *LTL* formulae with a single exponential blowup, or if the gap in Theorem 35 can be further widened.[22]

REMARK 36 *The succinctness gap between LTL + Past and LTL is important for model checking but not for validity. There exists a succinct satisfiability-preserving reduction from LTL + Past to LTL (left as an exercise), using fresh propositions that we call "history variables" in the programming community. However, introducing polynomially-many fresh propositions leads to an exponential blowup when we look at the size of Kripke structures, so that this translation is not practical for model checking.*

5.3 $CTL + Past$ and $CTL^* + Past$

Past-time modalities can be added to branching-time logics too. Here we concentrate on proposals where future is branching but past is linear. While this may seem like just one more arbitrary (and awkward) choice, it is actually the one more consistent with the view that a Kripke structure actually describes a computation tree, and that temporal logic is a logic for describing properties of this tree.

In classical temporal logic, this is referred to as "unpreventability of the past" and leads to so-called Ockhamist temporal logics [78, 12]. Logics with a branching past do not adhere to this view and really talk about something other than the behavior, most often the internal structure of the system.[23]

Syntactically, one defines $PCTL^*$, or $CTL^* + Past$, with the following grammar:

$$\varphi, \psi ::= \varphi \, \mathsf{S} \, \psi \mid \mathsf{X}^{-1} \varphi \mid \cdots \text{ usual } CTL^* \text{ syntax } \cdots \quad (PCTL^* \text{ syntax})$$

The defining abbreviations for F^{-1} and G^{-1} (2) still apply. The semantics is just obtained by combining clauses (S7) and (S8) from Section 5.1 with (S1–6), the semantics of CTL^*. $PCTL$ is the fragment of $PCTL^*$ where every future-time modality (U or X) is immediately under the scope of path quantifier (A or E).[24]

---

[22] Gabbay's effective procedure is not known to be elementary, however an elementary upper bound on formula size can be obtained by combining the standard translation from $PLTL$ to counter-free Büchi automata and the elementary translation from these automata to LTL using results from [102].

[23] The interested, or unconvinced, reader will find a longer argumentation in [66]. Proposals with branching past can be found e.g. in [83, 105, 89, 49, 51].

[24] These definitions for $PCTL$ and $PCTL^*$ are from [65]. These logics are equivalent to the CTL_{lp} and CTL^*_{lp} (*lp* for "linear past") from [51]. Our $PCTL^*$ further coincides with the OCL logic from [107]. The $PCTL^*$ from [45] differs from our $PCTL^*$ since its path quantifiers always forget the past (see [65, 66] for a formalism allowing both cumulative and forgettable past).

As far as expressive power is concerned, and comparing logics with the same *initial equivalence* criterion we used in Section 5.2, the following results are proved in [65]:

- $PCTL^*$ can be translated to CTL^*,
- $PCTL$ is strictly more expressive than CTL,
- the $CTL + \mathsf{F}^{-1}$ fragment of $PCTL$ can be translated to CTL.

The translations are not succinct: $CTL + \mathsf{F}^{-1}$ can be exponentially more succinct than CTL (a consequence of [102]) and $PCTL^*$ can be exponentially more succinct than CTL^* (a consequence of Theorem 35).

5.4 Model checking $LTL + Past$

While $LTL + Past$ is more succinctly expressive than LTL, this does not entail obviously harder model checking problems.

THEOREM 37 [86] *The model checking problem for PLTL is* PSPACE-*complete.*

In fact [86] directly gave their small model theorem for $PLTL$, and the proof of Theorem 9 can be extended to $PLTL$.

The automata-theoretic approach of Vardi and Wolper extends as well, and with a $PLTL$ formula φ, one can associate a Büchi automaton with $2^{O(|\varphi|)}$ states (see [64, 100]), so that $PLTL$ model checking can be done in time $2^{O(|\varphi|)} O(|S|)$ as for LTL and CTL^*.

An empirical observation is that, in most cases, linear-time logic with past is not more difficult than its pure-future fragment. For example, Theorem 21 can be extended to:

THEOREM 38 [68] *The model checking problem for $L(\mathsf{F}, \mathsf{F}^{-1})$ is* coNP-*complete.*

Here too the coNP algorithm given in [68] relies on a small model theorem: satisfiable $L(\mathsf{F}, \mathsf{F}^{-1})$ formulae admit a linear-sized ultimately periodic model.

5.5 Model checking $CTL + Past$ and $CTL^* + Past$

"Very probably" $CTL+Past$ does not allow polynomial-time model checking algorithms:

THEOREM 39 [66] *The model checking problem for PCTL is* PSPACE-*hard.*

This can be shown via a direct reduction from QBF (a.k.a. Quantified Boolean Formula, a well-known PSPACE-complete problem). We illustrate this on an example: consider a QBF formula θ of the form

$$\exists x_1 \forall x_2 \exists x_3 \forall x_4 \cdots \exists x_n \left[(x_1 \wedge \overline{x_2} \wedge \overline{x_4}) \vee (\overline{x_1} \wedge \cdots) \vee \cdots \right] \quad (\theta)$$

Then obviously θ is valid iff

$$S_n \models \mathsf{EX}\,\mathsf{AX}\,\mathsf{EX}\,\mathsf{AX}\cdots\mathsf{EX}\left[(\mathsf{F}^{-1}x_1 \wedge \mathsf{F}^{-1}\overline{x_2} \wedge \mathsf{F}^{-1}\overline{x_4}) \vee (\mathsf{F}^{-1}\overline{x_1} \wedge \cdots) \vee \cdots\right]$$

in the structure S_n (see Figure 5) we used in Section 4.1. The reduction applies to restricted fragments of *PCTL*: one future-time modality and one past-time modality is enough.

At the other end of the spectrum, model checking for $CTL^* + Past$ is in PSPACE:

THEOREM 40 [**52**] *Model checking for* $PCTL^*$ *is* PSPACE-*complete.*

As a corollary, we obtain PSPACE-completeness for *PCTL* too.

The results of Section 5 are summarized in Table 4.

LTL	PSPACE-complete	$LTL + Past$	PSPACE-complete
CTL	P-complete	$CTL + Past$	PSPACE-complete
CTL^*	PSPACE-complete	$CTL^* + Past$	PSPACE-complete

Table 4. Model checking temporal logics with past

6 The two parameters of model checking

Table 1 that concludes Section 3.3 shows that model checking *CTL* formulae is much easier than model checking *LTL* formulae, so that it seems *CTL* is perhaps the better choice when it comes to picking a temporal logic in which to state behavioral properties. Indeed this line of argument was widespread in the early days of model checking.

However, complexity results like the ones we surveyed in previous sections are not the most relevant for assessing the cost of model checking *in practical situations*. For these situations, it is sensible to distinguish between how much each of the two parameters contribute separately to the complexity of model checking. These two parameters are the Kripke structure S and the temporal formula φ. In practical situations, $|S|$ is usually quite large and

$|\varphi|$ is often small, so that what matters most is the impact $|S|$ has on the overall cost.

When Lichtenstein and Pnueli published their *LTL* model checking algorithm that runs in time $2^{O(|\varphi|)}O(|S|)$ [63], they explained that the algorithm is *linear-time w.r.t. the Kripke structure*, so that it is comparable to *CTL* model checking. They further said that the $2^{O(|\varphi|)}$ factor is not so important in practice, most formulae being short anyway, and this is validated by current practice showing that *LTL* model checking is very feasible.

6.1 Program-complexity and formula-complexity

These considerations can be stated and studied in the complexity-theoretic framework. The *program-complexity* of a model checking problem $MC(L)$ is its computational complexity measured as a function of the Kripke structure S (the program) only, with the temporal formula being fixed. The *formula-complexity* of $MC(L)$ is its computational complexity measured as a function of φ only, with S fixed. For model checking, these notions were introduced in [100], inspired from similar ideas from database querying [94].

Characterizing the program- and the formula-complexity of model checking allows one to measure the impact each input must have on the overall cost. In such investigations, it is natural to look for specialized algorithms that are asymptotically optimal in their handling of one parameter, perhaps to the detriment of the other parameter.

As an illustration, let us mention the following two results based on algorithms that try to reduce the space required for *CTL* model checking:

THEOREM 41 [56] *The model checking problem for CTL can be solved in space* $O(|\varphi| \times \log^2(|\varphi| \times |S|))$.

THEOREM 42 [87] *The model checking problem for CTL can be solved in space* $O(|S| \times \log|\varphi|)$.

6.2 A new view of the cost of model checking

The main results on the program-complexity and the formula-complexity of model checking are listed in the following theorem:

THEOREM 43 (See Table 5)
1. *The program-complexity of LTL model checking is* NLOGSPACE-*complete, its formula-complexity is* PSPACE-*complete.*
2. *The program-complexity of CTL model checking is* NLOGSPACE-*complete, its formula-complexity is in time* $O(|\varphi|)$ *and in space* $O(\log|\varphi|)$.

3. *The program-complexity of CTL^* model checking is* NLOGSPACE-*complete, its formula-complexity is* PSPACE-*complete*.

	program-complexity	formula-complexity	(overall complexity)
LTL	NLOGSPACE-complete	PSPACE-complete	PSPACE-complete
CTL	NLOGSPACE-complete	LOGSPACE	P-complete
CTL^*	NLOGSPACE-complete	PSPACE-complete	PSPACE-complete

Table 5. Three measures for the complexity of model checking

REMARK 44 *Saying that the program- (or formula-) complexity is $\mathcal{C}$-complete means that:*

1. *for any formula φ (resp. structure S) the problem of deciding whether $S \models \varphi$, a problem where S (resp. φ) is the only input, belongs to $\mathcal{C}$;*

2. *there exist formulae (resp. structures) for which the problem is $\mathcal{C}$-hard.*

Thus the overall complexity can be higher than the maximum of the program-complexity and the formula-complexity.

We now briefly explain how the results in Theorem 43 can be obtained:

Lower bound for program-complexity: In all three logics one can write a fixed formula stating that a given state is reachable (or is not reachable, in the case of LTL). This reachability problem is well-known to be NLOGSPACE-complete [47].

Upper bound for program-complexity: For a fixed LTL formula φ, telling whether some Kripke structure S satisfies φ reduces to a reachability problem in the product of S and the (fixed) Büchi automaton for $\neg\varphi$, a question that can be solved in NLOGSPACE.

For CTL^* and CTL, the same kind of reasoning applies [56] but one now reduces to an emptiness test for hesitating alternating tree-automata on a one-letter alphabet.

Formula-complexity for CTL: The $O(|\varphi|)$ time comes from Theorem 8, the $O(\log|\varphi|)$ space is from Theorem 42.

Formula-complexity for LTL and CTL^*: If we fix a finite set AP of atomic propositions, there exists a fixed Kripke structure S_{AP} where all possible AP-labeled runs can be found: if AP has k propositions, then S_{AP} is a clique with 2^k states, one for each Boolean valuation on AP. Now, checking whether $S_{AP} \models \varphi$ for an *LTL* formula φ amounts to deciding whether φ is valid, which is a PSPACE-hard problem.

Note that this uses the fact that *LTL* validity is PSPACE-hard even when AP is finite [23].

As a conclusion, the program-complexity of model checking is NLOGSPACE-complete, and this does not depend on whether we consider *LTL*, *CTL* or *CTL** formulae. There is a parallel here with the results from Section 3.1 where we saw that model checking for these three logics can be done in time that linearly depends on $|S|$. Finally, Table 1 is not a fair comparison of the relative merits of *CTL* and *LTL* [25].

7 The complexity of symbolic model checking

We already observed that, in practical situations, model checking mostly has to deal with large Kripke structures and small temporal formulae.

Kripke structures are frequently used to model the configurations of systems where several components interact, leading to a combinatorial explosion of the number of possible configurations. By *components*, we mean agents in a protocol, or logic gates in a circuit, or simply variables in a program (among many other possibilities). Each component can be identified and easily understood independently of the others, but the combination of their interacting behaviors makes the whole system hard to analyze without the help of a model checker.

Most model checkers let the user define his Kripke structures in a compositional way. But when n components interact together, and assuming that each of them only has a finite number of possible states, the resulting system has $2^{O(n)}$ possible configurations. It quickly becomes impossible to build it (e.g., to store its transitions) and one faces what is now called the *state explosion problem*.

There exist several ways to tackle the state explosion problem in model checking: the most prominent ones today are *compositional reasoning*, *abstraction methods*, *on the fly* model checking, and *symbolic* model checking [16, 4]. In the best approaches, all these methods are combined.

7.1 Symbolic model checking

By "symbolic model checking", we mean any model checking algorithm where the Kripke structure is not described in extension (by a description

---

[25] Even in terms of computational complexity, *LTL* behaves better than *CTL* on many verification problems (see [97, 98]) and it seems that model checking is the only place where *CTL* can hold its ground.

having size $|S|$, as in *enumerative* methods) but is rather handled via more succinct data structures, most often some kind of restricted logical formulae for which efficient constraint-solving techniques apply [26].

Symbolic algorithms can often verify systems that defy enumerative methods (see [5, 69]). But there are also systems on which they do not perform better than the naive non-symbolic approach (an approach that can be defined as "build the structure enumeratively and then use the best model checking algorithm at hand".)

7.2 A complexity-theoretic viewpoint

From a complexity-theoretic viewpoint, there is no reason why symbolic model checking could not be solved more efficiently, even in the "worst cases", than with the naive non-symbolic approach (that always builds the resulting structure). Indeed, symbolic model checkers only deal with a very special kind of huge structures, those that have a succinct representation as a combination of small components, while the naive non-symbolic approach is not so specific and, in particular, is tuned to perform as well as possible on all huge structures. (Observe that very few huge structures have a succinct representation, as a simple cardinality argument shows).

A possible formalization of this issue is as follows: given a sequence $S_1, \ldots, S_k$ of Kripke structures, we define their composition $S = S_1 \otimes \cdots \otimes S_k$ via some combining mechanism denoted $\otimes$. In practice, some kind of parallel combination with prescribed synchronization rules can be used: this mechanism is powerful enough to represent other ways of combining systems, like refinements *à la* Statecharts, or program with Boolean variables *à la* SMV. For the purposes of this survey, it is enough to know that in all these cases $|S|$ is in $O(\prod_{i=1}^{k} |S_i|)$, and S can be constructed in PSPACE.

Then, the *symbolic model checking problem* for a temporal logic L is the decision problem associated with the language

$$MC_{\text{symb}}(L) \stackrel{\text{def}}{=} \{\langle S_1, \ldots, S_k, \varphi\rangle \mid k \in \mathbb{N}, \varphi \in L \text{ and } (S_1 \otimes \cdots \otimes S_k) \models \varphi\}.$$

The cost of algorithms solving $MC_{\text{symb}}(L)$ must be evaluated in terms of the size n of its inputs, i.e. $n = |\varphi| + \sum_i |S_i|$.

Since $|\bigotimes_i S_i|$ is in $2^{O(\sum_i |S_i|)}$, the naive non-symbolic approach (Theorems 8, 10, and Corollary 14) provides upper bounds of time $2^{O(n)}$ for symbolic model checking of *CTL*, *LTL*, and *CTL**.

---

[26] The most famous example is of course the OBDD's (Ordered Binary Decision Diagrams) that made symbolic model checking so popular [5]. There had been earlier attempts at symbolic model checking but their choice of data structure and constraint-solving system was less successful.

Theorem 41 further provides a non-symbolic algorithm running in polynomial-space for *CTL* symbolic model checking, and this approach extends to *CTL* and *CTL** as well.

It can be proved that this upper bound is optimal:

THEOREM 45 [56] *For LTL, CTL and CTL**, *symbolic model checking is* PSPACE-*complete, and its program complexity is* PSPACE-*complete.*

REMARK 46 *Let us note that* PSPACE-*completeness is not the universal measure for symbolic model checking: for the branching-time µ-calculus, symbolic model checking is* EXPTIME-*complete [81].*

What seems to be universal is that, given our simplifying assumptions on the cost of algorithms, non-symbolic methods perform optimally on verification problems. Indeed, this has been observed in many instances, ranging from reachability problems to equivalence problems, and can now be called an empirical fact (see [67, 22] and the references therein). In other words, Kripke structures that admit a succinct representation are not simpler for model checking purposes than arbitrary Kripke structures.

Table 6 contrasts the cost of model checking with that of symbolic model checking. It provides one more argument against the view that model checking is easier for *CTL* than for *LTL*.

	model checking	symbolic model checking
LTL	PSPACE-complete	PSPACE-complete
CTL	P-complete	PSPACE-complete
CTL*	PSPACE-complete	PSPACE-complete

Table 6. Contrasting symbolic and non-symbolic model checking

8 Concluding remarks

So what does complexity theory have to say about model checking? A lot! For one thing, it helps identify barriers to the efficiency of model checking algorithms (e.g., symbolic model checking inherently requires the full power of PSPACE). It also helps compare different temporal logics (e.g., while much more expressive than *LTL*, *LTL* + *Past* or *CTL** are not essentially harder for model checking). Finally, it provides explanations of why there is no real difference in practice between the costs of *LTL* and *CTL* model checking: program-complexity is the important parameter! This argument applies equally to the enumerative, old-style, and the symbolic, new-style, approaches.

When does complexity theory miss the point? Many users (and most developers :-) of model checkers claim that computational complexity is only an abstract theory that bears little relationship with the actual difficulty of problems in practice. We see two situations where this criticism applies: (1) when costs must be measured precisely (i.e., when polynomial transformations are too brutal), and (2) when instances met in practice are nowhere like the hardest cases.

A fair answer to these objections is that, in principle, complexity theory is equipped with the concepts that are required in such situations (as we partly illustrated with our developments on program-complexity and complexity of symbolic model checking). The obstacles here are practical, not conceptual: fine-grained complexity and average complexity, are extremely difficult to measure.

Acknowledgements

This survey borrows a lot from investigations that I conducted with some of my colleagues: S. Demri, F. Laroussinie and N. Markey at LSV, and A. Rabinovich from Tel-Aviv.

BIBLIOGRAPHY

[1] A. Arnold and P. Crubillé. A linear algorithm to solve fixed-point equations on transition systems. *Information Processing Letters*, 29(2):57–66, 1988.

[2] M. Adler and N. Immerman. An $n!$ lower bound on formula size. In *Proc. 16th IEEE Symp. Logic in Computer Science (LICS'2001)*, pages 197–206. IEEE Comp. Soc. Press, 2001.

[3] A. Arnold and D. Niwiński. *Rudiments of μ-Calculus*, volume 146 of *Studies in Logic and the Foundations of Mathematics*. Elsevier Science, 2001.

[4] B. Bérard, M. Bidoit, A. Finkel, F. Laroussinie, A. Petit, L. Petrucci, and Ph. Schnoebelen. *Systems and Software Verification. Model-Checking Techniques and Tools*. Springer, 2001.

[5] J. R. Burch, E. M. Clarke, K. L. McMillan, D. L. Dill, and L. J. Hwang. Symbolic model checking: 10^{20} states and beyond. *Information and Computation*, 98(2):142–170, 1992.

[6] J. van Benthem. *Modal Logic and Classical Logic*. Bibliopolis, Naples, 1985.

[7] J. van Benthem. Time, logic and computation. In *Linear Time, Branching Time and Partial Order in Logics and Models for Concurrency*, volume 354 of *Lect. Notes in Comp. Sci.*, pages 1–49. Springer, 1989.

[8] R. Baumgartner and G. Gottlob. Propositional default logics made easier: computational complexity of model checking. *Theoretical Computer Science*, 289(1):591–627, 2002.

[9] M. Ben-Ari, A. Pnueli, and Z. Manna. The temporal logic of branching time. *Acta Informatica*, 20:207–226, 1983.

[10] P. Blackburn, M. de Rijke, and Y. Venema. *Modal Logic*, volume 53 of *Cambridge Tracts in Theoretical Computer Science*. Cambridge Univ. Press, 2001.

[11] J. C. Bradfield and C. Stirling. Modal logics and mu-calculi: an introduction. In J. A. Bergstra, A. Ponse, and S. A. Smolka, editors, *Handbook of Process Algebra*, chapter 4, pages 293–330. Elsevier Science, 2001.

[12] J. P. Burgess. Basic tense logic. In D. M. Gabbay and F. Guenthner, editors, *Handbook of Philosophical Logic, Vol. II: Extensions of Classical Logic*, chapter 2, pages 89–133. D. Reidel Publishing, 1984.

[13] E. M. Clarke and E. A. Emerson. Design and synthesis of synchronization skeletons using branching time temporal logic. In *Proc. Logics of Programs Workshop*, volume 131 of *Lect. Notes in Comp. Sci.*, pages 52–71. Springer, 1981.

[14] E. M. Clarke, E. A. Emerson, and A. P. Sistla. Automatic verification of finite-state concurrent systems using temporal logic specifications. *ACM Transactions on Programming Languages and Systems*, 8(2):244–263, 1986.

[15] E. M. Clarke and O. Grumberg. Research on automatic verification of finite-state concurrent systems. In *Ann. Rev. Comput. Sci.*, volume 2, pages 269–290. Annual Reviews Inc., 1987.

[16] E. M. Clarke, O. Grumberg, and D. A. Peled. *Model Checking*. MIT Press, 1999.

[17] R. Cleaveland and B. Steffen. A linear-time model-checking algorithm for the alternation-free modal mu-calculus. *Formal Methods in System Design*, 2(2):121–147, 1993.

[18] J. Castro and C. Seara. Complexity classes between Θ_k^p and Δ_k^p. *RAIRO Informatique Théorique et Applications*, 30(2):101–121, 1996.

[19] E. M. Clarke and B.-H. Schlingloff. Model checking. In A. Robinson and A. Voronkov, editors, *Handbook of Automated Reasoning, vol. 2*, chapter 24, pages 1635–1790. Elsevier Science, 2001.

[20] E. M. Clarke and J. M. Wing. Formal methods: state of the art and future directions. *ACM Computing Surveys*, 28(4):626–643, 1996.

[21] D. R. Dams. Flat fragments of CTL and CTL*: Separating the expressive and distinguishing powers. *Logic Journal of the IGPL*, 7(1):55–78, 1999.

[22] S. Demri, F. Laroussinie, and Ph. Schnoebelen. A parametric analysis of the state explosion problem in model checking. In *Proc. 19th Ann. Symp. Theoretical Aspects of Computer Science (STACS'2002)*, volume 2285 of *Lect. Notes in Comp. Sci.*, pages 620–631. Springer, 2002.

[23] S. Demri and Ph. Schnoebelen. The complexity of propositional linear temporal logics in simple cases. *Information and Computation*, 174(1):84–103, 2002.

[24] E. A. Emerson and J. Y. Halpern. Decision procedures and expressiveness in the temporal logic of branching time. *Journal of Computer and System Sciences*, 30(1):1–24, 1985.

[25] E. A. Emerson and J. Y. Halpern. "Sometimes" and "Not Never" revisited: On branching versus linear time temporal logic. *Journal of the ACM*, 33(1):151–178, 1986.

[26] E. A. Emerson and Chin-Laung Lei. Modalities for model checking: Branching time logic strikes back. *Science of Computer Programming*, 8(3):275–306, 1987.

[27] E. A. Emerson. Temporal and modal logic. In J. van Leeuwen, editor, *Handbook of Theoretical Computer Science, vol. B*, chapter 16, pages 995–1072. Elsevier Science, 1990.

[28] E. A. Emerson and J. Srinivasan. Branching time temporal logic. In *Linear Time, Branching Time and Partial Order in Logics and Models for Concurrency*, volume 354 of *Lect. Notes in Comp. Sci.*, pages 123–172. Springer, 1989.

[29] E. A. Emerson, T. Sadler, and J. Srinivasan. Efficient temporal satisfiability. *Journal of Logic and Computation*, 2(2):173–210, 1992.

[30] K. Etessami, M. Y. Vardi, and T. Wilke. First order logic with two variables and unary temporal logic. *Information and Computation*, 179(2):279–295, 2002.

[31] K. Etessami and T. Wilke. An until hierarchy and other applications of an Ehrenfeucht-Fraïssé game for temporal logic. *Information and Computation*, 160(1/2):88–108, 2000.

[32] N. Francez. *Fairness*. Springer, 1986.
[33] D. M. Gabbay. The declarative past and imperative future: Executable temporal logic for interactive systems. In *Proc. Workshop Temporal Logic in Specification*, volume 398 of *Lect. Notes in Comp. Sci.*, pages 409–448. Springer, 1989.
[34] R. Greenlaw, H. J. Hoover, and W. L. Ruzzo. *Limits to Parallel Computation: P-Completeness Theory*. Oxford Univ. Press, 1995.
[35] R. J. van Glabbeek. The linear time – branching time spectrum I. In J. A. Bergstra, A. Ponse, and S. A. Smolka, editors, *Handbook of Process Algebra*, chapter 1, pages 3–99. Elsevier Science, 2001.
[36] D. M. Gabbay, A. Pnueli, S. Shelah, and J. Stavi. On the temporal analysis of fairness. In *Proc. 7th ACM Symp. Principles of Programming Languages (POPL'80)*, pages 163–173, 1980.
[37] D. M. Gabbay, M. A. Reynolds, and M. Finger. *Temporal Logic: Mathematical Foundations and Computational Aspects. Vol. 2*, volume 40 of *Oxford Logic Guides*. Clarendon Press, 2000.
[38] Y. Gurevich and S. Shelah. The decision problem for branching time logic. *The Journal of Symbolic Logic*, 50(3):668–681, 1985.
[39] J. Y. Halpern. The effect of bounding the number of primitive propositions and the depth of nesting on the complexity of modal logic. *Artificial Intelligence*, 75(2):361–372, 1995.
[40] D. Harel. Recurring dominos: Making the highly undecidable highly understandable. *Annals of Discrete Mathematics*, 24:51–72, 1985.
[41] G. E. Hughes and M. J. Cresswell. *A new introduction to modal logic*. Routledge, London, 1996.
[42] G. J. Holzmann. *Design and Validation of Computer Protocols*. Prentice Hall Int., 1991.
[43] G. J. Holzmann. The model checker Spin. *IEEE Transactions on Software Engineering*, 23(5):279–295, 1997.
[44] J. Y. Halpern and J. H. Reif. The propositional dynamic logic of deterministic, well-structured programs. *Theoretical Computer Science*, 27(1–2):127–165, 1983.
[45] T. Hafer and W. Thomas. Computation tree logic CTL* and path quantifiers in the monadic theory of the binary tree. In *Proc. 14th Int. Coll. Automata, Languages, and Programming (ICALP'87)*, volume 267 of *Lect. Notes in Comp. Sci.*, pages 269–279. Springer, 1987.
[46] D. S. Johnson. A catalog of complexity classes. In J. van Leeuwen, editor, *Handbook of Theoretical Computer Science, vol. A*, chapter 2, pages 67–161. Elsevier Science, 1990.
[47] N. D. Jones. Space-bounded reducibility among combinatorial problems. *Journal of Computer and System Sciences*, 11(1):68–85, 1975.
[48] J. A. W. Kamp. *Tense Logic and the Theory of Linear Order*. PhD thesis, UCLA, Los Angeles, CA, USA, 1968.
[49] Michael Kaminski. A branching time logic with past operators. *Journal of Computer and System Sciences*, 49(2):223–246, 1994.
[50] O. Kupferman and O. Grumberg. Buy one, get one free!!! *Journal of Logic and Computation*, 6(4):523–539, 1996.
[51] O. Kupferman and A. Pnueli. Once and for all. In *Proc. 10th IEEE Symp. Logic in Computer Science (LICS'95)*, pages 25–35. IEEE Comp. Soc. Press, 1995.
[52] O. Kupferman, A. Pnueli, and M. Y. Vardi. Unpublished proof. Private communication with O. Kupferman, January 1998.
[53] M. W. Krentel. The complexity of optimization problems. *Journal of Computer and System Sciences*, 36(3):490–509, 1988.
[54] A. Kučera and J. Strejček. The stuttering principle revisited: On the expressiveness of nested X and U operators in the logic LTL. In *Proc. 16th Int. Workshop Computer Science Logic (CSL'2002)*, volume 2471 of *Lect. Notes in Comp. Sci.*, pages 276–291. Springer, 2002.

[55] R. P. Kurshan. *Computer-Aided Verification of Coordinating Processes.* Princeton Univ. Press, 1995.
[56] O. Kupferman, M. Y. Vardi, and P. Wolper. An automata-theoretic approach to branching-time model checking. *Journal of the ACM*, 47(2):312–360, 2000.
[57] L. Lamport. "Sometimes" is sometimes "Not Never". In *Proc. 7th ACM Symp. Principles of Programming Languages (POPL'80)*, pages 174–185, 1980.
[58] L. Lamport. What good is temporal logic? In *Information Processing'83. Proc. IFIP 9th World Computer Congress*, pages 657–668. North-Holland, 1983.
[59] F. Laroussinie. About the expressive power of CTL combinators. *Information Processing Letters*, 54(6):343–345, 1995.
[60] F. Laroussinie, N. Markey, and Ph. Schnoebelen. Model checking CTL^+ and $FCTL$ is hard. In *Proc. 4th Int. Conf. Foundations of Software Science and Computation Structures (FOSSACS'2001)*, volume 2030 of *Lect. Notes in Comp. Sci.*, pages 318–331. Springer, 2001.
[61] F. Laroussinie, N. Markey, and Ph. Schnoebelen. On model checking durational Kripke structures (extended abstract). In *Proc. 5th Int. Conf. Foundations of Software Science and Computation Structures (FOSSACS'2002)*, volume 2303 of *Lect. Notes in Comp. Sci.*, pages 264–279. Springer, 2002.
[62] F. Laroussinie, N. Markey, and Ph. Schnoebelen. Temporal logic with forgettable past. In *Proc. 17th IEEE Symp. Logic in Computer Science (LICS'2002)*, pages 383–392. IEEE Comp. Soc. Press, 2002.
[63] O. Lichtenstein and A. Pnueli. Checking that finite state concurrent programs satisfy their linear specification. In *Proc. 12th ACM Symp. Principles of Programming Languages (POPL'85)*, pages 97–107, 1985.
[64] O. Lichtenstein, A. Pnueli, and L. D. Zuck. The glory of the past. In *Proc. Logics of Programs Workshop*, volume 193 of *Lect. Notes in Comp. Sci.*, pages 196–218. Springer, 1985.
[65] F. Laroussinie and Ph. Schnoebelen. A hierarchy of temporal logics with past. *Theoretical Computer Science*, 148(2):303–324, 1995.
[66] F. Laroussinie and Ph. Schnoebelen. Specification in CTL+Past for verification in CTL. *Information and Computation*, 156(1/2):236–263, 2000.
[67] F. Laroussinie and Ph. Schnoebelen. The state explosion problem from trace to bisimulation equivalence. In *Proc. 3rd Int. Conf. Foundations of Software Science and Computation Structures (FOSSACS'2000)*, volume 1784 of *Lect. Notes in Comp. Sci.*, pages 192–207. Springer, 2000.
[68] N. Markey. Past is for free: on the complexity of verifying linear temporal properties with past. In *Proc. 9th Int. Workshop on Expressiveness in Concurrency (EXPRESS'2002)*, volume 68.2 of *Electronic Notes in Theor. Comp. Sci.* Elsevier Science, 2002.
[69] K. L. McMillan. *Symbolic Model Checking.* Kluwer Academic, 1993.
[70] R. Milner. *Communication and Concurrency.* Prentice Hall Int., 1989.
[71] Z. Manna and A. Pnueli. The anchored version of the temporal framework. In *Linear Time, Branching Time and Partial Order in Logics and Models for Concurrency*, volume 354 of *Lect. Notes in Comp. Sci.*, pages 201–284. Springer, 1989.
[72] Z. Manna and A. Pnueli. *The Temporal Logic of Reactive and Concurrent Systems: Specification.* Springer, 1992.
[73] S. Owicki and L. Lamport. Proving liveness properties of concurrent programs. *ACM Transactions on Programming Languages and Systems*, 4(3):455–495, 1982.
[74] H. Ono and A. Nakamura. On the size of refutation Kripke models for some linear modal and tense logics. *Studia Logica*, 39(4):325–333, 1980.
[75] C. H. Papadimitriou. *Computational Complexity.* Addison-Wesley, 1994.
[76] A. Pnueli. The temporal logic of programs. In *Proc. 18th IEEE Symp. Foundations of Computer Science (FOCS'77)*, pages 46–57, 1977.
[77] A. Pnueli. The temporal semantics of concurrent programs. *Theoretical Computer Science*, 13(1):45–60, 1981.

[78] A. N. Prior. *Past, Present, and Future*. Clarendon Press, Oxford, 1967.
[79] J.-P. Queille and J. Sifakis. Specification and verification of concurrent systems in CESAR. In *Proc. Int. Symp. on Programming*, volume 137 of *Lect. Notes in Comp. Sci.*, pages 337–351. Springer, 1982.
[80] M. O. Rabin. Decidability of second-order theories and automata on infinite trees. *Trans. Amer. Math. Soc.*, 141:1–35, 1969.
[81] A. Rabinovich. Symbolic model checking for μ-calculus requires exponential time. *Theoretical Computer Science*, 243(1–2):467–475, 2000.
[82] A. Rabinovich. Expressive power of temporal logics. In *Proc. 13th Int. Conf. Concurrency Theory (CONCUR'2002)*, volume 2421 of *Lect. Notes in Comp. Sci.*, pages 57–75. Springer, 2002.
[83] W. Reisig. Towards a temporal logic for causality and choice in distributed systems. In *Linear Time, Branching Time and Partial Order in Logics and Models for Concurrency*, volume 354 of *Lect. Notes in Comp. Sci.*, pages 603–627. Springer, 1989.
[84] A. Rabinovich and S. Maoz. An infinite hierarchy of temporal logics over branching time. *Information and Computation*, 171(2):306–332, 2001.
[85] A. Rabinovich and Ph. Schnoebelen. BTL_2 and expressive completeness for $ECTL^+$. Research Report LSV-00-8, Lab. Specification and Verification, ENS de Cachan, Cachan, France, October 2000.
[86] A. P. Sistla and E. M. Clarke. The complexity of propositional linear temporal logics. *Journal of the ACM*, 32(3):733–749, 1985.
[87] Ph. Schnoebelen. Spécification et vérification des systèmes concurrents. Mémoire d'habilitation à diriger des recherches, Université Paris 7, October 2001.
[88] Ph. Schnoebelen. Oracle circuits for branching-time model checking. In *Proc. 30th Int. Coll. Automata, Languages, and Programming (ICALP'2003)*, volume 2719 of Lect. Notes Comp. Sci., pages 790–801, Springer, 2003.
[89] C. Stirling. Modal and temporal logics. In S. Abramsky, D. M. Gabbay, and T. S. E. Maibaum, editors, *Handbook of Logic in Computer Science, vol.2. Background: Computational Structures*, pages 477–563. Oxford Univ. Press, 1992.
[90] L. J. Stockmeyer. The polynomial-time hierarchy. *Theoretical Computer Science*, 3(1):1–22, 1976.
[91] L. J. Stockmeyer. Classifying the computational complexity of problems. *The Journal of Symbolic Logic*, 52:1–43, 1987.
[92] A. P. Sistla, M. Y. Vardi, and P. Wolper. The complementation problem for Büchi automata with applications to temporal logic. *Theoretical Computer Science*, 49(2–3):217–237, 1987.
[93] W. Thomas. Automata on infinite objects. In J. van Leeuwen, editor, *Handbook of Theoretical Computer Science, vol. B*, chapter 4, pages 133–191. Elsevier Science, 1990.
[94] M. Y. Vardi. The complexity of relational query languages. In *Proc. 14th ACM Symp. Theory of Computing (STOC'82)*, pages 137–146, 1982.
[95] M. Y. Vardi. Alternating automata and program verification. In *Computer Science Today. Recent Trends and Developments*, volume 1000 of *Lect. Notes in Comp. Sci.*, pages 471–485. Springer, 1995.
[96] M. Y. Vardi. An automata-theoretic approach to linear temporal logic. In *Logics for Concurrency: Structure Versus Automata*, volume 1043 of *Lect. Notes in Comp. Sci.*, pages 238–266. Springer, 1996.
[97] M. Y. Vardi. Linear vs. branching time: A complexity-theoretic perspective. In *Proc. 13th IEEE Symp. Logic in Computer Science (LICS'98)*, pages 394–405. IEEE Comp. Soc. Press, 1998.
[98] M. Y. Vardi. Branching vs. linear time: Final showdown. In *Proc. 7th Int. Conf. Tools and Algorithms for the Construction and Analysis of Systems (TACAS'2001)*, volume 2031 of *Lect. Notes in Comp. Sci.*, pages 1–22. Springer, 2001.
[99] W. Visser and H. Barringer. Practical CTL* model checking: Should SPIN be extended? *Journal of Software Tools for Technology Transfer*, 2(4):350–365, 2000.

[100] M. Y. Vardi and P. Wolper. An automata-theoretic approach to automatic program verification. In *Proc. 1st IEEE Symp. Logic in Computer Science (LICS'86)*, pages 332–344. IEEE Comp. Soc. Press, 1986.

[101] K. W. Wagner. More complicated questions about maxima and minima, and some closures of NP. *Theoretical Computer Science*, 51(1–2):53–80, 1987.

[102] T. Wilke. Classifying discrete temporal properties. In *Proc. 16th Ann. Symp. Theoretical Aspects of Computer Science (STACS'99)*, volume 1563 of *Lect. Notes in Comp. Sci.*, pages 32–46. Springer, 1999.

[103] T. Wilke. CTL^+ is exponentially more succint than CTL. In *Proc. 19th Conf. Found. of Software Technology and Theor. Comp. Sci. (FST&TCS'99)*, volume 1738 of *Lect. Notes in Comp. Sci.*, pages 110–121. Springer, 1999.

[104] P. Wolper. Temporal logic can be more expressive. *Information and Control*, 56(1/2):72–99, 1983.

[105] P. Wolper. On the relation of programs and computations to models of temporal logic. In *Proc. Workshop Temporal Logic in Specification*, volume 398 of *Lect. Notes in Comp. Sci.*, pages 75–123. Springer, 1989.

[106] P. Wolper. Constructing automata from temporal logic formulas: A tutorial. In *Lectures on Formal Methods and Performance Analysis*, volume 2090 of *Lect. Notes in Comp. Sci.*, pages 261–277. Springer, 2001.

[107] A. Zanardo and J. Carmo. Ockhamist computational logic: Past-sensitive necessitation in CTL*. *Journal of Logic and Computation*, 3(3):249–268, 1993.

Ph. Schnoebelen
Laboratoire Spécification & Vérification (LSV),
École Normale Supérieure de Cachan
61, avenue du Prsident-Wilson
94235 CACHAN Cedex, France
E-mail: phs@lsv.ens-cachan.fr

20
Chronological future modality in Minkowski spacetime

ILYA SHAPIROVSKY AND VALENTIN SHEHTMAN

1 Introduction

The problem of logical foundations of contemporary physics was included by David Hilbert in the list of the most important mathematical problems and generated an interesting research area in nonclassical logic. Study of relativistic temporal logics is a natural topic within this area. Their investigation was initiated by Arthur Prior's proposal in [10], but at the early stage did not move fast — perhaps because relativistic time is both branching and dense, which is rather unusual for modal logic.

Let us recall that two basic relations in Minkowski spacetime are causal ($\prec$) and chronological ($\preceq$) accessibility. The causal future of a point-event x consists of all those points y, to which a signal from x can be sent; $x \prec y$ if this signal is slower than light.

The first significant result in relativistic temporal logic was the theorem by Robert Goldblatt [6] identifying the ("Diodorean") modal logic of relation $\preceq$ as the well-known **S4.2**. Then in [12] the second author described modal logics of domains on Minkowski plane ordered by $\preceq$.

This paper makes the next essential step after the past twenty years. It solves one of three problems put by R. Goldblatt in [6] (see also [7]): to axiomatize the modal logic of the frame $(\mathbb{R}^n, \prec)$.

For this logic $\mathbf{L}_2$ we present an axiom system. Its axioms are widely known in modal logic, except for the specific axiom of 2-density, first introduced in [6]. The logic $\mathbf{L}_2$ is rather standard, but the completeness proof for the intended interpretation is not so straightforward. The main technical problem is the proof of the finite model property. As 2-density is not preserved under filtration in the Lemmon–Segerberg style (when some worlds are identified), we use the Kripke–Gabbay method of selective filtration instead. This method allows us to extract a finite submodel from an infinite model, see e.g. [2]. In our case selective filtration is applied to the canonical model in a combination with the method of maximal points, due to Kit Fine [3]. Similar arguments were used for various modal and intermediate logics by P. Miglioli, M. Zakharyaschev, F. Wolter and others. The finite model characterization is convenient for obtaining complexity bounds of $\mathbf{L}_1$, $\mathbf{L}_2$; this subject is postponed until a further publication.

The final part of the proof is a geometric construction of a p-morphism following the lines of [6] and [12].

In the last Section we discuss applications of our results to many-dimensional modal logics, such as products and interval logics.

2 Basic notions

In this paper all *modal logics* are normal monomodal propositional logics containing **K4**; as usual, modal logics are considered as certain sets of formulas. For a modal logic Λ and a modal formula A, the notation $\Lambda \vdash A$ means $A \in \Lambda$; $\Lambda + A$ denotes the smallest modal logic containing $\Lambda \cup \{A\}$.

We assume that $\Diamond$, $\rightarrow$, $\bot$ are the basic connectives, and $\Box$, $\neg$, $\vee$, $\wedge$, $\top$ are derived. PV denotes the set of propositional variables.

Here are the names for some particular modal formulas:

$A4 := \Diamond\Diamond p \rightarrow \Diamond p,$ $\qquad\qquad AD := \Diamond\top,$
$A2 := \Diamond\Box p \rightarrow \Box\Diamond p,$ $\qquad\qquad Ad := Ad_1 = \Diamond p \rightarrow \Diamond\Diamond p,$
$Ad_n := \Diamond p_1 \wedge \ldots \wedge \Diamond p_n \rightarrow \Diamond(\Diamond p_1 \wedge \ldots \wedge \Diamond p_n),$

and the names for some modal logics:

$\mathbf{K4} := \mathbf{K} + A4,$ $\qquad\qquad \mathbf{D4} := \mathbf{K4} + AD,$
$\mathbf{D4.2} := \mathbf{D4} + A2,$ $\qquad\qquad \mathbf{L}_1 := \mathbf{D4d}_2 := \mathbf{D4} + Ad_2,$
$\mathbf{L}_2 := \mathbf{D4.2d}_2 := \mathbf{L}_1 + A2.$

By *(Kripke) frame* we mean a non-empty set with a transitive relation (W, R). A *(Kripke) model* is a Kripke frame with a valuation: $M = (W, R, \theta)$, where $\theta : PV \longrightarrow 2^W$ (2^W denotes the power set of W).

For a Kripke model $M = (W, R, \theta)$, the notation $x \in M$ means $x \in W$. The sign $\vDash$ denotes the truth at a point of a Kripke model and also the validity in a Kripke frame.

For a class of frames $\mathcal{F}$, $\mathbf{L}(\mathcal{F})$ denotes the *modal logic determined by* $\mathcal{F}$, i.e., the set of all formulas that are valid in all frames from $\mathcal{F}$. For a single frame F, $\mathbf{L}(F)$ abbreviates $\mathbf{L}(\{F\})$. Recall that a *cluster* in (W, R) is an equivalence class under the relation

$$\sim_R := (R \cap R^{-1}) \cup I_W$$

(where I_W is the equality relation on W), a *degenerate cluster* is an irreflexive singleton. As usual, for $x \in W$, $V \subseteq W$ we denote $R(x) := \{y \mid xRy\}$, $R(V) := \bigcup_{x \in V} R(x)$. A cluster is called *maximal* if $R(C) \subseteq C$; a point

$x \in W$ is called *maximal* if its cluster is maximal. The associated relation between clusters

$$C \leq_R D := D \subseteq R(C), \ C <_R D := C \leq_R D \text{ and } C \neq D$$

are transitive and antisymmetric, and $<_R$ is irreflexive. For a frame $F = (W, R)$ let $F/\sim_R := (W/\sim_R, \leq_R)$.

A Kripke model $M_1 = (W_1, R_1, \theta_1)$ is a *(weak) submodel* of $M = (W, R, \theta)$ (notation: $M_1 \subseteq M$) if $W_1 \subseteq W$, $R_1 \subseteq R$, $\theta_1(q) = \theta(q) \cap 2^{W_1}$ for every $q \in PV$. A particular case of a submodel is when $R_1 = R \cap (W_1 \times W_1)$; in this case M_1 is called a *restriction of M to W_1* and denoted by $M|W_1$. A submodel $M|W_1$ is called *generated* if $R(W_1) \subseteq W_1$. A particular case of a generated submodel is a *cone* $M^x = M|W^x$, where $W^x = \{x\} \cup R(x)$.

It is well-known that formula $A4$ corresponds to transitivity of a Kripke frame and AD corresponds to seriality: $\forall x \exists y \ xRy$. The correspondents for $A2$, Ad_n are also easily described:

LEMMA 1. *For a frame F,*

- $F \vDash A2$ *iff F is confluent, i.e., satisfies*

$$\forall x \forall y_1 \forall y_2 \exists z (xRy_1 \ \& \ xRy_2 \Rightarrow y_1Rz \ \& \ y_2Rz);$$

- $F \vDash Ad_n$ *iff F is n-dense, i.e., the following holds:*

$$\forall x \forall y_1 \ldots \forall y_n \exists z (xRy_1 \ \& \ \ldots \ \& \ xRy_n \Rightarrow xRz \ \& \ zRy_1 \ \& \ \ldots \ \& \ zRy_n).$$

By Sahlqvist Theorem (cf. [2]), we also have completeness:

PROPOSITION 2. *The logics $\mathbf{D4.2}$, $\mathbf{L_1}$, $\mathbf{L_2}$ are canonical.*

So we obtain

PROPOSITION 3. *$\mathbf{L_1}$ is determined by the class of all serial 2-dense frames, $\mathbf{L_2}$ is determined by the class of all serial confluent 2-dense frames.*

LEMMA 4. $\mathbf{K4} + Ad_2 \vdash Ad_n$ *for all n.*

Proof. It is clear that $\mathbf{K4} + Ad_2 \vdash Ad_1$. A syntactic proof of $\mathbf{K4} + Ad_2 \vdash Ad_n$ for $n > 2$ is rather easy and is left for the reader. Another proof is based on completeness theorem for $\mathbf{K4} + Ad_n$ and the observation that for a transitive relation, 2-density implies n-density. ∎

3 The finite model property

Let us first recall a simple lemma on selective filtrations.

DEFINITION 5. Let M be a Kripke model, Ψ a set of formulas closed under subformulas.

A submodel $M_1 \subseteq M$ (with the relation R_1) is called a *selective filtration* of M through Ψ if for any $x \in M_1$, for any formula A

$$\Diamond A \in \Psi \ \& \ M, x \vDash \Diamond A \Rightarrow \exists y \in R_1(x) \ M, y \vDash A.$$

LEMMA 6. *If M_1 is a selective filtration of M through Ψ, then for any $x \in M_1$, for any $A \in \Psi$*

$$M, x \vDash A \Leftrightarrow M_1, x \vDash A.$$

Proof. By induction on the length of A. We consider only the case $A = \Diamond B$.

If $M, x \vDash \Diamond B$, then by Definition 5, $M, y \vDash B$ for some $y \in R_1(x)$. So $M_1, y \vDash B$ by the induction hypothesis, and thus $M_1, x \vDash \Diamond B$.

Conversely, if $M_1, x \vDash \Diamond B$, then $M_1, y \vDash B$ for some $y \in R_1(x) \subseteq R(x)$. Hence $M, y \vDash B$ by the induction hypothesis, and thus $M, x \vDash \Diamond B$. ∎

Now let us prove some useful properties of canonical models. First we recall the maximality property of canonical models, cf. [3], [2]:

LEMMA 7. *Let $\mathfrak{M}$ be the canonical model of a modal logic Λ, and assume that a formula B is satisfied in some $x \in \mathfrak{M}$. Consider the set of all those clusters in $\mathfrak{M}^x$, in which B is satisfied:*

$$\Gamma := \{C \subseteq \mathfrak{M}^x \mid \exists y \in C \ B \in y\}.$$

Then $\mathfrak{M} | \bigcup \Gamma$ contains a maximal cluster (with respect to the relation $\leq_R$).

Proof. By assumption, $\Gamma \neq \varnothing$, and so due to Zorn Lemma, it suffices to check that every $\leq_R$-chain Σ of clusters from Γ has an upper bound in Γ. Let

$$S := \{A \mid \exists C \in \Sigma \ \forall y \in C \ \Box A \in y\} \cup \{B\},$$

and consider two cases.

(1) Assume that S is Λ-inconsistent. Then there exist clusters $C_1, \ldots, C_n \in \Sigma$ and points $y_1 \in C_1, \ldots, y_n \in C_n$ such that $B \in y_i$ for every i, and for some formulas $\Box A_1 \in y_1, \ldots, \Box A_n \in y_n$ we have

$$(*) \quad \neg(A_1 \wedge \ldots \wedge A_n \wedge B) \in \Lambda.$$

Since Σ is a chain, $C_1,\ldots,C_n$ are $\leq_R$-comparable, and we may assume that C_1 is the $\leq_R$-greatest among them. So for every i, $y_i R y_1$ or $y_i = y_1$. But in the canonical model we have $y_i \vDash \Box A_i$, and thus we obtain $y_1 \vDash \Box A_1 \wedge \ldots \wedge \Box A_n$. Then it follows that C_1 is an upper bound of Σ. In fact, otherwise for some $y_0 \in R(y_1)$ we have $y_0 \vDash A_1 \wedge \ldots \wedge A_n \wedge B$, which contradicts (*).

(2) Now assume that S is Λ-consistent. Then by Lindenbaum Lemma, $S \subseteq z$ for some $z \in \mathfrak{M}$. Let Z be the cluster of z. By definition of S, $Z \in \Gamma$, and for any $y \in \bigcup \Sigma$ we have yRz. Thus Z is an upper bound of Σ. ∎

LEMMA 8. *Let Λ be a modal logic containing $\mathbf{L}_1$, $\mathfrak{M}$ a canonical model of Λ. Let $x \in \mathfrak{M}$, and assume that $\Diamond A_1, \ldots, \Diamond A_n \in x$. Let*

$$Y := \{y \mid xRy \ \& \ \mathfrak{M}, y \vDash \Diamond A_1 \wedge \ldots \wedge \Diamond A_n\}.$$

Then $\mathfrak{M}|Y$ contains a maximal cluster, which is non-degenerate.

Proof. Let $B := \Diamond A_1 \wedge \ldots \wedge \Diamond A_n$. By Lemma 4, $Ad_n \in \mathbf{L}_1$, and thus $\mathfrak{M} \vDash B \to \Diamond B$. So we have $\mathfrak{M}, x \vDash \Diamond B$, which implies $Y \neq \varnothing$ (Fig. 1).

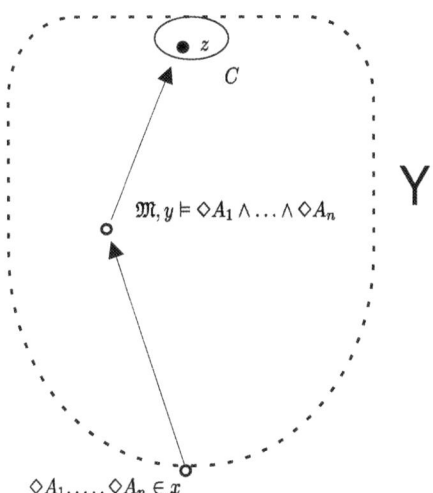

Figure 1.

By Lemma 7, $\mathfrak{M}|Y$ has a maximal cluster C. Since $\mathfrak{M} \vDash B \to \Diamond B$, for any $z \in C$ we have $\mathfrak{M}, z \vDash \Diamond B$, and so there exists $t \in Y$ such that zRt. But $t \in C$, by maximality of C. Therefore, C is non-degenerate. ∎

One can easily check the following

LEMMA 9. *A finite frame $F = (W, R)$ is 2-dense iff for any irreflexive $x \in W$ with $R(x) \neq \varnothing$, there exists a unique successor cluster, which is non-degenerate.*

Let $\mathcal{F}_1$ be the class of all finite 2-dense serial frames, and let

$$\mathcal{F}_2 := \{F + C \mid F \in \mathcal{F}_1, C \text{ is a finite non-degenerate cluster}\},$$

where $+$ denotes the ordinal sum. Then we have

THEOREM 10. $\mathbf{L}_1 = \mathbf{L}(\mathcal{F}_1)$, $\mathbf{L}_2 = \mathbf{L}(\mathcal{F}_2)$.

Proof. (I) First consider $\mathbf{L}_1$. Let μ_1 be the class of Kripke models over frames from $\mathcal{F}_1$. Given an $\mathbf{L}_1$-consistent formula B, we have to construct a a model $M \in \mu_1$, where B is satisfied.

Let $\mathfrak{M}$ be the canonical model of $\mathbf{L}_1$, R the accessibility relation in $\mathfrak{M}$. Let us say that $y \in \mathfrak{M}$ *realizes* a formula $\Diamond A$ if $\mathfrak{M}, y \vDash A$ (i.e., $A \in y$). Since B is consistent, we have $\mathfrak{M}, x_0 \vDash B$ for some x_0. Let Ψ be the set of all subformulas of B,

$$S := \{\Diamond A \mid \Diamond A \in \Psi\} \cup \{\Diamond\top\}.$$

For $x, y \in \mathfrak{M}$ let

$$x \approx y := (S \cap x = S \cap y).$$

Note that xRy implies $S \cap y \subseteq S \cap x$, and thus $x \sim_R y$ implies $x \approx y$.

For $x \in \mathfrak{M}$ let

$$Y_x := \{y \in R(x) \mid x \approx y\}.$$

Then Y_x is a union of clusters.

A point $x \in \mathfrak{M}$ is called *critical* if $\forall y \in R(x) \, (x \approx y \Rightarrow x \sim_R y)$, i.e., if the cluster of x is maximal in Y_x. By Lemma 8, for any $x \in \mathfrak{M}$ there exists a maximal and non-degenerate cluster in Y_x; so we can choose a critical point x' in Y_x. We also assume that $x' = x$ if x already is critical; thus $x'' = x'$ for every x.

Now we construct a required M by induction.

Stage 0. Put $M_0 := \mathfrak{M}|\{x_0, x_0'\}$.

If $S = \{\Diamond\top\}$, the construction terminates at this stage, i.e., we take $M = M_0$. Trivially, M_0 is a selective filtration of $\mathfrak{M}$ through Ψ, and so $M_0, x_0 \vDash B$ by Lemma 6. $M_0 \in \mu_1$, since x_0' is reflexive (in this case its cluster is maximal in $\mathfrak{M}$).

If $S \neq \{\Diamond\top\}$, we store the new point in the set: $X_0 := \{x_0'\}$, and continue the construction.

Stage (n+1). Assume that at stage n we have a model $M_n \subseteq \mathfrak{M}$, $M_n \in \mu_1$, and a non-empty subset $X_n \subseteq M_n$ such that for every $x \in M_n$ the following holds.

(1) $x' \in M_n$;

(2) if $x \in X_n$, then x is critical;

(3) if $x' \notin X_n$, $\Diamond A \in S \cap x$, then $\exists y \in R_n(x')\ A \in y$,

where R_n is the accessibility relation in M_n. Every formula $\Diamond A \in S \cap x$ is called *essential* (for x). For a critical x, let $S_1(x)$ be the set of all essential formulas that are realizable in the cluster of x:

$$S_1(x) := \{\Diamond A \in S \cap x \mid \exists t \sim_R x\ A \in t\},$$

and let

$$S_2(x) := (S \cap x) - S_1(x).$$

Obviously, $S_1(x) \neq \varnothing$, because $\top \in x$. Now for every $x \in X_n$ we proceed as follows.

If $S_1(x) = \{\Diamond A_1, \ldots, \Diamond A_m\}$, then for $i = 1, \ldots, m$ we choose $t_i \sim_R x$ such that $A_i \in t_i$. Of course, it may happen that $t_i \in M_n$ or $t_i = t_j$ for some different i, j.

Similarly, if $S_2(x) = \{\Diamond B_1, \ldots, \Diamond B_k\}$, then for $i = 1, \ldots, k$ we choose $z_1, \ldots, z_k \in R(x)$ such that $B_i \in z_i$. Let
$U_x := \{z_1, \ldots, z_k\}$ or $U_x = \varnothing$ (if $S_2(x) = \varnothing$),
$U'_x := \{z' \mid z \in U_x\}$,
$W_x := U_x \cup U'_x \cup \{t_1, \ldots, t_m\}$ (Fig. 2).

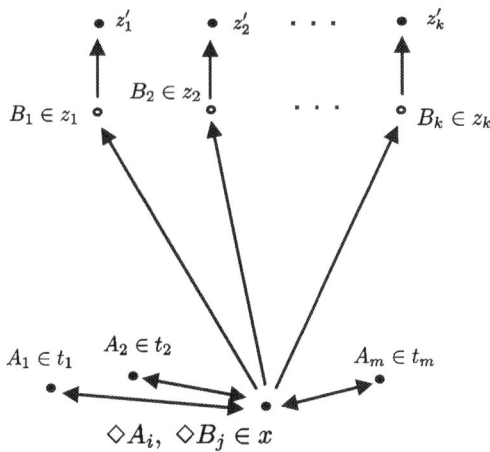

Figure 2.

We define M_{n+1} as the submodel of $\mathfrak{M}$ obtained by adding all points of $\bigcup_{x \in X_n} W_x$ to M_n. The relation in M_{n+1} is defined as follows.

- If a is reflexive then $R_{n+1}(a) := R(a) \cap M_{n+1}$.
- If a is irreflexive then $R_{n+1}(a) := R(a') \cap M_{n+1}$.

One can easily check that R_{n+1} is transitive; note that for irreflexive a, $aR_{n+1}bR_{n+1}c$ implies $a'RbRc$, and thus $a'Rc$, i.e., $aR_{n+1}c$.

R_{n+1} is serial, since $aR_{n+1}a'$.

By Lemma 9, R_{n+1} is 2-dense; in fact, if a is irreflexive, then the cluster of a' is the first in $R_{n+1}(a)$. Also note that $R_n \subseteq R_{n+1}$, since $R_n(a)$ is defined in the same manner as $R_{n+1}(a)$, and the case $n = 0$ is not exceptional.

Thus $M_{n+1} \in \mu_1$.

Let X_{n+1} be the set of new critical points:

$$X_{n+1} := \bigcup_{x \in X_n} U'_x.$$

Then properties (1)-(3) hold in M_{n+1}. In fact, (1), (2) are obvious.

To check (3), suppose $x \in M_{n+1}$, $x' \notin X_{n+1}$. Then by construction, $x \in M_n$. Since (3) holds for M_n and $R_n \subseteq R_{n+1}$, it remains to consider the case $x' \in X_n$. But then by construction, every essential formula is realizable in M_{n+1} (in one of the points $t_1, \ldots, t_m, z_1 \ldots, z_k$).

If $X_{n+1} = \varnothing$, the construction terminates at this stage, and we put $M := M_{n+1}$.

According to the construction, if $x \in X_n$, $\Diamond B_i \in S_2(x)$, we have $z_i \not\prec_R x$, and thus $S \cap z'_i = S \cap z_i \subset S \cap x$, because x is critical. So the number of essential formulas decreases at every step, more precisely:

$$\max\{|S \cap x| \mid x \in X_{n+1}\} < \max\{|S \cap x| \mid x \in X_n\}.$$

Therefore the number of steps in the construction does not exceed the cardinality of S, and we obtain a finite model M.

Due to property (3), the resulting model M is a selective filtration of $\mathfrak{M}$ through Ψ. Since $\mathfrak{M}, x_0 \vDash B$, we obtain $M, x_0 \vDash B$ by Lemma 6.

(II) Similarly, in the case of $\mathbf{L}_2$, consider the canonical model $\mathfrak{M}$, an $\mathbf{L}_2$-consistent formula B; take a world $x \in \mathfrak{M}$, where B is satisfied. By Lemma 8, the cone $\mathfrak{M}^x$ has a maximal and non-degenerate cluster C. By Lemma 1 and Proposition 2, $\mathfrak{M}$ is confluent, and thus C is its final cluster.

As we are interested only in subformulas of B, consider the equivalence relation on C:

$$x \approx y := (\Psi \cap x = \Psi \cap y)$$

and take a finite cluster $C_1 \subseteq C$ containing all representatives under $\approx$. Then the submodel $M' \subseteq \mathfrak{M}$ obtained by adding the cluster C_1 to M (constructed in the proof of (I)) is in μ_1. In fact, the successor cluster of an irreflexive x in M also fits for M'.

To show that M' is a selective filtration of $\mathfrak{M}$ through Ψ, first note that for $x \in M$, all essential formulas can be realized within M. And for $x \in C_1$, if $\Diamond A \in x \cap \Psi$, then $A \in y$ for some $y \in C$ (since C is maximal), and by the choice of C_1, there exists $y_1 \approx y$ in C_1.

Thus B is satisfiable in M'.

Let F' be the frame of M', and put the copy of C_1 on the top of F'. The new frame F'' is in $\mathcal{F}_2$, since $F' \in \mathcal{F}_1$. We also obtain that B is satisfiable in F'', since it is satisfiable in F', and there exists an obvious p-morphism from F'' onto F', which identifies two copies of C_1.

Therefore we obtain the inclusion $\mathbf{L}(\mathcal{F}_2) \subseteq \mathbf{L}_2$. The converse follows easily, because every frame in $\mathcal{F}_2$ is confluent, serial and 2-dense. ∎

DEFINITION 11. A *reflexive tree* is a rooted poset (W, R), in which every subset $R^{-1}(x)$ is a chain; a *tree* (in this paper) is frame, whose reflexive closure is a reflexive tree. A frame F is called a *quasitree* if its cluster frame $F/\sim_R$ is a tree.

Now let $\mathcal{G}_1$ be the class of all finite 2-dense serial quasitrees, and let

$$\mathcal{G}_2 := \{F + C \mid F \in \mathcal{G}_1, \ C \text{ is a finite non-degenerate cluster}\}.$$

We can give more specific characterizations of $\mathbf{L}_1$, $\mathbf{L}_2$:

LEMMA 12. $\mathbf{L}_1 = \mathbf{L}(\mathcal{G}_1)$; $\mathbf{L}_2 = \mathbf{L}(\mathcal{G}_2)$.

Proof. The inclusions $\mathcal{G}_1 \subseteq \mathcal{F}_1$ and $\mathcal{G}_2 \subseteq \mathcal{F}_2$ imply $\mathbf{L}(\mathcal{G}_1) \supseteq \mathbf{L}_1$, $\mathbf{L}(\mathcal{G}_2) \supseteq \mathbf{L}_2$. Thus to prove the first statement of Lemma, it is sufficient to show that every rooted frame $H = (W, R) \in \mathcal{F}_1$ is a p-morphic image of some frame from $\mathcal{G}_1$. First let us consider the cluster frame $H_1 = H/\sim_R$. By standard unravelling argument, we can present H_1 as a p-morphic image of a finite tree H' consisting of all paths in H_1, cf. [2]. In more detail, $H' = (W', R')$, where

$W' := \{(C_0, \ldots, C_k) \mid C_0 <_R C_1 <_R \cdots <_R C_k\}$,

C_0 is the initial cluster in H;

$\alpha R' \beta$ iff either β is a continuation of α and $\alpha \neq \beta$, or $\alpha = \beta$ and the last cluster of α is non-degenerate.

Then the map $f : (C_0, \ldots, C_k) \mapsto C_k$ is a p-morphism $H' \twoheadrightarrow H_1$. Now consider a quasitree $H^* = (W^*, R^*)$, in which

$W^* := \{(C_0, \ldots, C_k, w) \mid (C_0, \ldots, C_k) \in W', \ w \in C_k\}$,
$(C_0, \ldots, C_k, w) R^* (C_0, \ldots, C'_l, v)$ iff $(C_0, \ldots, C_k) R' (C_0, \ldots, C'_l)$.

Note that $H^* \in \mathcal{G}_1$, and we obtain a p-morphism $g : H^* \twoheadrightarrow H$ such that $g(C_0, \ldots, C_k, w) = w$.

Next, for any finite non-degenerate cluster C, we can extend g to a p-morphism $g' : H^* + C \twoheadrightarrow H + C$. Since $H^* + C \in \mathcal{G}_2$, this yields the second statement. ∎

4 Further completeness results

Let $T_\mathbb{Z}$ be the set of finite sequences of integers, T_n the set of finite sequences of numbers $\{1, 2, \ldots, n\}$. The sequences are ordered in a standard way: $\sigma_1 \sqsubseteq \sigma_2$ iff σ_1 is an initial part of σ_2. Let $\sigma_1 \sqsubset \sigma_2$ iff $\sigma_1 \sqsubseteq \sigma_2$ and $\sigma_1 \neq \sigma_2$; $\mathbf{T}_\mathbb{Z} := (T_\mathbb{Z}, \sqsubseteq)$, $\mathbf{T}_2 := (T_2, \sqsubseteq)$.

λ denotes the empty sequence; $\sigma_1 \sigma_2$ denotes the concatenation of sequences σ_1 and σ_2.

Now let us extend the tree $\mathbf{T_2}$ by inserting an extra irreflexive point at every edge. More precisely, this means the following.

For a non-empty $\tau \in T_2$ let $\tau^- := (\tau, 0)$. Consider the set

$$I_2 := T_2 \cup \{\tau^- \mid \tau \in T_2, \tau \neq \lambda\}.$$

Let $\lessdot$ be the minimal transitive relation on I_2 satisfying the conditions:

- $\lessdot | T_2 = \sqsubset$;

- if $\sigma \in T_2$, $i \in \{0, 1\}$, $\tau = \sigma i$, then $\sigma \lessdot \tau^- \lessdot \tau$.

Let $\mathbf{I_2} := (I_2, \lessdot)$, $\mathbf{I_2^+} := \mathbf{I_2} + C_\mathbb{Z}$, where $C_\mathbb{Z}$ is a countable cluster (Fig. 3). Since $\mathbf{I_2}$ is 2-dense, we have
$\mathbf{L}(\mathbf{I_2}) \supseteq \mathbf{L}_1$, $\mathbf{L}(\mathbf{I_2^+}) \supseteq \mathbf{L}_2$.

THEOREM 13. $\mathbf{L}(\mathbf{I_2}) = \mathbf{L}_1$, $\mathbf{L}(\mathbf{I_2^+}) = \mathbf{L}_2$.

Proof. For a frame F let $F_0 := E + F$, where E is a reflexive singleton. For $i = 1, 2$ let $\mathcal{K}_i := \{F_0 \mid F \in \mathcal{G}_i\}$. Then $\mathbf{L}_i = \mathbf{L}(\mathcal{K}_i)$. In fact, $\mathcal{K}_i \subseteq \mathcal{G}_i$ implies $\mathbf{L}_i \subseteq \mathbf{L}(\mathcal{K}_i)$ (by Lemma 12). On the other hand, for any $F \in \mathcal{G}_i$ there exists a p-morphism $F_0 \twoheadrightarrow F$ sticking E to the initial cluster of F (which is non-degenerate); therefore $\mathbf{L}_i \subseteq \mathbf{L}(\mathcal{K}_i)$.

Now to prove the inclusion $\mathbf{L}(\mathbf{I_2}) \subseteq \mathbf{L}_1 = \mathbf{L}(\mathcal{K}_1)$, let us construct a p-morphism from $\mathbf{I_2}$ onto an arbitrary frame $G = (W, R) \in \mathcal{K}_1$.

Let F be the restriction of G to the set of reflexive points. It is well-known [6], [12] that there exists a p-morphism $f : \mathbf{T_2} \twoheadrightarrow F$.

For a cluster $C \subseteq F$, let M_C be the set of all minimal elements in $f^{-1}(C)$. Obviously, $M_C \neq \emptyset$.

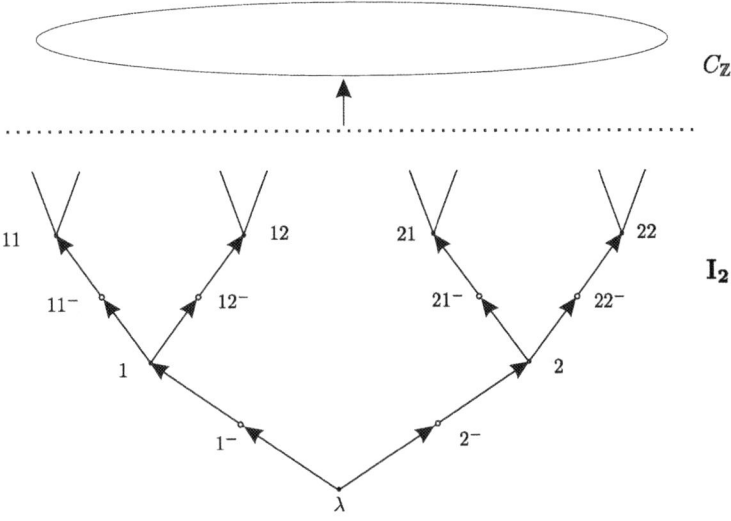

Figure 3.

Consider $w \in F$, and let $C(w)$ be its cluster. If w is not a root of G, the cluster $C(w)$ has a single $<_R$-predecessor. Choose an element $C(w)^-$ in this preceding cluster.

Let us extend f to $f' : \mathbf{I_2} \to G$ as follows: if $f(\alpha) = w$, $\alpha \neq \lambda$, we put

$$f'(\alpha^-) = \begin{cases} C(w)^-, & \text{if } \alpha \in M_{C(w)} \text{ and } C(w)^- \text{ is irreflexive,} \\ w, & \text{otherwise.} \end{cases}$$

(Informally: $f'(\alpha^-) = f(\alpha)$, whenever this is possible.)

Let us show that f' is surjective. Due to the surjectivity of f, it is sufficient to show that every irreflexive $w \in G$ is in the range of f'. So let C be the (unique) successor cluster of w, $x \in M_C$. Then $C^- = w$ and by definition $f'(x) = w$.

The monotonicity of f' obviously follows from the monotonicity of f and from the observation that $f'(\alpha^-) R f(\alpha)$.

Let us prove the lift property for f'. Assuming $f'(x) R w$, we have to find $y > x$ such that $f'(y) = w$. Consider different cases.

(1) $x \in T_2$, $w \in F$.

Then we can apply the lift property of f.

(2) $x \in T_2$, w is irreflexive.

Let C be the successor cluster of w,
$$Y := \{y \mid x \sqsubseteq y, \ y \in f^{-1}(C)\}.$$

Since f has the lift property, Y is non-empty. Choose a minimal $\alpha \in Y$. Then $x \lessdot \alpha^-$, $f'(\alpha^-) = w$.

(3) x is irreflexive, $f'(x) \in F$.

Let $x = \alpha^-$. By definition, $f'(x) = f'(\alpha)$. By the lift property of f, there exists $y > \alpha > x$ such that $f'(y) = w$.

(4) Both x, $f'(x)$ are irreflexive.

Let C be the successor cluster of $f'(x)$, $x = \alpha^-$. Then by definition, $f'(\alpha) = f(\alpha) \in C$. Thus $f(\alpha) R w$ and by the lift property of f, we have $y > \alpha > x$ such that $f'(y) = w$.

Therefore $f' : \mathbf{I_2} \twoheadrightarrow G$, and the first assertion of Theorem follows from P-morphism Lemma and Lemma 12.

Finally, since a finite cluster C is a p-morphic image of the infinite cluster $C_{\mathbb{Z}}$, we can extend f' to a p-morphism $\mathbf{I_2^+} \twoheadrightarrow G + C$. This yields the inclusion $\mathbf{L}(\mathbf{I_2^+}) \subseteq \mathbf{L}(\mathcal{K}_2) = \mathbf{L_2}$. ∎

Now recall the definitions of causal and chronological future relations ($\preceq$, $\prec$) in Minkowski spacetime $\mathbb{R}^n$ (where $n \geq 2$):

$$(x_1, \ldots, x_n) \preceq (y_1, \ldots, y_n) \Leftrightarrow \sum_{i=1}^{n-1}(y_i - x_i)^2 \leq (x_n - y_n)^2 \ \& \ x_n \leq y_n,$$

$$(x_1, \ldots, x_n) \prec (y_1, \ldots, y_n) \Leftrightarrow \sum_{i=1}^{n-1}(y_i - x_i)^2 < (x_n - y_n)^2 \ \& \ x_n < y_n.$$

Let $MF_D^n = (\mathbb{R}^n, \preceq)$, $MF_C^n = (\mathbb{R}^n, \prec)$.

From [6], [12] it is known that $\mathbf{L}(MF_D^n) = \mathbf{S4.2}$ for any $n \geq 2$. Our next aim is to describe $\mathbf{L}(MF_C^n)$.

The following soundness lemma is proved easily [6]:

LEMMA 14. $\mathbf{L}(MF_C^n) \supseteq \mathbf{L_2}$ for $n \geq 2$.

To prove the converse, we begin with the two-dimensional case.

Let us introduce some notation. pr_1 and pr_2 denote the standard projections $\mathbb{R}^2 \longrightarrow \mathbb{R}$; $\mathrm{pr}_i(x_1, x_2) := x_i$.

Let $P, Q \in \mathbb{R}^2$, $P \neq Q$, $\mathrm{pr}_2(P) = \mathrm{pr}_2(Q)$. Take the lower isosceles right triangle[1] PRQ with the right angle at R and subtract its hypotenuse $[PQ]$. Let
$\nabla(PQ) := PRQ \setminus [PQ]$,
$K(PQ) := [RQ] \cup [RP]$.

For $m \in \mathbb{Z}$ define the points $S_m(PQ)$ on the segment $[PQ]$ as follows:
$S_0(PQ)$ is the middle of $[PQ]$,
$S_m(PQ)$ is the middle of $[PS_{m+1}(PQ)]$ for $m < 0$,
$S_m(PQ)$ is the middle of $[S_{m-1}(PQ)Q]$ for $m > 0$.

For $x \in \mathbb{R}^2$, $M \subseteq \mathbb{R}^2$ let

$$h(x) := |\mathrm{pr}_2(x)|, \quad h(M) := \sup\{h(x) \mid x \in M\}.$$

Now fix the points $A := (-1, 0)$, $B := (1, 0)$, $C := (0, -1)$, and the corresponding open triangle together with the vertex C:

$$\nabla := (\nabla(AB) \setminus K(AB)) \cup \{C\}.$$

LEMMA 15. [2] *There exists a p-morphism $(\nabla, \prec) \twoheadrightarrow \mathbf{I_2}$.*

Proof. Our construction is a modification of those from [6], [12].

For $\sigma \in T_\mathbb{Z}$ we define points A^σ, B^σ by induction on the length $|\sigma|$. Put
$A^\lambda := A$, $B^\lambda := B$,
$A^{\sigma m} := S_m(A^\sigma B^\sigma)$, $B^{\sigma m} := S_{m+1}(A^\sigma B^\sigma)$, where $m \in \mathbb{Z}$;

$$\nabla^\sigma := \nabla(A^\sigma B^\sigma), \quad K^\sigma := K(A^\sigma B^\sigma), \quad W^\sigma := \bigcup_{m \in \mathbb{Z}} \nabla^{\sigma m},$$

and let C^σ be the lower vertex of ∇^σ.

One can easily check the following:

(1) $\sigma_1 \sqsubseteq \sigma_2 \Leftrightarrow \nabla^{\sigma_1} \supseteq \nabla^{\sigma_2}$.

(2) $\nabla^{\sigma_1} \cap \nabla^{\sigma_2} \neq \varnothing \Rightarrow \sigma_1 \sqsubseteq \sigma_2$ or $\sigma_2 \sqsubseteq \sigma_1$.

Note that
$h(\nabla^\lambda) = 1$,
$h(\nabla^{\sigma m}) \leq h(\nabla^{\sigma 0}) = \frac{1}{4} h(\nabla^\sigma)$, and thus

$$h(\nabla^\sigma) \leq 4^{-|\sigma|}, \quad h(W^\sigma) = h(\nabla^{\sigma 0}) \leq 4^{-|\sigma|-1}.$$

Since $h(x) > 0$ for any $x \in \nabla$, there exists the longest sequence σ_x such that $x \in \nabla^{\sigma_x}$; then $x \notin W^{\sigma_x}$. Due to (2), (1), σ_x is unique. Since $x \in \nabla^{\sigma_x}$, we have

---
[1] All triangles are considered as closed domains on the plane.
[2] A more accurate notation for $(\nabla, \prec)$ should be $(\nabla, \prec |\nabla)$.

(3) $\prec (x) \subseteq \nabla^{\sigma_x}$.

Now we define the map $f : \nabla \to I_2$ as follows.

If $x \neq C$ and $\sigma_x = m_1 \ldots m_l$, $l \geq 0$ [3], put $\tau_x := \tilde{m}_1 \ldots \tilde{m}_l$, where $\tilde{m} \in \{1, 2\}$, $\tilde{m} \equiv m \pmod{2}$. Let

$$f(x) := \begin{cases} \tau_x & \text{if } x \text{ is an interior point of } \nabla^{\sigma_x}; \\ \lambda & \text{if } x = C; \\ \tau_x^- & \text{otherwise.} \end{cases}$$

For $x \in K^{\sigma_x}$ put $f(x) = \tau_x^-$, otherwise $f(x) = \tau_x$ (Fig. 4).

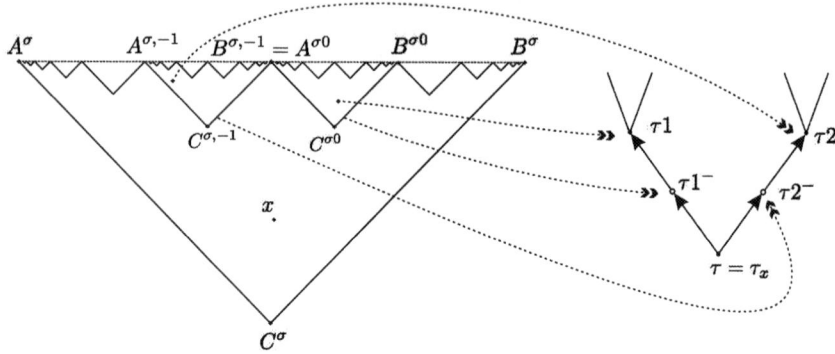

Figure 4.

Let us show that f is a p-morphism.

To check monotonicity, assume $x \prec y$. (3) implies that $y \in \nabla^{\sigma_x}$. Then $y \in \nabla^{\sigma_x} \cap \nabla^{\sigma_y}$, and so $\sigma_x \sqsubseteq \sigma_y$ or $\sigma_y \sqsubseteq \sigma_x$, by (2).

If $\sigma_y \sqsubset \sigma_x$, then $W^{\sigma_y} \supseteq \nabla^{\sigma_x}$, which contradicts $y \in \nabla^{\sigma_x}$.

If $\sigma_x \sqsubset \sigma_y$, then $\tau_x \sqsubset \tau_y$, and thus $\tau_x^- \triangleleft \tau_x \triangleleft \tau_y^- \triangleleft \tau_y$, which implies $f(x) \triangleleft f(y)$.

If $\sigma_x = \sigma_y$, then $f(x) = \tau_x$ or $f(x) = \tau_x^-$, $f(y) = \tau_x$, and again $f(x) \triangleleft f(y)$.

To check the lift property for f, assume $f(x) \triangleleft v$, where $v = \alpha$ or $v = \alpha^-$ for some $\alpha \in T_2$, $\sigma_x = m_1 \ldots m_k$, $\alpha = i_1 \ldots i_l$.

We have to find $y \succ x$ such that $f(y) = v$. Now there are two cases: $l = k$ or $l > k$.

Suppose $l = k$. Since α^- is irreflexive, we have $v = \alpha = \tau_x$.

Let $[xz_1]$ be the perpendicular to $[AB]$, $]xz_2[:=]xz_1[\setminus W^{\sigma_x}$, and let y be the middle of $]xz_2[$ (Fig. 5).

---

[3] If $l = 0$, $m_1 \ldots m_l$ means λ.

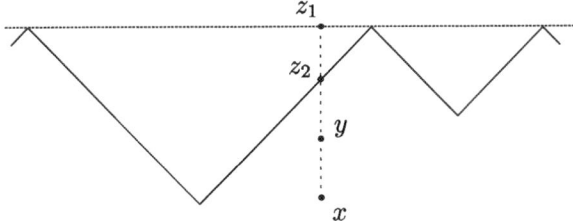

Figure 5.

Then obviously $x \prec y$. Since $y \in \nabla^{\sigma_x}$, $y \notin W^{\sigma_x}$, we obtain $\sigma_y = \sigma_x$, $f(y) = \tau_x = v$.

Now suppose $l > k$. Due to the transitivity of $\prec$ and $\lessdot$, it is sufficient to consider only the case $l = k+1$. Then $\alpha = \tau_x i_l$.

Since $x \notin W^{\sigma_x}$, for some division point $Q = A^{\sigma_x m}$ we have $x \prec Q$ (otherwise $\prec(x) \subseteq \nabla^{\sigma_x m}$ for some m). Put

$$m' := \begin{cases} m & \text{if } m \equiv i_l \pmod{2}, \\ m-1 & \text{otherwise}, \end{cases}$$

and let $\rho := \sigma_x m'$. Then Q is the right or the left vertex of ∇^ρ. Since $x \prec Q$, for some $u \in K^\rho$, we have $x \prec u$.

Let $[uz_1]$ be the perpendicular to $[AB]$, $]uz_2[:=]uz_1[\setminus W^\rho$, and let y be the middle of $]uz_2[$ (Fig. 6).

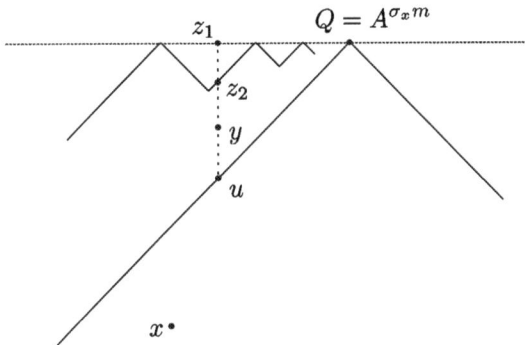

Figure 6.

Then we have $x \prec u \prec y$, $\tau_u = \tau_y = \alpha$, and thus $f(u) = \alpha^-$, $f(y) = \alpha$. Therefore $f(u) = v$ or $f(y) = v$. ∎

THEOREM 16. $\mathbf{L}(MF_C^2) = \mathbf{L_2}$.

Proof. Let $F = MF_C^2$. Since $\mathbf{L}(F) = \bigcap_{x \in \mathbb{R}^2} \mathbf{L}(F^x)$ and all cones F^x are isomorphic, we have $\mathbf{L}(F) = \mathbf{L}(F^x)$.

So consider the cone $G = F^{(0,-1)}$. In view of Theorem 13, it is sufficient to construct a p-morphism $G \twoheadrightarrow \mathbf{I}_2^+ = \mathbf{I_2} + C_{\mathbb{Z}}$.

We extend the p-morphism $f : \nabla \twoheadrightarrow \mathbf{I_2}$ from Lemma 15 to $f' : G \twoheadrightarrow \mathbf{I}_2^+$ as follows.

$$f'(x) := \begin{cases} f(x) & \text{if } x \in \nabla, \\ c_{m(x)} & \text{if } x \notin \nabla. \end{cases}$$

Here we assume that $C_{\mathbb{Z}} := \{c_n \mid n \in \mathbb{Z}\}$; for $x \in G$ $m(x)$ denotes the integer part of $\mathrm{pr}_1(x)$. Then one can easily check that f' is a p-morphism; note that if $x \notin \nabla$, then $\prec (x)$ contains points y with arbitrary $m(y)$. ∎

THEOREM 17. $L(MF_C^n) = \mathbf{L_2}$ *for any* $n \geq 2$.

Proof. Similar to [6]. Let p be the projection $\mathbb{R}^n \times \mathbb{R} \longrightarrow \mathbb{R} \times \mathbb{R}$:

$$p(x_1, \ldots, x_{n+1}) := (x_1, x_{n+1}).$$

It is easily seen that p maps a $\prec$-cone onto a $\prec$-cone, and thus $p : MF_C^n \twoheadrightarrow MF_C^2$, whence

$$\mathbf{L}(MF_C^n) \subseteq \mathbf{L}(MF_C^2) = \mathbf{L_2}.$$

On the other hand, $\mathbf{L}(MF_C^n) \supseteq \mathbf{L_2}$ by Lemma 14. ∎

Now let us consider logics of some domains on the plane with the relation $\prec$. Let us first prove a generalization of Lemma 15.

Let PRQ be a right triangle, in which $\mathrm{pr}_1(P) < \mathrm{pr}_1(Q)$, $ang(RP) = ang(RQ) = \pi/4$, where for a straight line l, $ang(l)$ denotes the (smallest) angle between l and the x-axis.

Let F be a real-valued differentiable function whose domain contains $[\mathrm{pr}_1(P), \mathrm{pr}_1(Q)]$, and assume that

$$\forall x \in \,]\mathrm{pr}_1(P), \mathrm{pr}_1(Q)[\ \ |F'(x)| < 1.$$

Also assume that γ is the graph of F, and $P, Q \in \gamma$. Then from Lagrange theorem we obtain that $\gamma \cap \,]RP[\, = \gamma \cap \,]RQ[\, = \varnothing$.

Let D be the open domain bounded by $[RP]$, $[RQ]$ and γ, and let

$$\nabla_F(PQ) := D \cup \{R\}, \ \widehat{\nabla}_F := D \cup \{R\} \cup \,]PR[\, \cup \,]RQ[.$$

LEMMA 18. *There exists a p-morphism* $(\nabla_F(PQ), \prec) \twoheadrightarrow \mathbf{I_2}$.

Proof. Let $A = P$, $B = Q$, and for $\sigma \in T_\mathbb{Z}$ consider the same division points $A^\sigma, B^\sigma \in [PQ]$ as in the proof of Lemma 15. Take the corresponding points on the curve γ:

$$A_F^\sigma := (\mathrm{pr}_1(A^\sigma), F(\mathrm{pr}_1(A^\sigma))), \ B_F^\sigma := (\mathrm{pr}_1(B^\sigma), F(\mathrm{pr}_1(B^\sigma))),$$

and let

$$\nabla^\sigma := \widehat{\nabla}_F(A_F^\sigma B_F^\sigma), \ W^\sigma := \bigcup_{m \in \mathbb{Z}} \nabla^{\sigma m}.$$

Similarly to Lemma 15, we obtain that for every $x \in \nabla_F(PQ)$ there exists a unique sequence σ_x such that $x \in \nabla^{\sigma_x}$ and $x \notin W^{\sigma_x}$.

So we can define a map $f : \nabla_F(PQ) \longrightarrow I_2$ in the same way as in Lemma 15. Namely, for $\sigma_x = m_1 \ldots m_l$, put $\tau_x := \tilde{m}_1 \ldots \tilde{m}_l$. The definition of f is as follows (Fig. 7).

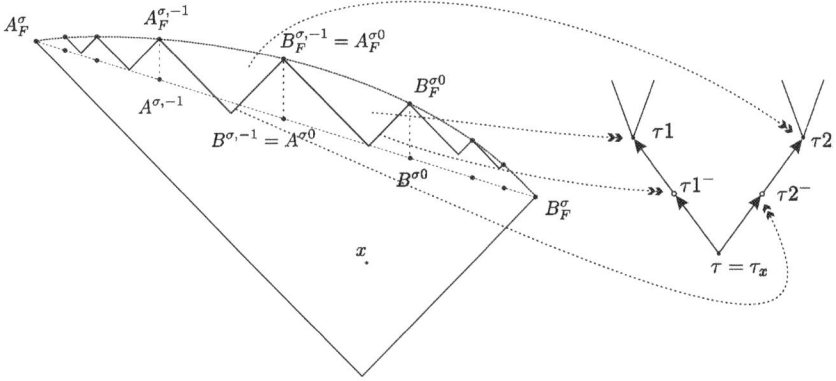

Figure 7.

$$f(x) := \begin{cases} \tau_x & \text{if } x \text{ is an interior point of } \nabla^{\sigma_x}; \\ \lambda & \text{if } x = R; \\ \tau_x^- & \text{otherwise.} \end{cases}$$

By the same argument as in Lemma 15, one can show that f is a p-morphism. ∎

THEOREM 19. *Assume that X is an open convex polygon in $\mathbb{R}^2$. Then $\mathbf{L}(X, \prec)$ is either $\mathbf{L}_1$ or $\mathbf{L}_2$.*

Proof. (Cf. [12]). Let V be the highest vertex of X, and let A and B be the vertices of X that are adjacent to V.

There may be two cases:

(1) $ang(AV) < \frac{\pi}{4}$ or $ang(BV) < \frac{\pi}{4}$;

(2) $ang(AV) \geq \frac{\pi}{4}$ and $ang(BV) \geq \frac{\pi}{4}$.

Consider the first case. Assume that for example, $ang(AV) < \frac{\pi}{4}$ (Fig. 8). Take the linear function $F : \mathbb{R} \longrightarrow \mathbb{R}$ with the graph AV. Take two

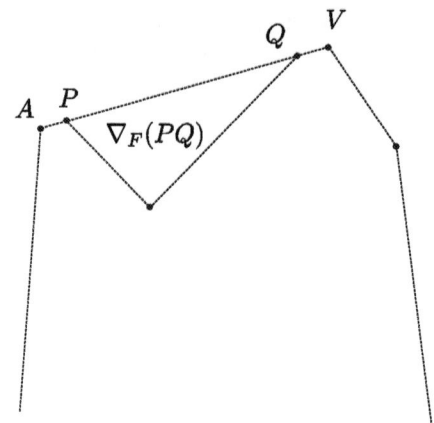

Figure 8.

points $P, Q \in [AV]$ such that $\nabla_F(PQ) \subseteq X$. Then $\nabla_F(PQ)$ is a generated subframe in $(X, \prec)$, and thus by Lemma 18,

$$\mathbf{L}(X, \prec) \subseteq \mathbf{L}(\nabla_F(PQ), \prec) \subseteq \mathbf{L}_1.$$

One can easily see that $(X, \prec)$ is serial and 2-dense, and thus we obtain $\mathbf{L}(X, \prec) = \mathbf{L}_1$.

In the case (2), take an interior point R in the triangle AVB (Fig. 9). Then there exist points $P \in [AV]$, $Q \in [BV]$ such that $ang(RP) = ang(RQ) = \frac{\pi}{4}$. The cone $Y := X \cap \prec (R)$ can be presented as $Y = \nabla(PQ) \cup D$, where $D = PQV \setminus ([PV] \cup [QV])$.

Let us show that Y can be p-morphically mapped onto $\mathbf{I_2} + C_{\mathbb{N}}$, where $C_{\mathbb{N}}$ is a countable cluster. By Lemma 18, there exists $h : \nabla(PQ) \twoheadrightarrow \mathbf{I_2}$, so let us prolong h to Y. Assume that (A_n) is a sequence of points from D converging to V. Consider the infinite sequence of natural numbers taking

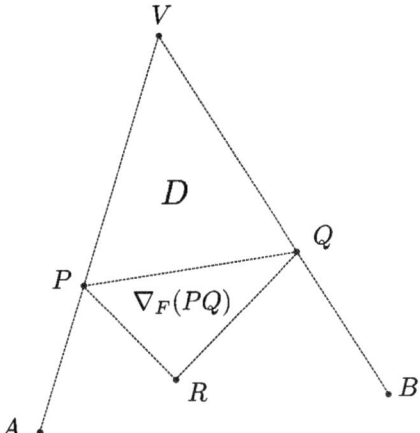

Figure 9.

every value infinitely many times, for instance $(s_n) = 1121231234\ldots$. Let $C_{\mathbb{N}} = \{c_0, c_1, c_2, \ldots\}$. Then put

$$h(A_n) := c_{s_n}.$$

For all other points $B \in D$ put $h(B) := c_0$. A straightforward argument shows that $p : Y \twoheadrightarrow \mathbf{I_2} + C_{\mathbb{N}}$. Therefore,

$$\mathbf{L}(X, \prec) \subseteq \mathbf{L}(Y) \subseteq \mathbf{L}_2.$$

Since X is convex, the condition (2) implies that the frame $(X, \prec)$ is confluent, and thus $\mathbf{L}(X, \prec) \supseteq \mathbf{L}_2$. ∎

THEOREM 20. *Let X be an open connected domain in $\mathbb{R}^2$ bounded by a closed smooth curve. Then $\mathbf{L}(X, \prec) = \mathbf{L}_1$.*

Proof. Let γ be the boundary of X. Due to smoothness, it can be presented as $\{(x(t), y(t)) \mid t \in [0, 1]\}$, where $x(t)$ and $y(t)$ are smooth functions on $[0, 1]$ such that for any t, $(\dot{x}(t), \dot{y}(t)) \neq (0, 0)$. Let $C = (x(t_0), y(t_0))$ be the highest point of γ (which may be not unique). Then $\dot{y}(t_0) = 0$, $\dot{x}(t_0) \neq 0$, and without any loss of generality, we may assume that $\dot{x}(t_0) > 0$, $t_0 \neq 0$, $t_0 \neq 1$. Since the derivatives $\dot{x}$ and $\dot{y}$ are continuous, there exists an interval $]t_1, t_2[\ni t_0$, where $\dot{x}(t) > \varepsilon$, $|\dot{y}(t)| < \varepsilon$ for some $\varepsilon > 0$. Then $x(t)$ is invertible in this interval, and the corresponding part of γ is a graph of some function $F :]x_1, x_2[\longrightarrow \mathbb{R}$, where $x_1 = x(t_1)$, $x_2 = x(t_2)$. Now it follows that $|F'(x)| = |\dot{y}(t)/\dot{x}(t)| < 1$ for $x \in]x_1, x_2[$.

We can choose points P, Q in this part of γ such that $\nabla_F(PQ) \subseteq X$. Since $\nabla_F(PQ)$ is a cone in $(X, \prec)$ and by Lemma 18, we have

$$\mathbf{L}(X, \prec) \subseteq \mathbf{L}(\nabla_F(PQ), \prec) \subseteq \mathbf{L}_1.$$

It remains to note that $\mathbf{L}(X, \prec)$ is 2-dense. In fact, for any $x \in X$ there exists an open disk $U \subseteq X$ containing x. Then every two points $y, z \in \prec(x)$ are $\prec$-accessible from some $u \in U \cap \prec(x)$.

Thus $\mathbf{L}(X, \prec) \supseteq \mathbf{L}_1$. ∎

5 Conclusion

Logics of relativistic time studied so far are examples of many-dimensional modal logics that are rather simple from the computational viewpoint. To say more on that topic, let us recall the connection between relativistic modal logics and modal products [13].

Consider the product frame
$(\mathbb{R}, <)^2 := (\mathbb{R}, <) \times (\mathbb{R}, <) := (\mathbb{R}^2, R_1, R_2)$,
where
$(x, y) R_1 (x', y') \Leftrightarrow x < x' \,\&\, y = y'$;
$(x, y) R_2 (x', y') \Leftrightarrow x = x' \,\&\, y < y'$.

The logic of $\mathbf{L}((\mathbb{R}, <)^2)$ is known to be Π_1^1-complete, and therefore it is not recursively axiomatizable [11], [4]. But our $\mathbf{L}_2$ is its rather natural decidable fragment. In fact, the frames $(\mathbb{R}^2, R_1 \circ R_2)$ and $(\mathbb{R}^2, \prec)$ are isomorphic; an isomorphism is given by rotation. Thus the corresponding logics can be identified by translation φ from 1-modal to 2-modal formulas reading $\square$ as $\square_1 \square_2$:

COROLLARY 21. $\mathbf{L}_2 = \{A \mid (\mathbb{R}, <)^2 \vDash \varphi(A)\}$.

The proofs in Section 4 easily transfer to the rational case, and in the same way we obtain:

COROLLARY 22. $\mathbf{L}_2 = \{A \mid (\mathbb{Q}, <)^2 \vDash \varphi(A)\}$.

Note that the whole logic $\mathbf{L}((\mathbb{Q}, <)^2)$ is undecidable [11], [4], but unlike the real case, it is recursively enumerable and coincides with the corresponding product of modal logics $\mathbf{L}(\mathbb{Q}, <) \times \mathbf{L}(\mathbb{Q}, <)$ [13], [4].

It remains a serious open problem, whether all "φ-fragments" of products of linear modal logics are decidable. Probably, the most interesting unknown case is $\mathbf{L}(\mathbb{Z}, <) \times \mathbf{L}(\mathbb{Z}, <)$. A related question [6] is about properties of the logic of discrete Minkowski spacetime (which coincides with $\mathbf{L}((\mathbb{Z}, <)^2)$).

The above two corollaries can be reformulated in terms of classical first-order theories in the style of [5]. Viz., consider the first-order language $\mathcal{L}$ with binary predicates $<, P_1, P_2, \ldots$ Every 1-modal propositional formula

A translates into an $\mathcal{L}$-formula $A^*(x, y)$:
$p_n^*(x, y) := P_n(x, y)$,
$\perp^* := \perp$,
$(A \to B)^* := A^* \to B^*$,
$\Diamond A^*(x, y) := \exists x_1 \exists y_1 \, (x < x_1 \wedge y < y_1 \wedge A^*(x_1, y_1))$.

Let $Th^2(W, <)$ be the $\mathcal{L}$-theory of all structures $(W, <, \ldots)$ with fixed $(W, <)$ and varying interpretations of $P_1, P_2, \ldots$ Then we obtain

COROLLARY 23. $\mathbf{L}_2 = \{A \mid Th^2(\mathbb{R}, <) \vdash A^*\} = \{A \mid Th^2(\mathbb{Q}, <) \vdash A^*\}$.

One can show that the same logics arise in the many-dimensional case. In fact, consider the n-dimensional product $(\mathbb{R}, <)^n = (\mathbb{R}^n, R_1, \ldots, R_n)$, where $(x_1, \ldots, x_n) R_i (y_1, \ldots, y_n)$ iff $x_i < y_i$ & $\forall j \neq i \; x_j = y_j$. Then a standard projection is a p-morphism $p : (\mathbb{R}, <)^n \twoheadrightarrow (\mathbb{R}, <)^2$.

Let φ_n be translation from 1-modal to n-modal formulas interpreting $\square$ as $\square_1 \ldots \square_n$. Then similarly to Corollaries 21, 22, we have

COROLLARY 24. $\mathbf{L}_2 = \{A \mid (\mathbb{R}, <)^n \vDash \varphi_n(A)\} = \{A \mid (\mathbb{Q}, <)^n \vDash \varphi_n(A)\}$.

This Corollary also has a classical analogue similar to Corollary 23; we leave the precise details for the reader.

The logics $\mathbf{L}_1$, $\mathbf{L}_2$ can also be interpreted as fragments of interval temporal logics. Logics of intervals have motivations in Computer Science, Linguistics, and Philosophy; the reader is addressed to [9] for further references and a brief overview of this area. Let us recall that there are 13 basic relations between intervals in a linearly ordered set, and the corresponding full modal logic of these accessibility relations is undecidable, according to the result by Halpern and Shoham [8]. This happens in most cases, in particular, for integers, rationals, and reals.

On the other hand, for rational and real time, as stated in [13], some natural fragments of interval logics are equivalent to the well-known systems **S4**, **S4.2**. A similar property holds for $\mathbf{L}_1$, $\mathbf{L}_2$. In fact, let $I(W, <)$ be the set of all nontrivial closed intervals (segments) in a linearly ordered set $(W, <)$. Obviously, we can identify $I(W, <)$ with the half-plane $\{(x, y) \in W^2 \mid x < y\}$.

Consider the following relations between intervals.

$$(x_1, y_1) \sqsubset (x_2, y_2) := x_2 < x_1 \; \& \; y_1 < y_2 \; (\text{``during''});$$

$$(x_1, y_1) < (x_2, y_2) := x_1 < x_2 \; \& \; y_1 < y_2 \; (\text{``weakly earlier''}),$$

and their converses $\sqsupset, >$. (Note that $<$ is not among the basic 13 relations, because $>$ is the union of "later than" and "overlaps".) Fig. 10 shows the points of $I(W, <)$ accessible from a certain point by these relations. So we can see that on the real plane the cone $\sqsubset$ is isomorphic to the triangle

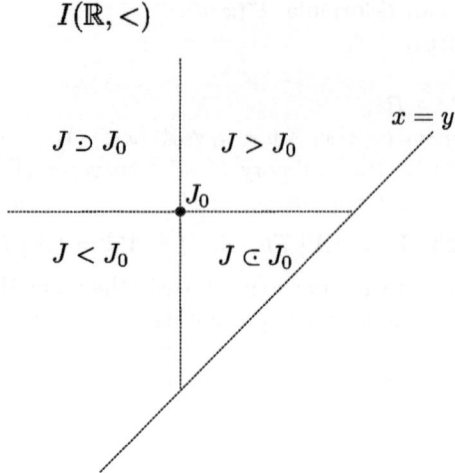

Figure 10.

$(\nabla, \prec)$ considered in Section 3, and thus its logic is $\mathbf{L_1}$. Three other cones are confluent, and their logic is $\mathbf{L_2}$. The case of rationals is completely analogous. So we obtain

COROLLARY 25. *Let $P = (\mathbb{R}, <)$ or $(\mathbb{Q}, <)$. Then*
$\mathbf{L}(I(P), \supset) = \mathbf{L_1},$
$\mathbf{L}(I(P), <) = \mathbf{L}(I(P), >) = \mathbf{L}(I(P), \subset) = \mathbf{L_2}.$

Finally let us make some remarks on the computational complexity of $\mathbf{L_1}$, $\mathbf{L_2}$. It is well-known that their reflexive analogues $\mathbf{S4}$, $\mathbf{S4.2}$ are PSPACE-complete [2]. Our proof of the f.m.p. in Section 3 motivates the same complexity bound for $\mathbf{L_1}$, $\mathbf{L_2}$. The corresponding result recently proved by the first author will be published in the sequel.

Also note that there probably exists an alternative proof of the f.m.p. based the method from [2] (Theorem 11.52). However this method does not provide a good complexity bound.

There remain many open problems in the field. For example, nothing is known about decidability, complexity and expressive power of relativistic polymodal logics. The same questions certainly make sense in the context of general relativity theory (a brief discussion can be found in [1]) or for many-dimensional analogues of interval logics. So there is enough room for further investigations of spacetime logics.

Acknowledgements

The authors thank the anonymous referees for very useful and detailed comments on the earlier versions of the paper.

The work on this paper was partly supported by Russian Foundation for Basic Research and ESPRC project on many-dimensional modal logic.

BIBLIOGRAPHY

[1] J. Burgess. Logic and time. Journal of Symbolic Logic, v. 44(1979), 566-582.
[2] A. Chagrov, M. Zakharyaschev. Modal logic. Oxford University Press, 1997.
[3] K. Fine. Logics containing K4. I. Journal of Symbolic Logic, v. 39(1974), 31-42.
[4] D. Gabbay, A. Kurusz, F. Wolter, M. Zakharyaschev. Many-dimensional modal logics: theory and applications. Studies in Logic, Elsevier, 2003.
[5] D. Gabbay, V. Shehtman. Products of modal logics. II. Relativised quantifiers in classical logic. Logic Journal of the IGPL, v. 8 (2000), No. 2, p. 165-210.
[6] R. Goldblatt. Diodorean modality in Minkowski spacetime. Studia Logica, v. 39 (1980), 219-236.
[7] R. Goldblatt. Logic of time and computation. CSLI Lecture Notes, No. 7. Stanford, 1987.
[8] J. Halpern, Y. Shoham. A propositional modal logic of time intervals. Journal of the ACM, v. 38(1991), 935-962.
[9] M. Marx, Y. Venema. Multi-dimensional modal logic. Applied Logic Series, v.4. Kluwer Academic Publishers, 1997.
[10] A.N. Prior. Past, present and future. Clarendon Press, Oxford, 1967.
[11] M. Reynolds, M. Zakharyaschev. On the products of linear modal logics. Journal of Logic and Computation, v. 11 (2001), 909-931.
[12] V.B. Shehtman. Modal logics of domains on the real plane. Studia Logica, v. 42 (1983), 63-80.
[13] V.B. Shehtman. On some two-dimensional modal logics. In: 8th International Congress on Logic, Methodology, and Philosophy of Science, Moscow, 1987. Abstracts, v.1, 326-330, 1987.

Ilya Shapirovsky
Department of mathematical logic
Faculty of mathematics and mechanics
Moscow State University
Moscow, Russia
E-mail: ilshapir@lpcs.math.msu.ru

Valentin Shehtman
Institute of Infomation Transmission Problems
B. Karetny 19
101447 Moscow, Russia
E-mail: shehtman@lpcs.math.msu.ru

An Incompleteness Result for Predicate Extensions of Intermediate Propositional Logics

D. Skvortsov

ABSTRACT. For a continuum of predicate extensions of intermediate propositional logics of small height or of small width we establish incompleteness in the semantics of Kripke bundles introduced in [7].

The semantics KB of *Kripke bundles* was introduced in [7]. It generalizes the standard Kripke semantics for predicate superintuitionistic logics. Its further strengthening, the *functor semantics* (the semantics of SET-valued functors over categories, or so-called C-*sets*) is considered in S. Ghilardi's paper [2]. An essential point in studying different semantics is the comparison of their strength.

This can be done in two ways. First, we can compare the completeness of logics: in a stronger semantics, more logics are complete. Second, we can compare separability of certain formulas from certain logics. Clearly, if a formula A is unseparable from a logic $\mathbf{L}$, then every extension $\mathbf{L}'$ of $\mathbf{L}$ which does not contain A is incomplete. Most of incompleteness results are proved in this way.

In this paper we establish KB-unseparability of some formulas introduced in Ghilardi's paper [3] (these formulas describe, in some sense, binary trees of finite height) from superintuitionistic predicate logics of small height or small width. This implies, in particular, the KB-incompleteness of the predicate version **Q-L** of any intermediate propositional logic **L** of small height or width, and the incompleteness of many its extensions (with the same propositional fragment).

Note that the predicate version **Q-L** of a *canonical* propositional logic **L** is complete in functor semantics [4]. On the other hand, for any propositional logic **L** of finite height (except for classical logic) its predicate version **Q-L** is incomplete in standard Kripke semantics [6].

REMARK 1. The result presented in this paper was obtained almost ten years ago, in the early 1990s and was motivated by Ghilardi's papers [2, 3] and by constructions presented there. At that time it was probably the first explicit example of KB-incompleteness; an indirect KB-incompleteness proof based on non-axiomatizability of KB-completions of some finitely axiomatizable logics was sketched in [8].

In 1997, K. Nagaoka and E. Isoda [5] proved very strong results which imply the KB-incompleteness of predicate versions **Q-L** for all propositional logics **L** of finite height or of finite width except classical logic and Dummett's logic (in the latter two cases **Q-L** are Kripke-complete). However, the idea of the proof and constructions presented in [5] essentially differs from ours. Let us also point out that the KB-unseparability is proved in [5] for another kind of formulas, different from Ghilardi's ones. Thus the ranges of KB-incompleteness are definitely incomparable. For example, our construction can be used to show the KB-incompleteness of [**Q-L** + Z], where $Z = \forall x, y \, (P(x) \supset \neg\neg P(y))$, for some logics of finite width, while the C-set constructed in [5] (in its part referring to the width) refutes the formula Z.

Thus we present this paper almost in its original form, only with some minimal re-compositions and re-formulations.

In Section 1 we give necessary definitions and formulate the Main Theorem on KB-unseparability (and its consequences on KB-incompleteness). In Section 2 we formulate Technical Lemma and prove the Main Theorem basing on this Lemma. Finally, in Section 3 we prove this Technical Lemma.

1 Formulation of KB-incompleteness theorem

1.1 *Predicate superintuitionistic logics* are understood as usual (cf. e.g. [7, 1]), i.e., as sets of predicate formulas (without equality and function symbols), containing all axioms of Heyting predicate calculus and closed under modus ponens, generalization, and substitution of arbitrary formulas instead of atomic ones.

The *propositional fragment* of a predicate logic is the set of all its propositional formulas.

The *predicate version* **Q-L** of a propositional (intermediate) logic **L** is the minimal predicate logic, the propositional fragment of which equals to **L**. In particular, Heyting predicate logic **Q-H** is the predicate version of Heyting propositional logic **H**.

Let us consider propositional formulas Q_n and P_n for any $n \in \omega$:

$$Q_n = \bigvee_{i=0}^{n} [q_i \supset \bigvee_{j \neq i} q_j]$$

(in particular, $Q_0 = \neg q_0$, and $Q_1 = (q_0 \supset q_1) \vee (q_1 \supset q_0)$, i.e., (**H** + Q_1) is the propositional Dummett's logic),

$$P_0 = \bot, P_{n+1} = q_n \vee (q_n \supset P_n),$$

$q_0, q_1, q_2, \ldots$ being a list of different propositional symbols.

Let also
$$J = \neg q_0 \vee \neg\neg q_0.$$

A propositional intermediate logic **L** is *of width n* if $Q_n \in \mathbf{L}, Q_{n-1} \notin \mathbf{L}$; a logic **L** is *of height n* if $P_n \in \mathbf{L}, P_{n-1} \notin \mathbf{L}$.

It is well-known that **L** is a logic of finite height iff $\mathbf{L} \not\subseteq (\mathbf{H} + Q_1)$.

Let $(W, \leq)$ be a pre-ordered set, i.e., $\leq$ is a reflexive and transitive, but not necessarily antisymmetric relation on W. As usual, the *propositional logic* $\mathbf{PL}(W, \leq)$ *of* $(W, \leq)$ is defined as the logic of the Heyting algebra of all $\leq$-stable (upward closed) subsets of W. It is well-known that if $(W, \leq)$ has the least element, then $\mathbf{PL}(W, \leq)$ is of width (or of height) $\leq n$ iff W has no antichains (or respectively, $<$-chains) of more than n elements. On the other hand, every propositional logic **L** of height $\leq n$ can be presented as $\mathbf{PL}(\Omega) = \bigcap(\mathbf{PL}(W, \leq) : (W, \leq) \in \Omega)$ for a family Ω of pre-ordered sets of height $\leq n$, i.e., of pre-ordered sets without $(n+1)$-element chains.

1.2 A *Kripke bundle* [7, Sect. 5] and [1] is a triple $\mathbb{F} = \langle W, D, \pi \rangle$ formed by non-empty pre-ordered sets W and D (R and ϱ denote corresponding pre-orders), and by a p-morphism π. This means that $\pi : D \to W$ is a surjective map such that

$$(i) a \varrho b \Rightarrow \pi(a) R \pi(b), \text{ and } (ii) \pi(a) R v \Rightarrow \exists b [(a \varrho b) \& (\pi(b) = v)].$$

We set:
$$D_u = \pi^{-1}(u) \text{ for } u \in W;$$

thus $D = \bigcup_{u \in W} D_u, D_u \neq \emptyset$ for each $u \in W$, and $D_u \cap D_v = \emptyset$ if $u \neq v$.

We also set $D_n = \bigcup_{u \in W} (D_u)^n$ for $n > 0$. D_n is pre-ordered by the relation:

$$(a_1, \ldots, a_n) \leq_n (b_1, \ldots, b_n) \Leftrightarrow \bigwedge_i (a_i \varrho b_i) \& \bigwedge_{i \neq j} [(a_i = a_j) \Rightarrow (b_i = b_j)].$$

In particular, $(D_1, \leq_1) = (D, \varrho)$. Finally we set: $(D_0, \leq_0) = (W, R)$.

Note that:

(π) for any $a_0, a_1, \ldots, a_n \in D_u$ and $b_1, \ldots, b_n \in D_v (n \geq 0)$
such that $(a_1, \ldots, a_n) \leq_n (b_1, \ldots, b_n)$ (hence $u \leq_0 v$)
there exists $b_0 \in D_v$ such that $(a_0, a_1, \ldots, a_n) \leq_{n+1} (b_0, b_1, \ldots, b_n)$

(cf. (ii)).

A *valuation* ξ in a bundle $\mathbb{F}$ (see [7]) can be defined as the forcing relation: $u \vDash A(\mathbf{a})$ for predicate formulas $A(x_1, \ldots, x_n)$, $u \in W$, $\mathbf{a} = (a_1, \ldots, a_n) \in (D_u)^n$, satisfying the *monotonicity*:

$$(u \leq_0 v) \& (\mathbf{a} \in (D_u)^n) \& (\mathbf{b} \in (D_v)^n) \& (\mathbf{a} \leq_n \mathbf{b}) \& (u \vDash A(\mathbf{a})) \Rightarrow (v \vDash A(\mathbf{b}))$$

and the following inductive clauses:

$u \nvDash \bot; u \vDash (B \wedge C) \Leftrightarrow (u \vDash B \text{ and } u \vDash C);$
$u \vDash (B \vee C) \Leftrightarrow (u \vDash B \text{ or } u \vDash C);$
$u \vDash (B \supset C)(\mathbf{a}) \Leftrightarrow \forall v \geq u \forall \mathbf{b} \in (D_v)^n [(\mathbf{a} \leq_n \mathbf{b}) \& (v \vDash B(\mathbf{b})) \Rightarrow (v \vDash C(\mathbf{b}))];$
$u \vDash \forall y B(\mathbf{a}, y) \Leftrightarrow \forall v \geq u \forall \mathbf{b} \in (D_v)^n \forall c \in D_v [(\mathbf{a} \leq_n \mathbf{b}) \Rightarrow (v \vDash B(\mathbf{b}, c))];$
$u \vDash \exists y B(\mathbf{a}, y) \Leftrightarrow \exists c \in D_u (u \vDash B(\mathbf{a}, c)).$

As usual, to obtain a valuation, it is sufficient to know $u \vDash A(\mathbf{a})$ only for atomic A, and the monotonicity for atomic A implies the monotonicity for arbitrary A, by induction.

A predicate formula $A(x_1, \ldots, x_n)$ is called *valid* in a Kripke bundle $\mathbb{F}$ if $u \vDash A(\mathbf{a})$ for any valuation in $\mathbb{F}$, for any $u \in W$, and any $\mathbf{a} \in (D_u)^n$. A formula A is called *strongly valid* in $\mathbb{F}$ if all its substitution instances are valid in $\mathbb{F}$. The set $\mathbf{QL}(\mathbb{F})$ of all predicate formulas strongly valid in $\mathbb{F}$ is a predicate logic [7, Sect. 5]. A predicate logic $\mathbf{L}$ is called *KB-complete* if it can be presented as $\mathbf{L} = \bigcap (\mathbf{QL}(\mathbb{F}) : \mathbb{F} \in \Omega)$ for some family Ω of Kripke bundles.

The following simple lemma is analogous to Lemma 12 from [4] (its proof is postponed until Section 3):

LEMMA 2. *For any Kripke bundle $\mathbb{F}$, the propositional fragment of $\mathbf{QL}(\mathbb{F})$ is $\bigcap_{n \in \omega} \mathbf{PL}(D_n, \leq_n)$.*

1.3 Let $T_h^2 = \{(d_1, \ldots, d_k) : 0 \leq k \leq h, \forall i (d_i \in \{0, 1\})\}$ be the binary tree of height $(h+1)$ (for $h \in \omega, h > 0$). Recall the definition of the predicate formula $F_\mathbf{d}^h$ corresponding to an arbitrary $\mathbf{d} \in T_h^2$ given in [3, Sect. 5]. This is a formula with two monadic predicate symbols R_0, R_1 and with k parameters $x_1, \ldots, x_k$ (where k is the length of $\mathbf{d}$):

$$F_\mathbf{d}^h = \bigvee_{i=1}^{h} R_{d_i}(x_i) \text{ if } k = h,$$
$$F_\mathbf{d}^h = \forall x_{k+1}[(R_0(x_{k+1}) \supset F_{(\mathbf{d},1)}^h) \vee (R_1(x_{k+1}) \supset F_{(\mathbf{d},0)}^h)] \text{ if } k < h.$$

In [3] it is proved that $\mathbf{Q}\text{-}\mathbf{L} \nvdash \neg\neg F_\varepsilon^h$ for any propositional logic $\mathbf{L}$ such that $J \in \mathbf{L}$, $Q_1 \notin \mathbf{L}$, and for any $h > 0$, where ε is the empty sequence from T_h^2. More precisely, [3] describes a C-set separating $\neg\neg F_\varepsilon^h$ from **Q-L**. Actually the same C-set separates $\neg\neg F_\varepsilon^h$ from $[\mathbf{Q\text{-}L} + E + Z]$, where

$$E = \neg\neg \exists x P(x) \supset \exists x \, \neg\neg P(x),$$
$$Z = \forall x, y \, (P(x) \supset \neg\neg P(y)),$$

and from many other extensions of **Q-L**.

THEOREM 3 (on KB-unseparability). *Let $\mathbb{F}$ be a Kripke bundle such that $(J \wedge P_m) \in \mathbf{QL}(\mathbb{F})$ for some $m \in \omega$, and $\neg\neg F_\varepsilon^3 \notin \mathbf{QL}(\mathbb{F})$. Then the propositional fragment of $\mathbf{QL}(\mathbb{F})$ is a logic of height ≥ 5 and of width ≥ 5.*

In other words, if the formula $\neg\neg F_\varepsilon^3$ is KB-separable from a predicate logic **L** containing $J \wedge P_m$, then the propositional fragment of **L** has height ≥ 5 and width ≥ 5.

COROLLARY 4. *Let **L** be a propositional logic of finite height, $J \in \mathbf{L}$, $Q_1 \notin \mathbf{L}$. If the predicate version **Q-L** is KB-complete, then **L** is a logic of height ≥ 5 and of width ≥ 5.*

Hence the KB-incompleteness of **Q-L** follows for continuum of logics **L** of height 4, for $\mathbf{L} = (\mathbf{H} + P_k \wedge J)$, where $3 \leq k \leq 4$, and for $\mathbf{L} = (\mathbf{H} + P_n \wedge Q_m \wedge J)$, where $(2 \leq m \leq 4, n \geq 3)$ or $(3 \leq n \leq 4, m \geq 2)$. Similarly we obtain the KB-incompleteness of the logics $[\mathbf{Q\text{-}L} + E]$, $[\mathbf{Q\text{-}L} + Z]$, $[\mathbf{Q\text{-}L} + E + Z]$ (for the same **L**), etc.

The bound 5 for height and width in Theorem on KB-unseparability cannot be increased, and we cannot say anything about the width without some additional restrictions referring to height. In fact, let us consider the p.o. sets W_1 and W_2 as in Fig.1: W_1 is of width and height 5, and W_2 is of width 2 (and of infinite height). For $s = 1, 2$ there exist Kripke bundles $\mathbb{F}_s$ (Fig.2) such that the propositional fragment of $\mathbf{QL}(\mathbb{F}_s)$ is $\mathbf{PL}(W_s)$ and $\neg\neg F_\varepsilon^h \notin \mathbf{QL}(\mathbb{F}_s)$ for all $h > 0$. In other words, all formulas $\neg\neg F_\varepsilon^h$ are separable from **Q-L** in the semantics of Kripke bundles KB if $\mathbf{L} = \mathbf{PL}(W_s), s = 1, 2$ (here $J \in \mathbf{L}, Q_1 \notin \mathbf{L}$). The proof for $\mathbb{F}_1$ is analogous to that of Theorem 5.6 in [3], the proof for $\mathbb{F}_2$ is similar.

If the formula $\neg\neg F_\varepsilon^3$ is replaced with a simpler formula $\neg\neg F_\varepsilon^2$, then the bound becomes 4 for height and 3 for width. The corresponding statement can be proved similarly. This result is also the optimal, because $\neg\neg F_\varepsilon^2$ is KB-separable from **Q-L** if $\mathbf{L} = \mathbf{PL}(W_3)$ (Fig.3). A corresponding bundle $\mathbb{F}_3$ is more complicated than $\mathbb{F}_1$, but it can be constructed in a straightforward way using the formulation of Technical Lemma (Lemma 5 below) for the case $h = 2$.

As for $h = 1$, the formula

$$\neg\neg F_\varepsilon^1 = \neg\neg \forall x[(R_0(x) \supset R_1(x)) \vee (R_1(x) \supset R_0(x))]$$

W_1: 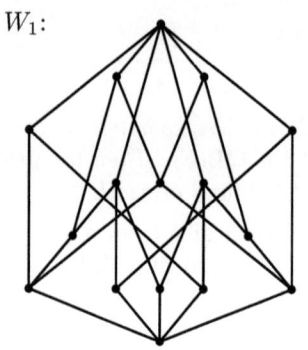 W_2: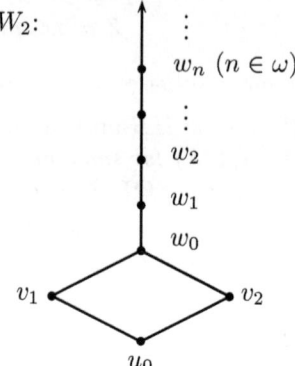

Figure 1.

is KB-separable from **Q-L** whenever $Q_1 \notin \mathbf{L}$; Fig. 4 shows the corresponding bundles.

If $\mathbf{L} \supseteq (\mathbf{H} + Q_1)$, then **Q-L** contains F_ε^1, leave alone $\neg\neg F_\varepsilon^1$.

Finally let us point out that the condition '$\exists m(P_m \in \mathbf{QL}(\mathbb{F}))$' in the statement of Theorem actually can be weakened to '$\mathbf{PL}(\mathbb{F}) \not\subseteq \mathbf{PL}(W_2)$', where $\mathbf{PL}(\mathbb{F})$ denotes the propositional fragment of $\mathbf{QL}(\mathbb{F})$.

2 Proof of KB-unseparability theorem

2.1 We use the following notation. For $\mathbf{d} = (d_1, \ldots, d_k) \in T_h^2$, where $0 \leq k \leq h$, we put: $k^* = k+1$ if $k < h$, and $k^* = h$ if $k = h$, $\mathbf{d}_i = (d_1, \ldots, d_i)$ for $1 \leq i \leq k$, $\mathbf{d}_0 = \varepsilon$, and $\mathbf{d}^- = \mathbf{d}_{k-1} = (d_1, \ldots, d_{k-1})$ if $0 < k \leq h$ ($\mathbf{d}^-$ for $\mathbf{d} = \varepsilon$ is not defined).

A routine proof of the following lemma will be given in Section 3.

LEMMA 5 (Technical Lemma). *Let $\mathbb{F} = \langle W, D, \pi \rangle$ be a Kripke bundle such that $(J \wedge P_m) \in \mathbf{QL}(\mathbb{F})$ for some $m \in \omega$. Then the following conditions are equivalent (for any $h > 0$):*

(i) $\neg\neg F_\varepsilon^h \notin \mathbf{QL}(\mathbb{F})$;

(ii) *there exists $n \in \omega$, and for all $\mathbf{d} = (d_1, \ldots, d_k) \in T_h^2$ there exist $u^{\mathbf{d}} \in W, \mathbf{c}^{\mathbf{d}} = (c_1^{\mathbf{d}}, \ldots, c_n^{\mathbf{d}}) \in (D_{u^{\mathbf{d}}})^n, \mathbf{a}^{\mathbf{d}} = (a_1^{\mathbf{d}}, \ldots, a_{k^*}^{\mathbf{d}}) \in (D_{u^{\mathbf{d}}})^{k^*}$, such that*

(0) $\forall \mathbf{c} \in D_n(\mathbf{c}^\varepsilon \leq_n \mathbf{c} \Rightarrow \mathbf{c} \leq_n \mathbf{c}^\varepsilon)$
  (and thus $\forall v \in W(u^\varepsilon \leq_0 v \Rightarrow v \leq_0 u^\varepsilon)$);

$\mathbb{F}_1:\ W = \{u_0\} \quad D = D_{u_0}$

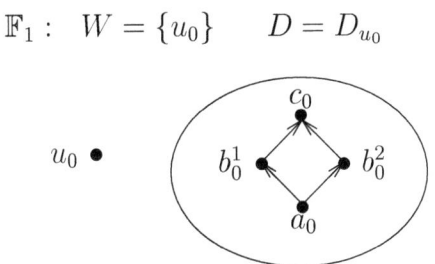

$\mathbb{F}_2:\ W = \{u_n : n \in \omega\}$

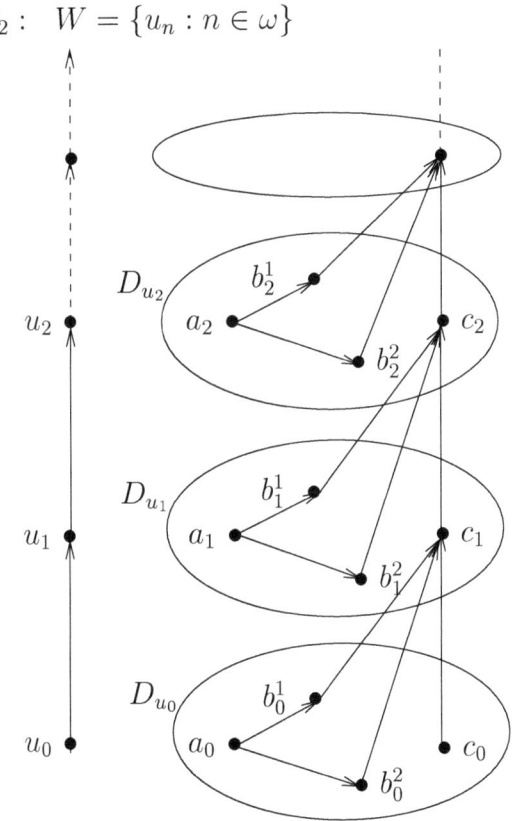

Figure 2.

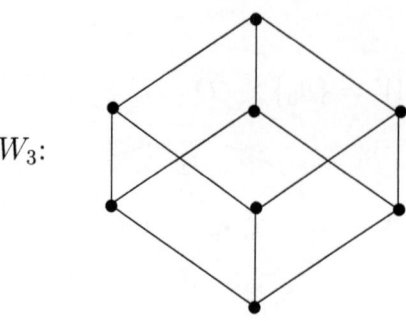

Figure 3.

(1) $(\mathbf{c^{d^-}}, \mathbf{a^{d^-}}) \leq_{n+k} (\mathbf{c^d}, (\mathbf{a^d})^-)$ for $\mathbf{d} = (d_1, \ldots, d_k), 0 < k \leq h$,
where $(\mathbf{a^d})^- = (a_1^d, \ldots, a_k^d)$
(and thus $u^\varepsilon \leq_0 u^{\mathbf{d}}, c^\varepsilon \leq_n c^{\mathbf{d}}$ for any $\mathbf{d} \in T_h^2$);

(2) $(\mathbf{c^d}, a_i^{\mathbf{d}}) \not\leq_{n+1} (\mathbf{c^{d'}}, a_j^{\mathbf{d'}})$ if $\mathbf{d} = (d_1, \ldots, d_i, \ldots, d_k)$,
$\mathbf{d'} = (d'_1, \ldots, d'_j, \ldots, d'_{k'}), d_i \neq d'_j$
(here $1 \leq i \leq k \leq h, 1 \leq j \leq k' \leq h$);

(Warning: $u^{\mathbf{d}}, a_i^{\mathbf{d}}, c_i^{\mathbf{d}}$ are not necessarily different)

(iii) there exist $n \in \omega$, $u^{\mathbf{d}} \in W$, $\mathbf{c^d} \in (D_{u^{\mathbf{d}}})^n$, $\mathbf{a^d} \in (D_{u^{\mathbf{d}}})^{k^*}$ as in (ii), and also $b^{\mathbf{d}} \in D_{u^\varepsilon}$ for all $\mathbf{d} = (d_1, \ldots, d_k) \in T_{h-1}^2, 0 \leq k < h$, such that

(3) $(\mathbf{c}^\varepsilon, b^{\mathbf{d}}) \not\leq_{n+1} (\mathbf{c^{d'}}, a_i^{\mathbf{d'}})$
for all $\mathbf{d'} = (d'_1, \ldots, d'_j) \in T_h^2, 1 \leq i \leq j \leq h$;

(4) $(\mathbf{c^{d'}}, a_1^{\mathbf{d'}}, \ldots, a_{k+1}^{\mathbf{d'}}) \leq_{n+k+1} (\mathbf{c}^\varepsilon, b^{\mathbf{d}_0}, \ldots, b^{\mathbf{d}_k})$
for any $\mathbf{d'} = (d_1, \ldots, d_k, d'_{k+1}, \ldots, d'_h)$.

2.2 Proof of KB-unseparability theorem We use Technical Lemma for $h = 3$. Let $u = u^{(0,1)}, (\mathbf{c}, a_1, a_2, a_3) = (\mathbf{c}^{(0,1)}, \mathbf{a}^{(0,1)}) \in (D_u)^{n+3}, u^d = u^{(0,1,d)}, (\mathbf{c}^d, a_1^d, a_2^d, a_3^d) = (\mathbf{c}^{(0,1,d)}, \mathbf{a}^{(0,1,d)}) \in (D_{u^d})^{n+3}$, for $d \in \{0,1\}$. Let $\mathbf{c} = (c_1, \ldots, c_n), \mathbf{c}^d = (c_1^d, \ldots, c_n^d)$. Take $b_i \in D_u, b_i^d \in D_{u^d}$ for $i \in \{1,2,3\}, d \in \{0,1\}$, such that

$$(\mathbf{c}^\varepsilon, b^\varepsilon, b^{(0)}, b^{(0,1)}) \leq_{n+3} (\mathbf{c}, b_1, b_2, b_3) \leq_{n+3} (\mathbf{c}^d, b_1^d, b_2^d, b_3^d)$$

for $d \in \{0,1\}$. Then $(\mathbf{c}, a_1, a_2, a_3) \leq_{n+3} (\mathbf{c}^d, a_1^d, a_2^d, a_3^d) \leq_{n+3} (\mathbf{c}, b_1, b_2, b_3)$ for $d \in \{0,1\}$ and $(\mathbf{c'}, a') \not\leq_{n+1} (\mathbf{c''}, a''), (\mathbf{c''}, a'') \not\leq_{n+1} (\mathbf{c'}, a')$ for any

$$(\mathbf{c'}, a') \in \{(\mathbf{c}, a_1), (\mathbf{c}^0, a_1^0), (\mathbf{c}^1, a_1^1), (\mathbf{c}^0, a_3^0)\},$$

$W = \{u_0\}$ $D = D_{u_0}$

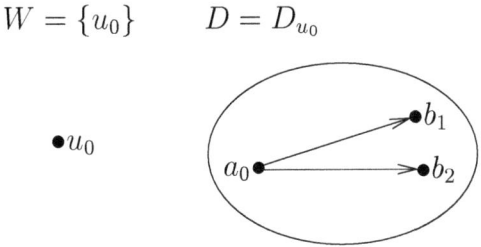

$W = \{u_0, v_0\}$

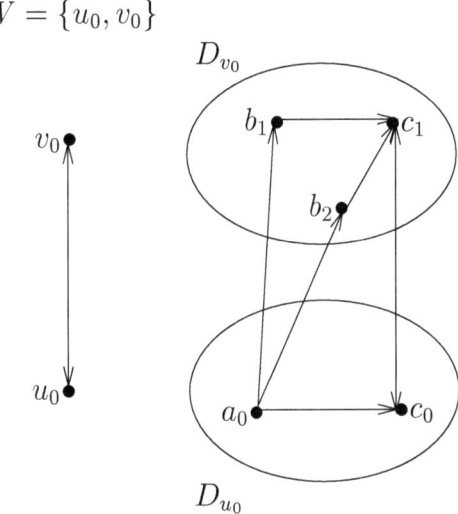

Figure 4.

$(\mathbf{c}'', a'') \in \{(\mathbf{c}, a_2), (\mathbf{c}^0, a_2^0), (\mathbf{c}^1, a_2^1), (\mathbf{c}^1, a_3^1)\}.$

Also $(\mathbf{c}, b_i) \not\leq_{n+1} (\mathbf{c}^d, a_j^d)$ for any $i, j \in \{1, 2, 3\}, d \in \{0, 1\}$ (see Fig.5, where directed arrows correspond to $<$, dashed lines in the upward direction correspond to $\leq$, and different signs near a^i, a_i^d mean that the corresponding $(\mathbf{c}, a_i), (\mathbf{c}^d, a_i^d)$ are incomparable).

So it follows that $a_i \notin \{c_1, \ldots, c_n, b_1, b_2, b_3\}$ for $i \in \{1, 2, 3\}$ — since $(\mathbf{c}, a_i) \leq_{n+1} (\mathbf{c}, b_i)$ and $(\mathbf{c}, b_i) \not\leq_{n+1} (\mathbf{c}, a_j)$ for $i, j \in \{1, 2, 3\}$. Similarly we have $a_i^d \notin \{c_1^d, \ldots, c_n^d, b_1^d, b_2^d, b_3^d\}$ for $d \in \{0, 1\}, i \in \{1, 2, 3\}$. Also $a_i \neq a_j$ for any $i \neq j (\in \{1, 2, 3\})$ — since $(\mathbf{c}, a_1) \not\leq_{n+1} (\mathbf{c}, a_2), (\mathbf{c}, a_1) \not\leq_{n+1} (\mathbf{c}^1, a_3^1), (\mathbf{c}, a_2) \not\leq_{n+1} (\mathbf{c}^0, a_3^0)$, and thus $(\mathbf{c}, a_1) \not\leq_{n+1} (\mathbf{c}, a_3), (\mathbf{c}, a_2) \not\leq_{n+1}$

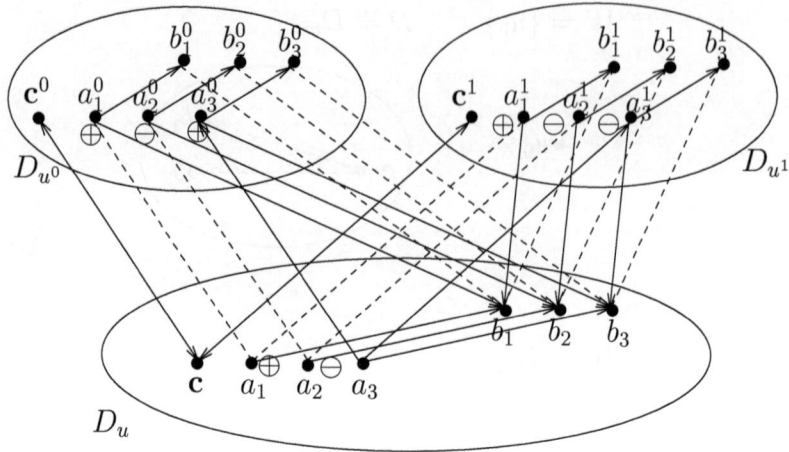

Figure 5.

$(\mathbf{c}, a_3)$.

We obtain that $(\mathbf{c}, a_1, a_2, a_3) <_{n+3} (\mathbf{c}, b_1, a_2, a_3) <_{n+3} (\mathbf{c}, b_1, b_2, a_3) <_{n+3} <_{n+3} (\mathbf{c}^0, b_1^0, b_2^0, a_3^0) <_{n+3} (\mathbf{c}^0, b_1^0, b_2^0, b_3^0)$, taking into account that $(\mathbf{c}^0, a_3^0)$
$\not<_{n+1} (\mathbf{c}, a_3)$ since $(\mathbf{c}^0, a_3^0) \not<_{n+1} (\mathbf{c}^1, a_3^1)$. Hence $\mathbf{PL}(D_{n+3}, \leq_{n+3})$ is a logic of height ≥ 5.

Also $(\mathbf{c}, a_1, a_2, a_3) \leq_{n+3} (\mathbf{c}', \mathbf{a}')$ for any $(\mathbf{c}', \mathbf{a}')$ from

$$X = \{(\mathbf{c}^0, a_1^0, a_2^0, b_3^0), (\mathbf{c}^0, a_1^0, b_2^0, a_3^0), (\mathbf{c}^0, b_1^0, a_2^0, a_3^0), \\ (\mathbf{c}^1, a_1^1, b_2^1, a_3^1), (\mathbf{c}^1, b_1^1, a_2^1, a_3^1)\},$$

and X is a $\leq_{n+3}$-antichain in D_{n+3} (since $(\mathbf{c}^0, a_3^0) \not<_{n+1} (\mathbf{c}^1, a_3^1), (\mathbf{c}^1, a_3^1)$
$\not<_{n+1} (\mathbf{c}^0, a_3^0)$, and $(\mathbf{c}^d, b_i^d) \not<_{n+1} (\mathbf{c}^e, a_i^e)$ for any $d, e \in \{0, 1\}, i \in \{1, 2, 3\})$.
Hence $\mathbf{PL}(D_{n+3}, \leq_{n+3})$ is a logic of width ≥ 5. ■

Note that Technical Lemma actually gives more information on Kripke bundles that refute the formulas $\neg\neg F_\varepsilon^h$ (for $h > 0$) than we need for the proof of our theorem. Thus one can obtain some versions of KB-unseparability theorem using other parameters than height or width. There are also versions formulated in terms of 'embeddings' in frames of certain logics.

3 Proof of technical lemma

3.1 Let A be a predicate formula, $(P_1^{(n_1)}, \ldots, P_m^{(n_m)})$ be the list of predicate symbols occurring in A, and $(z_1, \ldots, z_n), (Q_1^{(n+n_1)}, \ldots, Q_m^{(n+n_m)})$ be lists of different individual variables and predicate symbols non-occurring in A. Then $A^{[n]}$ is a substitution instance of A obtained by replacing every occurrence of $P_i(y_1,\ldots,y_{n_i})$ in A with $Q_i(z_1,\ldots,z_n,y_1,\ldots,y_{n_i})$.

LEMMA 6. *A formula A is strongly valid in a bundle $\mathbb{F}$ iff all $A^{[n]}$ are valid in $\mathbb{F}$ for $n \in \omega$.*

Proof. Let B be a substitution instance of A obtained by replacing every occurrence of $P_i(y_1,\ldots,y_{n_i})$ in A with $B_i(z_1,\ldots,z_n,y_1,\ldots,y_{n_i})$ $(1 \leq i \leq m)$ and $(x_1,\ldots,x_k),(z_1,\ldots,z_n,x_1,\ldots,x_k)$ be lists of free variables of formulas A and B respectively. Let ξ be a valuation in $\mathbb{F}$ such that $u \not\models B(c_1,\ldots,c_n,a_1,\ldots,a_k)$. Consider a valuation (for $Q_1^{(n+n_1)},\ldots,Q_m^{(n+n_m)}$) in $\mathbb{F}$ such that $v \models Q_i(d_1,\ldots,d_n,b_1,\ldots,b_{n_i})$ iff

$$(v \models B_i(d_1,\ldots,d_n,b_1,\ldots,b_{n_i}) \text{ under the valuation } \xi)$$

for $1 \leq i \leq m$. Then $u \not\models A^{[n]}(c_1,\ldots,c_n,a_1,\ldots,a_k)$. ∎

Note that Lemma 2 follows immediately from Lemma 6, since for any propositional formula A, $A^{[n]}$ is valid in $\mathbb{F}$ iff $A \in \mathbf{PL}(D_n, \leq_n)$.

A pre-ordered set $(W, \leq)$ is called *directed* if:

$$\forall u, v_1, v_2 \in W[(u \leq v_1, u \leq v_2) \Rightarrow \exists w \in W(v_1 \leq w, v_2 \leq w)].$$

It is well-known that $J \in \mathbf{PL}(W, \leq)$ iff $(W, \leq)$ is directed.

The following lemma is quite obvious:

LEMMA 7. *Let $(W, \leq)$ be a directed pre-ordered set. Let $u \leq w$, and w is a maximal element of W (i.e., $\forall v \geq w(v \leq w)$). Then $\forall v \geq u(v \leq w)$.*

3.1 Proof of Lemma 5, $(i) \Rightarrow (iii)$.

Assume that $\neg\neg F_\varepsilon^h \notin \mathbf{QL}(\mathbb{F})$. Take a valuation ξ in $\mathbb{F}, n \in \omega, u \in W, \mathbf{c} \in (D_u)^n$ such that $u \models \neg(F_\varepsilon^h)^{[n]}(\mathbf{c})$. Since $(D_n, \leq_n)$ is a pre-ordered set of a finite height $(\leq m)$, we can assume that $\mathbf{c}$ is a $\leq_n$-maximal element of D_n (and thus u is maximal in W).

LEMMA 8. *For any $\mathbf{d} = (d_1,\ldots,d_k) \in T_h^2, 0 \leq k \leq h$, there exist $u^{\mathbf{d}} \in W, (\mathbf{c}^{\mathbf{d}}, \mathbf{a}^{\mathbf{d}}) \in D_{u^{\mathbf{d}}}^{(n+k^*)}$ such that:*

(a) $u \leq_0 u^{\mathbf{d}}, \mathbf{c} \leq_n \mathbf{c}^{\mathbf{d}}$;

(b) $u^{\mathbf{d}} \not\models R_{d_i}(\mathbf{c}^{\mathbf{d}}, a_i^{\mathbf{d}}), u^{\mathbf{d}} \models R_{(1-d_i)}(\mathbf{c}^{\mathbf{d}}, a_i^{\mathbf{d}})$ *for* $1 \leq i \leq k$
(here $\mathbf{a}^{\mathbf{d}} = (a_1^{\mathbf{d}},\ldots,a_{k^*}^{\mathbf{d}})$);

(c) $u^{\mathbf{d}} \not\models (R_0(\mathbf{c}^{\mathbf{d}}, a_{k+1}^{\mathbf{d}}) \supset (F_{(\mathbf{d},1)}^h)^{[n]}(\mathbf{c}^{\mathbf{d}}, \mathbf{a}^{\mathbf{d}})) \vee$
$(R_1(\mathbf{c}^{\mathbf{d}}, a_{k+1}^{\mathbf{d}}) \supset (F_{(\mathbf{d},0)}^h)^{[n]}(\mathbf{c}^{\mathbf{d}}, \mathbf{a}^{\mathbf{d}}))$ *for* $0 \leq k < h$;

(d) $(\mathbf{c}^{\mathbf{d}^-}, \mathbf{a}^{\mathbf{d}^-}) \leq_{n+k} (\mathbf{c}^{\mathbf{d}}, (\mathbf{a}^{\mathbf{d}})^-)$,
where $\mathbf{d}^- = (d_1, \ldots, d_{k-1}), (\mathbf{a}^{\mathbf{d}})^- = (a_1^{\mathbf{d}}, \ldots, a_k^{\mathbf{d}})$, for $0 < k \leq h$.

Proof. By induction on k.
For $k = 0, \mathbf{d} = \varepsilon$ the claim holds, since

$$u \not\models \forall x_1[(R_0(\mathbf{c}, x_1) \supset (F_{(1)}^h)^{[n]}(\mathbf{c}, x_1)) \vee (R_1(\mathbf{c}, x_1) \supset (F_{(0)}^h)^{[n]}(\mathbf{c}, x_1))].$$

Let $k > 0$. By the inductive hypothesis,
$u^{\mathbf{d}^-} \not\models [R_{(1-d_k)}(\mathbf{c}^{\mathbf{d}^-}, a_k^{\mathbf{d}^-}) \supset (F_{\mathbf{d}}^h)^{[n]}(\mathbf{c}^{\mathbf{d}^-}, \mathbf{a}^{\mathbf{d}^-})]$, where $\mathbf{d}^- = (d_1, \ldots, d_{k-1})$.
If $k < h$, we have:

$$(F_{\mathbf{d}}^h)^{[n]}(\mathbf{c}^{\mathbf{d}^-}, \mathbf{a}^{\mathbf{d}^-}) = \forall x_{k+1}[(R_0(\mathbf{c}^{\mathbf{d}^-}, x_{k+1}) \supset (F_{(\mathbf{d},1)}^h)^{[n]}(\mathbf{c}^{\mathbf{d}^-}, \mathbf{a}^{\mathbf{d}^-}, x_{k+1})) \vee$$
$$\vee (R_1(\mathbf{c}^{\mathbf{d}^-}, x_{k+1}) \supset (F_{(\mathbf{d},0)}^h)^{[n]}(\mathbf{c}^{\mathbf{d}^-}, \mathbf{a}^{\mathbf{d}^-}, x_{k+1}))].$$

Hence there exists $(\mathbf{c}^{\mathbf{d}}, \mathbf{a}^{\mathbf{d}}) \in D^{n+k+1}$, satisfying the conditions $(c), (d)$, and $(b') : u^{\mathbf{d}} \models R_{(1-d_k)}(\mathbf{c}^{\mathbf{d}}, a_k^{\mathbf{d}})$. Also $u^{\mathbf{d}} \models R_{(1-d_i)}(\mathbf{c}^{\mathbf{d}}, a_i^{\mathbf{d}})$ for $i < k$, since $u^{\mathbf{d}^-} \models R_{(1-d_i)}(\mathbf{c}^{\mathbf{d}^-}, a_i^{\mathbf{d}^-}), (\mathbf{c}^{\mathbf{d}^-}, a_i^{\mathbf{d}^-}) \leq_{n+1} (\mathbf{c}^{\mathbf{d}}, a_i^{\mathbf{d}})$.
Finally, if $k = h$, then $(F_{\mathbf{d}}^h)^{[n]}(\mathbf{c}^{\mathbf{d}^-}, \mathbf{a}^{\mathbf{d}^-}) = \bigvee_{i=1}^{h} R_{d_i}(\mathbf{c}^{\mathbf{d}^-}, a_i^{\mathbf{d}^-})$.

Hence there exists $(\mathbf{c}^{\mathbf{d}}, \mathbf{a}^{\mathbf{d}})$ satisfying $(b), (d)$. Therefore, $u^{\mathbf{d}'} \not\models R_{d_i}(\mathbf{c}^{\mathbf{d}'}, a_i^{\mathbf{d}'})$ for any $i \leq j < h$ and $\mathbf{d}' = (d_1, \ldots, d_j)$, since $(\mathbf{c}^{\mathbf{d}'}, a_i^{\mathbf{d}'}) \leq_{n+1} (\mathbf{c}^{\mathbf{d}}, a_i^{\mathbf{d}})$. ∎

The condition (ii) of Lemma 5 is obviously satisfied for our $(\mathbf{c}^{\mathbf{d}}, \mathbf{a}^{\mathbf{d}})$. Namely, $(\mathbf{c}^{\mathbf{d}}, a_i^{\mathbf{d}}) \not\leq_{n+1} (\mathbf{c}^{\mathbf{d}'}, a_j^{\mathbf{d}'})$ for any $\mathbf{d} = (d_1, \ldots, d_i, \ldots, d_k)$, $\mathbf{d}' = (d_1', \ldots, d_j', \ldots, d_{k'}')$ ($i \leq k \leq h, j \leq k' \leq h$) such that $d_i \neq d_j'$, i.e., $d_j' = 1 - d_i$, since $u^{\mathbf{d}} \models R_{d_i'}(\mathbf{c}^{\mathbf{d}}, a_i^{\mathbf{d}}), u^{\mathbf{d}'} \not\models R_{d_j'}(\mathbf{c}^{\mathbf{d}'}, a_j^{\mathbf{d}'})$. Also $\mathbf{c}^{\varepsilon}$ is maximal in D_n since $\mathbf{c} \leq_n \mathbf{c}^{\varepsilon}$. Without any loss of generality we can assume that $\mathbf{c} = \mathbf{c}^{\varepsilon}$ (and $u = u^{\varepsilon}$).

Let $(\mathbf{c}'^{\mathbf{d}}, \mathbf{e}^{\mathbf{d}})$ be a $\leq_{n+k+1}$-maximal element of D_{n+k+1} such that $(\mathbf{c}^{\mathbf{d}}, \mathbf{a}^{\mathbf{d}}) \leq_{n+k+1} (\mathbf{c}'^{\mathbf{d}}, \mathbf{e}^{\mathbf{d}})$, for $\mathbf{d} = (d_1, \ldots, d_k) \in T_{h-1}^2, 0 \leq k < h$ (recall that D_{n+k+1} is a pre-ordered set of finite height). Then

$$(\mathbf{c}^{\mathbf{d}'}, a_1^{\mathbf{d}'}, \ldots, a_{k+1}^{\mathbf{d}'}) \leq_{n+k+1} (\mathbf{c}'^{\mathbf{d}}, \mathbf{e}^{\mathbf{d}})$$

for any $\mathbf{d} = (d_1, \ldots, d_k), \mathbf{d}' = (d_1, \ldots, d_k, d_{k+1}', \ldots, d_h'), \mathbf{a}^{\mathbf{d}'} = (a_1^{\mathbf{d}'}, \ldots, a_h^{\mathbf{d}'})$, by Lemma 7 (since $(\mathbf{c}^{\mathbf{d}}, \mathbf{a}^{\mathbf{d}}) \leq_{n+k+1} (\mathbf{c}^{\mathbf{d}'}, a_1^{\mathbf{d}'}, \ldots, a_{k+1}^{\mathbf{d}'}))$.

LEMMA 9. *There exist $b^{\mathbf{d}} \in D_u$ (for $\mathbf{d} \in T_{h-1}^2$) such that $(\mathbf{c}'^{\mathbf{d}}, \mathbf{e}^{\mathbf{d}}) \leq_{n+k+1} \leq_{n+k+1} (\mathbf{c}, b^{\mathbf{d}_0}, \ldots b^{\mathbf{d}_k})$ for any $\mathbf{d} = (d_1, \ldots, d_k) \in T_{h-1}^2, 0 \leq k < h$ (here $\mathbf{d}_i = (d_1, \ldots, d_i)$ for $i \leq k$, cf. Sect. 1.4).*

Proof. By induction on k (the length of $\mathbf{d}$).

For $k = 0, \mathbf{d} = \varepsilon$ we have $\mathbf{c}'^{\varepsilon} \leq_n \mathbf{c}$, since $\mathbf{c}$ is $\leq_n$-maximal in D_n; thus there exists $b_{\varepsilon} \in D_u$ such that $(\mathbf{c}'^{\varepsilon}, \mathbf{e}^{\varepsilon}) \leq_{n+1} (\mathbf{c}, b_{\varepsilon})$ (by (π), Sect. 1.2).

Let $k > 0$. By the inductive hypothesis, $(\mathbf{c}'^{\mathbf{d}^-}, \mathbf{e}^{\mathbf{d}^-}) \leq_{n+k} (\mathbf{c}, b^{\mathbf{d}_0}, \ldots, b^{\mathbf{d}_{k-1}})$, where $\mathbf{d}^- = (d_1, \ldots, d_{k-1})$. On the other hand, $(\mathbf{c}'^{\mathbf{d}}, (\mathbf{e}^{\mathbf{d}})^-) \leq_{n+k} (\mathbf{c}'^{\mathbf{d}^-}, \mathbf{e}^{\mathbf{d}^-})$, where $\mathbf{e}^{\mathbf{d}} = (e_1^{\mathbf{d}}, \ldots e_k^{\mathbf{d}}, e_{k+1}^{\mathbf{d}})$, $(\mathbf{e}^{\mathbf{d}})^- = (e_1^{\mathbf{d}}, \ldots, e_k^{\mathbf{d}})$ (by Lemma 7, since $(\mathbf{c}^{\mathbf{d}^-}, \mathbf{a}^{\mathbf{d}^-}) \leq_{n+k} (\mathbf{c}^{\mathbf{d}}, (\mathbf{a}^{\mathbf{d}})^-) \leq_{n+k} (\mathbf{c}'^{\mathbf{d}}, (\mathbf{e}^{\mathbf{d}})^-)$). Hence, by (π), there exists $b^{\mathbf{d}} \in D_u$ such that $(\mathbf{c}'^{\mathbf{d}}, \mathbf{e}^{\mathbf{d}}) \leq_{n+k+1} (\mathbf{c}, b^{\mathbf{d}_0}, \ldots, b^{\mathbf{d}_{k-1}}, b^{\mathbf{d}})$. ∎

The condition (iii) of Lemma 5 is satisfied for these $b^{\mathbf{d}} \in D_u$. Namely, $(\mathbf{c}^{\mathbf{d}'}, a_1^{\mathbf{d}'}, \ldots, a_{k+1}^{\mathbf{d}'}) \leq_{n+k+1} (\mathbf{c}'^{\mathbf{d}}, \mathbf{e}^{\mathbf{d}}) \leq_{n+k+1} (\mathbf{c}, b^{\mathbf{d}_0}, \ldots, b^{\mathbf{d}_k})$ for $\mathbf{d} = (d_1, \ldots, d_k), \mathbf{d}' = (d_1, \ldots, d_k, d'_{k+1}, \ldots, d'_h), 0 \leq k < h$.

Finally, assume that $(\mathbf{c}, b^{\mathbf{d}}) \leq_{n+1} (\mathbf{c}^{\mathbf{d}'}, a_i^{\mathbf{d}'})$ for some $\mathbf{d} = (d_1, \ldots, d_k)$, $\mathbf{d}' = (d'_1, \ldots, d'_i, \ldots, d'_j)(k < h, 1 \leq i \leq j \leq h)$.

Take $\mathbf{d}'' = (d_1, \ldots, d_k, 1 - d'_i, \ldots, d''_h)$. Then $(\mathbf{c}^{\mathbf{d}''}, a_{k+1}^{\mathbf{d}''}) \leq_{n+1} (\mathbf{c}, b^{\mathbf{d}}) \leq_{n+1}$
$\leq_{n+1} (\mathbf{c}^{\mathbf{d}'}, a_i^{\mathbf{d}'})$, which contradicts the condition $(ii)(2)$. ∎

Note that if instead of finiteness of height (i.e., the validity of some P_m) we assume only that $\mathbf{PL}(\mathbb{F}) \not\subseteq \mathbf{PL}(W_2)$, then we have to construct $b^{\mathbf{d}}$ in a more accurate way. But this can be done, using the directedness of all D_k.

3.2 Proof of Lemma 5, $(ii) \Rightarrow (i)$.

Consider a valuation ξ in $\mathbb{F}$ such that (for $d \in \{0, 1\}$):

$$u \models R_d(\mathbf{c}, a) \text{ iff } [(\mathbf{c}^{\mathbf{d}}, a_k^{\mathbf{d}}) \leq_{n+1} (\mathbf{c}, a) \text{ for some }$$
$$\mathbf{d} = (d_1, \ldots, d_{k-1}, 1 - d) \in (T_2^h \backslash \{\varepsilon\})].$$

LEMMA 10. *For any* $\mathbf{d} = (d_1, \ldots, d_k) \in T_h^2$:

$$u^{\mathbf{d}} \not\models (F_{\mathbf{d}}^h)^{[n]}(\mathbf{c}^{\mathbf{d}}, \mathbf{a}_k^{\mathbf{d}})$$

(here $\mathbf{a}_k^{\mathbf{d}} = (a_1^{\mathbf{d}}, \ldots, a_k^{\mathbf{d}})$ if $\mathbf{a}^{\mathbf{d}} = (a_1^{\mathbf{d}}, \ldots, a_{k^*}^{\mathbf{d}})$).

Proof. By induction on $(h - k)$.

If $k = h$, we have $u^{\mathbf{d}} \not\models R_{d_i}(\mathbf{c}^{\mathbf{d}}, a_i^{\mathbf{d}})$ for $i \leq h$, by the condition (2).

If $k < h$, we have

$$u^{\mathbf{d}} \not\models \forall x_{k+1}[(R_0(\mathbf{c}^{\mathbf{d}}, x_{k+1}) \supset (F_{(\mathbf{d},1)}^h)^{[n]}(\mathbf{c}^{\mathbf{d}}, \mathbf{a}_k^{\mathbf{d}}, x_{k+1})) \vee$$
$$\vee (R_1(\mathbf{c}^{\mathbf{d}}, x_{k+1}) \supset (F_{(\mathbf{d},0)}^h)^{[n]}(\mathbf{c}^{\mathbf{d}}, \mathbf{a}_k^{\mathbf{d}}, x_{k+1}))],$$

since $u^{(\mathbf{d},d)} \models R_{1-d}(\mathbf{c}^{(\mathbf{d},d)}, a_{k+1}^{(\mathbf{d},d)}), u^{(\mathbf{d},d)} \not\models (F_{(\mathbf{d},d)}^h)^{[n]}(\mathbf{c}^{(\mathbf{d},d)}, \mathbf{a}_k^{(\mathbf{d},d)}, a_{k+1}^{(\mathbf{d},d)})$, and $(\mathbf{c}^{\mathbf{d}}, \mathbf{a}^{\mathbf{d}}) = (\mathbf{c}^{\mathbf{d}}, \mathbf{a}_k^{\mathbf{d}}, a_{k+1}^{\mathbf{d}}) \leq_{n+k+1} (\mathbf{c}^{(\mathbf{d},d)}, \mathbf{a}_k^{(\mathbf{d},d)}, a_{k+1}^{(\mathbf{d},d)})(= (\mathbf{c}^{(\mathbf{d},d)}, \mathbf{a}_{k+1}^{(\mathbf{d},d)}))$ for $d \in \{0, 1\}$, by the condition (1) (recall that $k^* = k+1$ for $k < h$). ∎

Hence, for $\mathbf{d} = \varepsilon$ we conclude: $u^\varepsilon \not\models (F_\varepsilon^h)^{[n]}(\mathbf{c}^\varepsilon)$. Therefore, by the condition (0):

$$\forall u \geq u^\varepsilon \forall \mathbf{c} \in (D_u)^n [(\mathbf{c}^\varepsilon \leq_n \mathbf{c}) \Rightarrow u \not\models (F_\varepsilon^h)^{[n]}(\mathbf{c})].$$

Thus $u^\varepsilon \models (\neg F_\varepsilon^h)^{[n]}(\mathbf{c}^\varepsilon)$ and $u^\varepsilon \not\models (\neg\neg F_\varepsilon^h)^{[n]}(\mathbf{c}^\varepsilon)$. ∎

Acknowledgements

The author is grateful to anonymous referees who drew his attention to the paper by K. Nagaoka and E. Isoda.

The work on this paper was partly supported by Russian Foundation for Basic Research.

BIBLIOGRAPHY

[1] D. M. Gabbay, V. Shehtman and D. Skvortsov. *Quantification in Nonclassical Logic*, forthcoming.
[2] S. Ghilardi. Presheaf semantics and independence results for some non-classical first-order logics. *Archive for Math. Logic*, **29**, 125–136, 1989.
[3] S. Ghilardi. Incompleteness results in Kripke semantics. *Journal of Symb. Logic*, **56**, 517–538, 1991.
[4] S. Ghilardi. Quantified extensions of canonical propositional intermediate logics. *Studia Logica*, **51**, 195-214, 1992.
[5] K. Nagaoka and E. Isoda. Incompleteness results in Kripke bundle semantics. *Math. Logic Quarterly*, **43**, 485-498, 1997.
[6] H. Ono. Model extension theorem and Craig's interpolation theorem for intermediate predicate logics. *Reports on Math. Logic*, **15**, 41–58, 1983.
[7] V. Shehtman and D. Skvortsov. Semantics of non-classical first order predicate logics. In *Mathematical Logic*, P. Petkov, ed., 105–116. Plenum Press, N.Y., 1990 (Proc. of Summer school and conference in math. logic Heyting'88).
[8] D. Skvortsov. An incompleteness result for intermediate predicate logics. *Journ. Symb. Logic*, **56**, 1145-1146, 1991.

All-Russian Institute of Scientific and Technical Information,
Molodogvardejskaja 22, korp.3, kv. 29
121351, Moscow
Russia
E-mail skvortsov@viniti.ru, skvortsov@lpcs.math.msu.ru

22
On IF Modal Logic and its Expressive Power

TERO TULENHEIMO

ABSTRACT. *Independence-friendly (IF) modal logic* is defined and its expressive power is compared with that of basic modal logic. It is shown that for all $k \geq 1$, **IFML**$[k]$ (IF modal logic speaking of k accessibility relations) is relative to arbitrary k-ary modal structures (models with k accessibility relations) strictly more expressive than the corresponding basic modal logic **ML**$[k]$ (Theorem 5). An analogous result holds for the tense logics **TL**$[k]$ and **IFTL**$[k]$ (Theorem 7). On the other hand, over the class of unary temporal structures with linear frames, the tense logics **IFTL**$[1]$ and **TL**$[1]$ have the same expressive power (Theorem 9). Likewise the expressive powers of **IFML**$[1]$ and **ML**$[1]$ coincide on unary linear modal structures (Theorem 10). By contrast, for $k \geq n \geq 2$ and for k-ary models of whose accessibility relations n are linear, IF modal (*resp.* tense) logic is more expressive than the basic modal (*resp.* tense) logic (Theorem 11).

1 Introduction

My aim in this paper is to study whether allowing a modal operator to be logically independent of a specified subset of syntactically superordinate operators results in a modal language with extra expressive power as compared with basic modal logic.

First I recall the definitions of basic modal and basic tense logics of k accessibility relations (**ML**$[k]$ *resp.* **TL**$[k]$), and sketch their semantics in game-theoretical terms (*Section* 2). I then move on to obtain a language **IFML**$[k]$ from **ML**$[k]$ by allowing formulae with operator prefixes $O_1...O_{n-1}(O_n/W)$, where the intended interpretation of the indication "$/W$" is that the operator O_n is *logically (informationally) independent* of those preceding operators that are identified by indices in the set $W \subseteq \{1,...,n-1\}$. This intuition is made precise by making use of games of imperfect information in defining the semantics for the language **IFML**$[k]$ (*Section* 3).

In *Section* 4 it is shown that the expressive power of **IFML**$[k]$ is strictly greater than that of **ML**$[k]$ relative to the class of all k-ary modal structures.

A similar result about tense logics is proven for all k-ary temporal structures, whose accessibility relations are, as is usual, required to be in particular irreflexive and transitive. Further, it is shown that IF tense logic and basic tense logic evaluated relative to one *linear* accessibility relation have the same expressive power. From the proof of this result the analogous result follows for modal logics **IFML**[1] and **ML**[1]. Finally, it is shown that these coincidence results are *not* generalizable to modal (temporal) structures with at least two linear accessibility relations.

2   Basic modal logic

We shall consider the two basic modalities for each *modality type* $i < \omega$: $\Diamond_i$, $\Box_i$. Occurrences of such operators are marked with unique indices (positive integers), as for instance in the string $\Diamond_{3,1}\Box_{1,2}\Box_{1,3}\Diamond_{2,4}p$. The formulae of *basic modal logic* **ML**[k] are generated from a fixed class **prop** of propositional atoms, and negations of propositional atoms, by the rules of closure under conjunction ($\wedge$), disjunction ($\vee$), and application of any of the (unary) modal operators $\Diamond_i$ and $\Box_i$ for $i < k$.

The semantics for **ML**[k] is defined in terms of k-ary modal structures $\mathcal{M} = (D, R_0, ..., R_{k-1}, \mathfrak{h})$, where D is a non-empty domain, the R_i are binary relations on D, and $\mathfrak{h}$ is a function assigning a subset of D to each propositional atom from **prop**. A pair $(\mathcal{M}, d)$ consisting of a modal structure and an element of its domain is called a *pointed modal structure*. The formulae of **ML**[k] are in effect evaluated relative to such pointed modal structures. We write $\mathcal{M} \models^+ \varphi[d]$ to make the judgement that the formula $\varphi \in $ **ML**[k] is true in $\mathcal{M}$ at d; and we write $\mathcal{M} \models^- \varphi[d]$ to say that $\varphi \in $ **ML**[k] is not true in $\mathcal{M}$ at d. The semantic clauses for formulae of **ML**[k] are completely standard:

- $\mathcal{M} \models^+ p[d] \iff d \in \mathfrak{h}(p)$
- $\mathcal{M} \models^+ \neg p[d] \iff d \notin \mathfrak{h}(p)$
- $\mathcal{M} \models^+ \varphi \vee \psi[d] \iff$ for some $\theta \in \{\varphi, \psi\}$: $\mathcal{M} \models^+ \theta[d]$
- $\mathcal{M} \models^+ \varphi \wedge \psi[d] \iff$ for every $\theta \in \{\varphi, \psi\}$: $\mathcal{M} \models^+ \theta[d]$

For each modality type $i < k$:

- $\mathcal{M} \models^+ \Diamond_i(\varphi)[d] \iff$ for some $c \in D$ with $R_i(d, c)$: $\mathcal{M} \models^+ \varphi[c]$.
- $\mathcal{M} \models^+ \Box_i(\varphi)[d] \iff$ for every $c \in D$ with $R_i(d, c)$: $\mathcal{M} \models^+ \varphi[c]$.

ML[0] evaluated over degenerate modal structures $(\{d_0\}, \mathfrak{h})$ is simply *Propositional logic*.

The semantics for a formula $\varphi \in \mathbf{ML}[k]$ can alternatively be given in terms of a semantical game $G(\varphi, \mathcal{M}, d)$ between two players, *Abélard* ($\forall$) and *Héloïse* ($\exists$). A play of this game is defined in a straightforward fashion:

- If $\varphi \in \{p, \neg p\}$, no move is made. If $\varphi = p$ and $d \in \mathfrak{h}(p)$ or if $\varphi = \neg p$ and $d \notin \mathfrak{h}(p)$, then *Héloïse* wins the play and *Abélard* loses. Otherwise *Abélard* wins and *Héloïse* loses.

- If $\varphi = (\theta \wedge \psi)$ [*resp.* $\varphi = (\theta \vee \psi)$], then *Abélard* picks out a conjunct $\chi \in \{\theta, \psi\}$ [*resp. Héloïse* picks out a disjunct $\chi \in \{\theta, \psi\}$], and the play goes on as $G(\chi, \mathcal{M}, d)$.

- Let $i < k$. If $\varphi = \Box_i(\psi)$ [*resp.* $\varphi = \Diamond_i(\psi)$], then *Abélard* [*resp. Héloïse*] picks out, if possible, a state d' with $R_i(d, d')$; and the play continues as $G(\psi, \mathcal{M}, d')$. If however such a choice is not possible (i.e. if d is R_i-maximal), then *Héloïse* [*resp. Abélard*] wins and *Abélard* [*resp. Héloïse*] loses.

A *strategy* of the player $j \in \{\forall, \exists\}$ in the game $G(\varphi, \mathcal{M}, d)$ is any function providing a move for this player j in any situation where it is according to the rules of the game j's turn to move. A strategy is *winning* (in short, a *w.s.*) for the player whose strategy it is, if using this strategy the player wins any play of the game $G(\varphi, \mathcal{M}, d)$ - irrespectively, that is, of the moves of his or her opponent.

Truth and falsity for $\mathbf{ML}[k]$-formulae are then defined game-theoretically as follows:

- $\mathcal{M} \models^+_{\text{GTS}} \varphi[d] \iff$ there exists a w.s. for *Héloïse* in $G(\varphi, \mathcal{M}, d)$.

- $\mathcal{M} \models^-_{\text{GTS}} \varphi[d] \iff$ there exists a w.s. for *Abélard* in $G(\varphi, \mathcal{M}, d)$.

It is easy to check that the above recursively defined semantics to $\mathbf{ML}[k]$ and the game-theoretical semantics coincide:[1]

$\mathcal{M} \models^+ \varphi[d] \iff \mathcal{M} \models^+_{\text{GTS}} \varphi[d]$

$\mathcal{M} \models^- \varphi[d] \iff \mathcal{M} \models^-_{\text{GTS}} \varphi[d]$

The semantics to $\mathbf{ML}[k]$ could well be defined more generally than is done above, by allowing evaluation relative to *tuples* of elements from a specified Cartesian product $D_0 \times ... \times D_{N_k} - 1$ with $1 \leq N_k \leq k$. Then for each (binary) accessibility relation R_i a component $D_{\pi(i)}$ of the Cartesian product would

---

[1] In general, the directions from left (the recursive definition) to right (the GTS-definition) require *Axiom of Choice*.

have to be associated by some surjection $\pi : \{0, ..., k-1\} \to \{0, ..., N_k - 1\}$. With an eye to many applications such a more flexible semantics would be useful. If for instance we'd wish to consider two modality types such as physical necessity and temporality, we would probably find it desirable to have one co-ordinate for possible worlds and another for time in our semantics. On the other hand, for instance in multiagent epistemic logic, k equals the number of the agents while the parameter N_k has the value 1, since formulae of that logic are evaluated relative to single possible worlds.

The semantics of *basic tense logic* $\mathbf{TL}[k] = \mathbf{ML}[2k]$ employs $2k$-ary modal structures
$$\mathcal{M} = (D, R_0, ..., R_{2k-1}, \mathfrak{h})$$
with the characteristic features that: **(i)** for each $j < k$, the relation R_{j+k} is the *converse* of the relation R_j; **(ii)** the accessibility relations are *irreflexive partial orders*, in other words irreflexive and transitive binary relations. The motivation for the restriction **(i)** is that the inverse of an operator O (which by definition always exists in a tense logic) must make use of the converse of precisely the accessibility relation associated with O. The feature **(ii)** is required, because at least irreflexivity and transitivity are thought of as essential to the temporal *earlier than* -relation. In connection with $\mathbf{TL}[k] = \mathbf{ML}[2k]$, we shall by convention write $\Diamond_i^{-1}$ resp. $\Box_i^{-1}$ for the operators $\Diamond_{i+k}$ resp. $\Box_{i+k}$ ($i < k$).

$2k$-ary modal structures $\mathcal{M}$ satisfying the above conditions **(i)** and **(ii)** are termed k-ary *temporal* structures. By contrast, we shall call $2k$-ary modal structures meeting the requirement **(i)** but not the requirement **(ii)** *quasi-temporal* structures.

3 IF modal logic and its uniformity interpretation

I proceed to define syntax and semantics to an extension of basic modal logic I call *IF modal logic of k modality types*, or $\mathbf{IFML}[k]$. Thereby more dependencies and independencies among modal operators will become expressible than are expressible within $\mathbf{ML}[k]$.

3.1 The language

The language of $\mathbf{IFML}[k]$ is simply this:

$$\mathbf{IFML}[k] := \mathbf{ML}[k] \cup \{O_1...O_{n-1}(O_n/W)\varphi : \varphi \in \mathbf{ML}[k], n \geq 1\},$$

where:

- for all $j \in \{1, ..., n\}$, O_j is one of the modal operators $\Diamond_{i,j}$, $\Box_{i,j}$ for some $i < k$.

- W is a (possibly empty) subset of the interval $[1, n-1]$ of natural numbers.

The intuitive reading of the expression $(\Diamond_{i,n}/W)$ is "there is a state s_n such that $R_i(s_{n-1}, s_n)$, *independently* of the states s_j with $j \in W$". The intuitive reading of $(\Box_{i,n}/W)$ is analogous.

The class of well-formed formulae for IF modal logic could well be defined more generally. In principle for every modal operator (and even for every propositional connective), its own independence indication "/W" could be added, where W would be an (empty or nonempty) set of indices of other operators or connectives appearing in the same formula. In the present paper, however, we stick to the definition of **IFML**[k] given above, as already this class turns out to be interesting as compared with the basic modal logic **ML**[k].

IF *tense* logic of k temporal modality types, or **IFTL**[k], is defined to be the logic **IFML**[$2k$] with the same stipulations about notation as were introduced above in connection with basic tense logic **TL**[k] = **ML**[$2k$].

3.2 Semantics intuitively

The semantics of **IFML**[k] is going to be given game-theoretically. For each formula $\varphi \in$ **IFML**[k], a semantical game is defined [relative to a k-ary pointed modal structure $(\mathcal{M}, d)$] by extending the above game-theoretical definition of the semantics for **ML**[k] to cover also formulae of the form $O_1...O_{n-1}(O_n/W)\varphi$. The key semantic idea in dealing with these new formulae is to make use of the game-theoretical notion of *informational independence* in defining how to evaluate the operator (O_n/W). Roughly, it will be required that a winning strategy of a player be *uniform* in those components of its arguments that correspond to operators O_j with $j \in W$, i.e. that distinct moves made for these operators O_j - of which the operator O_n is declared to be *independent* - in different possible plays of the semantical game never suffice by themselves for prompting a distinct move for O_n. Whichever move is made for O_n after *one* distribution of moves corresponding to the operators O_j with $j \in W$, is made - other things being equal, i.e. as long as the choices for the operators with $j \notin W$ are the same - for *all* distributions that can come up in the course of a play of the game. Different plays corresponding to evaluating the block $O_1...O_{n-1}$ are partitioned into equivalence classes so that plays h and h' are in the same class if and only if they differ at most for moves corresponding to operators with indices from W (that is, agree at least for indices from $\{1, ..., n-1\}\backslash W$); and the relevant player's strategy can only be winning if it respects these information sets in the sense that it gives histories from the same the set a common value.

We now move on to define the semantics in a precise way. We need to have the notion of substring of an **IFML**[k]-formula available. The class $Sub(\varphi)$ of *proper substrings* of an **IFML**[k]-formula is defined recursively for strings φ including all **IFML**[k]-formulae - but *not all* of which are formulae of this logic. (We cannot simply define here the class of sub*formulae* of **IFML**[k], since substrings $O_i...O_{n-1}(O_n/W)\varphi$ - which we wish to have always in the class $Sub(\varphi)$ - are by the syntax *not* formulae of **IFML**[k], if W is not contained in $\{i,...,n-1\}$.) The clauses for (negated) propositional atoms and Boolean connectives are standard: for an atom $p \in$ **prop**, $Sub(p) = Sub(\neg p)$ is empty; and $Sub(\varphi \land \psi) = Sub(\varphi \lor \psi) = \{\varphi, \psi\} \cup Sub(\varphi) \cup Sub(\psi)$. If $W \subseteq \{1,...,n-1\}$ and $1 \leq i \leq n$, we define $Sub(O_i...O_{n-1}(O_n/W)\varphi) = \{O_{i+1}...O_{n-1}(O_n/W)\varphi\} \cup Sub(O_{i+1}...O_{n-1}(O_n/W)\varphi)$. Finally, by stipulation the string FAIL is in $Sub(\varphi)$ for all $\varphi \in$ **IFML**[k]. The strings in the class $Sub(\varphi) \cap$ **IFML**[k] are said to be *subformulae* of φ.

3.3 The semantical games

A semantical game $G_A(\varphi, \mathcal{M}, d) =$

$$\langle \{\forall, \exists\}, H, Z, P, \{u_\forall, u_\exists\}, \{I_\forall, I_\exists\} \rangle$$

in extensive normal form is associated with each triple $(\varphi, \mathcal{M}, d)$ consisting of a formula φ of **IFML**[k]; a k-ary modal structure $\mathcal{M} = (D, R_0, ..., R_{k-1}, \mathfrak{h})$; and an element $d \in D$. *Positions* in this game are pairs (ψ, a) from the set

$$A = (Sub(\varphi) \cup \{\varphi\}) \times (dom(\mathcal{M}) \cup \{\star\}).$$

The object $\star$ goes together with the substring FAIL; the pair (FAIL,$\star$) is a position that can be chosen in situations where there were otherwise no choice complying with the game rules available at all. The game $G_A(\varphi, \mathcal{M}, d)$ is a game between two players, *Héloïse* (or $\exists$) and *Abélard* (or $\forall$). The players are said to be *opponents* of each other. The set H of *plays* (or, *histories*) of $G_A(\varphi, \mathcal{M}, d)$ is a set of *finite sequences of positions* (ψ, a), defined recursively (on the subformula structure of φ) as follows:[2]

- $(\varphi, d) \in H$.

- If the last position in $h \in H$ is $(\theta \lor \psi, a)$, then $h^\frown(\theta, a) \in H$ and $h^\frown(\psi, a) \in H$.
  It is *Héloïse* who makes a choice from $\{(\theta, a), (\psi, a)\}$ to extend h. We write $P(h) = \exists$ to indicate that the move corresponding to the history h is made by $\exists$.

---

[2] If $h = (\alpha_1, ..., \alpha_m)$ and $\alpha \in A$, we write "$h^\frown \alpha$" (read: h *extended* by α) to denote the sequence $(\alpha_1, ..., \alpha_m, \alpha)$.

- The case where the last position of $h \in H$ is $(\theta \wedge \psi, a)$ is defined analogously, the only difference being that here it is *Abélard* who makes the choice: $P(h) = \forall$.

For $i < k$:

- If the last position in $h \in H$ is from $\{((\Diamond_{i,n}/W)\psi, a), (\Diamond_{i,n}\psi, a)\}$, then for all $a' \in dom(\mathcal{M})$ with $R_i(a, a')$, $h^\frown(\psi, a') \in H$; if no such $a' \in dom(\mathcal{M})$ exists, then $h^\frown(\texttt{FAIL}, \star) \in H$. Further, $P(h) = \exists$.

- The definitions for $\Box_{i,n}/W$ and $\Box_{i,n}$ are like those for $\Diamond_{i,n}/W$ and $\Diamond_{i,n}$, except that here it is $\forall$'s turn to move.

All sequences in the set H are finite. The set Z of *terminal histories* is defined simply as the set of maximally long histories, i.e. sequences from H which cannot be extended by any position so as to yield a sequence in H. By definition, then, the last position in a terminal history has as its substring-component a propositional atom, the negation of a propositional atom, or the label $\texttt{FAIL}$. The function $P : H\backslash Z \to \{\forall, \exists\}$ constructed above simultaneously with the definition of H will be called the *player function*.

The *utility functions* $u_\forall$ and $u_\exists$ for the two players of the game are maps from the class Z of terminal histories to the values 1 (a win) and -1 (a loss) as follows. For any $h \in Z$, $u_\exists(h) = 1$ and $u_\forall(h) = -1$, if the last position of h is of the form (p, a) and $a \in \mathfrak{h}(p)$, or is of the form $(\neg p, a)$ and $a \notin \mathfrak{h}(p)$, or is the position $(\texttt{FAIL}, \star)$ chosen by $\forall$. Otherwise $u_\forall(h) = 1$ and $u_\exists(h) = -1$.

The *information partitions* $I_\forall$ and $I_\exists$ for the two players still need be defined. For an arbitrary history $h = ((\varphi_0, a_0), ..., (\varphi_m, a_m))$, we call the sequence $pr_1(h) = (\varphi_0, ..., \varphi_m)$ its *left projection*; and the sequence $pr_2(h) = (a_0, ..., a_m)$ its *right projection*. If $(s_0, ..., s_n)$ is any finite sequence and $i \leq n$, we write $(s_0, ..., s_n)[i]$ for its member s_i. Now the sets $P^{-1}(\{\exists\})$ and $P^{-1}(\{\forall\})$ are partitioned into equivalence classes under the following equivalence relations $\sim_\exists$ and $\sim_\forall$, respectively.

- $h_1 \sim_\exists h_2 \iff$
  (for some $a_1, a_2 \in dom(\mathcal{M})$, the last position in h_1 is $((\Diamond_{i,n}/W)\theta, a_1)$ and the last position in h_2 is $((\Diamond_{i,n}/W)\theta, a_2)$; and $pr_1(h_1) = pr_1(h_2)$; and for all $j \notin W$, $pr_2(h_1)[j] = pr_2(h_2)[j]$) **or**
  ($h_1 = h_2$ and its last position is of the form (ψ, a) for some $\psi \in \{\Diamond_{i,n}\theta, (\theta \vee \chi)\}$ and $a \in dom(\mathcal{M})$).

The condition under which $h_1 \sim_\forall h_2$ holds is similar to the condition for $\sim_\exists$ and is obtained from it by replacing the occurrences of the symbols

"◇" and "∨" by the symbols "□" and "∧", respectively. The promised partitions of $P^{-1}(\{\exists\})$ and $P^{-1}(\{\forall\})$ then are:

$$I_\exists = \{[h]_{\sim_\exists} : h \in H\backslash Z\} \text{ and } I_\forall = \{[h]_{\sim_\forall} : h \in H\backslash Z\}.$$

The sets $I_\exists$ and $I_\forall$ themselves are called *information partitions*, and their members (i.e. the cells of these partitions) are called *information sets*.

The definition of the semantical game

$$G_A(\varphi, \mathcal{M}, d) = \langle \{\forall, \exists\}, H, Z, P, \{u_\forall, u_\exists\}, \{I_\forall, I_\exists\}\rangle$$

in its extensive form has become completed. By definition it is a zero-sum game. And it is a game of *imperfect* information.

3.4 Imperfect information modelling logical independence

Intuitively the idea with information partitions is that histories in the same cell of the information partition of the player $j \in \{\exists, \forall\}$ are indistinguishable to this player: unless there is some way of ruling out a subset of such an information set as histories that definitely cannot have been reached in the course of a play of the game, any move j makes after one history belonging to this set, j must be able to make after *any* history from this set, in order for j's strategy for making these moves to be winning: j's winning strategy has to agree on all these histories. To get a better grasp of these partitions, let us take an example.

Example Let $\mathcal{M} = (\{0,1,2,3,4\}, <, \mathfrak{h})$ be a modal structure, where $<$ is the ordering of the natural numbers 0,1,2,3,4 by magnitude, the assignment $\mathfrak{h}$ being arbitrary. Consider the formulae:

(a) $\varphi := \Box_1 \Diamond_2 \Diamond_3 / \{1\} p.$
(b) $\psi := \Box_1 \Diamond_2 \Diamond_3 / \{2\} p.$

Up to indicating the relevant information partitions, the extensive form of both semantical games, $G_A(\varphi, \mathcal{M}, 0)$ and $G_A(\psi, \mathcal{M}, 0)$, is depicted as below.

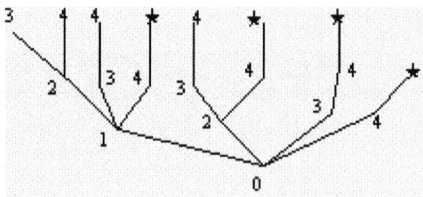

Let the histories of length 2 that appear in the picture be listed as $h_1, ..., h_7$ from left to right. Now in the case of the formula **(a)**, the information partition for *Héloïse* contains, in addition to four singletons each consisting of a history of length 1, the following cells whose members are histories of length 2:

$$\{h_1\}, \{h_2, h_4\}, \{h_3, h_5, h_6\} \text{ and } \{h_7\}.$$

In the same cell are those histories of length 2 that have their last member in common. By contrast, in the case of the formula (b), the cells of *Héloïse*'s information partition consisting of histories of length 2 are these:

$$\{h_1, h_2, h_3\}, \{h_4, h_5\}, \{h_6\} \text{ and } \{h_7\}.$$

Here the histories in the same cell share their second member (i.e. the first member chosen after the initial position $(\psi, 0)$).

3.5 Uniformity and plans of action

To make use of the games $G_A(\varphi, \mathcal{M}, d)$ in defining semantics to **IFML**[k], we need the notion of strategy of a player. A *strategy* of the player $j \in \{\exists, \forall\}$ is simply any function

$$f_j : P^{-1}(\{j\}) \to A.$$

Observe that strategies are *not* components in the extensive form of a semantical game. Given that $h \in H$ is a history on which a strategy f_j is defined, we say that the move $f_j(h)$ given by the strategy is *legal*, if $h \frown f_j(h) \in H$, i.e. when the extension to h determined by f_j remains inside the set H of histories of the relevant game. Otherwise the move is said to be *illegal*.

We shall employ the following terminology. If s, s_1 and s_2 are sequences such that $s = s_1 \frown s_2$, then s_1 is said to be an *initial segment* of s. If s_2 is non-empty, s_1 is a *proper* initial segment of s. When S is a set of sequences, we write $Cl(S)$ for the set of initial segments of members of S, i.e. for the closure of S under forming initial segments.

A strategy f_j of the player j is *winning* (is a w.s.), if there exists a subset $W \subseteq Z$ of terminal histories satisfying the following four conditions:

- **(a)** If $h \in Cl(W)$ and $P(h)$ is j, then $h \frown f_j(h) \in Cl(W)$.

- **(b)** If $h \in Cl(W)$ and $P(h)$ is the opponent of j, then for every $u \in A$ such that $h \frown u \in H$, $h \frown u \in Cl(W)$.

- **(c)** For every $h, h' \in Cl(W)$: if $h, h' \in I \in I_j$, then $f_j(h) = f_j(h')$.

- **(d)** Every $h \in W$ is a win for j.

The condition **(a)** simply says that $Cl(W)$ is closed under applications of the strategy f_j; and **(c)** expresses the 'uniformity condition' for this strategy. The condition **(b)** in meant to ensure that j wins against any move of his opponent. Together these conditions are meant to ensure that only the plays that are "reached" given that j uses the strategy f_j are relevant for its being winning: the only plays in which the strategy f_j ever needs be used, are those from $Cl(W)$.

A set $W \subseteq Z$ which thus establishes that a strategy is winning, is called a *plan of action*. We shall say that f_j is a winning strategy *based on* W. We observe that W guarantees the existence of a minimal (though not necessarily unique) partial function from $P^{-1}(\{j\})$ to A which specifies a move for all combinatorially possible moves of the opponent of j and always yields a win for j.

If there exists a set $W \subseteq Z$ satisfying the conditions **(a)**, **(b)** and **(c)** but not necessarily **(d)**, we say that a corresponding *strategy* f_j is *based on the plan of action* W. But unless the condition **(d)** is satisfied, such a strategy is *not* winning.

Fact 1 *At most one of the players has a w.s. in a game* $\Gamma = G_A(\varphi, M, d)$.

Proof. Assume for contradiction that strategies $f_\exists$ and $f_\forall$ are both winning in Γ. Fix W resp. W' as corresponding plans of action. We show that there is a history h such that $h \in W$ and $h \in W'$, which yields a contradiction by the condition **(d)**.

We construct h in steps. For the initial position $\alpha_1 = (\varphi, d)$, we trivially have that $(\alpha_1) \in Cl(W)$ and $(\alpha_1) \in Cl(W')$. Assume then that we have reached $(\alpha_1, ..., \alpha_m)$ such that $(\alpha_1, ..., \alpha_m) \in Cl(W)$ and $(\alpha_1, ..., \alpha_m) \in Cl(W')$. If $P((\alpha_1, ..., \alpha_m)) = \forall$, then by **(a)**, $(\alpha_1, ..., \alpha_m, f_\forall(\alpha_1, ..., \alpha_m)) \in Cl(W')$. But then by the clause **(b)** $(\alpha_1, ..., \alpha_m, f_\forall(\alpha_1, ..., \alpha_m)) \in Cl(W)$. The same reasoning applies if $P((\alpha_1, ..., \alpha_m)) = \exists$. Hence in a finite number of steps a terminal history $h = (\alpha_1, ..., \alpha_N)$ is reached that is both in W and in W'; and so by **(d)** h is a win for both players, which is impossible. ∎

For instance Ariel Rubinstein ([9], p. 66) suggests that it is natural to define strategies in terms of plans of action — i.e. as functions which essentially specify the actions of a player only for such histories that he or she may actually confront when applying this plan. This is in contrast to the typical game-theoretical practice of requiring that strategies be defined for all histories of the game, irrespective of whether such a history can be reached when following that strategy.

3.6 Semantics to IFML

Now that the plethora of requisite definitions all are given, we can state the semantics for **IFML**[k]. Let $\mathcal{M} = (D, R_0, ..., R_{k-1}, \mathfrak{h})$ be any k-ary modal structure; and let $d \in D$. Then:

- $\varphi \in \textbf{IFML}[k]$ is true in $\mathcal{M}$ at d, in symbols $\mathcal{M} \models^+ \varphi[d]$, iff there exists a w.s. for *Héloïse* in $G_A(\varphi, \mathcal{M}, d)$.

- $\varphi \in \textbf{IFML}[k]$ is false in $\mathcal{M}$ at d, in symbols $\mathcal{M} \models^- \varphi[d]$, iff there exists a w.s. for *Abélard* in $G_A(\varphi, \mathcal{M}, d)$.

- $\varphi \in \textbf{IFML}[k]$ is non-determined in $\mathcal{M}$ at d, in symbols $\mathcal{M} \models^0 \varphi[d]$, iff $\mathcal{M} \not\models^+ \varphi[d]$ and $\mathcal{M} \not\models^- \varphi[d]$.

The semantics just given to **IFML**[k] will be referred to as its *uniformity interpretation*.

It is clear that for the subclass **ML**[k] of **IFML**[k], the game-theoretical semantics defined in *Section* 2 coincides with the above game-theoretical semantics.

For $k = 0$, **IFML**[k] and **ML**[k] are the same languages. Hence, **IFML**[0] = **ML**[0] evaluated over degenerate modal structures $(\{d_0\}, \mathfrak{h})$ is indeed *Propositional logic*. For $k \geq 1$, **ML**[k] is a proper subclass of **IFML**[k].

The formulae of IF *tense logic* **IFTL**[k] = **IFML**[2k] are evaluated relative to k-ary temporal structures. (Recall that k-ary temporal structures are by definition a subclass of all $2k$-ary modal structures.)

4 Comparing the Expressive Powers of ML and IFML

If $\mathcal{L}$ and $\mathcal{L}'$ are modal logics (IF or not) and $\mathcal{C}$ is a class of modal structures on which the semantics of these logics is defined, it is said that $\mathcal{L}$ is *embeddable* in $\mathcal{L}'$ over $\mathcal{C}$ (symbolically $\mathcal{L} \leq_\mathcal{C} \mathcal{L}'$), if for each formula $\varphi \in \mathcal{L}$ there exists a formula $\psi_\varphi \in \mathcal{L}'$ such that for all $\mathcal{M} \in \mathcal{C}$ and all $d \in dom(\mathcal{M})$,

$$\varphi \text{ is true in } \mathcal{M} \text{ at } d \iff \psi_\varphi \text{ is true in } \mathcal{M} \text{ at } d.$$

If $\mathcal{L}$ is embeddable in $\mathcal{L}'$ over $\mathcal{C}$ but not *vice versa*, it is said that $\mathcal{L}'$ has *greater expressive power* than $\mathcal{L}$ over $\mathcal{C}$ (in symbols $\mathcal{L} <_\mathcal{C} \mathcal{L}'$). If $\mathcal{L}$ is embeddable in $\mathcal{L}'$ and *vice versa*, then it is said that they have *the same expressive power* over $\mathcal{C}$ (symbolically $\mathcal{L} =_\mathcal{C} \mathcal{L}'$).

4.1 Basic facts about expressive power

Let us write $\mathcal{C}_k$ for the class of all k-ary modal structures. For an arbitrary $k < \omega$, we have completely trivially the following embeddability result:

Fact 2 *Over C_k, $\mathbf{ML}[k]$ is embeddable in $\mathbf{IFML}[k]$.*

This is just because $\mathbf{ML}[k]$ is a subclass of $\mathbf{IFML}[k]$: any formula $\varphi \in \mathbf{ML}[k]$ is its own translation in $\mathbf{IFML}[k]$.

We say that the class $\mathbf{IFML}_{det}[k] :=$

$$\{\varphi \in \mathbf{IFML}[k] : for\ all\ \mathcal{M} \in \mathcal{C}_k\ and\ all\ d \in dom(\mathcal{M}),$$

$$\mathcal{M} \models^+ \varphi[d]\ or\ \mathcal{M} \models^- \varphi[d]\}$$

is the *determined fragment* of the logic $\mathbf{IFML}[k]$. It is straightforward to show that relative to the class $\mathcal{C}_k$ of all k-ary modal structures, the expressive powers of $\mathbf{IFML}_{det}[k]$ and $\mathbf{ML}[k]$ are the same:

THEOREM 1. $\mathbf{IFML}_{det}[k] =_{\mathcal{C}_k} \mathbf{ML}[k]$.

Proof. All formulae of $\mathbf{ML}[k]$ are determined (everywhere either true or false), and they are also (directly by syntactical criteria) formulae of $\mathbf{IFML}[k]$, so $\mathbf{ML}[k] \leq_{\mathcal{C}_k} \mathbf{IFML}_{det}[k]$. For the other direction, let $\varphi \in \mathbf{IFML}_{det}[k]$, $\mathcal{M} \in \mathcal{C}_k$ and $d \in dom(\mathcal{M})$ all be arbitrary. We have to find a formula $\psi_\varphi \in \mathbf{ML}[k]$, satisfying that

there is a *w.s.* for $\exists$ in $G_A(\varphi, \mathcal{M}, d) \iff \mathcal{M} \models^+_{\mathrm{GTS}} \psi_\varphi[d]$.

Now if in particular $\varphi \in \mathbf{ML}[k]$, then we may clearly take ψ_φ to be φ itself. If, again, φ is of the form $O_1...O_{n-1}(O_n/W)\chi$ with $\chi \in \mathbf{ML}[k]$, then put $\psi_\varphi := O_1...O_{n-1}O_n\chi$. What can be done in greater ignorance can be done in lesser ignorance: clearly if some $f_\exists$ is a w.s. for $\exists$ in $G_A(\varphi, \mathcal{M}, d)$, the same strategy is a w.s. for her in $G_A(\psi_\varphi, \mathcal{M}, d)$. Assume then that there is no w.s. for $\exists$ in $G_A(\varphi, \mathcal{M}, d)$. But then, by the assumption of determinacy, there is a w.s. for $\forall$ in $G_A(\varphi, \mathcal{M}, d)$. And a w.s. for $\forall$ in $G_A(\varphi, \mathcal{M}, d)$ is clearly also a w.s. for him in $G_A(\psi_\varphi, \mathcal{M}, d)$. This means, in turn, that there is no w.s. for $\exists$ in $G_A(\psi_\varphi, \mathcal{M}, d)$. ∎

We shall need subsequently the notion of *bisimulation* and the simple *Invariance Lemma* relating the relation of bisimilarity to the capacity of $\mathbf{ML}[k]$-formulae to distinguish k-ary modal structures.

DEFINITION 2. (*Bisimulation*) Let $\mathcal{M} = (D, R_0, ..., R_{k-1}, \mathfrak{h})$ and $\mathcal{N} = (D', R'_0, ..., R'_{k-1}, \mathfrak{h}')$ be k-ary modal structures. A *bisimulation* between the modal structures $\mathcal{M}$ and $\mathcal{N}$ is a binary relation $\equiv_{D,D'} \subseteq D \times D'$, satisfying the following three conditions **(1)**, **(2)** and **(3)**.

(1) *Atomic Harmony:*
$d \equiv_{D,D'} d' \implies$ for all $p \in \mathbf{prop}$: $d \in \mathfrak{h}(p) \iff d' \in \mathfrak{h}'(p)$.

(2) *Zigzag Forwards:* for all $i < k$,
$d \equiv_{D,D'} d'$ and $R_i(d, c) \implies \exists c'(R'_i(d', c')$ and $c \equiv_{D,D'} c')$.

(3) *Zigzag Backwards:* for all $i < k$,
$d \equiv_{D,D'} d'$ and $R'_i(d', c') \implies \exists c(R_i(d, c)$ and $c \equiv_{D,D'} c')$.

A *bisimilarity* $B_{D,D'}$ between modal structures $\mathcal{M}$ and $\mathcal{N}$ is the *union* of all bisimulations between $\mathcal{M}$ and $\mathcal{N}$. The pointed modal structures $(\mathcal{M}, d)$ and $(\mathcal{N}, d')$ are said to be *bisimilar*, if there is a bisimilarity $B_{D,D'}$ such that $(d, d') \in B_{D,D'}$. Further, we say that the pointed modal structures $(\mathcal{M}, d)$ and $(\mathcal{N}, d')$ are **ML**[k]*-equivalent*, if for all $\varphi \in \mathbf{ML}[k]$,

$$\mathcal{M} \models^+ \varphi[d] \iff \mathcal{N} \models^+ \varphi[d'].$$

The notion of **IFML**[k]-equivalence of $(\mathcal{M}, d)$ and $(\mathcal{N}, d')$ is defined similarly.

For formulae of **ML**[k] the following *Invariance Lemma* holds. The lemma says that bisimilarity preserves truth in **ML**[k] in the sense that if $(\mathcal{M}, d)$ and $(\mathcal{N}, d')$ are bisimilar, then these pointed modal structures are **ML**[k]-equivalent.

LEMMA 3. *(Invariance Lemma) Let* $\mathcal{M} = (D, R_0, ..., R_{k-1}, \mathfrak{h})$ *and* $\mathcal{N} = (D', R'_0, ..., R'_{k-1}, \mathfrak{h}')$ *be k-ary modal structures. Then for all* $\varphi \in \mathbf{ML}[k]$, *all $d \in D$ and all $d' \in D'$:*

$$(d, d') \in B_{D,D'} \implies (\mathcal{M} \models^+ \varphi[d] \iff \mathcal{N} \models^+ \varphi[d']).$$

Proof. A straightforward induction on the complexity of the formula $\varphi \in \mathbf{ML}[k]$. ∎

4.2 Expressive power over arbitrary modal structures

Let us next consider the modal logics **ML**[1] and **IFML**[1], which hence both are evaluated relative to precisely one accessibility relation. Let $\mathcal{C}_1$ be the class of all unary modal structures. We prove that **IFML**[k] has a greater expressive power than **ML**[k] relative to $\mathcal{C}_1$:

LEMMA 4. $\mathbf{ML}[1] <_{\mathcal{C}_1} \mathbf{IFML}[1]$.

Proof. Construct unary modal structures $\mathcal{M}$ and $\mathcal{M}'$ as follows. Put $\mathcal{M} = (D, R, \mathfrak{h})$ and $\mathcal{M}' = (D', R', \mathfrak{h}')$, where
- $D = D' = \{a, b, c, d, e, f\}$ (6 distinct elements)
- $R = R' = \{(a, b), (a, c), (b, d), (b, e), (c, e), (c, f)\}$

- $\mathfrak{h}(q) = \{e\}$, $\mathfrak{h}'(q) = \{d, f\}$

Define $\equiv \, \in Pow(D \times D')$ to be the following relation:

$$\{(a,a), (b,b), (c,c), (b,c), (c,b)(d,e), (f,e)(e,d), (e,f)\}.$$

It is easy to check that $\equiv$ is a bisimulation between the unary modal structures $\mathcal{M}$ and $\mathcal{M}'$. Hence, by *Invariance Lemma*, the pointed modal structures $(\mathcal{M}, a)$ and $(\mathcal{M}', a)$ are **ML**[1]-equivalent. On the other hand, we observe that the **IFML**[1]-formula $\Box_{0,1} \Diamond_{0,2} / \{1\} q$ is true in $\mathcal{M}$ at a, but is non-determined in $\mathcal{M}'$ at a. It follows that **IFML**[1] is not embeddable in **ML**[1]. By appealing to Fact 2 the proof is completed. ∎

Notice that the frames (D, R) and (D', R') of the modal structures employed in the proof of Lemma 4 are isomorphic, in fact identical.[3] Hence, by the proof, **IFML**[1] is able to distinguish properties properly of models (as opposed to properties of frames) which are not distinguishable using **ML**[1].

Remark. The result, provided by Lemma 4, that **IFML**[1] has greater expressive power than **ML**[1] over the class $\mathcal{C}_1$ of all unary modal structures, is *not* a corollary to the completely trivial fact that there exist **IFML**[1]-formulae (such as the formula $\Box_{0,1} \Diamond_{0,2} / \{1\} q$ employed in the proof of Lemma 4) that are non-determined in certain pointed modal structures. Obviously for such a formula φ there cannot exist an **ML**[1]-formula ψ_φ sharing the truth-value with φ in all structures $(\mathcal{M}, d)$ with $\mathcal{M} \in \mathcal{C}_1$, $d \in dom(\mathcal{M})$ - this is because all **ML**[1]-formulae are everywhere determined but φ by definition isn't. Had we defined embeddability of a logic $\mathcal{L}$ to a logic $\mathcal{L}'$ over a class $\mathcal{C}$ by the condition that for each $\varphi \in \mathcal{L}$ there must exist $\psi \in \mathcal{L}'$ such that for all $\mathcal{M} \in \mathcal{C}$ and for all $d \in dom(\mathcal{M})$:

φ and ψ are both true, both false, or both non-determined in $\mathcal{M}$ at d

and not by the condition we actually used (i.e. that for all $\varphi \in \mathcal{L}$ there is $\psi \in \mathcal{L}'$ that is *true* in precisely the same structures $(\mathcal{M}, d)$ in which φ is *true*), this trivial fact would indeed suffice for showing that **IFML**[1] is strictly more expressive than **ML**[1] over $\mathcal{C}_1$. But it does *not* follow from the fact that the truth-value of a given non-determined $\varphi \in$ **IFML**[1] never coincides in all structures $(\mathcal{M}, d)$ with the truth-value of any $\psi_\varphi \in$ **ML**[1] that there could not exist a formula $\psi_\varphi \in$ **ML**[1] such that φ and ψ_φ have all their *verifying* models in common - it being an additional question what

---

[3]Frames $(D, R_0, ..., R_{k-1})$ and $(D', R'_0, ..., R'_{k-1})$ are said to be *isomorphic* when there exists a bijection $f : D \to D'$ such that for all $d_1, d_2 \in D$ and for all $i < k$, $R_i(d_1, d_2) \iff R'_i(f(d_1), f(d_2))$.

would happen with those structures in which $\varphi \in$ **IFML**[1] were not true (i.e. were false or non-determined). Hence - our definition of embeddability being what it is - the sole existence of non-determined **IFML**[1]-formulae is by itself an insufficient ground for inferring Lemma 4.

Lemma 4 has the straightforward consequence that for any positive $k < \omega$, **IFML**[k] is strictly more expressive than **ML**[k] over the class $\mathcal{C}_k$ of arbitrary k-ary modal structures:

THEOREM 5. *For any positive* $k < \omega$, **ML**[k] $<_{\mathcal{C}_k}$ **IFML**[k].

By Theorem 5, we have for all positive $k < \omega$ that **TL**[k] $<_{\mathcal{C}_{2k}}$ **IFTL**[k]. By the proof of Lemma 4, this result can be improved to hold for the proper subclass QT_k of $\mathcal{C}_{2k}$ consisting of all k-ary *quasi-temporal* structures:

COROLLARY 6. *For any positive* $k < \omega$, **TL**[k] $<_{QT_k}$ **IFTL**[k].

Proof. The statement of the corollary will trivially follow, if this statement is established for the value $k = 1$. To do so, let $\mathcal{M} = (M, R, \mathfrak{h})$ and $\mathcal{M}' = (M', R', \mathfrak{h}')$ be the unary modal structures defined in the proof of Lemma 4 above. Then the structures $\mathcal{M}^+ = (M, R, R^{-1}, \mathfrak{h})$ and $\mathcal{M}'^+ = (M', R', R'^{-1}, \mathfrak{h}')$, obtained from $\mathcal{M}$ and $\mathcal{M}'$ by adding to these the converses of their accessibility relations, are in the class QT_1. But in fact the relation $\equiv$ defined in the proof of Lemma 4 is a bisimulation between the binary modal structures $\mathcal{M}^+$ and $\mathcal{M}'^+$. And the formula $\Box_{0,1} \Diamond_{0,2}/\{1\}q$ is true in $\mathcal{M}^+$ at a, but non-determined in $\mathcal{M}'^+$ at a. Hence the statement follows. ∎

However, the structures $\mathcal{M}^+$ and $\mathcal{M}'^+$ appearing in the proof of Corollary 6 are *not* genuine temporal structures - in particular the relations R and R' are not transitive. But it is possible to prove a genuine tense logical analogue to Theorem 5. Furthermore, the proof can be given so that the structures witnessing the result have isomorphic frames. Let us write $\mathcal{T}_k$ for the class of all k-ary temporal structures. (Recall that these are quasi-temporal strctures whose all accessibility relations are irreflexive partial orders.)

THEOREM 7. *For any positive* $k < \omega$, **TL**[k] $<_{\mathcal{T}_k}$ **IFTL**[k].

Proof. Let $(C_1, <_1)$ be the set of positive rationals ordered by magnitude. Write $C_2 := C_1 \times \{0\}$, and let $(C_2, <_2)$ be an isomorphic copy of $(C_1, <_1)$. Finally, let t, t_1, t_2 be three *irrational* numbers. Define a frame (D, R) as follows:
- $D := \mathbb{Q} \cup C_2 \cup \{t, t_1, t_2\}$;
- $R := < \cup [\mathbb{Q} \times \{t_1\}] \cup [(\mathbb{Q} \backslash C_1) \times (C_2 \cup \{t_2\})] \cup <_2 \cup$

$[C_2 \times \{t_2\}] \cup \{(t, t_1), (t, t_2)\}$;
where $\mathbb{Q}$ stands for the set of all rational numbers, and $<$ for their order by magnitude. Finally, let **prop** $= \{q\}$, and define assignments $\mathfrak{h}$ and $\mathfrak{h}'$ by:
- $\mathfrak{h}(q) = \{1, 1'\}$; $\mathfrak{h}'(q) = \{0\}$;

where $1'$ is the image of 1 under a fixed isomorphism that establishes the isomorphism between $(C_1, <_1)$ and $(C_2, <_2)$. The relation R is clearly irreflexive and transitive. The picture below illustrates the temporal structures $\mathcal{M} = (D, R, R^{-1}, \mathfrak{h})$ and $\mathcal{N} = (D, R, R^{-1}, \mathfrak{h}')$.

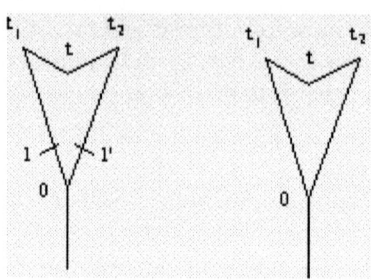

The structures $\mathcal{M}$ and $\mathcal{N}$ have a *common* frame (D, R, R^{-1}); and so *a fortiori* their frames are isomorphic. Furthermore, the temporal structures in question are bisimilar, as is witnessed by the relation

$\{(t, t)\} \cup [\{t_1, t_2\} \times \{t_1, t_2\}] \cup \{(1, 0), (1', 0)\} \cup$
$[\{x : 1 <_1 x <_1 t_1 \text{ or } 1' <_2 x <_2 t_2\} \times \{x : 0 <_1 x <_1 t_1 \text{ or } 0 <_2 x <_2 t_2\}] \cup$
$[\{x : x <_1 1 \text{ or } x <_2 1'\} \times \{x : x <_1 0\}]$.

Hence the pointed temporal structures $(\mathcal{M}, t)$ and $(\mathcal{N}, t)$ cannot be distinguished by any **TL**[1]-formula. On the other hand, the **IFTL**[1]-formula

$$\varphi := \Box_1 \Diamond_2^{-1} / \{1\} q$$

is non-determined in $\mathcal{M}$ at t; but it is true in $\mathcal{N}$ at t. Hence we may conclude that **IFTL**[1] is more expressive than **TL**[1] over the class of all unary temporal structures. The statement of the theorem then trivially follows. ∎

4.3 Expressive power and linear frames

We shall say that a k-ary modal structure $\mathcal{M}$ is *linear in n dimensions*, if $\mathcal{M}$ has $n \leq k$ accessibility relations which are irreflexive linear orders. A k-ary *temporal* structure is said to be linear in n dimensions, if it has $2n \leq 2k$ accessibility relations which are irreflexive linear orders. We adopt the convention of writing **LO**[k, n] for the class of k-ary modal structures linear in n dimensions. By stipulation we write $\mathbf{LO}[k] := \mathbf{LO}[k,k]$. The class of k-ary temporal structures linear in n dimensions we denote by writing $\mathbf{LO}_{temp}[k,n]$; and we stipulate that $\mathbf{LO}_{temp}[k] := \mathbf{LO}_{temp}[k,k]$. In what follows we always assume that the linear orders considered are *irreflexive*.

It can be shown that over unary temporal structures $(D, R, R^{-1}, \mathfrak{h})$ with an arbitrary irreflexive linear order R, the expressive powers of the tense logics $\mathbf{TL} := \mathbf{TL}[1]$ and $\mathbf{IFTL} := \mathbf{IFTL}[1]$ coincide. The following lemma is useful in the proof. (It holds for *all* logics $\mathbf{IFML}[k]$ and *all* k-ary modal structures.)

LEMMA 8. *Let* $W \subseteq [1, n-1]$; *and let* $U = [K+1, n-1] \subseteq W$, *where* $K \notin W$ $(0 \le K \le n-1)$. *For all* k-*ary modal structures* $\mathcal{M}$; *all* $d \in \text{dom}(\mathcal{M})$; *and all* $\varphi \in \mathbf{IFML}[k]$, *we have:*

Héloïse has a w.s. in $G(O_1...O_{n-1}(O_n/W)\varphi, \mathcal{M}, d) \iff$
Héloïse has a w.s. in $G(O_1...O_{n-1}(O_n/U)\varphi, \mathcal{M}, d)$.

Proof. The set $W \subseteq [1, n-1]$ clearly has a unique presentation as a disjoint union $W = V' \cup [K+1, n-1]$, where $K \notin W$. The implication from left to right is trivial. (What can be done in greater ignorance can be done in lesser ignorance.) Assume then that there is a w.s. (f) for *Héloïse* in $\Gamma = G_A(O_1...O_{n-1}(O_n/U)\varphi, \mathcal{M}, d)$. We observe that for any $i \in \{1, ..., n-1\}$, if *Héloïse* knows the state s_i chosen for the operator O_i (which is the case exactly when $i \notin W$), then in order to evaluate O_n, the choices made to evaluate operators with indices $j < i$ are of no relevance. The 'route' by which the play of the game has proceeded to the choice of this state s_i is irrelevant. And K precisely is *not* contained in W, nor in $U = [K+1, n-1]$. This is why *Héloïse*'s w.s. (f^*) in $\Gamma^* = G_A(O_1...O_{n-1}(O_n/W)\varphi, \mathcal{M}, d)$ can be constructed using her w.s. (f) in Γ in such a way that for each cell I of the information partition corresponding to O_n in Γ^* we choose some such subset I' of I which is a cell of the information partition corresponding to O_n in Γ (in general *Axiom of Choice* is needed for choosing I'); and then we define f^* on I by the value f gives to the histories from I'. By the observation just made, this value really works also for all other cells of Γ contained in the cell I of Γ^*, and thereby legally extends all histories from I. ∎

THEOREM 9. $\mathbf{TL} =_{\mathbf{LO}_{temp}[1]} \mathbf{IFTL}$.

Proof. $\mathbf{TL}$ is trivially embeddable in $\mathbf{IFTL}$ over any class of structures. For the other direction, consider $\mathbf{IFTL}$-formulae of the form $\psi = O_1...O_{n-1}(O_n/W)\varphi$. By Lemma 8 the set W referred to in ψ can be assumed to be of the form $[K+1, n-1]$. Indeed, we may suppose that $W = \{1, ..., n-1\}$, since clearly

$$O_1...O_K O_{K+1}...O_{n-1}(O_n/\{K+1, ..., n-1\})\varphi \iff O_1...O_K \chi,$$

if χ is a **TL**-translation of the **IFTL**-formula $O_{K+1}...O_{n-1}(O_n/\{K+1, ..., n-1\})\varphi$. The cases $W = \emptyset$ and $O_n \in \{\Box, \Box^{-1}\}$ are trivial. The laborious part of the proof is to show that the **IFTL**-formulae of the form $O_1...O_{n-1}(O_n/\{1, ..., n-1\})\varphi$ (where $O_n \in \{\Diamond, \Diamond^{-1}\}$, $n \geq 2$) can be translated to **TL**. By symmetry it suffices to consider formulae of the form $O_1...O_{n-1}(\Diamond_n/\{1, ..., n-1\})\varphi$. On their part the proof is by dividing the proof into subcases according to the distribution of operators in the prefix $O_1...O_{n-1}$ of the formula ψ, and taking into account different properties the linear order in the model may have. The relevant subcases are:

- **Case 1.** For all $i \in \{1, ..., n-1\}$: $O_i \in \{\Diamond_i\}$.
- **Case 2.** For all $i \in \{1, ..., n-1\}$: $O_i \in \{\Diamond_i, \Box_i^{-1}\}$; and $\Box^{-1}$ appears at least once in the block.
- **Case 3.** There is last occurrence of $\Diamond^{-1}$ in the block, and after it all operators are from $\{\Diamond, \Box^{-1}\}$.
- **Case 4.** There is last occurrence of $\Box$ in the block, and after it all operators are from $\{\Diamond, \Box^{-1}\}$.

Case 1 is trivial and cases 3 and 4 can be handled by reducing them to case 2. The case 2 is proven by showing that the (negation normal form of the) formula $\theta_\psi :=$

$$O_1...O_{n-1}\Diamond_n\varphi \wedge \neg\Diamond^{-1}(\neg\varphi \wedge \Box\neg\varphi \wedge \Box^{-1}\Diamond\varphi) \wedge \neg(\neg\varphi \wedge \Box\neg\varphi \wedge \Box^{-1}\Diamond\varphi)$$

counts as a required **TL**-translation of ψ. (The string θ_ψ is not in **TL**, as it contains $\neg$ prefixed to non-atomic formulae; it is a formula of the logic obtained from **TL** by closing **TL** under negation, and it has a unique negation normal form which is in **TL**.) The idea with the two latter conjuncts in θ_ψ is that they ban the possibility that the first conjunct ($O_1...O_{n-1}\Diamond_n\varphi$) would be true at t while there was a point $s \leq t$ such that

- φ is false at s and at all points later than s; but
- for all $s' < s$ there is s'' such that $s' < s'' < s$ and φ is true at s''.

The properties of the order that need be taken into account when proving the four cases are whether or not the point of evaluation is discrete to the left *resp.* to the right, and if yes, how many discrete moves can be made in the respective direction; and also whether the domain has or does not have a minimum *resp.* a maximum relative to its linear order. The relevant different cases concerning the order of course prompt fresh disjuncts to the translation in question. ∎

The proof of Theorem 9, when spelled out in detail, shows that the expressive powers of **ML**[1] and **IFML**[1] coincide on unary modal structures with an arbitrary linear order:

THEOREM 10. **ML**[1] $=_{\mathbf{LO}[1]}$ **IFML**[1].

Let us finally consider binary modal structures linear in both dimensions. We may observe that on the one hand, Theorem 9 shows that **IFML**[2] does *not* have more expressive power than **ML**[2] over the *subclass* $\mathbf{LO}_{temp}[1]$ of **LO**[2] consisting of *such* binary modal structures linear in both dimensions whose two accessibility relations are *converses to each other* (i.e. over unary linear *temporal* structures). On the other hand, by Theorem 5 we know that **IFML**[2] *has* a greater expressive power than **ML**[2] over *arbitrary* binary modal structures. The case we now move on to consider is hence an intermediate one: since

$$\mathbf{LO}_{temp}[1] \subset \mathbf{LO}[2] \subset \mathcal{C}_2$$

is a chain of proper inclusions, the theorems 5 and 9 do not suffice for deciding the question whether the relation **ML**[2] $<_{\mathbf{LO}[2]}$ **IFML**[2] holds or not. The following theorem implies that in fact **IFML**[2] is already over **LO**[2] more expressive than **ML**[2]. More generally, the theorem states that the equalities **ML**[k] $=_{\mathbf{LO}[k,n]}$ **IFML**[k] and **TL**[k] $=_{\mathbf{LO}_{temp}[k,n]}$ **IFTL**[k] which by the theorems 9 and 10 are established for the value $k = n = 1$, cannot be generalized to any values $k \geq n \geq 2$.

THEOREM 11. *For all $n \geq 2$ and for all $k \geq n$, we have:*
(a) **ML**[k] $<_{\mathbf{LO}[k,n]}$ **IFML**[k].
(b) **TL**[k] $<_{\mathbf{LO}_{temp}[k,n]}$ **IFTL**[k].

Proof. We shall show that each of the claims **(a)** and **(b)** holds for the values $k := 2$ and $n := 2$. The results then trivially generalize to all $k \geq n \geq 2$. Now let $\mathcal{M} = (D, <_0, <_1, \mathfrak{h})$ and $\mathcal{M}' = (D', <'_0, <'_1, \mathfrak{h}')$ be binary modal structures whose components are given as follows.
- D is the set of rational numbers, and $<_0 = <_1$ is the order of rationals by magnitude.

Let $C = D \times \{0\}$, and let $(C, <_0^+, <_1^+)$ be an isomorphic copy of $(D, <_0, <_1)$. By construction $C \cap D = \emptyset$. Then put:
- $D' := D \cup C$
- $<'_0 := <_0 \cup <_0^+ \cup (D \times C)$
- $<'_1 := <_1 \cup <_1^+ \cup (C \times D)$

Finally, assume for simplicity that **prop** $= \{\top\}$ and put $\mathfrak{h}(\top) = D$ and $\mathfrak{h}'(\top) = D'$. The structures $\mathcal{M}$ and $\mathcal{M}'$ can respectively be depicted as below.

It is clear that the relation $D \times D'$ is a bisimulation between $\mathcal{M}$ and $\mathcal{M}'$. Fix then points $\zeta \in D$ and $\xi \in D' \cap C$. It follows by *Invariance Lemma* that the pointed modal structures $(\mathcal{M},\zeta)$ and $(\mathcal{M}',\xi)$ cannot be distinguished by any formula of **ML**[2]. But we observe that the **IFML**[2]-formula $\square_{0,1} \lozenge_{1,2}/\{1\}\top$ is true in $\mathcal{M}'$ at ξ (*Héloïse* can choose for $\lozenge_{1,2}$ any $\xi'' \in D' \cap D$, as any such ξ'' will be a $<'_1$-successor of any possible ξ' with $\xi <'_0 \xi'$ chosen by *Abélard*); but on the other hand this formula is non-determined in $\mathcal{M}$ at ζ (not true, since no $\zeta'' \in D$ is a $<_1$-successor to all $\zeta' \in D$ with $\zeta <_0 \zeta'$; not false, as for all $\zeta' \in D$ there is some $\zeta'' \in D$ such that $\zeta <_0 \zeta' <_1 \zeta''$). It follows that **IFML**[2] is not embeddable in **ML**[2]. By appealing to Fact 2, the claim **(a)** is hence shown for the value $k := 2$; and the generalized claim **(a)** can be inferred.

Then observe that as both $<_0$ and $<_1$ are irreflexive and transitive relations, the structures

$$\mathcal{M}^+ = (D, <_0, <_1, <_0^{-1}, <_1^{-1}, \mathfrak{h}) \text{ and } \mathcal{M}'^+ = (D', <'_0, <'_1, <'^{-1}_0, <'^{-1}_1, \mathfrak{h}')$$

are temporal structures. But the relation $D \times D'$ is a bisimulation even between these binary temporal structures $\mathcal{M}^+$ and $\mathcal{M}'^+$. Hence $(\mathcal{M}^+,\zeta)$ and $(\mathcal{M}'^+,\xi)$ are **TL**[2]-equivalent but can of course be distinguished by the formula $\square_{0,1}\lozenge_{1,2}/\{1\}\top$. Hence the claim **(b)** follows. ∎

Summary of the relations between the logics considered

The relative expressive powers of the logics considered in this paper are seen to be as follows.

Arbitrary modal structures:

$$\textbf{ML}\,[0] =_{c_0} \textbf{IFML}\,[0] <_{c_1} \textbf{ML}\,[1] <_{c_1} \textbf{IFML}\,[1] \ldots$$
$$\ldots <_{c_k} \textbf{ML}\,[k] <_{c_k} \textbf{IFML}\,[k] \ldots$$

After the first members **ML**[0] $=_{c_0}$ **IFML**[0], this is a hierarchy of ever increasing expressive power. However, once the relation **ML**[1] $<_{c_1}$ **IFML**[1] is shown to hold, the fact that the other relations hold follows trivially.

Arbitrary temporal structures:

$$\mathbf{TL}\,[0] =_{\mathcal{T}_0} \mathbf{IFTL}\,[0] <_{\mathcal{T}_1} \mathbf{TL}\,[1] <_{\mathcal{T}_1} \mathbf{IFTL}\,[1]\ldots$$
$$<_{\mathcal{T}_k} \mathbf{TL}\,[k] <_{\mathcal{T}_k} \mathbf{IFTL}\,[k]\ldots$$

This hierarchy of tense logics is completely analogous to the above hierarchy among modal logics; and holds relative to genuine temporal structures.

By the definitional identities $\mathbf{TL}[k] = \mathbf{ML}[2k]$ and $\mathbf{IFTL}[k] = \mathbf{IFML}[2k]$ (and the transitivity of the relation "of greater expressive power than"), the above hierarchy of modal logics includes a hierarchy of tense logics evaluated relative to *modal* structures. The above tense logical hierarchy is relative to *temporal* structures, and thereby it is an improvement of the result already known from the modal logical case.

Modal logics and linear structures:

$$\mathbf{ML}\,[1] =_{\mathbf{LO}[1]} \mathbf{IFML}\,[1] <_{\mathbf{LO}[2,n_2]} \mathbf{ML}\,[2] <_{\mathbf{LO}[2,n_2]} \mathbf{IFML}\,[2]\ldots$$
$$\ldots <_{\mathbf{LO}[k,n_k]} \mathbf{ML}\,[k] <_{\mathbf{LO}[k,n_k]} \mathbf{IFML}\,[k]\ldots,$$

where for all $i \geq 2, n_i \geq 2$.

Tense logics and linear structures:

$$\mathbf{TL}\,[1] =_{\mathbf{LO}_{temp}[1]} \mathbf{IFTL}\,[1] <_{\mathbf{LO}_{temp}[2,n_2]} \mathbf{TL}\,[2] <_{\mathbf{LO}_{temp}[2,n_2]}$$
$$\mathbf{IFTL}\,[2]\ldots <_{\mathbf{LO}_{temp}[k,n_k]} \mathbf{TL}\,[k] <_{\mathbf{LO}_{temp}[k,n_k]} \mathbf{IFTL}\,[k]\ldots,$$

where for all $i \geq 2, n_i \geq 2$.

5 Related work

The basic framework of this paper stems from the independence-friendly (IF) first-order logic of Hintikka and Sandu [5, 10], which allows presenting arbitrary logical priorities among logical expressions (quantifiers and propositional connectives) in first-order formulae. Henkin quantifier formulae are examples of formulae expressible in IF first-order logic but not in the usual first-order logic (**FO**). (For Henkin quantifiers, see [3, 4, 7, 12].) They are however by no means the only representatives for patterns of logical expressions that cannot be equivalently expressed in usual first-order logic. Other examples are provided by the partially ordered connectives (conjunctions, disjunctions) studied by Sandu and Väänänen [11]; and mutually dependent quantifiers discussed recently on various occasions by Hintikka (see e.g. [6]).

The systematical motivation for the present paper has been the following: as it is possible to define a first-order logic capable of exhibiting more freedom (in fact, full freedom) in relations of logical priority among its logical expressions than usual first-order logic can (a fact witnessed by the existence of IF first-order logic), we may ask whether the same could be done in the case of modal logic. Is there a reasonable way to define 'IF modal logic' in which scope relations among modal operators are relaxed as compared with 'basic modal logic', in a way analogous to the way in which relations of priority scope are relaxed among quantifiers in IF first-order logic? And if so, what can be said about the relative expressive powers of such an IF modal logic and its traditional sibling?

The possibility of defining IF modal logic has been studied in Bradfield [1] and Bradfield & Fröschle [2]. The way IF modal logic is formulated in these papers results in a very expressive logic. In fact Bradfield's general framework makes his IF modal logic more expressive than the logics I have studied in the present paper: his 'Henkin-modalities' are not expressible in the logics studied here. (It can be shown that **IFML**[k] can be translated into usual first-order logic.) The motivation of the studies by Bradfield and Fröschle has been to investigate whether the logical notion of independence appearing in IF logics could prove useful for *concurrency theory* in computer science. My approach in defining IF modal logic in the present paper proceeds from the traditional modal logical setting in a more straightforward way than does Bradfield's approach. In both cases informational independence is made use of when defining the semantics. Bradfield, however, wishes to further require that those transitions of which a given transition is to be independent, must all be concurrent with that independent transition.

Thinking of further conceivable applications to IF modal logic, linguistic theorizing and in particular the analysis of tenses might be a case in point. There is a tradition stemming from Prior [8] of construing natural language tenses as sentential operators. Most contemporary linguists (e.g. Enç, Hornstein, Kamp) oppose, in one way or another, to construing tenses as operators. On the other hand, all that would be needed for an ineliminable theoretical role for operators in the case of tenses is that it be possible that the interpretation of one tense be logically (functionally) dependent on the interpretation of another tense, in analogy to the way in which the interpretation of the quantifier $\exists y$ is dependent on the interpretation of the quantifier $\forall x$ in a formula of the form $\forall x \exists y \; \varphi$. If no such functional dependencies ever occur with tenses, then their sole role is to introduce times and they may as well be taken for pronominal expressions, or for adverbs (which act only very locally). But if such dependency can occur, then there are instances of natural language tenses as operators. When suitably elaborated, IF tense logic offers a natural framework for discussing the logical role of tenses in natural language as opposed to other features connected with tense semantics, notably the mechanism of choosing the temporal interpretation of a tense as temporally related (via a relation such as *earlier than* or *contemporaneousness*) to the interpretation of a syntactically superordinate tense.

6 Conclusion

We have shown that modifying basic modal logic by dissociating the syntactical substring-relation and the relation of logical dependence results in a modal logic whose expressive power is greater than that of basic modal logic over arbitrary modal structures (Theorem 5); an analogous fact holds true of corresponding tense logics and genuine temporal structures (Theorem 7). A case especially worthy of notice is the case of unary *linear* structures. The expressive power of IF tense logic does *not* add to that of basic tense logic in the case of unary linear temporal structures (Theorem 9). Similarly, the expressive powers of IF modal logic and basic modal logic coincide over unary linear modal structures (Theorem 10). It was proven that adding more linear accessibility relations renders the resulting IF modal (tense) logic again more expressive than its basic sibling (Theorem 11).

Acknowledgements

I am most greatly indebted to Dr. Tapani Hyttinen and Prof. Gabriel Sandu for invaluable help during my research for this paper. I am also thankful to Dr. Julian Bradfield for profitable discussions. Finally, I wish to thank the anonymous referees for highly useful comments.

BIBLIOGRAPHY

[1] J. Bradfield. "Independence: logics and concurrency", *Lecture Notes in Computer Science* vol. 1862, 2002. Available electronically as http://www.dcs.ed.ac.uk/home/jcb/Research/cs100.ps.gz

[2] J. Bradfield and S. Fröschle, Sibylle. "Independence-friendly modal logic and true concurrency" in *Nordic Journal of Computing*, vol. 9, no. 2, pp. 102–117, 2002.

[3] H. B. Enderton. "Finite partially ordered quantifiers", *Zeitschrift für Matematische Logik und Grundlagen der Mathematik*, vol. 16, pp. 393–397, 1970.

[4] L. Henkin. "Some remarks on infinitely long formulas", in (no ed. given): *Infinitistic Methods*, Pergamon Press, Oxford, pp. 167–183, 1961.

[5] J. Hintikka and G. Sandu. "Informational Independence as a Semantical Phenomenon" in J. E. Fenstad et alii (eds.): *Logic, Methodology and Philosophy of Science VIII*, North-Holland, Amsterdam, pp. 571–589, 1989.

[6] J. Hintikka. "Quantum Logic as a Fragment of Independence-Friendly Logic", *Journal of Philosophical Logic* 31, pp. 197–209, 2002.

[7] M. Krynicki and M. Mostowski. "Henkin Quantifiers" in Krynicki, Mostowski, Szczebra (eds.): *Quantifiers: Logics, Models and Computation*, vol. 1, Kluwer Academic Publishers, Dordrecht, pp. 193–262, 1995.

[8] A. Prior. *Past, Present and Future*, Oxford University Press, Oxford, 1967.

[9] A. Rubinstein. *Modelling Bounded Rationality*, The MIT Press, Cambridge, Mass., 1998.

[10] G. Sandu. "On the Logic of Informational Independence and its Applications", *Journal of Philosophical Logic* 22, pp. 29–60, 1989.

[11] G. Sandu and J. Väänänen. "Partially Ordered Connectives", *Zeitschrift für Mathematische Logik und Grundlagen der Mathematik*, 38, pp. 361–372, 1992.

[12] W. J. Walkoe, Jr. "Finite partially ordered quantification", *Journal of Symbolic Logic*, vol. 35, pp. 535–55, 1970.

Tero Tulenheimo
Department of Philosophy
00014 University of Helsinki
Finland
E-mail: tero.tulenheimo@helsinki.fi

23
Modal Definability in Languages with a Finite Number of Propositional variables and a New Extension of the Sahlqvist's Class

DIMITER VAKARELOV

ABSTRACT. Modal definability in languages with a finite number of propositional variables is introduced and the related notion of minimal modal equivalent is studied. Some applications to the Sahlqvist theory are given: it is proved that the class of Sahlqvist formulas needs infinitely many propositional variables and that there are infinitely many Sahlqvist formulas with minimal modal equivalents which are not Sahlqvist formulas. This suggests a new extension of the class of Sahlqvist formulas, called complex Sahlqvist formulas.

Introduction

In order to keep things as simpler as possible we will work in the standard modal language, denoted here by L_∞, containing an infinite number of propositional variables, the logical constants and operations $\bot, \top, \wedge, \vee$ and the modalities $\Box$ and $\Diamond$ all taken as primitives. We assume also that the notions of modal definability (local and global) for this language are as in [1]. In this paper we will be interested only with the local version of modal definability and from now on "modal definability" will mean "local modal definability".

Let L_n, $n = 0, 1, \ldots$ be the restriction of L_∞ with n different propositional variables. Our first aim in this paper is to study the notion of modal definability in L_n and its connection with the famous Sahlqvist's theorem. For this theorem the reader may consult [7] and also [1]. A very short topological proof of the Sahlqvist's theorem is given in [8]. Nice presentation of the same proof is given also in the book [3]. Extensions for polyadic languages are given in [6] and in a very readable form in the book [2]. The last generalization which extends those given in [6, 2] is in [5].

Probably the only result in modal definability with restricted number of propositional variables is that the first-order condition

$$F(x): \quad (\forall yz)(xRy \& xRz \to (yRz \vee zRy \vee y = z))$$

is modally definable in $L2$ by the formula with two propositional variables

$$\Box(\Box p \wedge p \to q) \vee \Box(\Box q \to p)$$

but it is not modally definable in L_1 [1].

Let F be a first-order formula with one free variable in the first-order language with the only predicate letters R and $=$. If F is modally definable by some modal formula B then F is said to be the *first-order equivalent* of B and B is said to be the *modal equivalent* of F. If F is modally definable then obviously there exists a natural number n such that F is modally definable in L_n but not in L_{n-1}. This means that there exists a modal formula $A(p_1, \ldots, p_n)$ with n different propositional variables $p_1, \ldots, p_n$ such that $A(p_1, \ldots, p_n)$ is a modal equivalent of F and that there is no modal formula with less than n variables which is a modal equivalent of F. The modal formula $A(p_1, \ldots, p_n)$, which modally defines F in L_n, will be called a *minimal modal equivalent* of F. If B is any modal formula which modally defines F then $A(p_1, \ldots, p_n)$ will also be called a minimal modal equivalent of B.

We will be interested in the following general problem: given a first-order formula F corresponding to some Sahlqvist formula B to find a minimal modal equivalent A of B.

The following two theorems are in this direction and are among the main results in the paper.

THEOREM 1 *For each natural number n there exists a Sahlqvist formula A_n, containing $n+1$ different propositional variables, such that A_n is a minimal modal equivalent of the first-order condition corresponding to A_n.*

THEOREM 2 *There exist infinitely many Sahlqvist formulas with minimal modal equivalents which are not Sahlqvist formulas.*

As a corollary of Theorem 1 we obtain that the Sahlqvist's class of formulas needs infinitely many propositional variables. Theorem 2 has a more interesting consequence: the Sahlqvist's class is not closed under taking minimal equivalents of their members. This fact suggests to consider a new extension of the class of Sahlqvist formulas, called now *complex Sahlqvist formulas*, which is the second aim of this paper. Complex Sahlqvist formulas violate some of the forbidden combinations in (classical) Sahlqvist formulas: *boxes over disjunctions* in (simple) Sahlqvist antecedents. Note that violating this condition leads in general outside of first-order definability ([1, 2]).

We found an effective translation τ transforming each complex Sahlqvist formula A into a deductively equivalent in K (classical) Sahlqvist formula $\tau(A)$ (Translation Lemma). In general $\tau(A)$ has much more propositional variables than A. As a corollary of the Translation Lemma we obtain an analog of the Sahlqvist's theorem: complex Sahlqvist formulas are (locally) first-order definable and canonical (the Main Theorem). Let us note that the Sahlqvist-van Benthem substitution algorithm for computing the first-order conditions of Sahlqvist formulas is not applicable to complex Sahqvist formulas. At the end of the paper we give an extension of this algorithm for complex Sahlqvist formulas.

1 Some preliminary technical lemmas

In this section we will list some technical lemmas needed later on.

LEMMA 3 (Combinatorial Lemma) *Let $\|W\| > 2^n$ and let $A_1, \ldots, A_n$ be subsets of W. Then there exist two different elements $x, y \in W$ such that for any $j = 1, \ldots, n$ either $x, y \in A_j$ or $x, y \notin A_j$.*

Proof. Let $A^0 =_{def} W \setminus A$ and $A^1 =_{def} A$. By a pure Boolean reasoning we have that

$$\bigcup_{i_1,\ldots,i_n} A_1^{i_1} \cap \ldots \cap A_n^{i_n} = W$$

Note that the number of the summands in this union is equal to 2^n and that the different summands are disjoint. Then if for all indices $i_1 \ldots i_n$ we have $\|A_1^{i_1} \cap \ldots \cap A_n^{i_n}\| \leq 1$, we will have $\|W\| \leq 2^n$ contrary to the assumption that $\|W\| > 2^n$. Hence there exist indices $i_1 \ldots i_n$ such that $\|A_1^{i_1} \cap \ldots \cap A_n^{i_n}\| > 1$. This means that there exist two different elements $x, y \in A_1^{i_1} \cap \ldots \cap A_n^{i_n}$. Let $j \in \{1, \ldots, n\}$. If $i_j = 1$ then both x, y are in A_j and if $i_j = 0$ then both x, y are not in A_j, which proves the Combinatorial Lemma. ∎

In order to formulate the next lemma we will remind the reader of some definitions and notations from the classical propositional logic PL. If A, B are formulas of PL then $A \vdash B$ means that the implication $A \Rightarrow B$ is a theorem of PL. We define $A \equiv B$ iff $A \vdash B$ and $B \vdash A$. By a substitution in the set of propositional variables of PL we mean any homomorphism S in the set of formulas of PL.

Let p be a propositional variable. Then $p^0 =_{def} \neg p$ and $p^1 =_{def} p$. Let $p_1, \ldots, p_n$ be a list of different variables. Formulas of the form $D = p_1^{i_1} \vee \ldots \vee p_n^{i_n}$ (in this order of the variables) will be called elementary disjunctions. In a similar way one can define elementary conjunctions. It is known that there are exactly 2^n non-equivalent elementary disjunctions (conjunctions) of $p_1, \ldots, p_n$. Sometimes we will consider a fixed order of all elementary disjunctions (conjunctions) $D_1, \ldots, D_{2^n}$. If we consider the strings $(i_1, \ldots, i_n)$ as binary codes of the integers then there are two natural orderings of the sequence D_i: the *increasing order* – starting from $(0, \ldots, 0) = 0$ and ending with $(1, \ldots, 1) = 2^n - 1$, and the *decreasing order* which is just the opposite of the increasing order.

LEMMA 4 (Substitution lemma 1) *Let $p_1 \ldots p_n$ be a list of different propositional variables and $D_1 \ldots D_{2^n}$ be a fixed list of all elementary disjunctions of them. Let $A_1 \ldots A_{2^n}$ be an arbitrary list of propositional formulas. Then:*

(A) The following two conditions are equivalent:

(i) There exists a substitution S acting on the variables $p_1 \ldots p_n$ such that the following equations hold:

(#1)
$$\left\|\begin{array}{rcl} A_1 & \equiv & S(D_1) \\ A_2 & \equiv & S(D_2) \\ \ldots & & \ldots \\ A_{2^n} & \equiv & S(D_{2^n}). \end{array}\right.$$

(ii) The following two conditions hold for any $i, j \leq 2^n$:
  (a) If $i \neq j$ then $A_i \vee A_j \equiv \top$,
  (b) $\bigwedge_{i=1}^{2^n} A_i = \bot$.

(B) If (ii) is fulfilled then the substitution S is uniquely determined by the following equations:

(#2) $\quad S(p_k) = \bigwedge \{A_l : p_k \in D_l, l \leq 2^n\}$, $k = 1, \ldots, n$.

We consider this lemma as a folklore result and left it without proof. The statements for $n = 2, 3$ are quite easy and can be obtained by standard Boolean manipulations. We invite the reader to extend these proofs for arbitrary n.

The following lemma can be considered as a non-trivial statement in the Propositional Logic. We do not know if it is a new one.

LEMMA 5 (Substitution lemma 2) *Let $p_1, \ldots, p_n$, be a sequence of different propositional variables, let $D_1, \ldots, D_{2^n}$ be a sequence of all elementary disjunctions of these variables and let $q_1, \ldots, q_{2^n-1}$ be a sequence of propositional formulas. Then:*

(i) There exists a substitution S (depending on the given order of the disjunctions D_i) satisfying the following conditions:

(*)
$$\left\|\begin{array}{rcl} q_1 & \equiv & S(D_1) \\ q_1 \wedge q_2 & \equiv & S(D_1 \wedge D_2) \\ \ldots & & \ldots \\ q_1 \wedge q_2 \wedge \ldots \wedge q_{2^n-1} & \equiv & S(D_1 \wedge D_2 \wedge \ldots \wedge D_{2^n-1}). \end{array}\right.$$

(ii) The conditions $(*)$ uniquely determine S (up to logical equivalence) and S can be effectively computed from them.

(iii) The substitution S satisfies also the following conditions:

$$(**) \quad \left\| \begin{array}{ll} q_1 & \vdash \ S(D_1) \\ q_2 & \vdash \ S(D_2) \\ \ldots & \ldots \\ q_{2^n-1} & \vdash \ S(D_{2^n-1}). \end{array} \right.$$

Proof. (i). Define the following sequence of formulas: $A_1 = q_1$, $A_2 = \neg q_1 \vee q_2$, $A_3 = \neg q_1 \vee \neg q_2 \vee q_3$,..., $A_{2^n-1} = \neg q_1 \vee \ldots \vee \neg q_{2^n-2} \vee q_{2^n-1}$ and $A_{2^n} = \neg q_1 \vee \ldots \vee \neg q_{2^n-2} \vee \neg q_{2^n-1}$. It is easy to see that the conditions (a) and (b) for $A_1, \ldots, A_{2^n}$ from the Substitution lemma 1 are fulfilled and hence there exists a substitution S satisfying the equations (#1).

One can see by pure propositional reasoning that $A_1, \ldots, A_{2^n}$ satisfy also the following equations:

(#3) $\quad A_1 \wedge \ldots \wedge A_k \equiv B_1 \wedge \ldots \wedge B_k$, $k = 1, \ldots, 2^n$.

Now from (#1), (#3) and the fact that S is a homomorphism one can easily derive the equations $(*)$.

(ii). Suppose now that S exists and satisfies $(*)$. Using the special structure of the elementary disjunctions one can see that they satisfy the conditions (a) and (b):

(a) $\quad$ If $i \neq j$ then $D_i \vee D_j \equiv \top$, $i, j \leq 2^n$,

(b) $\quad \bigwedge_{i=1}^{2^n} D_i = \bot$.

Applying the condition (a) one can see that the following equations are true

(#4) $\quad \neg(D_1 \wedge \ldots \wedge D_k) \vee D_{k+1} \equiv D_{k+1}$, $k < 2^n$.

Now we transform the equations $(*)$ as follows – negate both sides of the k-th row and add it disjunctively to the $(k+1)$-th row:

$\neg(q_1 \wedge \ldots \wedge q_k) \vee (q_1 \wedge \ldots \wedge q_k \wedge q_{k+1}) \equiv S(\neg(D_1 \wedge \ldots \wedge D_k) \vee (D_1 \wedge \ldots \wedge D_k \wedge D_{k+1}))$.

The left side of this equation is equivalent to A_{k+1} and by (#4) the right side is equivalent to $S(D_{k+1})$. In this way, using also (b), we equivalently transform the system $(*)$ into the system (#1). Then by the part B of Substitution Lemma 1 the substitution S is uniquely determined and can be computed effectively.

(iii). From the definition of A_i, $i < 2^n$ one can see that the following is true:

(#5) $\quad q_i \vdash A_i$.

Then applying (#5) to the system (#1) we obtain the conditions $(**)$. ∎

Let for brevity AB and $\overline{A}$ denote the conjunction $A \wedge B$ and the negation $\neg A$ respectively and consider the decreasing order of the elementary disjunctions of the variables p_1, p_2 (p_1, p_2, p_3). In the next two examples we show the explicit form of the substitution S for $n = 2$ and $n = 3$.

Example 1: ($n = 2$).

$S(p_1) = q_1 q_2,$
$S(p_2) = (q_1 \overline{q_2}) \vee (q_1 q_3).$

Example 2: ($n = 3$).

$S(p_1) = q_1 q_2 q_3 q_4,$
$S(p_2) = q_1 q_2 \overline{q_3} \overline{q_4} \vee q_1 q_2 q_5 q_6,$
$S(p_3) = q_1 \overline{q_2} \vee q_1 q_3 \overline{q_4} \vee q_1 q_3 q_5 \overline{q_6} \vee q_1 q_3 q_5 q_7.$

In the next lemma we will consider systems of equations over an arbitrary Boolean algebra B in the following forms:

$$(+) \quad \left\| \begin{array}{rcl} q_1 & = & D_1 \\ q_1 \wedge q_2 & = & D_1 \wedge D_2 \\ \ldots & & \ldots \\ q_1 \wedge q_2 \wedge \ldots \wedge q_k & = & D_1 \wedge D_2 \wedge \ldots \wedge D_k \end{array} \right.$$

and

$$(++) \quad \left\| \begin{array}{rcl} q_1 & \leq & D_1 \\ q_2 & \leq & D_2 \\ \ldots & & \ldots \\ q_k & \leq & D_k. \end{array} \right.$$

In (+) and (++) $q_1, \ldots, q_k$ denote given elements from B. A solution of the system (+) (system (++) or both together) is any valuation of the variables $p_1, \ldots, p_n$ in B which satisfies the system.

LEMMA 6 (Boolean lemma) *Let $B = (B, 0, 1, \wedge, \vee, \neg)$ be a Boolean algebra and $D_1, \ldots, D_{2^n}$ be a list of all elementary disjunctions of the different variables $p_1, \ldots, p_n$ and let $q_1, \ldots, q_k$, $1 \leq k \leq 2^n - 1$, be given elements of B. Then the system consisting of (+) and (++) together has a solution in B and if $k = 2^n - 1$ then this solution is unique.*

Proof. Using the connection between Boolean algebras and PL one can easily derive the Boolean Lemma from the Substitution Lemma 2. ∎

2 Proof of Theorem 1

Let (W, R) be a frame and consider the following first-order condition:

$Alt_{k-1}(x) \quad (\forall y_1 \ldots y_k)(xRy_1 \& \ldots \& xRy_k \to \|y_1, \ldots, y_k\| < k), k > 1.$

It is known that this condition is modally definable by the following formula ([9]):

$\mathbf{Alt}_{k-1}$ $\quad \Box q_1 \vee \Box(q_1 \Rightarrow q_2) \vee \ldots \vee \Box(q_1 \wedge \ldots \wedge q_{k-1} \Rightarrow q_k)$.

We will us this formula in the following equivalent form:

$\Box q_1 \vee \Box(\neg q_1 \vee q_2) \vee \ldots \vee \Box(\neg q_1 \vee \ldots \vee \neg q_{k-1} \vee q_k)$.

Note. If in the above formula we substitute q_k with $\bot$ we obtain an equivalent in K formula to $\mathbf{Alt}_{k-1}$ which we denote by $\mathbf{Alt}'_{k-1}$.
Note also that $\mathbf{Alt}_{k-1}$ and $\mathbf{Alt}'_{k-1}$ can easily be rewritten into a Sahlqvist form. However neither of these equivalents of $\mathbf{Alt}_{k-1}$ is a minimal modal equivalent of $Alt_{k-1}(x)$. Now we will construct another Sahlqvist formula which is a minimal equivalent of $Alt_{k-1}(x)$.

Let $p_1, \ldots, p_{n+1}$ be a sequence of $n+1$ different propositional variables and let $Q_1, \ldots, Q_{2^{n+1}}$ be a fixed sequence of all 2^{n+1} elementary conjunctions of the variables p_i (for instance in the increasing order). We introduce the following Sahlqvist formula of $n+1$ propositional variables:

$\mathbf{ALT}^{min}_{k-1}$ $\quad \Diamond Q_1 \wedge \ldots \wedge \Diamond Q_k \Rightarrow \bot$.

We will use this formula also in the following equivalent form:

$\Box D_1 \vee \ldots \vee \Box D_k$,

where the formula D_i $(i = 1, \ldots, k)$ is the elementary disjunction equivalent to $\neg Q_i$.

Our first aim now is to show that $\mathbf{ALT}^{min}_{k-1}$ is a modal equivalent of $Alt_{k-1}(x)$. For each concrete k and n one may use for this purpose the Sahlqvist's Theorem but for the general case it is more convenient just to prove that $\mathbf{ALT}^{min}_{k-1}$ and $\mathbf{Alt}_{k-1}$, are deductively equivalent in K. So we have the following lemma.

LEMMA 7 *Let $k \leq 2^{n+1}$. Then:*
(i) The formulas $\mathbf{ALT}^{min}_{k-1}$ and $\mathbf{Alt}_{k-1}$ are deductively equivalent in K.
(ii) $\mathbf{ALT}^{min}_{k-1}$ modally defines $Alt_{k-1}(x)$.

Proof. (i). Let $k \leq 2^{n+1}$ and let $D_1, \ldots, D_{2^{n+1}}$ be the list of the elementary disjunctions equivalent to the negations of the elementary conjunctions $Q_1, \ldots, Q_{2^{n+1}}$.

For the implication from $\mathbf{Alt}_{k-1}$ to $\mathbf{ALT}^{min}_{k-1}$ substitute q_i in $\mathbf{Alt}_{k-1}$ with D_i, $i = 1, \ldots, k$. Using the fact that $D_i \vee D_j \equiv \top$ for $i \neq j$ one can easily prove that the following equivalences are true in the Propositional Logic:

$\neg D_1 \vee D_2 \equiv D_2$, $\neg D_1 \vee \neg D_2 \vee D_3 \equiv D_3$, $\ldots \neg D_1 \vee \ldots \vee \neg D_{k-1} \vee D_k \equiv D_k$.

Then the result of the substitution is just equivalent to $\mathbf{ALT}^{min}_{k-1}$.

For the implication from $\mathbf{ALT}_{k-1}^{min}$ to $\mathbf{Alt}_{k-1}$ define the following sequence of formulas:

$A_1 = q_1$, $A_2 = \neg q_1 \vee q_2$,..., $A_{k-1} = \neg q_1 \vee \ldots \vee \neg q_{k-2} \vee q_{k-1}$,
$A_k = \neg q_1 \vee \ldots \vee \neg q_{k-2} \vee \neg q_{k-1}$, $A_{k+1} = \ldots = A_{2^n+1} = \top$.

This definition is possible because of the inequality $k \leq 2^n+1$. It can easily be seen that the formulas A_i satisfy the conditions (a) and (b) of the Substitution Lemma 1. Then there is a substitution S such that

$$A_1 \equiv S(D_1)$$
$$A_2 \equiv S(D_2)$$
$$\ldots \quad \ldots$$
$$A_k \equiv S(D_k).$$

By this substitution we obtain

$$S(\mathbf{ALT}_{k-1}^{min}) = \Box A_1 \vee \ldots \vee \Box A_k$$

which is just the formula $\mathbf{Alt}'_{k-1}$.

(ii). Condition (ii) is a corollary of (i). ∎

PROPOSITION 8 *Let $2^n < k \leq 2^{n+1}$. Then:*

(i) There is no modal formula of n variables which modally defines the condition $Alt_{k-1}(x)$.

(ii) The Sahlqvist formula $\mathbf{ALT}_{k-1}^{min}$ of $n+1$ variables is a minimal modal equivalent of $Alt_{k-1}(x)$.

Proof. (i). Let $2^n < k \leq 2^{n+1}$ and let $p_1, \ldots, p_n$ be a list of different propositional variables. Suppose that such a formula $\alpha = \alpha(p_1, \ldots, p_n)$ exists and proceed to obtain a contradiction. This can be done realizing the following

PROGRAM

(1) First we construct a frame $F = (W, R)$ which is not a frame for α and consequently we define a model $M = (W, R, v)$ falsifying α.
The next steps are:
(2) to construct a frame $F' = (W', R')$ which is a frame for α and
(3) to construct a p-morphism f from F onto F'.
(4) Then we define a valuation v' in F' such that f is a p-morphism from the model M onto the model $M' = (W', R', v')$.
(5) Now the contradiction can be obtained as follows. By (1), (4) and p-morphism lemma M' falsifies α which contradicts (2).
Let us start realizing the above program.

Step (1). Define the frame $F = (W, R)$ as follows. Let $0, a_1, \ldots, a_k$ be $k+1$ different elements. Then put

$W = \{0, a_1, \ldots, a_k\}$ and $R = \{(0, a_1), \ldots, (0, a_k)\}$.

Obviously F does not satisfy the condition $Alt_{k-1}(x)$ because 0 has just k different R-successors and hence F is not a frame for α. Then there exists a valuation v in F falsifying α.

Step (2). Define $A_i = v(p_i) \cap \{a_1, \ldots, a_k\}$, $i = 1, \ldots, n$. Since $k > 2^n$ then by the Combinatorial Lemma there exist $x \neq y \in \{a_1, \ldots, a_k\}$ such that for any $i = 1, \ldots, n$ we have either $x, y \in A_i$ or $x, y \notin A_i$. Then for some $l \neq m \in \{1, \ldots, k\}$ we have $x = a_l$ and $y = a_m$. Consequently we obtain the following equivalence:

(#) $a_l \in v(p_i)$ iff $a_m \in v(p_i)$, $i = 1, \ldots, n$.

Now define F' in the same way as F assuming only that $a_l = a_m$. Now $\|a_1, \ldots, a_k\| = k - 1$ and consequently F' satisfies $Alt_{k-1}(x)$. Hence F' is a frame for α.

Step (3). Define f from W onto W' as follows: $f(0) = 0$ and for $i \neq l$ put $f(a_i) = a_i$ and finally put $f(a_l) = a_m$. Obviously f is a p-morphism from W onto W' "gluing" a_l and a_m.

Step (4). The valuation v' in W' is defined by $v'(p_i) = v(p_i) \cap W'$. We will prove that f is a p-morphism from the model M onto the model $M' = (W', R', v')$. The only thing we have to show is the equivalence:

$(\forall x \in W)(x \in v(p_i)$ iff $f(x) \in v'(p_i))$, $i = 1, \ldots, n$.

For $x \neq a_l$ this is obvious by the definition of f and v'.
For $x = a_l$ we proceed as follows:

$a_l \in v(p_i)$ iff (by (#)) $a_m \in v(p_i)$ iff $a_m \in v(p_i) \cap W'$ iff (by the definitions of f and v') $f(a_l) \in v'(p_i)$, $i = 1, \ldots, n$.

The proof of (i) is finished.

(ii). Let $2^n < k \leq 2^{n+1}$. By (i) there is no modal formula of n variables which modally defines $Alt_{k-1}(x)$. By 7 the formula $\mathbf{ALT}_{k-1}^{min}$ of $n+1$ propositional variables modally defines $Alt_{k-1}(x)$. Hence $\mathbf{ALT}_{k-1}^{min}$ is a minimal modal equivalent of $Alt_{k-1}(x)$. ∎

Proof of Theorem 1. Let n be a given natural number and take any k such that $2^n < k \leq 2^{n+1}$. Then by Proposition 8 the Sahlqvist formula $A_n = \mathbf{ALT}_{k-1}^{min}$ contains $n+1$ propositional variables and is a minimal modal equivalent of the corresponding first-order condition $Alt_{k-1}(x)$. ∎

COROLLARY 9 *The class of Sahlqvist formulas needs an infinite number of propositional variables.*

REMARK 10 (i) The formula $\mathbf{ALT}_{k-1}^{min}$ depends on the chosen order of the string $< Q_1, \ldots, Q_{2^{n+1}} >$ of the elementary conjunctions. Different orders present different formulas but on the base of lemma 7 all are deductively equivalent in K.

(ii) The Segerberg's formula $\mathbf{Alt}_{k-1}$ is based on the fact that the difference between k points in a frame can be expressed by means of k (or $k-1$) different propositional variables. In $\mathbf{ALT}_{k-1}^{min}$ the same thing is done by means of a smaller number of variables, namely by the smallest number $n+1$ such that $k \leq 2^{n+1}$.

3 Proof of Theorem 2

In order to prove Theorem 2 we introduce the following sequence of first-order conditions which generalize the well-known Churc-Rosser confluence condition:

$CR_{k-1}(x)$ $\quad (\forall y_1 \ldots y_k)(xRy_1 \& \ldots \& xRy_k \to (\exists z)(y_1 Rz \& \ldots \& y_k R_k z))$.

LEMMA 11 *The following Sahlqvist formula is a modal equivalent of the Churc-Rosser confluence condition $CR_{k-1}(x)$:*

$\mathbf{CR}_{k-1}$ $\quad \Diamond \Box q_1 \wedge \ldots \wedge \Diamond \Box q_{k-1} \Rightarrow \Box \Diamond (q_1 \wedge \ldots \wedge q_{k-1})$.

Proof. Use directly the Sahlqvist's theorem. ∎

To define a minimal equivalent of the Churc-Rosser confluence condition $CR_{k-1}(x)$ let $2^n < k \leq 2^{n+1}$, let $p_1, \ldots, p_{n+1}$ be a sequence of different propositional variables and let $D_1, \ldots, D_{2^{n+1}}$ be a sequence of all non-equivalent elementary disjunctions of these variables. Then we define the formula:

$\mathbf{CR}_{k-1}^{min}$ $\quad \Diamond \Box D_1 \wedge \ldots \wedge \Diamond \Box D_{k-1} \Rightarrow \Box \Diamond (D_1 \wedge \ldots \wedge D_{k-1})$.

Example 3:

$\Diamond \Box (p_1 \vee p_2) \wedge \Diamond \Box (p_1 \vee \neg p_2) \wedge \Diamond \Box (\neg p_1 \vee p_2) \Rightarrow \Box \Diamond (p_1 \wedge p_2)$.

(Here $p_1 \wedge p_2$ is substituted for its equivalent $(p_1 \vee p_2) \wedge (p_1 \vee \neg p_2) \wedge (\neg p_1 \vee p_2)$.)

LEMMA 12 *Let $2^n < k \leq 2^{n+1}$. Then:*
(i) The formulas $\mathbf{CR}_{k-1}$ and $\mathbf{CR}_{k-1}^{min}$ are deductively equivalent in K.
(ii) The formula $\mathbf{CR}_{k-1}^{min}$ is canonical and a modal equivalent of the first-order formula $CR_{k-1}(x)$.

Proof. (i). The direction from $\mathbf{CR}_{k-1}$ to $\mathbf{CR}_{k-1}^{min}$ is easy – simply substitute q_i with D_i, $i = 1, \ldots, k-1$.

For the converse direction let S be the substitution guaranteed by the Substitution Lemma 2. Then we obtain:

(#) $\quad \Diamond \Box S(D_1) \wedge \ldots \wedge \Diamond \Box S(D_{k-1}) \Rightarrow \Box \Diamond S(D_1 \wedge \ldots \wedge D_{k-1})$.

By the Substitution Lemma 2 we have:

$$(**)\quad \begin{array}{rl} q_1 & \vdash S(D_1) \\ q_2 & \vdash S(D_2) \\ \ldots & \ldots \\ q_{k-1} & \vdash S(D_{k-1}) \end{array}$$

and also

$(*)\quad q_1 \wedge q_2 \wedge \ldots \wedge q_{k-1} \equiv S(D_1 \wedge D_2 \wedge \ldots \wedge D_{k-1})$.

Using the monotonicity of $\Diamond$ and $\Box$ and the replacement of equivalents we obtain from (#), (**) and (*) the formula

$$\Diamond\Box q_1 \wedge \ldots \wedge \Diamond\Box q_{k-1} \Rightarrow \Box\Diamond(q_1 \wedge \ldots \wedge q_{k-1}).$$

This is just the formula $\mathbf{CR}_{k-1}$ and the proof is finished.

(ii). The proof of (ii) follows from (i). ∎

PROPOSITION 13 *Let $2^n < k \leq 2^{n+1}$. Then:*

(i) There is no modal formula of n variables which modally defines the condition $CR_{k-1}(x)$.

(ii) The non-Sahlqvist formula $\mathbf{CR}_{k-1}^{min}$ (with $n+1$ variables) is a canonical minimal modal equivalent of $CR_{k-1}(x)$.

Proof. (i). The proof is similar to the proof of Proposition 8. Let $2^n < k \leq 2^{n+1}$ and suppose that such a formula $\alpha = \alpha(p_1, \ldots, p_n)$ exists and proceed to obtain a contradiction. The contradiction can be obtained realizing the same PROGRAM of 5 steps as in Proposition 8 but now the constructions are more complicated.

Step (1). The frame $F = (W, R)$ from which the program starts for this case is the following:

$W = \{0, 1, \ldots, k, \omega, a_1, \ldots, a_k\}$, where $a_1, \ldots, a_k$ are different elements and different from $0, 1, \ldots, k, \omega$.

$R = \{(0, i)/i = 1, \ldots, k\} \cup \{(a_i, \omega)/i = 1, \ldots, k\} \cup \{(i, a_j)/i \neq j, i, j = 1, \ldots, k\} \cup \{(\omega, \omega)\}$.

We invite the reader to draw this frame in order to follow more easily the next reasonings.

It can be seen that F does not satisfy the condition $CR_{k-1}(x)$, because we have $0R1, \ldots, 0Rk$ but there is no $z \in W$ such that $1Rz, \ldots, kRz$. Hence F is not a frame for α. Then there is a valuation v in F falsifying α.

Step (2). Define $A_i = v(p_i) \cap \{a_1, \ldots, a_k\}$, $i = 1, \ldots, n$. Since $k > 2^n$ then by the Combinatorial Lemma there exist $x \neq y \in \{a_1, \ldots, a_k\}$ such that for any $i = 1, \ldots, n$ we have either $x, y \in A_i$ or $x, y \notin A_i$. Then for some $l \neq m \in \{1, \ldots, k\}$ we have $x = a_l$ and $y = a_m$. Consequently we obtain the following equivalence:

(#) $a_l \in v(p_i)$ iff $a_m \in v(p_i)$, $i = 1, \ldots, n$.

Define F' in the same way as F assuming only that $a_l = a_m$. It is easy to see now that F' satisfies the condition $CR_{k-1}(x)$. Hence F' is a frame for α.

Step (3). Define f from W into W' as follows: $f(i) = i$ for $i = 0, 1, \ldots, k, \omega$ and for $i \neq l$ put $f(a_i) = a_i$ and finally put $f(a_l) = a_m$. Obviously f is a p-morphism from W onto W' "gluing" a_l and a_m.

Step (4). The valuation v' in W' is defined by $v'(p_i) = v(p_i) \cap W'$. The proof that f is a p-morphism from the model M onto the model $M' = (W', R', v')$ is similar to the corresponding proof in Proposition 8.

The PROGRAM has been realized and the expected contradiction can be obtained as in Proposition 8: since α is falsified in the model (F, v) then by the p-morphism lemma it is falsified also in the model (F', v'), which is impossible, because F' is a frame for α. The proof of (i) is finished.

(ii). Let $2^n < k \leq 2^{n+1}$. By Lemma 12 the formula $\mathbf{CR}_{k-1}^{min}$ containing $n+1$ propositional variables is a modal equivalent of $CR_{k-1}(x)$. It follows from (i) that it is a canonical minimal modal equivalent of $CR_{k-1}(x)$. ∎

Proof of Theorem 2. Theorem 2 follows just from the Proposition 13 and the remark that $\mathbf{CR}_{k-1}^{min}$ are not Sahlqvist formulas. ∎

REMARK 14 (i) Note that $\mathbf{CR}_{k-1}^{min}$ are not Sahlqvist formulas. They violate one of the forbidden conditions for the Sahlqvist antecedents: disjunctions in the scope of boxes.

(ii) The formula $\mathbf{CR}_{k-1}^{min}$ depends on the given order of the elementary disjunctions. But by Lemma 12 any order will give a formula deductively equivalent in K to the Sahlqvist formula $\mathbf{CR}_{k-1}$. Note that this equivalence is not based on easy and obvious transformations.

(iii) The formula $\mathbf{CR}_{k-1}$ is based on the fact that common R-successor of k points in a frame can be expressed by means of $k-1$ different propositional variables. The formula $\mathbf{CR}_{k-1}^{min}$ is based on the observation that the same thing can be done by a smaller number of variables, namely by the smallest number $n+1$ such that $k \leq 2^{n+1}$. Proposition 13 shows that this number is optimal.

4 A new extension of the Sahlqvist's class

The non-Sahlqvist formulas $\mathbf{CR}_{k-1}^{min}$ which are canonical and local correspondents of some first-order conditions – features typical for the formulas of the Sahlqvist's class, suggest this class to be extended. These formulas violate the forbidden condition disjunctions to be preceded by a box in the Sahlqvist antecedents. But in this case the disjunctions are of a special kind – elementary disjunctions. If we compare the equivalent formulas $\mathbf{CR}_{k-1}$ and $\mathbf{CR}_{k-1}^{min}$, we can see the following things:

(i) The number of the variables in the first formula is $k-1$, the number in the second $-n+1$, and the numbers k,n satisfy the following inequality $2^n < k \leq 2^{n+1}$. It follows from this inequality that for $n > 1$ the second formula always has a smaller number of variables.

(ii) The second thing we can see is that the elementary disjunctions D_i in the second formula just "code" the variables q_i from the first formula. This suggests to look at the elementary disjunctions D_i as a new kind of "complex" variables. Another important block in the second formula is the conjunction $D_1 \wedge \ldots \wedge D_{k-1}$ in the consequent. Looking at the Substitution Lemma 2 we can see similar blocks $D_1, D_1 \wedge D_2, D_1 \wedge D_2 \wedge D_3, \ldots$ and they will be also used in the generalization.

Let us start building the new definition.

Let $p = <p_1, \ldots, p_n>$ be a string of different propositional variables and let the elementary disjunctions built from p have the same order of the variables. Let $D(p) = <D_1, \ldots, D_{2^n-1}>$ be a fixed sequence of different elementary disjunctions (the last one D_{2^n} is missing). Consider also the sequence $D^\wedge(p) = <D_1, D_1 \wedge D_2, \ldots, D_1 \wedge \ldots \wedge D_{2^n-1}>$. Then the pair $P = <D(p), D^\wedge(p)>$ is called *propositional complex of dimension n*, the disjunctions D_i from $D(p)$ are called *simple blocks* or *complex variables* and the elements from $D^\wedge(p)$ are called *compound blocks*. We say that the complex $<D(p), D^\wedge(p)>$ is *disjoint* from the complex $<D(q), D^\wedge(q)>$ if the strings p and q do not share common variables.

By a *constant formula* we mean any formula not containing propositional variables. By a *positive formula* we mean any formula built from variables and constant formulas by means of the operations $\wedge, \vee, \Diamond, \Box$. Negation of a positive formula is called a *negative formula*.

Let Σ be a set of pairwise disjoined propositional complexes. The expression $\Box^n A$ where A is a simple block from Σ is called *complex boxed atom* ($n=0$ is included: $\Box^0 A = A$). Let $A = A(p_1, \ldots, p_k)$ be a positive (negative) formula and $p_1, \ldots, p_k$ be the list of all variables of A. If $B_1, \ldots, B_k$ is a list of compound blocks from Σ then $A(B_1, \ldots, B_k)$ is called *complex positive (negative) atom* in Σ. By a **complex Sahlqvist antecedent** in Σ we mean any formula composed from complex boxed atoms and negative complex atoms from Σ by means of $\wedge$ and $\Diamond$. Formulas of the form $\alpha \Rightarrow \beta$ where α is a complex Sahlqvist antecedent and β is a complex positive atom are called **complex Sahlqvist implications** in Σ.

Now the class of **complex Sahlqvist formulas** in Σ can be obtained from the complex Sahlqvist implications in Σ by freely applying boxes and conjunctions, and disjunctions only to formulas sharing no common variables. We allow Boolean simplifications of compound blocks as in the example 3.

Suppose that we have a string $p = \langle p_1 \rangle$ of dimension 1. Then D_1 is either p_1 or $\neg p_1$. So in the first case we have $D(p) = D^\wedge(p) = \langle p_1 \rangle$ which shows that if all complex variables are of dimension 1 then the notion of complex Sahlqvist formula coincides with the notion of Sahlqvist formula. Thus the class of complex Sahlqvist formulas contains the original Sahlqvist's class. The next lemma however shows that the new class do not extend the corresponding class of first-order equivalents.

LEMMA 15 (Translation Lemma) *There exists an effective translation τ such that every complex Sahlqvist formula A is deductively equivalent in K to the Sahlqvist formula $\tau(A)$.*

Proof. The translation τ is very simple – just consider complex variables in A as notations of new propositional variables ("new" here means "not included in A"). It is clear that if A is a complex Sahlqvist formula then $\tau(A)$ is an (ordinary) Sahlqvist formula. Note that if we prove the lemma for the complex Sahlqvist implications then the general case is straightforward.

Let $A = \alpha \Rightarrow \beta$ be a complex Sahlqvist implication. The proof for this case is quite similar to the proof of lemma 12. Since the propositional complexes in A are disjoined we may realized the translation τ step-by-step substituting the complex variables one by one (in arbitrary order). This means that we consider a "partial" translation τ_p for each propositional complex $P = \langle D(p), D^\wedge(p) \rangle$ just substituting in A only occurrences of complex variables from P.

So we will prove that A is deductively equivalent in K to $\tau_p(A)$.

For the implication from $\tau_p(A)$ to A note that $\tau_p(A)$ can be obtained from A by ordinary substitution, so this case is obvious.

For the implication from A to $\tau_p(A)$ we proceed as follows.

Let $p = \langle p_1, \ldots, p_n \rangle$ and let $D(p) = \langle D_1, \ldots, D_{2^n-1} \rangle$ be the string of the simple blocks from p. For the translation we need a string of new variables $\langle q_1, \ldots, q_{2^n-1} \rangle$. Let $Box_1(D_{i_1}), \ldots, Box_j(D_{i_j})$ be the list of complex boxed atoms occurring in α and let $C_1, \ldots, C_k$ be the complex negative atoms occurring in α. Then α is in the form

$$\alpha(Box_1(D_{i_1}), \ldots, Box_j(D_{i_j}); C_1, \ldots, C_k).$$

Note that complex boxed atoms in α occur only positively, compound blocks in α occur only negatively and that β is a complex positive atom.

Now we shall make use of the substitution lemma 2. This lemma guarantees existence of a substitution S such that

$$(*) \quad \left\| \begin{array}{lll} q_1 & \equiv & S(D_1) \\ q_1 \wedge q_2 & \equiv & S(D_1 \wedge D_2) \\ \ldots & & \ldots \\ q_1 \wedge q_2 \wedge \ldots \wedge q_{2^n-1} & \equiv & S(D_1 \wedge D_2 \wedge \ldots \wedge D_{2^n-1}), \end{array} \right.$$

$$(**) \quad \left\| \begin{array}{ccc} q_1 & \vdash & S(D_1) \\ q_2 & \vdash & S(D_2) \\ \ldots & & \ldots \\ q_{2^n-1} & \vdash & S(D_{2^n-1}). \end{array} \right.$$

It can be assumed that the substitution S acts only on the variables from p and leaves unchanged the other variables.

Using $(*)$ the following statement can be proved by an easy induction.

Statement. *If B is a complex positive or negative atom then the following equivalence is true: $S(B) \equiv \tau_p(B)$.*

Applying S to A we obtain:

(1) $S(A) = \alpha(Box_1(S(D_{i_1})),\ldots,Box_j(S(D_{i_j})); S(C_1),\ldots,S(C_k)) \Rightarrow S(\beta)$

Applying the above Statement to (1) we obtain

(2) $S(A) \equiv \alpha(Box_1(S(D_{i_1})),\ldots,Box_j(S(D_{i_j})); \tau_p(C_1),\ldots,\tau_p(C_k)) \Rightarrow \tau_p(\beta)$

Using the fact that complex boxed atoms occur in α only positively, then applying $(**)$ to (2) we obtain

(3) $S(A) \vdash \alpha(Box_1(q_{i_1}),\ldots,Box_j(q_{i_j}); \tau_p(C_1),\ldots,\tau_p(C_k)) \Rightarrow \tau_p(\beta)$.

But obviously we have

(4) $\alpha(Box_1(q_{i_1}),\ldots,Box_j(q_{i_j}); \tau_p(C_1),\ldots,\tau_p(C_k)) = \tau_p(\alpha)$.

Now from (4) and (3) we obtain

(5) $S(A) \vdash \tau_p(\alpha) \Rightarrow \tau_p(\beta)$

and consequently $S(A) \vdash \tau_p(A)$. This shows that starting from A we can derive in K $\tau_p(A)$. The proof is finished. ∎

The following statement is an analog of the Sahlqvist's Theorem.

THEOREM 16 (The Main Theorem) *The following conditions are true for each complex Sahlqvist formula A:*

(i) A defines locally a first-order condition $F_A(x)$ which can effectively be computed from A.

(ii) A is canonical.

Proof. The theorem is a corollary of the Translation Lemma. ∎

5 Extension of the Sahlqvist–van Benthem substitution method to the complex Sahlqvist formulas

The Translation Lemma gives an algorithm for finding the corresponding first-order condition for a given complex Sahlqvist formula A: first translate A to some original Sahlqvist formula $\tau(A)$ and then apply the standard

Sahlqvist-van Benthem substitution method [1]. Note that this method can not be directly applied to complex Sahlqvist formulas. The reason is that it is applicable only to formulas having the format of standard Sahlqvist formulas and strongly uses the fact that boxes in the boxed atoms are in front of propositional variables. If we look at the set-theoretical form of this method in [8] or in [3] we can see that the algorithm eliminates the second-order universal quantifiers on propositional variables (considered as ranging over subsets of the universe) *one by one in an arbitrary order*, making suitable "minimal" substitution of the corresponding variable. This feature characterizes the substitutions in Sahlqvist-van Benthem algorithm as *simple substitutions*, substitutions only for one given variable in each step. The fact that boxes in complex boxed atoms are in front of elementary disjunctions of several propositional variables suggests to consider "simultaneous substitutions" or substitutions of several variables simultaneously in a given step of the algorithm. This idea works and gives rise to a suitable extension of the Sahlqvist-van Benthem substitution method, now as a "method of simultaneous substitutions". This extended version of the substitution method works directly on complex Sahlqvist formulas for computing their first-order equivalents without going trough their translations into the original Sahlqvist's class.

We will demonstrate the new method by an example. Consider the formula from example 3:

$A = A(p,q) = \Diamond\Box(p \vee q) \wedge \Diamond\Box(p \vee \neg q) \wedge \Diamond\Box(\neg p \vee q) \Rightarrow \Box\Diamond(p \wedge q)$.

The corresponding first-order condition $F(x)$ is:

$(\forall y_1 y_2 y_3 y_4)(xRy_1 \& xRy_2 \& xRy_3 \& xRy_4 \to (\exists z)(y_1 Rz \& y_2 Rz \& y_3 Rz \& y_4 Rz))$

We consider $A(p,q)$ as a set-theoretical expression over the modal algebra of a given frame (W, R). The variables p, q range over subsets of W and the variables $x, y_1, \ldots, y_4$ – range over W. The (local) validity of $A(p,q)$ at a point $x \in W$ means the validity of the following second-order formula in (W, R):

$(\forall pq)(x \in A(p,q))$.

Rewriting it in an equivalent way using the known notations from Boolean modal set-algebras we obtain ($-p$ denotes the complement of p and $x \in \Box p$ is equivalent to $R(x) \subseteq p$, where $y \in R(x)$ iff xRy):

(1) $(\forall pq)(\forall y_1 y_2 y_3)(xRy_1 \& xRy_2 \& xRy_3 \& R(y_1) \subseteq (p \cup q) \& R(y_2) \subseteq (p \vee -q) \& R(y_3) \subseteq (-p \cup q) \to x \in \Box\Diamond(p \cap q))$.

By the Boolean Lemma there exist subsets $p, q \subseteq W$ such that the following equations are satisfied:

(2) $R(y_1) \subseteq (p \cup q)$, $R(y_2) \subseteq (p \cup -q)$, $R(y_3) \subseteq (-p \cup q)$, $R(y_1) \cap R(y_2) \cap R(y_3) = p \cap q$.

Substituting p and q from (2) in (1) we obtain:

(3) $(\forall y_1 y_2 y_3)(xRy_1 \& xRy_2 \& xRy_3 \to x \in \Box\Diamond(R(y_1) \cap R(y_2) \cap R(y_3)))$.

Rewriting $x \in \Box\Diamond(R(y_1) \cap R(y_2) \cap R(y_3))$ and substituting the result in (3) we obtain just the required first-order condition $F(x)$ which is the local correspondent of $A(p,q)$.

This shows that if $x \in A(p,q)$ holds in (W,R) for all subsets p,q then the first-order formula $F(x)$ holds in (W,R) (for that given $x \in W$). Note that to obtain $F(x)$ we have substantially used the Boolean Lemma to find the appropriate substitutions for p and q. This "simultaneous" substitution is in a sense " minimal": the obtained condition $F(x)$ from this "minimal" assumption guarantees the converse implication: if $F(x)$ holds in (W,R) (for that given $x \in W$) then we have $(\forall pq)(x \in A(p,q))$.

Indeed, let p,q be arbitrary subsets of W. In order to show that $x \in A(p,q)$ suppose

(4) $x \in \Diamond\Box(p \cup q) \cap \Diamond\Box(p \cup -q) \cap \Diamond\Box(-p \cup q)$

and proceed to show that

(5) $x \in \Box\Diamond(p \cap q)$.

The condition (4) can be rewritten as follows:

(6) $(\exists y_1 y_2 y_3)(xRy_1 \& xRy_2 \& xRy_3)$
such that

(7) $R(y_1) \subseteq (p \cup q)$, $R(y_2) \subseteq (p \vee -q)$ and $R(y_3) \subseteq (-p \cup q)$.

From (7) by simple Boolean transformations we obtain

(8) $R(y_1) \cap R(y_2) \cap R(y_3) \subseteq (p \cap q)$.

By the monotonicity of $\Box$ and $\Diamond$ we obtain from (8)

(9) $\Box\Diamond(R(y_1) \cap R(y_2) \cap R(y_3)) \subseteq \Box\Diamond(p \cap q)$.

From (6) and $F(x)$ (in the form of (3)) we obtain

(10) $x \in \Box\Diamond(R(y_1) \cap R(y_2) \cap R(y_3))$.

Then by (9) we obtain $x \in \Box\Diamond(p,q)$, i.e. (5), which had to be proved. Note that in this direction there were no need of the Boolean Lemma – the only thing from which the proof depends essentially is the monotonicity of the formula $\Box\Diamond(p \wedge q)$.

Let us note that the procedure demonstrated in the above example can be extended to an algorithm for obtaining the first-order equivalents for all complex Sahlqvist formulas which extends the Sahlqvist-van Benthem substitution method. The extension is based on the Boolean Lemma which guarantees the existence of a suitable simultaneous substitution of the complex variables.

REMARK 17 The method of simultaneous substitutions gives another motivation for complex Sahlqvist formulas: if the standard Sahlqvist-van Benthem method considered as a method of "simple substitutions" corresponds to the original Sahlqvist's class, then the method of "simultaneous substitutions" naturally corresponds to the class of complex Sahlqvist formulas. There is however a third kind of substitutions – compound or composed: the result of one substitution is used to define another substitution. Is there a class of "extended" Sahlqvist formulas corresponding to such a kind of substitutions? Yes, this is just the extension, given in [5]. In this sense the class of complex Sahlqvist formulas completes the picture: for each of the three kinds of substitutions – simple, compound and simultaneous – we have the most natural class of modal formulas for which the corresponding method for obtaining first-order equivalents works.

6 Concluding remarks

This paper has to be considered as an initial step to modal definability in languages with finitely many propositional variables with some applications to Sahlqvist theory. The notion of minimal modal equivalent of a first-order condition has been introduced and studied. Here an interesting phenomenon has been discovered: there are first-order conditions corresponding to Sahlqvist formulas with minimal modal equivalents not in the Sahlqvist's class. This fact suggests the idea of introducing a new extension of the class of Sahlqvist formulas, called "complex Sahlqvist formulas". For reasons of simplicity this extension is given for the simplest modal language but the definition and the obtained results can be easily extended to the case of Sahqvist formulas in the polyadic modal languages introduced in [5].

Complex Sahlqvist formulas violate one of the forbidden conditions for ordinary Sahlqvist formulas – disjunctions preceded by boxes in the Sahlqvist antecedents. Nevertheless complex Sahlqvist formulas possess two nice properties of the ordinary Sahlqvist formulas: they define locally first-order conditions and are canonical. A suitable extension of the Sahlqvist-van Benthem substitution method for obtaining (local) first-order equivalents is adapted for the class of complex Sahlqvist formulas called here the method of "simultaneous substitutions". The method of simultaneous substitutions can be considered also as a method of eliminating second-order quantifiers. In the literature there are other methods for eliminating second-order quantifiers, which have also applications in definability theory in modal logic. One for instance is the method SCAN described in [4]. It can be said that this paper is in a sense a result of experiments with SCAN: SCAN presented me first-order conditions for various non-Sahlqvist formulas for which there were no existing theory and methods to deal with. The present paper is an attempt for building of such a theory.

There are of course many natural open problems and questions which can be formulated by everyone interested in the field and in the present subject. I will mention only one: find enough invariants of the original Sahlqvist's class making possible to prove statements in the following form: "there is no Sahlqvist formula having such and such property". Example. We proved in Proposition 13 that the non-Sahlqvist formula $\mathbf{CR}_{k-1}^{min}$ having $n+1$ propositional variables is a minimal modal equivalent of the first-order condition $CR_{k-1}(x)$. It will be interesting to know if there exists a (classical) *Sahlqvist* formula with $n+1$ variables which is a minimal modal equivalent of $CR_{k-1}(x)$. For some cases such a formula exists. For instance for $k=3$ the Sahlqvist formula $\Diamond \Box q_1 \wedge \Diamond \Box q_2 \Rightarrow \Box \Diamond (q_1 \wedge q_2)$ of two variables is a minimal equivalent of $CR_{3-1}(x)$. But for $k=4$ I conjecture that such a Sahlqvist formula of two variables does not exist. At this stage however our knowledge about the structure of the Sahlqvist's class is too limited for proving such statements.

And one more question before the end: can we extend further the class of complex Sahlqvist formulas? Probably yes. Just before preparing the final version of this paper I have found several examples of formulas which are candidates for a such an extension. Here is one which defines (locally) a first-order condition and is canonical:

$\Diamond\Box(p_1 \vee \Box(q \vee r)) \wedge \Diamond\Box(\neg p_1 \vee \Box(q \vee r)) \wedge \Diamond\Box(p_2 \vee \Box(q \vee \neg r)) \wedge \Diamond\Box(\neg p_2 \vee \Box(q \vee \neg r)) \wedge \Diamond\Box(p_3 \vee \Box(\neg q \vee r)) \wedge \Diamond\Box(\neg p_3 \vee \Box(\neg q \vee r)) \wedge \Diamond\Box(p_4 \vee \Box(\neg q \vee \neg r)) \wedge \Diamond\Box(\neg p_4 \vee \Box(\neg q \vee \neg r)) \Rightarrow \bot.$

In this formula "boxed atoms" are not boxed pure elementary disjunctions but contain also boxes inside the elementary disjunctions. The proof of first-order definability and canonicity is based on some theory of "solving equations" in modal algebras similar to those in the Boolean Lemma.

Acknowledgements

I am grateful to Andreas Herzig for introducing me to SCAN and to Renate Schmidt and Valentin Goranko for discussing some problems related to SCAN and Sahlqvist's class.

BIBLIOGRAPHY

[1] Benthem, J. van 1983. *Modal Logic and Classical logic*. Naples: Bibliopolis.
[2] Blackburn, P., M. de Rijke, and Y. Venema. 2000. *Modal Logic*. Cambridg Tracts in Theoretical Computer Science.
[3] Chagrov, A., and M. Zakharyaschev. 1997. *Modal Logic*. Oxford Logic Guides, Vol. 35. Oxford: Oxford University Press.

[4] Gabbay, D., and H. J. Ohlbach. 1992. Quantifier elimination in second-order predicate logic. In: Proceedings of the KR'92, *Princiles of Knowledge Representation and Reasoning*, Morgan Kaufmann, 425-436.
[5] Goranko, V., and D. Vakarelov.2002. Sahlqvist Formulas Unleashed in Polyadic Modal Languages. In F. Wolter, H. Wansing, M. de Rijke, and M. Zakharyaschev, eds. *Advances in Modal Logic, vol. 3*, World Scientific, New Jersey, London, Singapore, Hong Kong, 221-240.
[6] de Rijke, M., and Y. Venema. 1995. Sahlqvist's Theorem for Boolean algebras with Operators with an Application to Cylindric Algebras. *Studia Logica* 54:61-78.
[7] Sahlqvist, H. 1975. Completeness and Correspondence in the First and Second Order Semantics for Modal Logic. In: Kanger, S. (ed.) *Proceedings of the Third Scandinavian Logic Symposium. Uppsala 1973*. Amsterdam, North-Holland.
[8] Sambin, G., and V. Vaccaro. 1989. A new Proof of Sahqvist's Theorem on Modal definability and Completeness. *Journal of Symbolic Logic* 54:992-999.
[9] Segerberg, K. 1971. *An Essay in Classical Modal Logic*, Filosofiska Studier 13, Department of Philosophy, University of Uppsala.

Dimiter Vakarelov
Department of Mathematical Logic
Faculty of Mathematics and Computer Science
Sofia University
blvd James Bouchier 5, 1126 Sofia, Bulgaria
E-mail: dvak@fmi.uni-sofia.bg

INDEX

abox, 292
action negation, 51
admissible, 272
aggregation function, 283
$\mathcal{ALC}(\mathcal{D})$, 269
alethic modality, 356
amalgamation property, 301

B-frame, 319
B-model, 319
basic path, 170
basic tense logic, 478
Basin, D., 243
behavioural structure, 396
Belardinelli, F., 346
Belnap, N. D., 46
Benthem, J. van, 52, 213
Beth properties, 297
Beth, E., 298
binary logics, 314
binders, 34
bisimilarity, 166
bisimulation, 98, 165, 486
Blackburn, P., 190, 199, 499
Boolean bisimulation, 117
Boolean matrices, 104
Boolean matrix multiplication, 105
Boolean vector space, 104
Boolean vectors, 102
bounded commutative residuated lattice, 338
Bourbaki, N., 186
Bradfield, J., 496, 497
Büchi automaton, 128, 424
Bull, R., 9, 11

bundle, 357
bundled frame, 357

canonical B-model, 320
canonical extension, 336
Chagrov, A., 197, 514
Chagrov, A. and Zakharyaschev, M., 499
characteristic formula, 134
Church–Rosser property, 222
coalgebra, 155, 156
commutativity, 222
Complex Sahlqvist formula, 511
complexity function, 80
computation trees, 398
computational structure, 396
concepts, 206
concrete domain, 268
concrete domain role constructor, 281
concurrency theory, 496
constnat formula, 511
context, 156
converse modality $\otimes$, 201
coproduct, 154
Craig, W., 297
cylindric modal logics, 235

$\mathcal{D}$-conjunction, 272
Dabrowski, A., 185, 201
Davidson, D., 31
definable modality, 176
Demri, S., 431
deontic logic, 51
description logic, 32, 265

diagonal constants, 233
diagonal modality, 201
diagonalisable algebra, 300
disjoint union, 155
Došen, K., 46

effort operator, 186
elimination rules, 37
Engelhardt, K, 10
epistemic modalities, 10
extraction function, 154

Fagin, R., 185
Fine, K., 9, 11, 12, 17, 29, 84
Fitting, M., 45
formula-complexity, 426
Fröschle, S., 496
functional completeness, 31
functor semantics, 461
fusion, 94, 221

Gabbay, D. M., 45, 189, 194, 201, 202, 420, 516
Gabbay-style IRR-rule, 358
generalized concrete domain constructor, 279
Gentzen systems, 45
Gentzen, G., 32
Georgatos, K., 185
Ghilardi, S., 461
Givant, S., 336
global consequence relation, 244
global modality, 201
global reduction function, 245
global tableau calculus, 246
Goldblatt, R., 319, 324, 437
Goré, R., 263
Goranko, V., 499, 516, 517
graded modalities, 17, 274
ground formula, 159
ground term, 157
guarded fragment, 205

Halpern, J. Y., 10, 71, 77, 185, 208, 457
Heinemann, B., 185, 198, 199
Hemaspaandra, E., 243, 251, 263
Henkin quantifiers, 495
Herzig, A., 517
hierarchical, 12
hierarchical structures, 9
Hilbert, D., 437
Hintikka, J., 495
historical necessity, 355
history, 357
Horn formula, 238
Hudelmaier, J., 243
hybrid logic, 31, 198
hybrid structured subset space, 198
Hyttinen, T., 497

i-local proposition, 12
independent combination, 94
information partitions, 481
information sets, 482
informational independence, 479
insertion function, 154
integral, 338
interior algebra, 300
interpolation, 297
interpolation properties, 298
interpretation function, 269
introduction rules, 37
inverse modality, 278
Isoda, E., 462, 474

Jipsen, P., 346

Kamp frames, 359
Kamp, H., 359
Kaplan, D., 9, 11, 17
Kesten, Y., 144
keys, 285
Klarlund, N., 137
knowledge operator, 186

Kolaitis, P., 29
Kozen, D., 124
Kracht, M., 84
Kripke bundles, 461, 463
Kripke, S., 9
Kripke–Gabbay method, 437
Kurtonina, N., 213
Kurucz, A., 194, 201

labelled deduction system, 45
Ladner, R. E., 71, 77
Lafont, Y., 351
Laroussinie, F., 431
Lemmon–Segerberg style, 437
limit closure, 362
linear programming, 288
local consequence relation, 244
local formula, 207
local reduction function, 245
local tableau calculus, 246

MacNeille completions, 335
Maehara, S., 346
many-dimensional formalisms, 221
Markey, N., 431
Marx, M., 47, 67, 186, 201, 202
Massaci, F., 263
Matthews, S., 243
McKenzie, R., 329
Meyden, R. van der, 10
Meyer, J.-J. Ch., 52, 57, 66
Miglioli, P., 437
minimal modal equivalent, 500
minimal models, 86
Minkowski spacetime, 437
modal action logic, 51
modal companion, 316
modal definability, 499
Modal equivalent, 500
modal logics of relations, 235
modalisation, 83
model checking problem, 404

monolithic formulas, 86
morphism of coalgebras, 155
Moses, Y., 10, 185
Moss, L. S., 185, 201
multi-agent systems, 9

Nagaoka, K., 462, 474
natural deduction, 31
negative formula, 511
Nguyen, L. A., 243, 259
nominals, 34, 278, 285
non-trivial type, 173
number restrictions, 274

obligation, 52
observable element, 155, 181
observable formula, 159
observable term, 157
observable type, 155
observation path, 170
observational ultrapower, 181
observationally indistinguishable, 166
observationally rich, 182
Ockhamist logic, 356
Ockhamist structure, 357
Ockhamist temporal logics, 423
Ohlbach, H. J., 516
Okada, M., 346, 351
one-variable, 71
ortho-complemented lattice, 313
ortholattice, 325
ortholattices, 313
orthologic, 325
Otto, M., 218

Parikh, R., 185, 201
partially ordered commutative monoid, 340
path, 169
path-free, 274
Peircean logic, 356

permission, 52
player function, 481
Pnueli, A., 144, 393
pointed modal structure, 476
pointed models, 214
polynomial functor, 155
positive formula, 511
Prawitz, D., 32
predicate superintuitionistic lgoics, 462
Prior, A., 31, 355, 356, 437, 496
product, 222
program-complexity, 426
projection function, 154
projective Beth property, 298
proposition, 11
PSPACE-complete, 71

quantification over propositions, 9
quantified propositional temporal logic, 127
quasi-completions, 335

Rabin automata, 128
Rabinovich, A., 431
Rasiowa–Sikorski lemma, 344
Reiter, R., 57, 65
relatinal translation, 205
relativised products, 224
representable cylindric algebras, 233
residual, 338
restricted amalgamation property, 301
rigid context, 157
rigid term, 157
rigid type, 156
Rijke, M. de, 190, 199, 213, 499
role hierarchy, 274
roles, 206
Rosen, E., 213, 214, 216, 218
Rubenstein, A., 484
run, 399

Sahlqvist's class, 500
Sahlqvist's Theorem, 513
Sahlqvist's theorem, 499
Sahlqvist, H., 499
Sahlqvist-van Benthem substitution method, 514
Sambin, G., 499, 514
Sandu, G., 495, 497
satisfaction operator, 34
Scherer, B. G., 196
Schmidt, R., 517
Schurz, G., 84
Segerberg, K., 52, 53, 505
selective filtration, 440
semantic web, 266
semantical game, 480
Shehtman, V., 189, 202
Shehtman,V., 241
Shoham, Y., 457
Spaan, E., 71, 72
spatial reasoning, 291
standard translation, 205
state path, 170
strategy, 483
strong amalgamation property, 301
strong epimorphs surjectivity, 305
structured subset frame, 187, 198
subcoalgebra, 155
subdirectly irreducible, 301
subframes, 223
subset frames, 186
substitution of terms, 162
subsumption, 270
subtype, 156
symbolic model checking, 428

TBox, 271, 274, 276
temporal action logic, 51
temporal reasoning, 289
temporalisation, 84
Tenney, R. L., 29

terminal histories, 481
Terui, K., 351
Thomason, S. K., 84
topoboolean algebra, 300
transition structure, 155
truth in a coalgebra, 164
Tsinakis, C., 346
two-dimensional modal lgoic, 186
two-variable, 71
Tzakova, M., 199

ultrafilter enlargement, 182
ultrapower, 181
universal modality, 35, 271
utility function, 481

Väänänen, J., 495
Vaccaro, V., 499, 514
Vakarelov, D., 499, 516
validity in a coalgebra, 164
van Benthem, J., 499, 514
Vardi, M., 185, 424
variable-free, 71
Venema, Y., 68, 186, 190, 199, 202, 336, 499
Viganò, L., 243

weakly idempotent, 338
Weiss, M. A., 185
Wolper, P., 424
Wolter, F., 84, 194, 201, 241, 263, 437

Zakharyaschev, M., 81, 194, 197, 201, 437, 514

www.ingramcontent.com/pod-product-compliance
Ingram Content Group UK Ltd.
Pitfield, Milton Keynes, MK11 3LW, UK
UKHW021315180426
11947UKWH00015B/1250